I0031466

Natural Occurrence and Biological Activities of Quinoline Derivatives

This latest volume in the *New Directions in Organic & Biological Chemistry* series provides a comprehensive overview of quinoline derivatives, focusing on their chemical structure, biological activities, and therapeutic benefits. It delves into the natural sources of these derivatives, covering their isolation, characterization, and potential therapeutic applications, particularly in areas such as anticancer, anti-inflammatory, antimicrobial, and antimalarial activities. With its emphasis on providing valuable reference material, this book is an indispensable resource for researchers, educators, and professionals in the fields of chemistry, biology, pharmacology, and drug discovery.

Dr Khadem is a chemist, holding B.Sc. and M.Sc. degrees from Ferdowsi University of Mashhad (Iran) and a Ph.D. from the University of Ottawa (Canada). With expertise gained from both academia and industry, he has gathered experience in diverse domains. Presently, he serves as a scientific evaluator and chemist at Health Canada, contributing to the field. His research focus spans organic synthesis, medicinal chemistry, and the exploration of natural products.

New Directions in Organic and Biological Chemistry

Series Editor: Philip Page, Emeritus Professor, School of Chemistry, University of East Anglia

For more information about this series, please visit: https://www.crcpress.com/
New-Directions-in-Organic–Biological-Chemistry/book-series/CRCNDOBCHE

Natural Occurrence and Biological Activities of Quinoline Derivatives

Shahriar Khadem

CRC Press
Taylor & Francis Group
Boca Raton London New York

CRC Press is an imprint of the
Taylor & Francis Group, an **informa** business

First edition published 2026
by CRC Press
2385 NW Executive Center Drive, Suite 320, Boca Raton FL 33431

and by CRC Press
4 Park Square, Milton Park, Abingdon, Oxon, OX14 4RN

CRC Press is an imprint of Taylor & Francis Group, LLC

© 2026 Shahriar Khadem

Reasonable efforts have been made to publish reliable data and information, but the author and publisher cannot assume responsibility for the validity of all materials or the consequences of their use. The author and publishers have attempted to trace the copyright holders of all material reproduced in this publication and apologize to copyright holders if permission to publish in this form has not been obtained. If any copyright material has not been acknowledged please write and let us know so we may rectify in any future reprint.

Except as permitted under U.S. Copyright Law, no part of this book may be reprinted, reproduced, transmitted, or utilized in any form by any electronic, mechanical, or other means, now known or hereafter invented, including photocopying, microfilming, and recording, or in any information storage or retrieval system, without written permission from the publishers.

For permission to photocopy or use material electronically from this work, access www.copyright. com or contact the Copyright Clearance Center, Inc. (CCC), 222 Rosewood Drive, Danvers, MA 01923, 978-750-8400. For works that are not available on CCC please contact mpkbookspermissions@ tandf.co.uk

For Product Safety Concerns and Information please contact our EU representative GPSR@ taylorandfrancis.com. Taylor & Francis Verlag GmbH, Kaufingerstraße 24, 80331 München, Germany.

Trademark notice: Product or corporate names may be trademarks or registered trademarks and are used only for identification and explanation without intent to infringe.

ISBN: 9781032876566 (hbk)
ISBN: 9781032878157 (pbk)
ISBN: 9781003534600 (ebk)

DOI: 10.1201/9781003534600

Typeset in Times
by codeMantra

Contents

Foreword

Quinoline derivatives are widespread in nature, being found naturally in plants, animals, fungi, and bacteria. While it is tempting to assume that their distribution across multiple kingdoms of life implies that the genes for these alkaloidal secondary metabolites are highly conserved, in fact, different lineages of organisms have evolved various biosynthetic pathways to create quinoline derivatives. Thus, their frequent occurrence in diverse organisms is more likely a case of convergent evolution. The biological activities of quinoline derivatives against viral, bacterial, protistan, and fungal pathogens, competing species, herbivores, and oxidative stress support a plausible hypothesis that these compounds confer survival advantages on their hosts.

Biologically active quinolines are also being investigated as leads for the synthesis of candidate drugs. Successful examples include Cilostazol (a phosphodiesterase-3 inhibitor); Aripiprazole (an antipsychotic); Carteolol (a non-selective beta-blocker); Vesnarinone (a cardiotonic agent investigational for HIV infections and Kaposi sarcoma treatment); Palonosetron (blocks the action of serotonin at 5-hydroxytryptamine type 3 receptors competitively, preventing nausea and vomiting associated with surgery and chemotherapy); Gliquidone (shows antioxidant and anti-inflammatory effects, lessening diabetic nephropathy); Cilostamide (a potent inhibitor of phosphodiesterase 3 enzyme for acute heart failure and cardiogenic shock therapy); Linomide (both an immunostimulant and an angiogenesis inhibitor); Rebamipide (used for mucosal protection, gastroduodenal ulcer healing, and gastritis treatment); and Brexpiprazole (for treatment of mental disorders such as depression and schizophrenia). These diverse therapeutic indications illustrate the great potential for further discovery of new drugs based on the quinoline structural motif.

Despite some notable successes in drug discovery and development, quinolines and their derivatives remain understudied. Recent reports about these compounds, scattered through the chemical, pharmacological, toxicological, and biological literature, have not been adequately compiled and evaluated. In the chapters that follow, the chemistry (including biosynthesis and synthetic approaches), natural occurrence, and biological activities of various chemical classes of quinoline derivatives are reviewed, with reflections on the implications of the current state of the science for drug discovery. Bringing together the latest advances in our knowledge and understanding of quinolines and their derivatives, with thoughtful analysis, this book provides a solid foundation for further research on this fascinating family of natural products.

Robin J. Marles, Ph.D., is Scientist Emeritus, Health Products and Food Branch, Health Canada, Ottawa, Ontario. He was a former Associate Professor of Botany, Brandon University, Brandon, Manitoba, Canada.

Preface

Quinoline derivatives have long stood at the forefront of medicinal chemistry due to their remarkable structural diversity and wide-ranging biological properties. This book has been developed as a consolidated scientific resource to provide a thorough and multidimensional overview of a subclass called semi-aromatic quinoline derivatives. Particular attention is given to their structural features, chemical identity, natural occurrence, synthesis and biosynthesis, pharmacological profiles, and therapeutic relevance. Revised and expanded content from a series of published and ongoing review articles has been integrated. In addition, newly written chapters have been included to deliver an in-depth analysis of quinoline-based molecules from their natural sources to their pharmacodynamic potential. The organization of this book reflects a deliberate attempt to guide the reader through the multifaceted landscape of semi-aromatic quinoline chemistry. Their structure–activity relationships and relevance to current drug development efforts are also given consideration. Emphasis has been placed throughout this book on interpreting experimental data and chemical modifications in the context of therapeutic efficacy. This work is intended to serve as a reliable scientific reference for researchers and professionals in drug discovery and development. It is particularly relevant to those working in the natural product chemicals and pharmaceutical sectors. The academic community is also addressed, including graduate students, postdoctoral researchers, and faculty members in chemistry, biology, and pharmacology. Access is provided to a harmonized, critically curated body of knowledge. This book is designed to bridge the gap between advanced research and practical applications. It is anticipated to be of particular value to those seeking to understand how quinoline-based molecular frameworks contribute to the design of new therapeutic agents. It is hoped that further investigation into this important class of compounds will be encouraged. Additionally, innovative approaches to address pressing medical challenges are expected to be inspired by this compilation.

Shahriar Khadem, Ph.D.
Ottawa, Canada

1 Introduction

1.1 FROM NATURE TO MEDICINE

Natural products (NPs) play an important role in drug discovery and development, originating from traditional medicine to modern medicinal chemistry. Figure 1.1 illustrates the distribution of Food and Drug Administration (FDA)-approved drugs by source from 1981 to 2019 (Newman and Cragg 2020). The largest proportion consists of purely synthetic drugs (S), followed by natural product derivatives (ND) and biologicals(B). NPs and their related forms (N, NB, ND, and S*) collectively account for a substantial portion, highlighting the continued importance of NP-based scaffolds in drug discovery. Pharmacology of NPs from plant origin and other microbial and animal sources has had a long history in traditional medicine systems available everywhere in the world, from plants, marine organisms, bacteria, and fungi. These classic uses provided a basis for understanding that is continually contributing to the development of modern pharmacology. The inherent complexity and structural diversity of these NPs also allow them to bind to a wide range of biological targets, which makes them invaluable starting points for developing new drugs. These naturally occurring metabolites still illustrated deep-seated therapeutic potential in several disease arenas, such as cancers, infectious diseases, and chronic diseases, such as cardiovascular and neurodegenerative diseases (Chopra and Dhingra 2021; Newman and Cragg 2020).

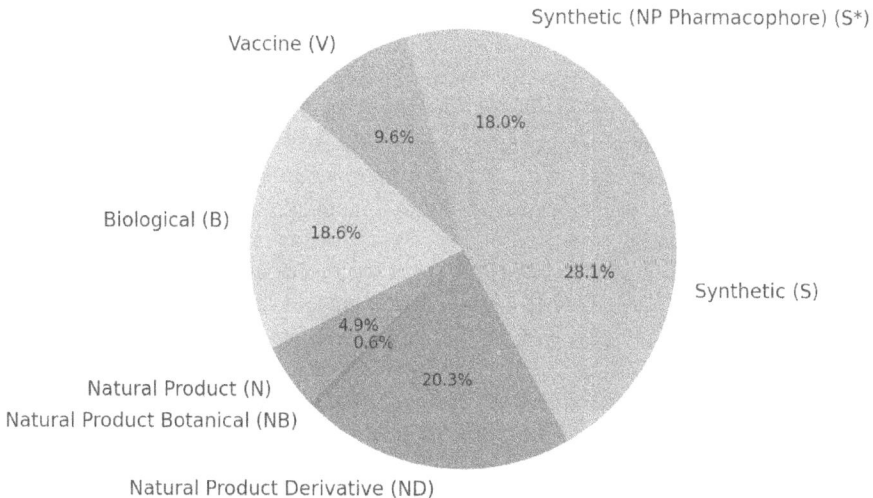

FIGURE 1.1 Distribution of approved drugs by source during 1981–2019. (Adopted from Newman and Cragg 2020.)

DOI: 10.1201/9781003534600-1

1

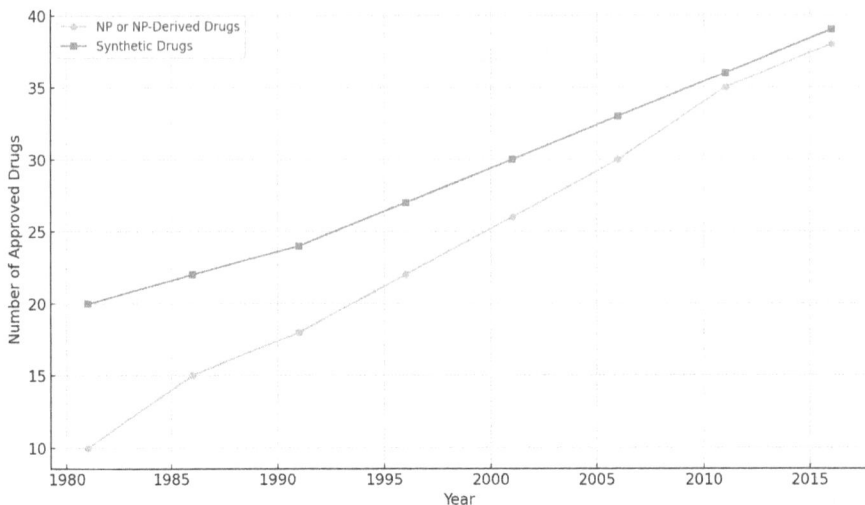

FIGURE 1.2 Drug approval over time (1981–2019): Natural products vs. Synthetic. (Adopted from Newman and Cragg 2020.)

Figure 1.2 shows the number of FDA-approved NPs or NP-derived drugs compared with synthetic drugs from 1981 to 2019. Both showed a steady increase over the years. However, the number of NP-based approvals remained smaller than that of synthetic drugs. By 2019, the gap between the two sources had narrowed. This reflects the growing interest and advancement in NP-based drug development.

The importance of NPs extends beyond their traditional applications; nowadays, a massive library of chemical diversity is widely used in modern drug discovery. The structural complexity of NPs tends to afford selective and potent interactions with their chosen biological targets, a feature that is much valued during drug discovery. At the molecular level, this selectivity leads to higher efficiency of the therapy, lower toxicity to non-target cells, and possibly fewer side effects. The chemical structure of NPs frequently contains privileged scaffolds that function as the coordinates of a core template for the construction of chemically enhanced derivatives. Chemical modifications to the structure of these natural compounds are applied to improve their physicochemical properties, bioavailability, and metabolic stability. These improvements have significantly expanded their potential therapeutic utility, enabling these NPs to progress from the laboratory to the clinical setting and into drug development. Newman and Cragg (2020), along with Chopra and Dhingra (2021), thoroughly discussed the NP-originated substances and their structural mimics in traditional and modern medicine. These researchers highlighted the uniqueness of these compounds for therapeutic diversity, especially for antibiotics, anticancer, immuno-modulating, and any Central Nervous System (CNS) disorder. The experts took into account specific illustrations like tumour treatment with paclitaxel in *Taxus* species and malaria treatment with artemisinin in *Artemisia annua*, thereby indicating that these plants have always stood as the source of effective drugs against disorders. The review also reported on marine-derived NPs, which are vital for drug discovery programmes due to their novel structural features and unique biological properties. Domingo-Fernandez et al. (2024) analysed NPs in clinical development and quantitatively demonstrated the clinical benefits of drugs based on NPs over synthetic drugs. In particular, the success rates in clinical studies of NPs, ND, and purely synthetic compounds were monitored by their study. An outstanding improvement in the success rates of NPs was observed in further clinical development steps. The authors found that the NP-derived drugs rose from about 35% of phase I studies to approximately 45% of phase III studies, and that pure synthetic drugs decreased in prevalence from 65% to 55% over this period. These trends underscore the poorer clinical efficacy and tolerability of synthetic versus ND. In addition, the *in vitro* and in silico toxicity prediction analyses revealed that NP-derived molecules had a lower toxicity profile, making them more desirable concerning safety, efficacy, and patient acceptance during clinical development. Newman and Cragg (2020) presented an in-depth longitudinal study on the contribution of NPs to drug approvals. Their large dataset showed that a significant portion of all marketed drugs had been directly invented based on NPs or synthetic compounds thereof. They singled out NPs and their derivatives that made up large portions of recently approved new drugs, particularly in areas such as oncology and infectious diseases. The authors acknowledged that despite the advances in synthetic chemistry and high-throughput screening, NPs remained broadly relevant because of their structural complexity and biological selectivity. They also highlighted the relatively poor track record of purely synthetic combinatorial chemistry in providing that paradigm of novel drug entities, noting that almost all successful synthetic approaches used NP scaffolds and/or pharmacophores as starting points. Additionally, they pointed out that although Lipinski's Rule of Five is commonly violated by NPs, they were shown to regularly hit large pharmacological target clusters with high therapeutic relevance. This finding contradicts current drug design paradigms and urges medicinal chemists to rethink established "drug-likeness" criteria for NPs. NP drug discovery has many challenges in sustainable sourcing and the isolation and characterization of bioactive compounds. However, such challenges are gradually being solved with technological advancements. Techniques, including genomic mining (e.g. ow-impact genetics), synthetic biology, and metabolic engineering, have been widely adopted in the production of compounds from nature in an environmentally friendly manner, overcoming the limitations of natural resources. These methods also make the NP's structural complexity, which was difficult or impossible to produce chemically, easier to identify and biosynthesize. Moreover, new computational tools and techniques such as machine learning approaches are now playing an important role in the NP field by providing quicker ways of target prediction, pharmacokinetics (PK)-properties optimization, and new drug target identification. The coupling of computational chemistry with bioinformatics has greatly advanced the studies of NPs and the process of drug discovery. Computational techniques like virtual screening, molecular docking, and Absorption, Distribution, Metabolism, Excretion, and Toxicity (ADMET) prediction have made lead discovery faster to discover and prioritize potential NP hits for further research. As a result of these strategies, time and cost towards drug development have been reduced, allowing the resources to focus on the best candidates. In addition, bioinformatics platforms help in the efficient processing and interpretation of multiple sets of data that emerge from NP research, thus facilitating data-driven decisions in drug discovery and development. Utilizing these computational tools hand in hand with classical organic and biological approaches further enhances the capacity and effectiveness of NP drug discovery.

Domingo-Fernández et al. (2024) compared the experimental toxicity profiles of synthetic, hybrid, and NP compounds across three endpoints: cytochrome P450 (CYP450) inhibition, hepatotoxicity, and carcinogenicity (Figure 1.3). Synthetic compounds exhibited the highest toxicity rates, particularly in CYP450 inhibition (~77%) and carcinogenicity (~62%). In contrast, NPs and hybrids showed significantly lower toxicity across all categories, with NPs displaying elevated hepatotoxicity but lower CYP450 inhibition and carcinogenic potential.

NPs are distributed across the biological kingdoms. Most of the NPs with pharmaceutical activity are derived from plants. These have contributed around 40%–60% of the known NPs. The contribution of these products is due to the maximum production of terpenoids, alkaloids, flavonoids, and many more. Approximately 20%–30% of bacteria make antibiotics, siderophores, and enzymes that help them survive and interact with other species. Fungi contribute about 15%–25% (Hamed et al. 2024), although fungal diversity is quite underexplored (Bar-on et al. 2018). They make diverse secondary metabolites, polyketides, alkaloids, and mycotoxins. Approximately 5%–10% of NPs come from animals, especially from marine sources; they include unique bioactive compounds such as toxins, enzymes, and hormones. However, the compounds from marine animals are not easily accessible (Princgenies et al. 2025). The above estimates indicate the bias of researchers favouring plants and culturable bacteria. Because of this, we have an incomplete but evolving understanding of NP diversity across biological kingdoms.

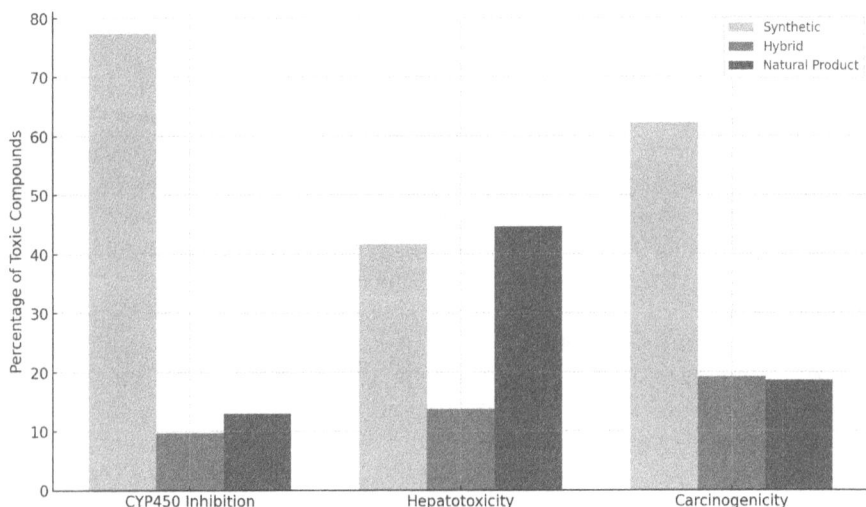

FIGURE 1.3 Experimental toxicity comparison by compound class. (Adopted from Domingo-Fernandez et al. 2024.)

1.2 A BRIEFING ON ALKALOIDS

The naturally occurring organic compounds known as alkaloids consist of nitrogen atoms, which are usually found inside heterocyclic ring structures. The biosynthesis of alkaloids primarily starts from the amino acids such as tyrosine, tryptophan, phenylalanine, ornithine, and lysine. However, their biosynthetic origins can stem from different pathways, including terpenoid and steroidal routes. Alkaloids exist in various structural forms throughout plants, fungi, bacteria, and certain animal species (e.g. marine animals). Alkaloids have typically complex structure types whereby they are capable of selectively reacting with biological macromolecules like nucleic acids, enzymes, as well as receptors, with a consequent powerful physiological effect. Alkaloids are classified based on their biosynthesis and structure into three types: true alkaloids, pseudo-alkaloids, and proto-alkaloids. True alkaloids contain nitrogen within a heterocyclic ring and are derived from amino acids. Pseudo-alkaloids, on the other hand, are nitrogenous compounds that originate from sources other than amino acids. Proto-alkaloids have nitrogen present but are not part of a ring structure. This diversity of chemistry of biosynthetic pathways forms the basis of extensive biological as well as pharmacological actions of the alkaloids (Daley and Cordell 2021; Letchuman et al. 2024; Rajput et al. 2022).

Alkaloids have many medical benefits. These include anticancer, antimicrobial, anti-inflammatory, antimalarial, antihypertensive, anti-pain, anti-diabetic, and immune system-boosting powers. Medicines like morphine, quinine, vincristine, atropine, and berberine were developed using these compounds. The complex structure of these compounds often makes them capable of selectively targeting biological targets, hence useful in drug discovery. The common foods we eat every day, like coffee, tea, tomato, and potato, contain alkaloids. Human physiology is influenced by alkaloids outside the therapeutic window as well. Currently, biomedicine examines the synergistic effects of alkaloids with other NPs (other than medications) for multi-target treatment. Integrating them into personalized medicine programs according to one's genetic variation is an emerging area of research that aims to enhance their effect while minimizing side effects. Although some alkaloids have benefits, they can also be toxic if used in large amounts. Thus, it is important to test the dose and toxicity of these alkaloids when using them in clinical and dietary uses (Daley and Cordell 2021; Letchuman et al. 2024; Rajput et al. 2022).

Researchers increasingly investigate alkaloids to meet global needs, such as dealing with Multidrug resistance (MDR) and neglected tropical diseases in the drug discovery of NPs. The structural diversity and biological potency of alkaloids qualify them as strategic candidates. While traditional pharma pipelines can be effective, they tend to prioritize varieties that are most likely to be profitable over others, which would ease the burden of global diseases. New techniques for discovering drugs harness the Fourth Industrial Revolution technologies, such as artificial intelligence, genomics, metabolomics, and bioinformatics. They facilitate the identification, screening, and optimization of alkaloid leads. The beneficial use of alkaloids is disadvantaged by sustainability constraints, biodiversity loss, dereplication complexity, and residual matrix effects in alkaloid isolation. To solve such issues, a collaborative framework such as the Quintuple Helix innovation model is recommended.

This model covers the interaction of civil society and the environment with higher education, industries, and government. In addition, the new strategies involve sourcing alkaloids from marine organisms, endophytes, and extremophiles, complemented by high-throughput biological screening and metabolite profiling databases. It means that in the future, alkaloids will remain significant with sustainable, ethical, and technologically driven discovery models.

Recent findings showed that alkaloids can also be used in personalized medicine and natural combination therapies. Ecological factors play an important role in determining the chemical variations of alkaloids in plants. The presence of soil, climate, altitude, and microbes are all factors that affect the biosynthesis and accumulation of alkaloids in plants. Alkaloids are traceable to amino acids and are known to exert a wide range of bioactivities. Alkaloids and several other NPs, including flavonoids and polyphenols, which are present in plants, increase bioactivity effectiveness. Supercritical fluid, microwave aqueous, and ultrasound aqueous extraction methods help in the proper isolation of alkaloids. Studies indicate that alkaloids alter the pathways that regulate oxidative stress, inflammation, apoptosis, and signal transduction. For instance, berberine modulates glucose metabolism and inflammatory pathways, Capsaicin, through transient receptor potential cation channel subfamily V member 1receptors, has analgesic effects, and ibogaine affects the neurochemical systems implicated in addiction. Using certain drugs to treat "diseases" for certain patients has gained much interest owing to genetic alterations. Studying less-known plants in biodiversity hotspots can help us find new alkaloids, which may offer unique treatments for complex diseases like cancer, neurodegeneration, and metabolic disorders that currently have unmet needs (Letchuman et al. 2024).

Some ingredients in medicines and foods are alkaloids. The amounts of active and toxic substances in these alkaloids can vary greatly. For instance, there are specific groups of alkaloids that contain indole and indolizidine structures. Important members include vincristine, reserpine, lycorine, and tylophorine. These alkaloids have good potential in the treatment of cancer, infectious diseases, and neurodegenerative diseases. Coffee, tea, bananas, and tomatoes contain beta-carboline alkaloids, which are biologically active molecules. The substances in the plant may protect the nerves, although high exposures cause neurotoxicity. Some alkaloids, for example, pyrrolizidine and tropane alkaloids, have been associated with adverse and serious effects, namely, hepatotoxicity, neurotoxicity, and carcinogenicity. Although widely used, caffeine, nicotine, and solanine are alkaloids that are toxic in high dosages. The breakdown of biosynthetic origins helps in the biological behaviour and safety evaluation of alkaloids. The frequent presence of alkaloids in commonly consumed food items underscores the need for regulatory standards and public health education. The aim should be to maximize their beneficial effects while mitigating potential harmful outcomes. Figure 1.4 illustrates the distribution of alkaloids in various components of the human food chain. Coffee and tea contribute the highest proportion at 30%, followed by vegetables and fruits (25%) and beverages (20%). Spices (15%) and dairy/meat products (10%) also contribute to alkaloid intake, indicating widespread dietary exposure across multiple food categories. However, more research is required to determine per-dose response, chronic toxicity profile and how their pharmacological potentials can be utilized safely (Rajput et al. 2022).

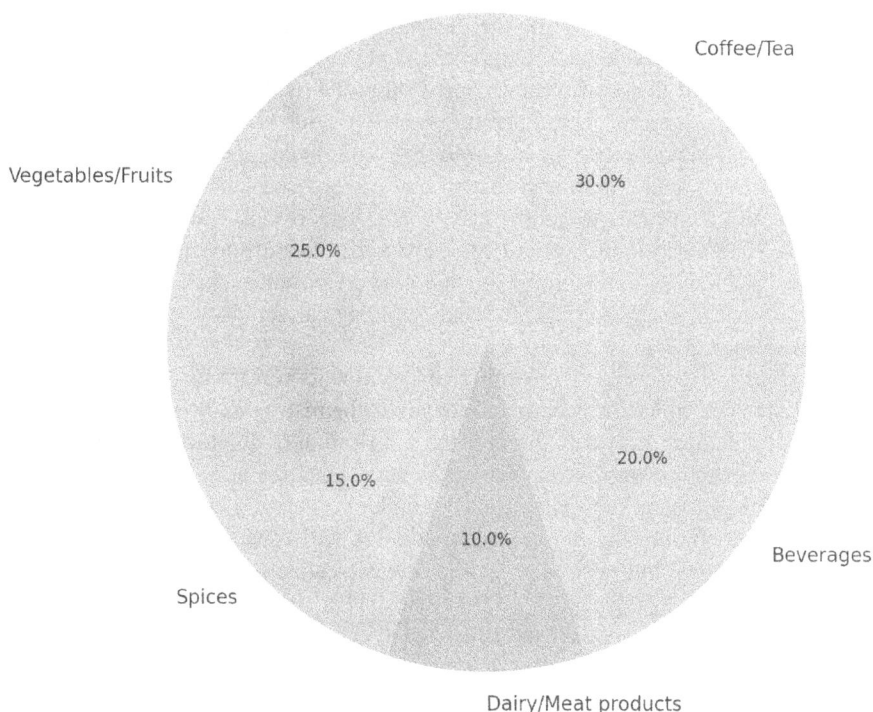

FIGURE 1.4 Alkaloids' presence in the human food chain. (Adopted from Rajput et al. 2022.)

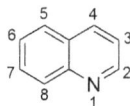

FIGURE 1.5 Quinoline and its numbering system.

1.3 QUINOLINES

Quinoline is an aromatic, bicyclic heterocyclic compound containing nitrogen. Its structure consists of a benzene and a pyridine that are fused (Figure 1.5). In 1834, during the distillation of coal tar, Friedlieb Ferdinand Runge first isolated quinoline, which turned out to be one of the first representatives of a significant class of nitrogenous heterocycles. Quinoline is a weak tertiary amine that can form salts with acids. It also undergoes electrophilic and nucleophilic substitution reactions. Somehow, nature managed to incorporate these motifs into several plants as key components of alkaloids. Significantly, the bark of the *Cinchona* tree has several quinoline-based alkaloids, including quinine, quinidine, cinchonine, and cinchonidine. They are historically important in the treatment of malaria. The camptothecin anticancer drug from *Camptotheca acuminata*, nitidine, a dimethoxylated quinoline from *Zanthoxylum nitidum*, and skimmianine, a furoquinoline from *Skimmia japonica*, are other natural quinoline derivatives. Besides plants, quinoline-based skeletons are found in a few microorganisms and marine organisms, further indicating the ubiquity of this scaffold in nature. Quinoline is described as a colourless, hygroscopic liquid at ambient temperatures and has a potent smell. It partially mixes with cold water but completely mixes with hot water and organic solvents like ethanol, diethyl ether, and chloroform. Light- and air-exposed quinoline first becomes colourless and then turns to yellow and brown. This reveals quinoline's oxidative instability. These subtle differences can have a significant impact on the chemical reactivity and biological functions of quinolines. Consequently, a lot of chemical modifications can be performed to enhance biological activity (Ajani et al. 2022; Yadav and Shah 2021).

Because of the structural flexibility and synthetic versatility of the quinoline scaffold, it is a privileged structure of medicinal chemistry. Many classical methods have been developed over the years for the preparation of quinoline and its derivatives. One of the first was Skraup synthesis, the cyclization of aniline with glycerol and an oxidizing agent (e.g. nitrobenzene) under strong acid conditions. Although the reaction conditions are severe, this method is still used to prepare quinolines. In the Friedländer synthesis, either acidic or basic catalysis allows for the condensation of 2-aminobenzaldehydes with ketones. This reaction has a much broader scope because it tolerates many substituents. Other notable classical methods are the Doebner–Miller reaction and the Combes synthesis, each with specific advantages in terms of substitution on the quinoline nucleus. The Povarov reaction has also been investigated with $[4+2]$-cycloadditions of aniline, aldehyde, and alkene to access fused quinoline. Recent improvements focus on finding sustainable and easy-to-use protocols. By utilizing methods from green chemistry, such as microwave-assisted syntheses, the reaction time is reduced and the yields are enhanced. The application of ionic liquids and catalysts supported by solids, such as magnetite, as well as ionic liquids supported by magnetite and nanoporous titanium dioxide (TiO_2), guarantees no waste and the ability to recycle catalysts. Plant extracts have been researched for the preparation of eco-friendly quinoline using biogenic catalysts.

Many compounds based on quinoline have been made into the pharmaceutical market, targeting many diseases. Chloroquine, primaquine, mefloquine, and amodiaquine are useful drugs in treating and prophylaxis malaria infection. The malaria-causing parasite dies because of the heme detoxification interference in *Plasmodium*. The fluoroquinolone antibiotics such as ciprofloxacin, levofloxacin, and gatifloxacin affect bacteria's DNA gyrase and topoisomerase IV. This action leads to the inhibition of DNA replication, thus causing the death of the bacterial cell. In another example, both irinotecan and topotecan (containing quinoline scaffold) damage DNA and trigger apoptosis in cancer cells by blocking topoisomerase I. They are also important cancer-fighting drugs. The quinoline-type antimycobacterial agent bedaquiline targets mycobacterial adenosine triphosphate (ATP) synthase through a novel mechanism of action for the treatment of multidrug-resistant tuberculosis.

Structural changes on the quinoline ring affect biological activity. When substituents are introduced at specific points, they can change the electronic distribution, steric factors, and hydrogen bonding ability. In other words, the substitute will be able to change and modulate interaction with the biological target. Using structure–activity relationship (SAR) analysis, quinoline derivatives have been classified into amino-quinolines, hydroxyquinolines, oxyquinolines, fused quinolines, metal-complexed quinolines, and alkyl-substituted quinolines. These structural classes can be used for pharmacological fine-tuning and thus enhance potency, selectivity, and/or pharmacokinetic profile. Noted from SAR studies, a Cl atom, an electron-withdrawing group, along with some hydrophobic side chains at a specific position, enhances potency, selectivity, and permeability. In addition, scientists have synthesized hybrid molecules with quinoline cores combined with other pharmacophores. They did this to check for synergistic effects and dual-target mechanisms. This expands the therapeutic applicability of this scaffold. Quinoline derivatives show their biological effect through several mechanisms of action. Some breast cancer drugs work by activating caspases, which cause apoptosis. Others suppress cancer cell proliferation through the inhibition of vascular endothelial growth factor signalling. However, many breast cancer drugs target heat shock protein 90. Recent studies have also revealed that some drugs can impair the toll-like receptor 4 signalling pathway, especially ones that block mechanistic (or mammalian) target of rapamycin and phosphoinositide 3-kinase. Molecular hybrids designed from quinoline scaffolds have enhanced their biological activities by engaging in multiple pathways. For example, ferroquine, a hybrid compound of 4-amino-7-chloroquinoline and ferrocene, is being tested clinically for its potential to block malaria, neonatally. Another compound, dactolisib, a dual phosphoinositide 3-kinase/mechanistic (or mammalian) target of rapamycin inhibitor with a quinoline nucleus, is being tested for its ability to kill solid tumours. Building new drugs based on the quinoline scaffold will always be required due to challenges like emerging resistance of drugs such as malaria and bacterial species, as well as the quality of new pathogens. Consequently, further investigation of quinoline derivatives concerning hybridization, new substitution patterns, and alternative mechanisms of action is one of the effective strategies to fill the unachieved clinical needs and subsequently a global approach for drug discovery (Yadav et al. 2022; Ajani et al. 2022; Yadav and Shah 2021).

1.4 SEMI-AROMATIC QUINOLINE DERIVATIVES

Quinoline derivatives can retain aromaticity when substituents are attached to positions 1 through 8 of the quinoline ring system since these modifications occur on the margin of the fused aromatic structure without disrupting the conjugated π-electron system. In other words, the core aromaticity of quinoline remains intact as long as the substituents do not introduce sp³-hybridized atoms into the ring backbone or participate in reactions that alter the electron delocalization (e.g. reduction or saturation of ring atoms). Semi-aromatic quinoline derivatives, named here for the first time, are characterized by partial hydrogenation of the pyridine ring (Figure 1.6). Such a change impacts the molecular geometry, aromaticity, electronic distribution, and chemical reactivity of the compound. Quinoline has full aromaticity through π-electron delocalization throughout both rings, but adding saturation at certain places (positions) in the pyridine ring breaks down this conjugation and excavates non-planar structures by introducing sp³-hybridized centres that decrease resonance stabilization.

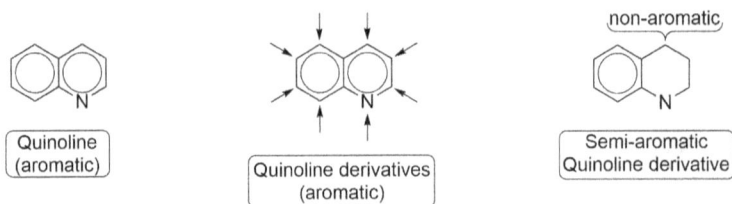

FIGURE 1.6 Quinoline, quinoline derivatives, and semi-aromatic quinoline derivatives.

This semi-aromaticity causes the formation of a new type of compound with localized aromaticity (in the benzene ring) and flexible structure, in contrast to the planar and rigid arrangement of the complete aromatic quinoline compound. Depending on the degree and position of saturation, aromaticity is affected negatively or positively. The non-aromatic six-membered ring with nitrogen in it (formerly pyridine) is generated by saturation, offers considerable conformational freedom, and gives rise to chiral centres that exhibit stereoisomerism. When there remains only one double bond in a nitrogenous ring system, partial unsaturation is maintained, although coupling between orbitals is somewhat reduced and geometric isomerism about the double bond is possible. Adding polar functional groups like carbonyls in strategic places makes conjugation weaker. This change leads to disturbances in the lactam structure. The saturation level and the presence of electrophilic functional groups determine the chemical reactivity of the semi-aromatic quinoline class. The benzene ring in these systems is intact and available for electrophilic aromatic substitution. Pyridine rings that are fully saturated in structures (piperidine) can be oxidized under some mild oxidizing conditions to regain their full aromaticity. It can have either an activating or deactivating effect on the reactivity patterns of the benzene ring. It is prone to oxidation and will undergo oxidative changes when exposed to air. The site of the reactive double bond results in nucleophilic additions, cycloadditions, and disproportionation reactions, facilitating a broad scope of synthetic manipulation and access to complex polycyclic structures.

Compounds having a carbonyl group in the ring, like lactam-containing compounds, show dual reactivity of aromatic and carbonyl compounds. Lactam groups are vulnerable to alkylation, acylation, and nucleophilic addition, whereas the aromatic element is subject to selective halogenation and transition-metal-catalysed C–H activation. The keto and enol forms can change into each other. They have tautomeric behaviour. This will offer additional reaction paths depending on the solvent mix and electronic environment. The derivatives containing two carbonyls within the quinoline system show a significant increase in their electrophilicity, which enables Michael-type additions and pericyclic reactions.

The semi-aromatic type of quinoline derivatives that possess a completely saturated nitrogen-containing ring and an aromatic benzene ring demonstrate considerable flexibility and the formation of multiple chiral centres. Due to this discrimination in stereochemistry, different biological interactions are created. The types of systems having only one double bond in the nitrogen ring become more reactive to oxidation and addition reactions and can be used as a synthon for further synthetic modifications. The possession of a carbonyl group at the second position in the structure brings a lactam functionality. This allows hydrogen bonding. As a result, it increases the polarity of the molecule. Moreover, this also enhances the reactivity of the overall molecule through nucleophilic and electrophilic reactions. Their ability to exist as tautomers further enhances their reactivity and biological activities. Derivatives having two carbonyl groups on quinoline constitute a class having remarkable electrophilicity. The π-systems of these compounds are highly electron-deficient. The dependence on nucleophilic reagents dictates their reaction reserves. The compounds also participate in conjugate addition. When they participate in cycloaddition reactions, they generate structurally complicated polycyclic systems. Moreover, due to their ring contraction reactions that occur in basic media, they are versatile in synthesis. In more complex systems with a semi-aromatic quinoline core fused to extra conjugated quinone components, the partially saturated amine flexibility is combined with the redox properties of quinones. These compounds can perform redox cycling and electron transfer and generate reactive oxygen species under the right circumstances. Furthermore, their chemical behaviour is much more complex than simpler quinolines or similar molecules.

1.5 SCOPE

This book provides comprehensive details on semi-aromatic quinoline derivatives, divided into different chapters dealing with individual subclasses of these important compounds. The chemical structure of the substances presented in this book is depicted in Figure 1.7.

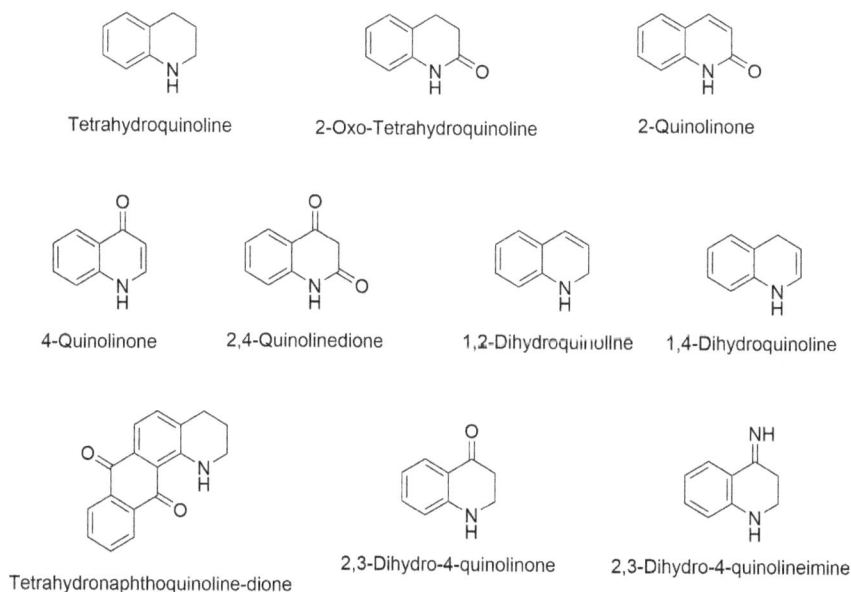

Tetrahydroquinoline   2-Oxo-Tetrahydroquinoline   2-Quinolinone

4-Quinolinone   2,4-Quinolinedione   1,2-Dihydroquinoline   1,4-Dihydroquinoline

Tetrahydronaphthoquinoline-dione   2,3-Dihydro-4-quinolinone   2,3-Dihydro-4-quinolineimine

FIGURE 1.7 Chemical structure of a few semi-aromatic quinoline derivatives.

Tetrahydroquinolines are cyclic organic compounds with a quinoline core in which the 1, 2, 3, and 4 positions are saturated. This partial reduction disrupts the aromaticity of the system and introduces sp³-hybridized carbon atoms, increasing conformational flexibility and the potential for chirality. The structural modification changes the electronic properties of the molecule, enhancing the chemical versatility of tetrahydroquinolines. This framework is found in a wide range of bioactive molecules and synthetic intermediates, especially in drug discovery, where their stereochemical complexity can influence the pharmacokinetics and target specificity (Khadem and Marles 2025f). 2-Quinolones (2-quinolinones) are structural isomers of quinoline characterized by a carbonyl group at position 2 and a hydrogen group at position 1, most commonly existing in the lactam form. This arrangement introduces polarity and enhances intermolecular interactions, impacting solubility and binding properties. A notable feature is their ability to tautomerize to the lactim form (2-hydroxyquinoline), with the equilibrium influenced by solvent environment, pH, and substitution. These compounds perform as pharmacophores in antimicrobial, anticancer, and enzyme-inhibiting agents (Khadem and Marles 2025a). 4-Quinolones (4-quinolinones) are quinoline derivatives bearing a carbonyl group at position 4. Similar to 2-quinolones, they exist in equilibrium between their keto (4-quinolone) and enol (4-hydroxyquinoline) tautomers. This tautomeric flexibility affects their hydrogen bonding, acidity, and chemical reactivity. The 4-quinolone scaffold is central to the fluoroquinolone class of antibiotics, where structural modifications on the core ring system modulate antibacterial potency and spectrum. Their biological activity is closely tied to the electron distribution influenced by the position of the carbonyl group. 2,4-Quinolinediones are dicarbonyl-substituted quinoline derivatives with carbonyl groups at both positions 2 and 4, forming a strongly electrophilic framework. The dual presence of electron-withdrawing groups enhances their reactivity towards nucleophiles and enables participation in a wide range of synthetic transformations. These structures often show pronounced biological activity, including antimicrobial and anticancer effects. This is attributed to their capacity to engage in multiple non-covalent interactions with biological targets and their propensity to undergo tautomeric and resonance stabilization (Khadem and Marles 2024). 2-Oxo-tetrahydroquinolines represent a hybrid structural class where the tetrahydroquinoline ring is retained, but a carbonyl group is introduced at the 2-position. This results in a cyclic amide (lactam), which combines the partially saturated core's stereochemical flexibility with a ketone's chemical functionality. These compounds are of significant interest because of their presence in NPs and synthetic bioactive compounds. They often serve as scaffolds in medicinal chemistry programs aiming to optimize hydrogen bonding, metabolic stability, and receptor affinity (Khadem and Marles 2025b, c, d). Tetrahydronaphthoquinoline-diones are polycyclic compounds incorporating a tetrahydroquinoline unit fused to a naphthoquinone moiety. This structural fusion produces a rigid and planar aromatic domain with multiple carbonyl functionalities and sites willing to substitution. The electronic features of the quinone and the conformational restriction imposed by the fused ring system make them valuable in the development of redox-active pharmaceuticals, enzyme inhibitors, and materials for electron transport applications. Their complex architecture allows precise tuning of both pharmacological and physicochemical properties (Khadem and

Marles 2025e). 1,2-Dihydroquinolines are produced by selective reduction of the double bond between positions 1 and 2 of the quinoline ring, resulting in a partially saturated system. This modification disrupts the extended aromaticity and introduces a sp³-hybridized centre at position 2, potentially creating chirality if the substitution is asymmetrical. The loss of full conjugation modifies the molecule's reactivity profile. It enhances nucleophilicity and makes the molecule responsive to further derivatization. 1,4-Dihydroquinolines are structural isomers of 1,2-dihydroquinolines in which the reduction occurs at positions 1 and 4. This alteration similarly breaks aromaticity and introduces sp³-hybridized centres that can serve as stereocentres. The resulting compounds can exhibit enhanced reactivity in electrophilic and radical reactions, and the non-planar nature of the dihydro ring can impact molecular recognition in biological systems. These features are exploited in medicinal chemistry and organic synthesis for the design of scaffolds with tailored pharmacological profiles. 2,3-Dihydroquinolin-4(1H)-imine is a heterocyclic compound where the double bond between positions 2 and 3 of the quinoline ring is reduced, while an imine group (C=NH) is present at position 4. This configuration interrupts aromaticity and introduces tautomeric behaviour between the imine and enamine forms, influencing its chemical reactivity. The presence of the imine also introduces electrophilic character at C-4, facilitating condensation and cyclization reactions. Such compounds are important in heterocyclic synthesis, offering pathways for forming nitrogen-containing fused ring systems. 2,3-Dihydroquinolin-4(1H)-one is a closely related structure in which the 4-imine group is replaced by a carbonyl group. The saturation at positions 2 and 3 removes aromaticity in part of the ring, while the carbonyl group at position 4 enhances the molecule's stability and capacity for hydrogen bonding. This compound is used as a synthetic intermediate and structural motif in pharmaceuticals, particularly for alkaloid synthesis, due to its rigidity, functional group compatibility, and pharmacologically relevant scaffold.

The chapters in this book give an overview of class-defining structural characteristics of semi-aromatic quinoline derivatives, including the full chemical identity, a summary of general synthetic strategies to access such frameworks, and relevant findings from SAR studies for biological applications. Moreover, natural occurrences and sources for each class are discussed in detail, linking their presence in animals, plants, fungi, or bacteria to their ecological and biochemical significance. Ultimately, the biological activities of every group have been extensively discussed, highlighting their role in medicinal chemistry. Through this chapter-wise arrangement, the book seeks to provide a systematic and detailed account of the chemical diversity, natural abundance, and biological potential of semi-aromatic quinoline derivatives.

REFERENCES

Ajani, O. O., Ayayi, K. T., & Ademosun, O. T. (2022). Recent advances in chemistry and therapeutic potential of functionalized quinoline motifs–A review. *RSC Advances, 12*(29), 18594–18614.

Bar-On, Y. M., Phillips, R., & Milo, R. (2018). The biomass distribution on earth. *Proceedings of the National Academy of Sciences of the United States of America, 115*(25), 6506–6511.

Chopra, B., & Dhingra, A. K. (2021). Natural products: A lead for drug discovery and development. *Phytotherapy Research, 35*(9), 4660–4702.

Daley, S., & Cordell, G. A. (2021). Alkaloids in contemporary drug discovery to meet global disease needs. *Molecules, 26*(13), 3800.

Domingo-Fernández, D., Gadiya, Y., Preto, A. J., Krettler, C. A., Mubeen, S., Allen, A., Healey, D., & Colluru, V. (2024). Natural products have increased rates of clinical trial success throughout the drug development process. *Journal of Natural Products, 87*(7), 1844–1851.

Hamed, A. A., Ghareeb, M. A., Soliman, N. R., Bakchiche, B., & Bardaweel, S. K. (2024). Insights into bioactive microbial natural products and drug discovery. *Egyptian Pharmaceutical Journal, 23*(1), 1–15.

Khadem, S., & Marles, R. J. (2024). 2, 4–quinolinedione alkaloids: Occurrence and biological activities. *Natural Product Research*, 1–12. https://doi.org/10.1080/14786419.2024.23 90611

Khadem, S., & Marles, R. J. (2025a). Biological activity of natural 2-quinolinones. *Natural Product Research, 39*(5), 1359–1373.

Khadem, S., & Marles, R. J. (2025b). Natural 3, 4-dihydro-2 (1H)-quinolinones–part III: Biological activities. *Natural Product Research, 39*(8), 2252–2259.

Khadem, S., & Marles, R. J. (2025c). Natural 3, 4-dihydro-2 (1H)-quinolinones-part I: Plant sources. *Natural Product Research, 39*(3), 593–608.

Khadem, S., & Marles, R. J. (2025d). Natural 3, 4-dihydro-2 (1H)-quinolinones-part II: Animal, bacterial, and fungal sources. *Natural Product Research, 39*(2), 374–387.

Khadem, S., & Marles, R. J. (2025e). The occurrence and bioactivity of tetrahydronaphtho-quinoline-diones (THNQ-dione). *Natural Product Research, 39*(6), 1622–1635.

Khadem, S., & Marles, R. J. (2025f). Tetrahydroquinoline-containing natural products discovered within the last decade: Occurrence and bioactivity. *Natural Product Research, 39*(1), 182–194.

Letchuman, S., Madhuranga, H. D., Kaushalya, M., Premarathna, A. D., & Saravanan, M. (2024). Alkaloids unveiled: A comprehensive analysis of novel therapeutic properties, mechanisms, and plant-based innovations. *Intelligent Pharmacy, 3*(4), 268–276.

Newman, D. J., & Cragg, G. M. (2020). Natural products as sources of new drugs over the nearly four decades from 01/1981 to 09/2019. *Journal of Natural Products, 83*(3), 770–803.

Pringgenies, D., Santosa, G. W., Susanto, A. B., & Nuraini, R. A. (2025). Bioactive compounds in some marine plants and invertebrates. *AACL Bioflux, 18*(1), 393–405.

Rajput, A., Sharma, R., & Bharti, R. (2022). Pharmacological activities and toxicities of alkaloids on human health. *Materials Today: Proceedings, 48*, 1407–1415.

Yadav, P., & Shah, K. (2021). Quinolines, a perpetual, multipurpose scaffold in medicinal chemistry. *Bioorganic Chemistry, 109*, 104639.

Yadav, V., Reang, J., Sharma, V., Majeed, J., Sharma, P. C., Sharma, K., Giri, N., Kumar, A., & Tonk, R. K. (2022). Quinoline-derivatives as privileged scaffolds for medicinal and pharmaceutical chemists: A comprehensive review. *Chemical Biology & Drug Design, 100*(3), 389–418.

2 Tetrahydroquinoline Alkaloids

2.1 INTRODUCTION

1,2,3,4-Tetrahydroquinoline (THQ) is an alkaloid and semi-hydrogenated quinoline derivative. It comprises a bicyclic structure containing a benzene fused ring with a saturated six-membered nitrogen ring (piperidine). The numbering within the THQ ring system starts at the nitrogen of the piperidine ring and goes around the fused ring system. The nitrogen atom is given the number 1, and the carbon atoms are numbered consecutively around the ring system, with the carbon atom next to the nitrogen receiving the number 2 (Figure 2.1). The nitrogen, C2, C3, and C4 are all sp^3-hybridized. The lone pair electrons associated with the nitrogen in the THQ structure provide an area for hydrogen bond acceptor activity. Also, when the nitrogen or carbon adjacent to the nitrogen in the saturated ring has hydrogen atoms extending out, to form possible hydrogen bonds. The electron density adjacent to the nitrogen allows it to interact with H-bond donors, enhancing its ability to participate in intermolecular hydrogen bonds. Influencing the biological activity of THQs is its ability to form hydrogen bonds with different biomolecules, which enables it to undergo various biological activity processes and can further attract the field of drug design and molecular interactions towards THQs.

THQs are found in numerous natural products exhibiting biological activities, including antimalarial, antiviral, antifungal, and anticancer properties (Sabale et al. 2013). A few examples are provided in Figure 2.2. Helquinoline, isolated from the bacterium *Janibacter limosus* Hel 1 (family Intrasporangiaceae), is a THQ antibiotic with high biological activity against bacteria and fungi (Asolkar et al. 2004). Galipinine is a THQ alkaloid from the trunk bark of *Galipea officinalis* (= *Angostura trifoliata*, *Rutaceae* family) was shown to have antimalarial properties and cytotoxic effects against certain cancer cell lines (Jacquemond-Collet et al. 2002). Produced by *Streptomyces* species, Virantmycin is an antiviral agent effective against DNA and RNA viruses (Omura and Nakagawa 1980). L-689560 is a potent and selective N-methyl-D-aspartate receptor antagonist, with potential neuroprotective properties (Leeson et al. 1992). Nicainoprol has been studied for its antiarrhythmic effects

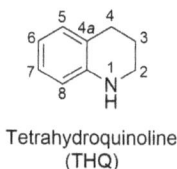

Tetrahydroquinoline
(THQ)

FIGURE 2.1 Chemical structure and numbering system of tetrahydroquinoline (THQ).

DOI: 10.1201/9781003534600-2

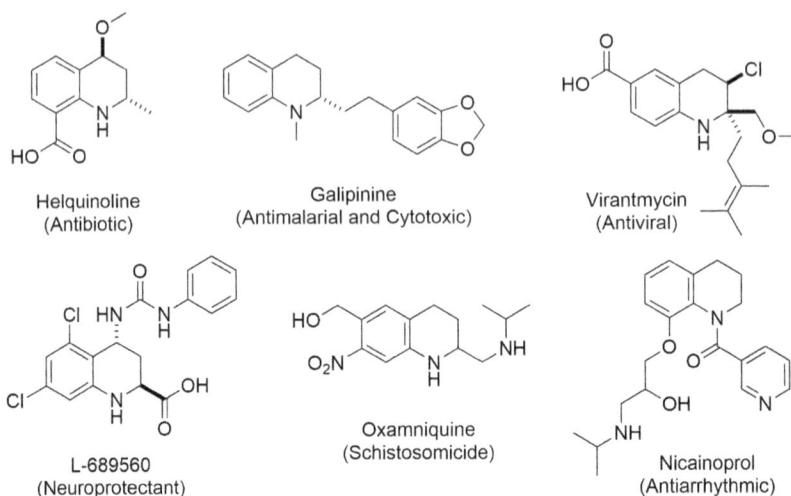

FIGURE 2.2 A few examples of THQ-containing drugs.

and possible use in treating cardiac disorders (Sen et al. 1986). As an anthelmintic drug, Oxamniquine is used to treat schistosomiasis, particularly infections caused by *Schistosoma mansoni* (Richards and Foster 1969). It works by interfering with the parasite's DNA, leading to its death (Pica-Mattoccia et al. 1989).

Many reviews on synthetic approaches for the preparation of THQs have been reported, such as Sridharan et al. (2011), Muthukrishnan et al. (2019), de Paiva et al. (2022), Nammalwar and Bunce (2013), and Limantseva et al. (2023). Synthetic biology has recently been indicated as a bioengineering platform to produce THQs (Klein et al. 2014). Nevertheless, in this particular insight, we intend to briefly review some efficient synthetic methods for generating the THQs (Katritzky et al. 1996; Khadem 2010).

The common traditional approach is the direct hydrogenation of unsaturated heterocycles. Several catalysts and reducing agents can be used, such as platinum catalysts for hydrogenation, cobalt stearate with triethylaluminium, palladium catalysts with formic acid and triethylamine, nickel-aluminium alloy for quinolinecarboxylic acid reductions, and sodium/ethanol for the reduction of 1,2-dihydroquinolines. Chiral catalysts, e.g., iridium catalysts with chiral bisphosphine ligands, have been developed to achieve high enantioselectivity in asymmetric hydrogenations. The above asymmetric reductions are performed under mild conditions, where Hantzsch esters deliver the hydrogen. These strategies have been utilized to synthesize natural products such as angustureine and galipinine (Muñoz and Dudley 2015).

The closure of the N–C (position No. 2) or the C4–C4a bond is also an approach to the synthesis of THQ. N–C bond formation can be accomplished via cyclocondensation of bromo compounds under basic conditions, reduction of nitro groups and subsequent cyclization or palladium-catalysed annulation of 1,4-dienes with ortho-iodoanilines. C4–C4a bond formation can be achieved by acid-catalysed cyclization of the corresponding precursors or intramolecular Diels–Alder cycloaddition of 2-substituted aminofurans.

THQs can be prepared in highly efficient one-pot, multi-step processes, such as tandem reactions. The examples include ozonolysis followed by catalytic hydrogenation for diastereoselective synthesis, as well as tandem Michael-aldol reactions of ortho-aminobenzaldehydes with α,β-unsaturated carbonyls. Other rearrangement reactions applicable to THQ scaffolds include the aza-Cope rearrangement and Mannich cyclization, and [3,3]-sigmatropic rearrangement of N-allyl-ortho-vinylanilines.

The formation of THQ can also be achieved using free radical cyclizations. They include tributyltin hydride-mediated cyclizations of ortho-iodoanilines in the presence of alkenes, as well as copper- or nickel-mediated atom transfer radical cyclizations. Another strategy for THQ synthesis involves transition metal-catalysed reactions. For example, palladium-catalysed carbonylative cyclization, ruthenium-catalysed ring-closing metathesis, and gold-catalysed cycloisomerization are typical intramolecular carbonylation processes.

Addition reactions to 1,2-dihydroquinolines have also been used. For instance, bromination of 1,2-dihydroquinolines with NBS provides 3-bromo-4-hydroxy-THQs, while chlorination followed by reduction can allow the formation of C3- and C4-substituents into THQs. Different types of cyclization reactions are introduced, including the catalytic hydrogenation and cyclization of ortho-nitrophenylpropanoic-acid derivatives to produce THQs. Similar to intramolecular Diels–Alder reactions of 2-aminofurans with tethered alkynes or alkenes, THQs can also be formed through oxa-bridge intermediates.

The 2,4-disubstituted THQs are obtained with high diastereoselectivity from ozonolysis of 2-nitrophenyl-butenes and subsequent catalytic hydrogenation. We also investigated the enantioselective synthesis of THQs; two routes to enantiomerically enriched 3-substituted THQs: asymmetric hydrogenation followed by catalytic reduction and protection, and asymmetric epoxidation, ring opening and reduction. We used Sharpless dihydroxylation and reductive cyclization for the simple and practical enantioselective synthesis of the THQ derivatives (Khadem et al. 2004, 2010).

Multicomponent reactions have evolved as relatively efficient strategies for THQ synthesis. Such multicomponent one-pot reactions are capable of generating highly elaborate architectures rapidly. One-pot condensation of anilines, aldehydes, and electron-rich alkenes, the so-called Povarov reaction, was popularized for the preparation of diverse THQ libraries. We made a one-pot, three-step synthesis of 2-amino-4-cyano-THQs from anilines, two different aldehydes, and trimethylsilyl cyanide (Khadem et al. 2009).

The basic structure of THQ alkaloids in plants is synthesized from anthranilic acid, a compound derived from shikimic acid. Anthraniloyl-coenzyme A (CoA) can react with malonyl-CoA to produce 4-hydroxy-2-quinolone, which can then be transformed into THQ or other substituted THQ alkaloids. Anthranilic acid is also involved in the biosynthesis of the amino acid L-tryptophan, which can be further metabolized back to anthranilic acid and used to synthesize other quinoline alkaloids. The pathways for alkaloid biosynthesis from anthranilic acid or L-tryptophan can be determined through isotopic labelling studies. In microbial biosynthesis, different pathways may be followed, such as the production of benzastatins from geranyl pyrophosphate and p-aminobenzoic acid, or the synthesis of 5-alkyl-1,2,3,4-THQ and streptoaminals from acetate units via polyketide synthase. The chemical identity (structure, name, and CAS number) of THQ alkaloids in this chapter is shown in Table 2.1.

2.2   NATURAL OCCURRENCE

While we will not discuss the THQ alkaloids discovered before 2012, we will briefly mention them here. Benzastatins C and D, as well as Virantmycin, were isolated from the cultural broth of *Streptomyces nitrosporeus* (the bacterial family Streptomycetaceae) (Kim et al. 1996; Omura and Nakagawa 1981). Martinelline and Martinellic acid were isolated from an extract of *Martinella iquitosensis* (Bignoniaceae) roots (Witherup et al. 1995). Jacquemond-Collet and coworkers (2001) study identified 15 alkaloids, among them 8 THQs, from the trunk bark of *Galipea officinalis* (Rutaceae). This Venezuelan shrubby tree is noted in folk medicine for its healing properties and is unique among the *Galipea* species for containing THQ alkaloids (Khan and Abourashed 2011). Figure 2.3 depicts the chemical structures of THQs from *G. officinalis*.

Table 2.2 illustrates the natural occurrence of THQ alkaloids across various biological kingdoms, highlighting their diverse origins. In the Animalia kingdom, these compounds are found in the beetle species *Allomyrina dichotoma* from the Scarabaeidae family (compounds 2.14–2.16). The Bacteria kingdom shows a significant presence, with numerous compounds (2.6-2.13, 2.17–2.20, and 2.28) produced by species in the Streptomycetaceae and Tsukamurellaceae families, particularly *Streptomyces nigrescens* and *Tsukamurella pulmonis* mixture. The Plantae kingdom also contributes several sources, with families such as Lamiaceae, Apiaceae, Orchidaceae, Apocynaceae, and Solanaceae. Notable genera include *Melodinus* (*Apocynaceae*) and *Lycium* (*Solanaceae*), which produce multiple compounds (2.21–2.27, 2.29). Additionally, compound 2.3 is derived from the fungus *Colletotrichum gloeosporioides* in the Fungi kingdom. This distribution underscores the extensive biosynthetic diversity leading to THQ alkaloids, which suggests diverse biological roles and potential applications in pharmaceuticals and agriculture. Moreover, it indicates these alkaloids' significant ecological and functional importance.

Figure 2.4 shows the distribution of occurrences across four kingdoms: Plantae, Bacteria, Animalia, and Fungi. Among these, plantae are the most common, accounting for 51.2%, followed by bacteria at 39.0%. Animalia comes in at 7.3%, and fungi make up the smallest portion at 2.5%.

2.3   BIOLOGICAL ACTIVITY

The various pharmacological effects of THQ alkaloids identified and reported before 2012 have been widely reviewed in many papers; hence, we will not elaborate on them here. Information on the various bioactivities of THQ alkaloids is summarised in Table 2.3 (also Khadem and Marles 2025). These compounds have broad therapeutic applications, such as antibacterial, antifungal, antiviral, anti-inflammatory, antioxidant, cytotoxic, and neuroprotective effects. They exhibited promising antibacterial

TABLE 2.1
Chemical Identity of Tetrahydroquinoline (THQ) Alkaloids

Structure (Number)	Systematic Name	Common Name	CAS Number
(2.1)	1,2,3,4-Tetrahydroquinoline	Tetrahydroquinoline	635-46-1
(2.2)	1,2,3,4-Tetrahydro-3-methylquinoline	–	20668-20-6
(2.3)	1,2,3,4-Tetrahydro-4,8-quinolinediol	–	1941224-59-4
(2.4)	1,2,3,4-Tetrahydro-2,2,4,6,7-pentamethylquinoline	–	49584-65-8
(2.5)	1,2,3,4-Tetrahydro-2-(1-methyl-1H-imidazol-4-yl)quinoline	Tetrahydromacrorine	2531-52-4
(2.6)	5-Heptyl-1,2,3,4-tetrahydroquinoline	5a-THQ-7n	1696430-98-4
(2.7)	1,2,3,4-Tetrahydro-5-octylquinoline	5a-THQ-8n	1696431-00-1
(2.8)	1,2,3,4-Tetrahydro-5-nonylquinoline	5a-THQ-9n	1696431-01-2

(Continued)

TABLE 2.1 (*Continued*)

Chemical Identity of Tetrahydroquinoline (THQ) Alkaloids

Structure (Number)	Systematic Name	Common Name	CAS Number
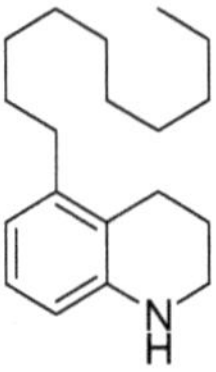 (2.9)	5-Decyl-1,2,3,4-tetrahydroquinoline	5a-THQ-10n	1696431-04-5
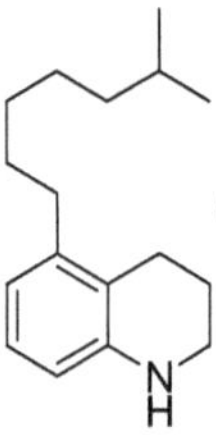 (2.10)	1,2,3,4-Tetrahydro-5-(6-methylheptyl)quinoline	5a-THQ-8i	1696430-99-5
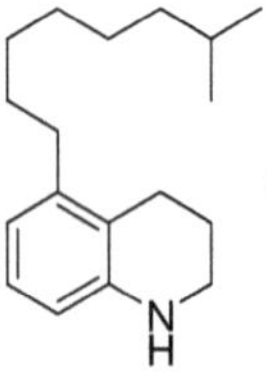 (2.11)	1,2,3,4-Tetrahydro-5-(7-methyloctyl)quinoline	5a-THQ-9i	1558031-18-7
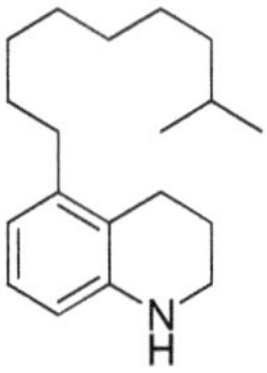 (2.12)	1,2,3,4-Tetrahydro-5-(8-methylnonyl)quinoline	5a-THQ-10i	1696431-03-4
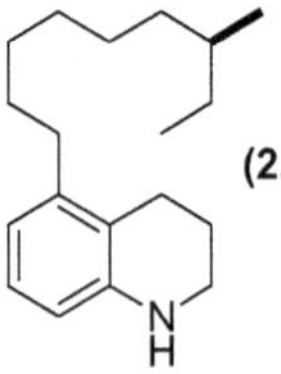 (2.13)	1,2,3,4-Tetrahydro-5-[(7S)-7-methylnonyl]quinoline	5a-THQ-10a	1696431-02-3
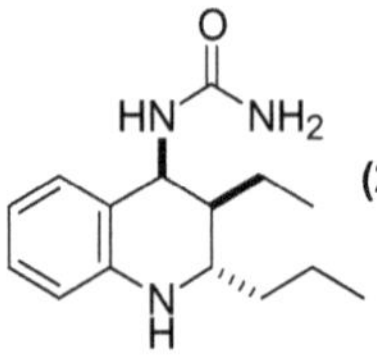 (2.14)	N-[(2S,3S,4S)-1,2,3,4-tetrahydro-2-propyl-3-ethyl-4-quinolinyl]-urea	Allomyrinaine A	–

(Continued)

TABLE 2.1 (*Continued*)
Chemical Identity of Tetrahydroquinoline (THQ) Alkaloids

Structure (Number)	Systematic Name	Common Name	CAS Number
(2.15)	N-[(2R,3S,4R)-1,2,3,4-tetrahydro-2-propyl-3-ethyl-4-quinolinyl]-urea	Allomyrinaine B	–
(2.16)	N-[(2S,3S,4R)-1,2,3,4-tetrahydro-2-propyl-3-ethyl-4-quinolinyl]-urea	Allomyrinaine C	–
(2.17)	(2R,3R)-2-(3,4-Dimethyl-3-penten-1-yl)-1,2,3,4-tetrahydro-3-hydroxy-N-(2-hydroxy-5-oxo-1-cyclopenten-1-yl)-2-(methoxymethyl)-6-quinolinecarboxamide	Virantmycin B	2299175-95-2
(2.18)	(7S,8aR,10aR)-5,6,7,8,8a,9,10,10a-octahydro-7-hydroxy-10a-(methoxymethyl)-7,8,8-trimethyl-2-acridinecarboxylic acid	Virantmycin D	2902718-97-0
(2.19)	(2R,3R)-3-chloro-1,2,3,4-tetrahydro-2-[3-(hydroxymethyl)-4-methyl-3-penten-1-yl]-2-(methoxymethyl)-6-quinolinecarboxylic acid	Virantmycin F	2902721-60-0
(2.20)	(2S,3S)-3-Chloro-2-(3,4-dimethyl-3-penten-1-yl)-1,2,3,4-tetrahydro-N-(2-hydroxy-5-oxo-1-cyclopenten-1-yl)-2-(methoxymethyl)-6-quinolinecarboxamide	Malamycin	2632983-21-0

(Continued)

TABLE 2.1 (*Continued*)
Chemical Identity of Tetrahydroquinoline (THQ) Alkaloids

Structure (Number)	Systematic Name	Common Name	CAS Number
(2.21) (R= COOMe)	–	Meloyunnanine A	2716939-73-8
(2.22) (R= COOMe)	–	Meloyunnanine B	2716939-75-0
(2.23) (R= COOMe)	–	Meloyunnanine C	2716939-77-2
(2.24) (R= COOMe)	–	Melognine	2664936-01-8
(2.25)	(2S,4S,5R)-4,5,6′,7′-Tetrahydro-4-hydroxy-5-(hydroxymethyl)spiro[furan-2(3H),2′(3′H)-[5H]pyrido[1,2,3-*de*][1,4]benzoxazine]-8′-carboxaldehyde	Lycibarbarine A	2701556-40-1
(2.26)	(2R,4S,5R)-4-hydroxy-5-(hydroxymethyl)-4,5,6′,7′-tetrahydro-3H,3′H,5′H-spiro[furan-2,2′-[1,4]oxazino[2,3,4-*ij*]quinoline]-8′-carbaldehyde	Lycibarbarine B	2701556-60-5

(*Continued*)

TABLE 2.1 (Continued)
Chemical Identity of Tetrahydroquinoline (THQ) Alkaloids

Structure (Number)	Systematic Name	Common Name	CAS Number
(2.27)	(2R,4S,5R)-3,4,5,6,6',7'-Hexahydro-4,5-dihydroxyspiro[2H-pyran-2,2'(3'H)-[5H]pyrido[1,2,3-de][1,4]benzoxazine]-8'-carboxaldehyde	Lycibarbarine C	2701556-61-6
(2.28)	1H-Imidazole-5-ethanaminium, α-carboxy-2-[[(2R,3R)-6-carboxy-2-(3,4-dimethyl-3-penten-1-yl)-1,2,3,4-tetrahydro-2-(methoxymethyl)-3-quinolinyl]thio]-N,N,N-trimethyl-, inner salt	JBIR-73	1342302-00-4
(2.29)	(6S,6aR,8R,12R,12aR,12bR)-12-ethyl-6,12b-dihydroxy-5,6,6a,7,9,10,11,12,12a,12b-decahydro-8H-6,12-ethanoindolizino[2,1-c]quinoline 8-oxide	Eugeniinaline A	–

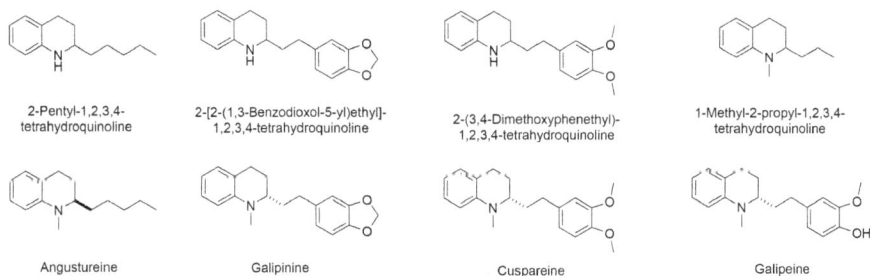

2-Pentyl-1,2,3,4-tetrahydroquinoline

2-[2-(1,3-Benzodioxol-5-yl)ethyl]-1,2,3,4-tetrahydroquinoline

2-(3,4-Dimethoxyphenethyl)-1,2,3,4-tetrahydroquinoline

1-Methyl-2-propyl-1,2,3,4-tetrahydroquinoline

Angustureine

Galipinine

Cuspareine

Galipeine

FIGURE 2.3 Chemical structures of THQ alkaloids isolated from *Galipea* officinalis.

activity against a wide range of pathogens. For instance, compound 2.2 was active against *Staphylococcus aureus*, *Bacillus subtilis*, *Escherichia coli*, and *Bacillus cereus*. Among these compounds, 2.6–2.8 and 2.11 exhibited antifungal activity, most notably against *Schizosaccharomyces pombe*. Compounds 2.17 and 2.19

TABLE 2.2

Natural Occurrence of Tetrahydroquinoline Alkaloids

Kingdom	Family	Genus	Species	Compound	References
Animalia	Scarabaeidae	*Allomyrina*	*dichotoma*	2.14	Park et al. (2020)
Animalia	Scarabaeidae	*Allomyrina*	*dichotoma*	2.15	Park et al. (2020)
Animalia	Scarabaeidae	*Allomyrina*	*dichotoma*	2.16	Park et al. (2020)
Bacteria	Streptomycetaceae and Tsukamurellaceae	*Streptomyces* and *Tsukamurella*	*nigrescens* HEK616 and *pulmonis* TP-B0596	2.6	Sugiyama et al. (2015)
Bacteria	Streptomycetaceae and Tsukamurellaceae	*Streptomyces* and *Tsukamurella*	*nigrescens* HEK616 and *pulmonis* TP-B0596	2.7	Sugiyama et al. (2015)
Bacteria	Streptomycetaceae and Tsukamurellaceae	*Streptomyces* and *Tsukamurella*	*nigrescens* HEK616 and *pulmonis* TP-B0596	2.8	Sugiyama et al. (2015)
Bacteria	Streptomycetaceae and Tsukamurellaceae	*Streptomyces* and *Tsukamurella*	*nigrescens* HEK616 and *pulmonis* TP-B0596	2.9	Sugiyama et al. (2015)
Bacteria	Streptomycetaceae and Tsukamurellaceae	*Streptomyces* and *Tsukamurella*	*nigrescens* HEK616 and *pulmonis* TP-B0596	2.10	Sugiyama et al. (2015)
Bacteria	Streptomycetaceae and Tsukamurellaceae	*Streptomyces* and *Tsukamurella*	*nigrescens* HEK616 and *pulmonis* TP-B0596	2.11	Sugiyama et al. (2015)
Bacteria	Streptomycetaceae and Tsukamurellaceae	*Streptomyces* and *Tsukamurella*	*nigrescens* HEK616 and *pulmonis* TP-B0596	2.12	Sugiyama et al. (2015)
Bacteria	Streptomycetaceae and Tsukamurellaceae	*Streptomyces* and *Tsukamurella*	*nigrescens* HEK616 and *pulmonis* TP-B0596	2.13	Sugiyama et al. (2015)
Bacteria	Streptomycetaceae	*Streptomyces*	sp. AM-2504	2.17	Kimura et al. (2019)
Bacteria	Streptomycetaceae	*Streptomyces*	*jiujiangensis* NBERC-24992	2.18	Liu et al. (2023)
Bacteria	Streptomycetaceae	*Streptomyces*	*jiujiangensis* NBERC-24992	2.19	Liu et al. (2023)

(Continued)

TABLE 2.2 (*Continued*)
Natural Occurrence of Tetrahydroquinoline Alkaloids

Kingdom	Family	Genus	Species	Compound	References
Bacteria	Streptomycetaceae	*Streptomyces*	*malaysiensis* SCSIO41397	2.20	Xie et al. (2021)
Bacteria	Streptomycetaceae	*Streptomyces*	sp. RI18	2.28	Motohashi et al. (2011)
Fungi	Asparagaceae	*Colletotrichum*	*gloeosporioides* strain B-142	2.3	Zhang et al. (2022)
Plantae	Lamiaceae	*Lagochilus*	*ilicifolius*	2.2	Dumaa et al. (2015)
Plantae	Apiaceae	*Anethum*	*Graveolens* (*sowa*)	2.4	Al Mansur et al. (2017)
Plantae	Orchidaceae	*Dendrobium*	*crepidatum*	2.5	Paudel et al. (2019)
Plantae	Apocynaceae	*Melodinus*	*yunnanensis* (*fusiformis*)	2.21	Wu et al. (2020)
Plantae	Apocynaceae	*Melodinus*	*Yunnanensis* (*fusiformis*)	2.22	Wu et al. (2020)
Plantae	Apocynaceae	*Melodinus*	*Yunnanensis* (*fusiformis*)	2.23	Wu et al. (2020)
Plantae	Apocynaceae	*Melodinus*	*fusiformis*	2.24	Zhou et al. (2021)
Plantae	Solanaceae	*Lycium*	*barbarum*	2.25	Chen et al. (2021)
Plantae	Solanaceae	*Lycium*	*barbarum*	2.26	Chen et al. (2021)
Plantae	Solanaceae	*Lycium*	*barbarum*	2.27	Chen et al. (2021)
Plantae	Apocynaceae	*Leuconotis*	*eugeniifolia*	2.29	Tan et al. (2024)

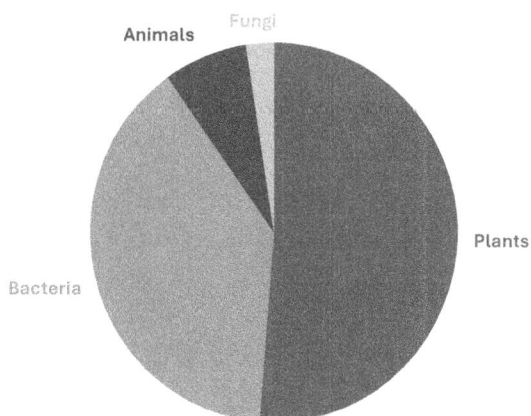

FIGURE 2.4 Distribution of THQ alkaloids in various life kingdoms.

TABLE 2.3
Biological Activity of Tetrahydroquinoline Alkaloids

Biological Activity	Target	Assay/Method	Results	Compound	References
Antibacterial	*Staphylococus aureus* (American Type Culture Collection (ATCC) 25923), *B. subtilis* (ATCC 6151), *Escherichia coli* (ATCC 25922), *B. cereus*	Agar disc diffusion method for extracts, fractions, and total alkaloids	Active at 500 μg/disc; low or inactive at 100 μg/disc	2.2	Dumaa et al. (2015)
Antifungal	Wild-type fission yeast cells (*Schizosaccharomyces pombe*)	—	MIC = 6.3 μM	2.6	Sugiyama et al. (2015)
Antifungal	Wild-type fission yeast cells (*S. pombe*)	—	MIC~ 100 μM	2.7	Sugiyama et al. (2015)
Antifungal	Wild-type fission yeast cells (*S. pombe*)	—	MIC~ 100 μM (20% growth)	2.8	Sugiyama et al. (2015)
Antifungal	Wild-type fission yeast cells (*S. pombe*)	—	MIC~ 100 μM (40% growth)	2.11	Sugiyama et al. (2015)
Anti-inflammatory	LPS-mediated human endothelial cells	—	Inhibited LPS-mediated barrier disruption by increasing barrier integrity; reduced neutrophils adhesion and migration towards human umbilical vein endothelial cells and inhibiting the expression of cell adhesion molecules; increase in the survival rate from 40% to 60%	2.14	Park et al. (2020)

(Continued)

TABLE 2.3 (*Continued*)
Biological Activity of Tetrahydroquinoline Alkaloids

Biological Activity	Target	Assay/Method	Results	Compound	References
Anti-inflammatory	LPS-mediated human endothelial cells	—	Inhibited LPS-mediated barrier disruption by increasing barrier integrity; reduced neutrophils adhesion and migration toward human umbilical vein endothelial cells and inhibiting the expression of cell adhesion molecules; increase in the survival rate from 40% to 60%	2.15	Park et al. (2020)
Anti-inflammatory	LPS-mediated human endothelial cells	—	Inhibited LPS-mediated barrier disruption by increasing barrier integrity; reduced neutrophils adhesion and migration toward human umbilical vein endothelial cells and inhibiting the expression of cell adhesion molecules; increased in the survival rate from 40% to 60%	2.16	Park et al. (2020)
Antioxidant	DPPH free radicals	DPPH assay	Ethanol and acetone extracts scavenged 94.7% and 93.4% of DPPH free radicals	2.5	Paudel et al. (2019)
Antiviral	Dengue virus type 2, New Guinea C strain	—	$IC_{50} = 43.2\ \mu M$	2.17	Kimura et al. (2019)
Antiviral	Pseudorabies virus	MTT assay	$EC_{50} = 1.74\ \mu g/mL$	2.19	Liu et al. (2023)
Cytotoxicity	HeLa (human cervical carcinoma) cells, U251 (human brain glioblastoma) cells	MTT Assay for Chloroform and Hexane extracts	At 800 µg/mL, the extracts inhibited the growth of 81.5% of HeLa cells and 76.5% of U251 cells	2.5	Paudel et al. (2019)
Cytotoxicity	Cancer cell lines C42B, 22Rv1, H446, MGC803, MDA-MB231	Cell viability, colony formation, caspase-3/7 activity and cell growth, qRT-PCR, and western blotting protocols	$IC_{50} = 0.067, 0.27, 0.07, 0.11, 0.20$ (respectively); affected the AR at both the mRNA and protein levels in a dose-dependent manner and suppressed the expression of androgen receptor target genes KLK2 and KLK3 in the C42B and 22RV1 cell lines	2.20	Xie et al. (2021)

(*Continued*)

TABLE 2.3 (*Continued*)

Biological Activity of Tetrahydroquinoline Alkaloids

Biological Activity	Target	Assay/Method	Results	Compound	References
Cytotoxicity	Human breast cancer BT549 cells	MTT assay	$IC_{50} = 1.49\,\mu M$; induces apoptosis in BT549 cells and upregulates the expression of caspase-3 and p53 while downregulating the expression of anti-apoptotic Bcl-2	2.24	Zhou et al. (2021)
Cytotoxicity	Human cervical carcinoma HeLa cell lines	—	$IC_{50} \sim 50\,\mu M$	2.28	Motohashi et al. (2011)
Cytotoxicity	Human colorectal adenocarcinoma HT-29 cancer cell line	—	$IC_{50} = 7.1\,\mu M$	2.29	Tan et al. (2024)
Neuroprotective (antidepressant)	Corticosterone-induced PC12 cell	In-cell western blotting	Within the range of $1-20\,\mu M$, demonstrated protective activity against injuries of PC12 cells; effectively reduced the expression of the cleavage level of caspase-3 and caspase-9	2.25	Chen et al. (2021)
Neuroprotective (antidepressant)	Corticosterone-induced PC12 cell	In-cell western blotting	Within the range of $1-20\,\mu M$, demonstrated protective activity against injuries of PC12 cells; effectively reduced the expression of the cleavage level of caspase-3 and caspase-9	2.26	Chen et al. (2021)
Neuroprotective (antidepressant)	Corticosterone-induced PC12 cell	In-cell western blotting	Within the range of $1-20\,\mu M$, demonstrated protective activity against injuries of PC12 cells; effectively reduced the expression of the cleavage level of caspase-3 and caspase-9	2.27	Chen et al. (2021)

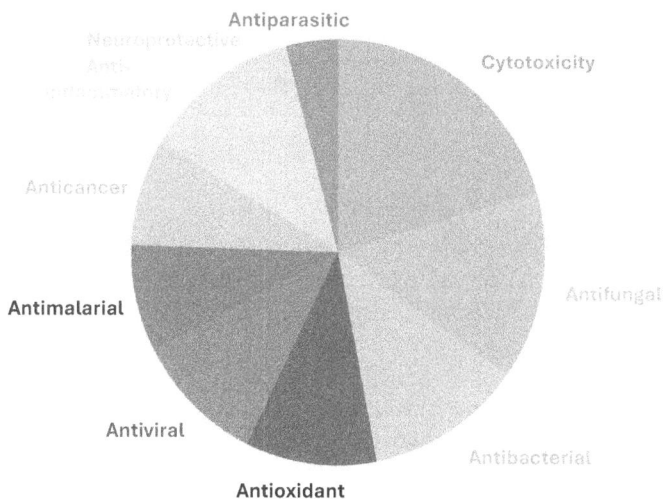

FIGURE 2.5 Biological activity of THQ alkaloids.

showed antiviral effects against the dengue virus and a pseudorabies virus, respectively. Some THQs have a particular interest due to the cytotoxic potential (2.20, 2.24, 2.28, and 2.29), presenting activity towards breast, cervical, and colorectal cell lines. Compound 2.20, for example, exhibits potent cytotoxicity in a broad range of cancer cell lines and induces androgen receptor expression. Compounds 2.14–2.16 also possess anti-inflammatory properties that inhibit lipopolysaccharide (LPS)-induced barrier disruption in human endothelial cells and reduce neutrophil adhesion and migration. Compound 2.5 demonstrates antioxidant activity through the effective scavenging of 2,2-diphenyl-1-picrylhydrazyl (DPPH) free radicals. Interestingly, compounds 2.25–2.27 have neuroprotective activity in corticosterone-induced PC12, which has antidepressant-like activity, by the ability to downregulate the expression of caspase-3 and -9.

The biological effects of THQs, including those identified before 2012, are shown in Figure 2.5. Bioactivities include cytotoxicity (20.4%), which represents killing and/or growth-inhibiting functions that are important in screening for anticancer therapeutics. Antifungal activity, which is 14.3%, allows the treatment of fungal diseases in human health and agriculture, improving food security. Antibacterial properties (12.2%), due to the increase in antibiotic resistance, can serve as the basis for the production of new antibiotics. The protection against oxidative stress (10.2%) or antioxidant activity highlights the development of nutraceuticals and functional foods for health and longevity. Likewise, antiviral activity (10.2%), which targets viral infections, is critical for fighting viral pathogens that have emerged. Antimalarials represent 8.2% of compounds through anti-Plasmodium drug discovery, targeting the pervasive global burden of malaria. Compounds with anticancer activity (8.2%) are selective for cancerous tissues, while anti-inflammatory drugs (6.1%) are useful for chronic inflammation, which is involved in the pathogenesis of many diseases. Neuroprotection (6.1%), like antidepressant activity, protects nerve cells and potentially is used for harmful neurodegenerative disorders. Antiparasitic activity (4.1%) specific for protozoan parasites is vital in developing treatment for target species, since there are currently limited advanced anti-parasitic options.

2.4   DISCUSSION

THQs have many different biological activities, showing their potential for use in various medical areas. Their natural occurrence in various organisms suggests that nature has optimized these compounds for specific biological functions. The structural diversity of THQs allows for extensive chemical modifications, which enable the optimization of their pharmacological properties. Researchers can design and synthesize new THQ derivatives with enhanced potency, selectivity, and reduced toxicity. The development of structure–activity relationship studies is crucial for understanding the key structural features responsible for their biological activities and guiding the design of more effective compounds. Despite their potential, several challenges remain in the development of THQ-based drugs. One major challenge is the synthesis and scalability of these compounds, as complex natural products often require complex and costly synthesis routes. Additionally, ensuring the safety and efficacy of THQ derivatives in clinical settings is crucial, which necessitates rigorous preclinical and clinical testing. Future research on THQs should focus on exploring their mechanisms of action, identifying novel natural sources, and developing new synthetic methodologies to access diverse derivatives. Advances in computational chemistry and high-throughput screening can accelerate the discovery of new THQ compounds with desirable biological activities. Moreover, investigating the synergistic effects of THQs with other bioactive compounds could lead to the development of combination therapies with improved therapeutic outcomes. Understanding the pharmacokinetics and pharmacodynamics of THQ derivatives will also be essential for optimizing their clinical applications.

REFERENCES

Al Mansur, M. A., Siddiqi, M. M. A., Akbor, M. A., & Saha, K. (2017). Phytochemical screening and GC-MS chemical profiling of ethyl acetate extract of seed and stem of *Anethum sowa* Linn. *Dhaka University Journal of Pharmaceutical Sciences, 16*(2), 187–194.

Asolkar, R. N., Schroeder, D., Heckmann, R., Lang, S., Wagner-Doebler, I., & Laatsch, H. (2004). Helquinoline, a new tetrahydroquinoline antibiotic from *Janibacter limosus* Hel 1. *The Journal of Antibiotics, 57*(1), 17–23.

Chen, H., Kong, J., Zhang, L., Wang, H., Cao, Y., Zeng, M., Li, M., Sun, Y., Du, K., & Xue, G. (2021). Lycibarbarines A–C, three tetrahydroquinoline alkaloids possessing a spiro-heterocycle moiety from the fruits of *Lycium barbarum*. *Organic Letters, 23*(3), 858–862.

de Paiva, W. F., de Freitas Rego, Y., de Fátima, Â, & Fernandes, S. A. (2022). The Povarov reaction: A versatile method to synthesize tetrahydroquinolines, quinolines and julolidines. *Synthesis, 54*(14), 3162–3179.

Dumaa, M., Gerelt-Od, Y., Javzan, S., Otgonhkishig, D., Doncheva, T., Yordanova, G., Philipov, S., & Selenge, D. (2015). GC-MS analysis and antibacterial activity of some fractions from *Lagochilus ilicifolius* Bge. grown in Mongolia. *Mongolian Journal of Chemistry, 16*, 39–43.

Jacquemond-Collet, I., Bessière, J., Hannedouche, S., Bertrand, C., Fourasté, I., & Moulis, C. (2001). Identification of the alkaloids of *Galipea officinalis* by gas chromatography-mass spectrometry. *Phytochemical Analysis: An International Journal of Plant Chemical and Biochemical Techniques, 12*(5), 312–319.

Jacquemond-Collet, I., Benoit-Vical, F., Valentin, M., Stanislas, A., Mallié, E., & Fourasté, M. (2002). Antiplasmodial and cytotoxic activity of galipinine and other tetrahydroquinolines from *Galipea officinalis*. *Planta Medica, 68*(01), 68–69.

Katritzky, A. R., Rachwal, S., & Rachwal, B. (1996). Recent progress in the synthesis of 1, 2, 3, 4,-tetrahydroquinolines. *Tetrahedron, 52*(48), 15031–15070.

Khadem, S. (2010). The development of stereocontrolled methods to obtain a diverse set of tetrahydroquinoline-based, natural product-inspired compounds (PhD Thesis), University of Ottawa, Canada.

Khadem, S., Joseph, R., Rastegar, M., Leek, D. M., Oudatchin, K. A., & Arya, P. (2004). A solution-and solid-phase approach to tetrahydroquinoline-derived polycyclics having a 10-membered ring. *Journal of Combinatorial Chemistry, 6*(5), 724–734.

Khadem, S., & Marles, R. J. (2025). Tetrahydroquinoline-containing natural products discovered within the last decade: Occurrence and bioactivity. *Natural Product Research, 39*(1), 182–194.

Khadem, S., Udachin, K. A., & Arya, P. (2010). Solution-and solid-phase synthesis of tetrahydroquinoline-based polycyclics having α, β-unsaturated γ-lactam and δ-lactone functionalities. *Synlett, 2010*(2), 199–202.

Khadem, S., Udachin, K. A., Enright, G. D., Prakesch, M., & Arya, P. (2009). One-pot construction of isoindolo [2, 1-a] quinoline system. *Tetrahedron Letters, 50*(48), 6661–6664.

Khan, I. A., & Abourashed, E. A. (2011). *Leung's encyclopedia of common natural ingredients: Used in food, drugs and cosmetics*. Hoboken, NJ: John Wiley & Sons.

Kim, W., Kim, J., & Ick-Dong, Y. (1996). Benzastatins A, B, C, and D: New free radical scavengers from *streptomyces nitrosporeus* 30643 II. Structure determination. *The Journal of Antibiotics, 49*(1), 26–30.

Kimura, T., Suga, T., Kameoka, M., Ueno, M., Inahashi, Y., Matsuo, H., Iwatsuki, M., Shigemura, K., Shiomi, K., & Takahashi, Y. (2019). New tetrahydroquinoline and indoline compounds containing a hydroxy cyclopentenone, virantmycin B and C, produced by streptomyces sp. AM-2504. *The Journal of Antibiotics, 72*(3), 169–173.

Klein, J., Heal, J. R., Hamilton, W. D., Boussemghoune, T., Tange, T. Ø, Delegrange, F., Jaeschke, G., Hatsch, A., & Heim, J. (2014). Yeast synthetic biology platform generates novel chemical structures as scaffolds for drug discovery. *ACS Synthetic Biology, 3*(5), 314–323.

Leeson, P. D., Carling, R. W., Moore, K. W., Moseley, A. M., Smith, J. D., Stevenson, G., Chan, T., Baker, R., Foster, A. C., & Grimwood, S. (1992). 4-amido-2-carboxytetrahydro-quinolines. structure-activity relationships for antagonism at the glycine site of the NMDA receptor. *Journal of Medicinal Chemistry, 35*(11), 1954–1968.

Limantseva, R. M., Savchenko, R. G., Odinokov, V. N., & Tolstikov, A. G. (2023). Povarov reaction in the synthesis of polycyclic nitrogen compounds containing a tetrahydroquinoline fragment. *Russian Journal of Organic Chemistry, 59*(7), 1102–1135.

Liu, M., Ren, M., Zhang, Y., Wan, Z., Wang, Y., Wu, Z., Wang, K., Fang, W., & Yang, X. (2023). Antiviral activity of benzo heterocyclic compounds from soil-derived *streptomyces jiujiangensis* NBERC-24992. *Molecules, 28*(2), 878.

Motohashi, K., Nagai, A., Takagi, M., & Shin-ya, K. (2011). Two novel benzastatin derivatives, JBIR-67 and JBIR-73, isolated from *Streptomyces* sp. RI18. *The Journal of Antibiotics, 64*(3), 281–283.

Muñoz, G. D., & Dudley, G. B. (2015). Synthesis of 1, 2, 3, 4-tetrahydroquinolines including angustureine and congeneric alkaloids. A review. *Organic Preparations and Procedures International, 47*(3), 179–206.

Muthukrishnan, I., Sridharan, V., & Menendez, J. C. (2019). Progress in the chemistry of tetrahydroquinolines. *Chemical Reviews, 119*(8), 5057–5191.

Nammalwar, B., & Bunce, R. A. (2013). Recent syntheses of 1, 2, 3, 4-tetrahydroquinolines, 2, 3-dihydro-4(1*H*)-quinolinones and 4(1*H*)-quinolinones using domino reactions. *Molecules, 19*(1), 204–232.

Omura, S., & Nakagawa, A. (1981). Structure of virantmycin, a novel antiviral antibiotic. *Tetrahedron Letters, 22*(23), 2199–2202.

Park, I., Lee, W., Yoo, Y., Shin, H., Oh, J., Kim, H., Kim, M., Hwang, J. S., Bae, J., & Na, M. (2020). Protective effect of tetrahydroquinolines from the edible insect *allomyrina dichotoma* on LPS-induced vascular inflammatory responses. *International Journal of Molecular Sciences, 21*(10), 3406.

Paudel, M. R., Chand, M. B., Pant, B., & Pant, B. (2019). Assessment of antioxidant and cytotoxic activities of extracts of *Dendrobium crepidatum*. *Biomolecules, 9*(9), 478.

Pica-Mattoccia, L., Cioli, D., & Archer, S. (1989). Binding of oxamniquine to the DNA of schistosomes. *Transactions of the Royal Society of Tropical Medicine and Hygiene, 83*(3), 373–376.

Richards, H. C., & Foster, R. (1969). A new series of 2-aminomethyltetrahydroquinoline derivatives displaying schistosomicidal activity in rodents and primates. *Nature, 222*(5193), 581–582.

Sabale, P. M., Patel, P., & Kaur, P. (2013). 1, 2, 3, 4–tetrahydroquinoline derivatives and its significance in medicinal chemistry. *Asian Journal of Research in Chemistry, 6*(6), 599–610.

Sen, S., Rettig, G., Özbek, C., Fröhlig, G., Schieffer, H., & Bette, L. (1986). Electrophysiological effects of a new antiarrhythmic agent, nicainoprol, in humans. *Journal of Cardiovascular Pharmacology, 8*(1), 144–150.

Sugiyama, R., Nishimura, S., Ozaki, T., Asamizu, S., Onaka, H., & Kakeya, H. (2015). 5-Alkyl-1, 2, 3, 4-tetrahydroquinolines, new membrane-interacting lipophilic metabolites produced by combined culture of Streptomyces nigrescens and Tsukamurella pulmonis. *Organic Letters*, *17*(8), 1918–1921.

Sridharan, V., Suryavanshi, P. A., & Menendez, J. C. (2011). Advances in the chemistry of tetrahydroquinolines. *Chemical Reviews, 111*(11), 7157–7259.

Tan, Y., Ng, M., Tan, C., Tang, W., Sim, K., Yong, K., Krishnan, P., Lim, K., Lim, S., & Low, Y. (2024). Quinoline, indole, and isogranatanine alkaloids from Malayan *Leuconotis eugeniifolia*. *Journal of Natural Products, 87*(2), 286–296.

Witherup, K. M., Ransom, R. W., Graham, A. C., Bernard, A. M., Salvatore, M. J., Lumma, W. C., Anderson, P. S., Pitzenberger, S. M., & Varga, S. L. (1995). Martinelline and martinellic acid, novel G-protein linked receptor antagonists from the tropical plant *Martinella iquitosensis* (bignoniaceae). *Journal of the American Chemical Society, 117*(25), 6682–6685.

Wu, J., Zhao, S., Shi, B., Bao, M., Schinnerl, J., & Cai, X. (2020). Cage-monoterpenoid quinoline alkaloids with neurite growth promoting effects from the fruits of *Melodinus yunnanensis*. *Organic Letters, 22*(19), 7676–7680.

Xie, Y., Guo, L., Huang, J., Huang, X., Cong, Z., Liu, Q., Wang, Q., Pang, X., Xiang, S., & Zhou, X. (2021). Cyclopentenone-containing tetrahydroquinoline and geldanamycin alkaloids from *Streptomyces malaysiensis* as potential anti-androgens against prostate cancer cells. *Journal of Natural Products, 84*(7), 2004–2011.

Zhang, B., Huang, Z., Si, Y., Zhang, K., Xu, X., Chen, J., Zhao, Q., & Zhang, X. (2024). A new 1, 4-benzoxazine derivative produced by endophytic *Colletotrichum gloeosporioides* B-142. *Natural Product Research, 38*(8), 1341–1346.

Zhou, J., Du, S., Fang, Z., & Feng, J. (2021). Melognine, a novel monoterpenoid indole alkaloid from *Melodinus fusiformis* that induce apoptosis in BT549 cells. *Natural Product Research, 35*(18), 3004–3010.

3 Tetrahydronaphtho-quinoline-diones (THNQ-dione)

3.1 INTRODUCTION

Anthraquinones are widely distributed in nature and represent the largest group of natural quinones (Malik & Muller, 2016; Diaz-Munoz et al. 2018). With approximately 700 compounds identified, this diverse group includes benzoquinones and naphthoquinones, mostly obtained from plants, lichens, bacteria, insects, and fungi. Anthraquinones share a common tricyclic aromatic organic structure known as 9,10-anthracenedione. Anthraquinones are known natural pigments, exhibiting various biological activities, including anti-tumour, anti-inflammatory, diuretic, laxative, anti-arthritic, antifungal, anti-bacterial, antimalarial, and antioxidant properties. In plants, they play a role in primary metabolism by affecting the electron transport chain and inhibiting energy transfer during photosynthesis. Moreover, they interact with DNA and inhibit the topoisomerase II enzyme, which leads to cell apoptosis. As a result, several DNA-intercalating compounds derived from anthraquinones are currently approved by the FDA for clinical use.

As seen in the previous chapter, the tetrahydroquinoline (THQ) structure emerges as particularly vital. It serves as a key heterocyclic framework present in a multitude of compounds, whether they are naturally occurring or synthetically derived (Muthukrishnan et al. 2019; Khadem and Marles 2025a). The significance of THQ lies in its association with compounds that possess remarkable biological properties, making it a focal point of interest in drug discovery.

1,2,3,4-Tetrahydronaphtho[2,3-*h*]quinoline-7,12-dione (simplified as tetrahydronaphthoquinoline-dione, THNQ-dione) [compound 3 in Figure 3.1] is a core structure found in various natural and synthetic compounds. The distinct characteristics of THNQ-diones may be attributed to the fusion of tetrahydroquinoline and anthraquinone moieties within their molecular framework, with phenyl as the bridge. This combination gives these compounds unique physical, chemical, and biological properties, making them interesting for scientific study. The numbering of the THNQ ring system begins at the nitrogen atom of the piperidine ring and proceeds around the fused ring system. The nitrogen atom is assigned the number 1, and the carbon atoms are numbered consecutively around the ring system, with the carbon atom next to the nitrogen atom receiving the number 2.

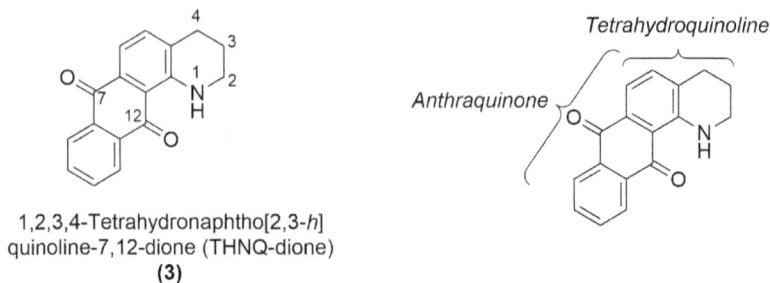

1,2,3,4-Tetrahydronaphtho[2,3-*h*]
quinoline-7,12-dione (THNQ-dione)
(3)

FIGURE 3.1   Chemical structure, numbering system, and structural features of THNQ-dione.

Anthraquinone-fused enediynes (AFEs), a subgroup of THQN-dione natural products, have received consistent attention since the discovery of dynemicin A (DYN A) (Konishi et al. 1989, 1990). One distinguishing feature is their 10-membered enediyne core, which provides a mechanism of action similar to that of other enediynes. Specifically, the Bergman or Myers-Saito rearrangement of the enediyne carbocycle forms a benzenoid diradical capable of extracting hydrogen atoms from DNA's deoxyribose structure. DNA-focused radicals can break either one or both strands of DNA, which can lead to cell death. On the other hand, the anthracycline family of anti-tumour antibiotics inserts itself into DNA, aided by their flat aromatic rings (Yan 2022).

We will briefly explore the methodologies employed for synthesizing THQN-diones in the literature, highlighting their significance within the field. Then, we will present a short review of the biosynthesis of AFEs. Finally, this chapter will focus on their natural occurrence across diverse organisms and their potential as a source of bioactive natural products. The chemical identity of THQN-dione alkaloids presented in this chapter is shown in Table 3.1.

3.2   SYNTHESIS AND BIOSYNTHESIS

3.2.1   SYNTHESIS

An earlier synthesis of THQN-dione compounds was accomplished using aromatic amines, formaldehyde, and electron-rich alkenes (Mellor et al., 1991). For instance, when 1-aminoanthraquinone, formaldehyde, and styrene were combined in an acidic medium, it resulted in a high yield of 4-phenyl-THQN-dione (Figure 3.2). Notably, the authors presented evidence suggesting that the reaction follows a nonconcerted, multi-step mechanism instead of a concerted [4+2] Diels–Alder process.

Albooyeh and Aghapour (2023) synthesized the THQN-dione system (3) by the aza-Claisen rearrangement of 1-(*N*-propargylamino)anthraquinone using zinc powder in 1-methylimidazolium tetrafluoroborate [Hmim]BF$_4$ as an ionic liquid in excellent yields. It is believed that the mechanism involves a reductive [3,3]-sigmatropic reaction followed by cyclization in a tandem manner (Figure 3.3).

Enediyne natural products have strong antibacterial and anti-tumour effects. Their effectiveness has led to the development of enediyne-based antibody-drug conjugates.

TABLE 3.1
The Chemical Identity of Tetrahydronaphtoquinoline-dione (THQN-dione) Alkaloids

Structure (Number)	Systematic Name	Common Name	CAS Number
(3.1)	4a,14a-Epoxy-4,14-[3]hexene[1,5]diynonaphtho[2,3-*c*]phenanthridine-2-carboxylic acid, 1,4,7,12,13,14-hexahydro-6,11-dihydroxy-3-methoxy-1-methyl-7,12-dioxo-, (1R*,4S*,4aS*,14R*,14aR*)-	Deoxydyne-micin A	130640-33-4
(3.2)	(1S,4R,4aR,14S,14aS,18Z)-1,4,7,12,13,14-Hexahydro-6,8,11-trihydroxy-3-methoxy-1-methyl-7,12-dioxo-4a,14a-epoxy-4,14-[3]hexene[1,5]diynonaphtho[2,3-c]phenanthridine-2-carboxylic acid	Dynemicin A (DYN A)	124412-57-3
(3.3)	4a,14a-Epoxy-4,14-[3]hexene[1,5]diynonaphtho[2,3-c]phenanthridine-2-carboxaldehyde, 1,4,7,12,13,14-hexahydro-6,8,11-trihydroxy-3-methoxy-1-methyl-7,12-dioxo-, (1S,4R,4aR,14S,14aS,18Z)-	DYN C	143250-23-1
(3.4)	9,8,14-[1]Buten[1]yl[4]ylideneanthra[1,2-b]benz[f]azocine-19-carboxylic acid, 5,8,9,14,15,16-hexahydro-1,4,6,17-tetrahydroxy-20-methoxy-18-methyl-5,16-dioxo-, (8S,9R,14S,17R,18R)-	DYN H	129985-03-1

(Continued)

TABLE 3.1 (*Continued*)
The Chemical Identity of Tetrahydronaphtoquinoline-dione
(THQN-dione) Alkaloids

Structure (Number)	Systematic Name	Common Name	CAS Number
(3.5)	9,8,14-[1]Buten[1]yl[4] ylideneanthra[1,2-b]benz [f]azocine-19-carboxylic acid, 8-chloro-5,8,9,14,15,16-hexahydro- 1,4,6,17-tetrahydroxy-20-methoxy- 18-methyl-5,16-dioxo-, (8R,9R,14S,17S,18S)-	DYN L	127032-74-0
(3.6)	9,8,14-[1]Butanyl[4]ylideneanthra [1,2-b]benz[f]azocine-5,16,19- trione, 8,9,14,15-tetrahydro- 1,4,6,8,17-pentahydroxy-20- methoxy-18-methyl-, (8R,9R, 14S,17S,18R)-	(DYN M)	127003-54-7
(3.7)	9,8,14-[1]Buten[1]yl[4] ylideneanthra[1,2-b]benz [f]azocine-19-carboxylic acid, 5,8,9,14,15,16-hexahydro- 1,4,6,8,17-pentahydroxy-20- methoxy-18-methyl- 5,16-dioxo-, (8R,9S,14S, 17S,18S)-	DYN N	127003-55-8
(3.8)	9,8,14-[1]Butanyl[4]ylideneanthra [1,2-b]benz[f]azocine-5,16,19- trione, 8,9,14,15-tetrahydro- 1,4,6,8,10,17-hexahydroxy- 20-methoxy-18-methyl-, (8R,9R,14S,17S,18R)-	DYN O	138370-12-4

(*Continued*)

TABLE 3.1 (*Continued*)
The Chemical Identity of Tetrahydronaphtoquinoline-dione
(THQN-dione) Alkaloids

Structure (Number)	Systematic Name	Common Name	CAS Number
(3.9)	9,8,14-[1]Butanyl[4]ylideneanthra [1,2-b]benz[f]azocine-5,16,19-trione, 8,9,14,15-tetrahydro-1,4,6,8,17,20-hexahydroxy-18-methyl-, (8R,9R,14S,17S,18R)-	DYN P	138370-13-5
(3.10)	9,8,14-[2]Buten[1]yl[4]ylideneanthra [1,2-b]benz[f]azocine-5,6,20-trione, 8,9,14,15-tetrahydro-1,4,6,8,17,19-hexahydroxy-18-methyl-, (8R,9S,14S,17S)-	DYN Q	138370-14-6
(3.11)	(2R,3R,4S,6R)-1,3,4,6-Tetrahydro-3-hydroxy-2,4,12-trimethyl-2,6-methano-2H-benz[5,6]anthra[2,1-c][1,5]oxazocine-9,16-dione	Marmycin A	1179526-18-1
(3.12)	(2R,3R,4S,6R)-8-Chloro-1,3,4,6-tetrahydro-3-hydroxy-2,4,12-trimethyl-2,6-methano-2H-benz[5,6]anthra[2,1-c][1,5]oxazocine-9,16-dione	Marmycin B	955119-60-5
(3.13)	1H-10,2,10-(Epoxymetheno)anthra[1,2-b]azacyclododecine-20-propanoic acid, 3,4,7,8-tetradehydro-2,9,13,18-tetrahydro-α,9,12-trihydroxy-β-methylene-13,18-dioxo-, methyl ester, (2S,5Z,9R,10S,20S)-	Sealutomicin A	2762229-09-2

(Continued)

TABLE 3.1 (*Continued*)
The Chemical Identity of Tetrahydronaphtoquinoline-dione (THQN-dione) Alkaloids

Structure (Number)	Systematic Name	Common Name	CAS Number
(3.14)	–	Sealutomicin B	2762229-13-8
(3.15)	Methyl (2E)-3-[(8S,9R,14S,17S)-5,8,9,14,15,16-hexahydro-6,8,9,17-tetrahydroxy-5,16-dioxo-8,14-methanoanthra[1,2-c][2]benzazocin-17-yl]-2-butenoate	Sealutomicin C	2254728-41-9
(3.16)	–	Sealutomicin D	2762229-16-1
(3.17)	(2S,5Z,9R,10S,20R)-3,4,7,8-Tetradehydro-2,9-dihydro-9,12,17-trihydroxy-20-[(1R)-1-hydroxyethyl]-16-methoxy-1H-10,2,10-(epoxymetheno)anthra[1,2-b]azacyclododecine-13,18-dione	Tiancimycin A (TNM A)	1965327-09-6

(*Continued*)

TABLE 3.1 (*Continued*)
The Chemical Identity of Tetrahydronaphtoquinoline-dione
(THQN-dione) Alkaloids

Structure (Number)	Systematic Name	Common Name	CAS Number
(3.18)	Methyl (2E)-3-[(1aS,5Z,9R,10S)-3,4,7,8-tetradehydro-2,9,13,18-tetrahydro-9,12-dihydroxy-13,18-dioxo-1H-10,2,10-(epoxymetheno)anthra[1,2-b]azacyclododecin-20-yl]-2-butenoate	TNM B	1965327-10-9
(3.19)	Methyl (βR,2S,5Z,9R,10S,20R)-3,4,7,8-tetradehydro-2,9,13,18-tetrahydro-α,β,9,12,16,17-hexahydroxy-β-methyl-13,18-dioxo-1H-10,2,10-(epoxymetheno)anthra[1,2-b]azacyclododecine-20-propanoate	TNM C	1965327-11-0
(3.20)	Methyl (αR,βR,2S,5Z,9R,10S,20R)-3,4,7,8-tetradehydro-2,9,13,18-tetrahydro-α,β,9,12,17-pentahydroxy-16-methoxy-β-methyl-13,18-dioxo-1H-10,2,10-(epoxymetheno)anthra[1,2-b]azacyclododecine-20-propanoate	TNM D	2254091-70-6
(3.21)	Methyl (αR,βR,2S,5Z,9R,10S,20R)-3,4,7,8-tetradehydro-2,9,13,18-tetrahydro-α,β,9,12-tetrahydroxy-β-methyl-13,18-dioxo-1H-10,2,10-(epoxymetheno)anthra[1,2-b]azacyclododecine-20-propanoate	TNM E	2254091-71-7

(Continued)

TABLE 3.1 (*Continued*)
The Chemical Identity of Tetrahydronaphtoquinoline-dione (THQN-dione) Alkaloids

Structure (Number)	Systematic Name	Common Name	CAS Number
(3.22)	Methyl (αR,βR,2S,5Z,9R,10S,20R)-3,4,7,8-tetradehydro-2,9,13,18-tetrahydro-α,β,9,12,16-pentahydroxy-β-methyl-13,18-dioxo-1H-10,2,10-(epoxymetheno)anthra[1,2-b]azacyclododecine-20-propanoate	TNM F	2254091-72-8
(3.23)	(2R,3R,4R,8′R,9′S,14′S)-3,4,8′,9′,14′,15′-Hexahydro-3,4,6′,9′-tetrahydroxy-3-methylspiro[furan-2(5H),17′-[8,14]methanoanthra[1,2-c][2]benzazocine]-5,5′,16′-trione	TNM G	2254728-70-4
(3.24)	(2R,3R,4R,8′S,9′R,14′S)-3,4,8′,9′,14′,15′-Hexahydro-3,4,6′,8′,9′,13′-hexahydroxy-3-methylspiro[furan-2(5H),17′-[8,14]methanoanthra[1,2-c][2]benzazocine]-5,5′,16′-trione	Tiancimycin congener (TNM congener)	2254728-68-0
(3.25)	(2R,3R,4R,8′S,9′R,14′S)-3,4,8′,9′,14′,15′-Hexahydro-3,4,6′,8′,9′-pentahydroxy-2′-methoxy-3-methylspiro[furan-2(5H),17′-[8,14]methanoanthra[1,2-c][2]benzazocine]-5,5′,16′-trione	TNM congener	2254728-38-4

(*Continued*)

TABLE 3.1 (*Continued*)
The Chemical Identity of Tetrahydronaphtoquinoline-dione (THQN-dione) Alkaloids

Structure (Number)	Systematic Name	Common Name	CAS Number
(3.26)	(2R,3R,4R,8′S,9′R,14′S)-3,4,8′,9′,14′,15′-Hexahydro-3,4,6′,8′,9′-pentahydroxy-3-methylspiro[furan-2(5H),17′-[8,14]methanoanthra[1,2-c][2]benzazocine]-5,5′,16′-trione	TNM congener	2254728-69-1
(3.27)	(2R,3R,4R,8′S,9′R,14′S)-3,4,8′,9′,14′,15′-Hexahydro-2′,3,4,6′,8′,9′-hexahydroxy-3-methylspiro[furan-2(5H),17′-[8,14]methanoanthra[1,2-c][2]benzazocine]-5,5′,16′-trione	TNM congener	2254728-71-5
(3.28)	(2*R*,3*R*,4*R*,8′*S*,9′*R*,14′*S*)-3,4,8′,9′,14′,15′-Hexahydro-1′,2′,3,4,6′,8′,9′-heptahydroxy-3-methylspiro[furan-2(5H),17′-[8,14]methanoanthra[1,2-*c*][2]benzazocine]-5,5′,16′-trione	TNM congener	2254728-72-6
(3.29)	(2R,3R,4R,8′S,9′R,14′S)-3,4,8′,9′,14′,15′-Hexahydro-1′,2′,3,4,6′,8′,9′,13′-octahydroxy-3-methylspiro[furan-2(5H),17′-[8,14]methanoanthra[1,2-c][2]benzazocine]-5,5′,16′-trione	TNM congener	2254728-75-9

(Continued)

TABLE 3.1 (*Continued*)

The Chemical Identity of Tetrahydronaphtoquinoline-dione (THQN-dione) Alkaloids

Structure (Number)	Systematic Name	Common Name	CAS Number
(3.30)	(2R,3R,4R,8′R,9′S,14′S)-3,4,8′,9′,14′,15′-Hexahydro-3,4,6′,9′-tetrahydroxy-2′-methoxy-3-methylspiro[furan-2(5H),17′-[8,14]methanoanthra[1,2-c][2]benzazocine]-5,5′,16′-trione	TNM congener	2254728-39-5
(3.31)	(2R,3R,4R,8′R,9′S,14′S)-3,4,8′,9′,14′,15′-Hexahydro-2′,3,4,6′,9′-pentahydroxy-3-methylspiro[furan-2(5H),17′-[8,14]methanoanthra[1,2-c][2]benzazocine]-5,5′,16′-trione	TNM congener	2254728-73-7
(3.32)	(2R,3R,4R,8′R,9′S,14′S)-3,4,8′,9′,14′,15′-Hexahydro-1′,2′,3,4,6′,9′-hexahydroxy-3-methylspiro[furan-2(5H),17′-[8,14]methanoanthra[1,2-c][2]benzazocine]-5,5′,16′-trione	TNM congener	2254728-74-8
(3.33)	(8R,9S,14S,17S)-6,8,17-trihydroxy-18-methyl-8,9,14,15-tetrahydro-8,14,9-(epibut[2]ene[1,1,4]triyl)anthra[1,2-b]benzo[f]azocine-5,16,20-trione	TNM congener	2254728-67-9

(*Continued*)

TABLE 3.1 (*Continued*)
The Chemical Identity of Tetrahydronaphtoquinoline-dione (THQN-dione) Alkaloids

Structure (Number)	Systematic Name	Common Name	CAS Number
(3.34)	Methyl (2E)-3-[(8R,9S,14S,17S)-5,8,9,14,15,16-hexahydro-6,9,17-trihydroxy-5,16-dioxo-8,14-methanoanthra[1,2-c][2]benzazocin-17-yl]-2-butenoate	TNM congener	2254728-63-5
(3.35)	ethyl (E)-3-((8S,9R,14S,17S)-6,8,9,17-tetrahydroxy-5,16-dioxo-5,8,9,14,15,16-hexahydro-8,14-methanoanthra[1,2-b]benzo[f]azocin-17-yl)but-2-enoate	TNM congener	–
(3.36)	(2R,3R,4R,8′R,9′S,14′S)-1′,3,4,6′,9′-pentahydroxy-2′-methoxy-3-methyl-3,4,8′,9′,14′,15′-hexahydro-5H-spiro[furan-2,17′-[8,14]methanoanthra[1,2-b]benzo[f]azocine]-5,5′,16′-trione	TNM congener	–
(3.37)	(2R,3R,4R,8′S,9′R,14′S)-1′,3,4,6′,8′,9′-hexahydroxy-2′-methoxy-3-methyl-3,4,8′,9′,14′,15′-hexahydro-5H-spiro[furan-2,17′-[8,14]methanoanthra[1,2-b]benzo[f]azocine]-5,5′,16′-trione	TNM congener	–

(*Continued*)

TABLE 3.1 (*Continued*)

The Chemical Identity of Tetrahydronaphtoquinoline-dione (THQN-dione) Alkaloids

Structure (Number)	Systematic Name	Common Name	CAS Number
(3.38)	(8R,9S,14S,17R)-1,6,9,17-tetrahydroxy-17-((R)-1-hydroxyethyl)-2-methoxy-8,9,14,15-tetrahydro-8,14-methanoanthra[1,2-b]benzo[f]azocine-5,16-dione	TNM congener	–
(3.39)	(8S,9R,14S,17R)-1,6,8,9,13,17-hexahydroxy-17-((R)-1-hydroxyethyl)-2-methoxy-8,9,14,15-tetrahydro-8,14-methanoanthra[1,2-b]benzo[f]azocine-5,16-dione	TNM congener	–
(3.40)	(8S,9R,14S,17R)-1,6,8,9,17-pentahydroxy-17-((R)-1-hydroxyethyl)-2-methoxy-8,9,14,15-tetrahydro-8,14-methanoanthra[1,2-b]benzo[f]azocine-5,16-dione	TNM congener	–
(3.41)	(2S,5Z,9R,10S,20R)-3,4,7,8-Tetradehydro-2,9-dihydro-9,12-dihydroxy-20-[(1R)-1-hydroxyethyl]-1H-10,2,10-(epoxymetheno)anthra[1,2-b]azacyclododecine-13,18-dione	Uncialamycin	870471-83-3

(Continued)

TABLE 3.1 (*Continued*)
The Chemical Identity of Tetrahydronaphtoquinoline-dione
(THQN-dione) Alkaloids

Structure (Number)	Systematic Name	Common Name	CAS Number
(3.42)	(8S,9R,14S,17R)-8,9,14,15-Tetrahydro-6,8,9,13,17-pentahydroxy-17-[(1R)-1-hydroxyethyl]-8,14-methanoanthra[1,2-c][2]benzazocine-5,16-dione	Unciaphenol	1818218-76-6
(3.43)	(2S,5Z,9R,10S,20R)-3,4,7,8-Tetradehydro-2,9-dihydro-9,12,17-trihydroxy-20-[(1R)-1-hydroxyethyl]-1H-10,2,10-(epoxymetheno)anthra[1,2-b]azacyclododecine-13,18-dione	Yangpumicin A (YPM A)	2243975-37-1
(3.44)	–	YPM B	2243975-38-2
(3.45)	–	YPM C	2243975-39-3

(Continued)

TABLE 3.1 (*Continued*)

The Chemical Identity of Tetrahydronaphtoquinoline-dione (THQN-dione) Alkaloids

Structure (Number)	Systematic Name	Common Name	CAS Number
(3.46)	–	YPM D	2243975-40-6
(3.47)	–	YPM E	2243975-41-7
(3.48)	1*H*-10,2,10-(Epoxymetheno) anthra[1,2-b]azacyclododecine-13,18-dione, 3,4,7,8-tetradehydro-20-[(1R)-1,2-dihydroxyethyl]-2,9-dihydro-9,12,17-trihydroxy-, (2S,5Z,9R,10S,20R)-	YPM F	2367015-13-0
(3.49)	1*H*-10,2,10-(Epoxymetheno) anthra[1,2-b]azacyclododecine-13,18-dione, 3,4,7,8-tetradehydro-20-[(1R)-1,2-dihydroxyethyl]-2,9-dihydro-9,12-dihydroxy-, (2S,5Z,9R,10S,20R)-	YPM G	2367015-14-1

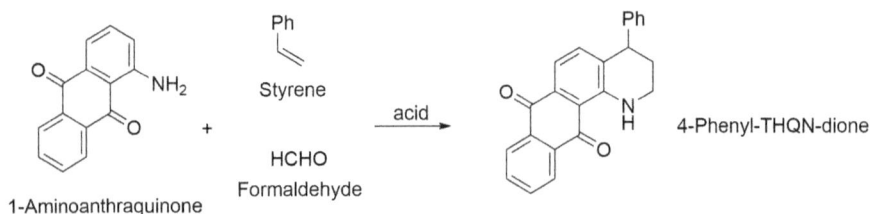

FIGURE 3.2 Synthesis of 4-phenyl-THNQ-dione.

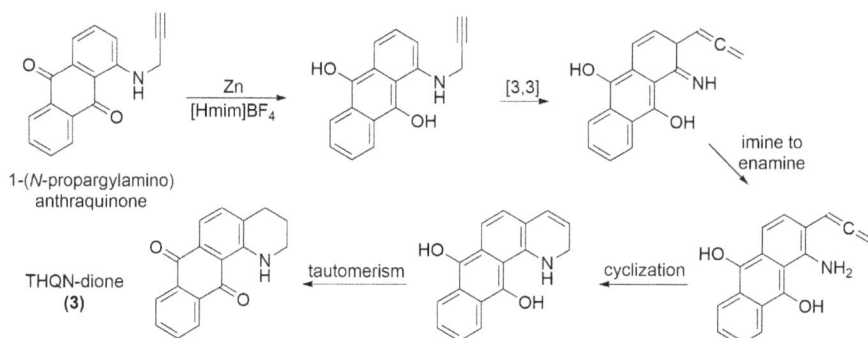

FIGURE 3.3 Synthesis of THNQ-dione.

As a result, researchers are focusing on creating these compounds and their derivatives through total synthesis (Gartman and Tambar 2022). Myers and colleagues (1995) achieved the total synthesis of DYN A and its analogues using a modified Diels–Alder cycloaddition/oxidation method, by the coupling of a substituted isobenzofuran moiety with a quinone-imine component. Further studies by Unno et al. (1997) showed that enhancing the water solubility of DYN A analogues increased their anti-tumour activity and reduced toxicity. Nicolaou and co-workers (2007, 2008) first synthesized uncialamycins (UCMs), determining that the natural UCM has an R configuration at C26. Nicolaou's team (2016) developed a more streamlined and enantioselective synthesis for UCM and its analogues, using a Noyori enantioselective reduction, Yamaguchi acetylide-pyridinium coupling, acetylide-aldehyde cyclization, and a Hauser-Kraus annulation. The Nicolaou group (2020) reported the total synthesis of tiancimycins A and B (TNM A, TNM B), yangpumicin A (YPM A), and their analogues using a similar streamlined synthetic strategy. The biological evaluation of these synthetic analogues revealed that novel AFEs, even at very low doses (sub-picomolar), are active against several cancer cell lines, including those resistant to multiple drugs. The structure–activity relationships suggested that the electronic properties of substituents on the A ring influence the redox potential of the anthraquinone moiety and its interaction with the DNA minor groove, while modifications near the epoxide ring could affect its activation. Future synthetic efforts are expected to yield more potent AFE analogues that can serve as ideal payloads for antibody-drug conjugates or toxins for other targeted delivery systems. However, additional factors such as monoclonal antibody specificity, linker properties, and drug-to-antibody ratio should be considered for developing these molecules into effective anticancer therapies (Flemming 2014).

3.2.2 BIOSYNTHESIS

The study of AFE biosynthesis accelerated after the discovery of DYN A (Konishi et al. 1989). Researchers found that DYN A is assembled from two distinct molecular chains, known as heptaketides. One of these chains forms the enediyne core, while the other forms the anthraquinone moiety. For many years, progress in understanding the biosynthetic pathways of these compounds was slow. However, later advancements in genetic engineering and molecular biology have greatly improved our understanding. In addition, cloning and analysing biosynthetic gene clusters (BGCs) and creating better genetic tools were found to be essential steps. Between 2016 and 2020, key studies in *Micromonospora chersina* and *Micromonospora* sp. MD118 provided crucial insights into the biosynthesis of the anthraquinone moiety and the complex coupling processes involved in AFE biosynthesis (Yan et al. 2017; Cohen and Townsend 2018b; Low et al. 2020).

3.2.2.1   Earlier Studies

Isotope labelling techniques were used to understand the biosynthetic pathways of AFEs. These techniques include the integration of specific isotopes into starting materials and tracking their incorporation into the final product. In 1992, researchers used isotope-labelled compounds to assign the carbon signals of DYN A (Tokiwa et al. 1992). This study revealed that specific carbon atoms in DYN A originate from different positions within acetate molecules, a common building block in many biosynthetic pathways. For example, carbons C5 and C30 are derived from the second carbon (C2) of acetate, whereas carbons C8 and C9 come from the first carbon (C1) of acetate. These findings indicated that DYN A is assembled from two heptaketide chains, each composed of seven acetate units, connected by a specific bond between two acetate-derived carbons. Further experiments indicated that the oxygen atoms in the anthraquinone moiety of DYN A come from molecular oxygen (O_2) rather than acetate. This suggested the involvement of a highly reducing type I polyketide synthase, which incorporates molecular oxygen into the structure (Low et al. 2020). BGCs are groups of genes that encode the enzymes responsible for producing complex natural products such as AFEs. Researchers cloned and characterized the BGC for DYN A, which spans approximately 54 kilobases (kb) and contains 47 genes (Gao and Thorson 2008). These genes are crucial for the production of the enediyne core and show high similarity to genes found in other enediyne-producing BGCs. Inactivation of the key gene *dynE8* resulted in the complete cessation of DYN A production. Comparative analyses of BGCs from other AFE-producing organisms, such as those producing tiancimycin (TNM), UCM, and yangpumicin (YPM), revealed that they share many homologous genes. This suggests a common biosynthetic mechanism across different species (Hindra et al. 2021).

The enediyne core is a highly reactive and critical component for the biological activity of AFEs. The synthesis of the enediyne core involves complex biochemical processes catalysed by polyketide synthase enzymes (PKSEs). These enzymes belong to the iterative type I polyketide synthase (PKS) family, having a specific set of features that are similar across different members of the group (Shen 2003). Studies on the biosynthesis of similar compounds, such as C-1027 and calicheamicin, have provided insights into the assembly of enediyne cores (Yan et al. 2017). These studies suggest that AFEs follow a comparable biosynthetic route, involving specific steps such as E/Z-isomerization and radical formation. The process begins with assembling a polyketide chain, followed by a series of modifications: cyclization, reduction, and radical formation. These steps conclude the formation of a 15-carbon enediyne skeleton, which is highly reactive and capable of generating diradicals that can cleave DNA (Belecki and Townsend 2012, 2013).

The origin and biosynthesis of the anthraquinone moiety in AFEs were initially unclear. Isotope labelling experiments suggested that a repetitive, highly reducing polyketide synthase might be involved in its synthesis. Early genomic analyses did not identify a typical PKS gene responsible for this process. As briefly mentioned above, this led researchers to hypothesize that the responsible gene might be located elsewhere in the genome (Gao and Thorson 2008). This puzzle was resolved in 2017 when the Townsend group identified a dual-role PKS, DynE8, in the genome of *M. chersina* (Cohen and Townsend 2018a). This enzyme was responsible for synthesizing both the enediyne core and the anthraquinone moiety in DYN A. The discovery of DynE8's dual role was surprising, as it challenged the traditional understanding of PKS enzymes, which typically specialize in synthesizing specific building blocks. The dynE13 gene, encoding a flavin-dependent oxidoreductase, was also essential for DYN A production. Inactivation of dynE13 resulted in the complete cessation of DYN A production and the accumulation of intermediate compounds. These intermediates, having different chemical structures, show a complex coupling process between the enediyne core and the anthraquinone moiety (Cohen and Townsend 2018b). This process likely involves radical reactions and specific bond formations that are critical for the assembly of the final AFE structure. The coupling of the enediyne core and the anthraquinone moiety is one of the most complex and critical steps in AFE biosynthesis. This process involves radical reactions and the formation of specific bonds that link the two distinct molecular components. The enzyme encoded by the dynE13 gene plays a central role in this process. When dynE13 was inactivated, researchers observed the accumulation of various intermediate compounds, confirming that dynE13 is involved in the coupling process. These intermediate compounds and their biosynthetic transformations provide valuable insights into the complex coupling and cyclization processes that lead to the final AFE structure.

3.2.2.2   Recent Studies

As indicated above, the origin of the C_{15} enediyne core is linked to a unique enzyme system, enediyne iterative type I PKS and its thioesterase. However, the exact initial product and its transformations were unclear. The compound C_{15}-heptaene (A) has been identified by Bhardwaj and coworkers (2023) as the initial product of PKSE and a precursor for the AFE enediyne core and the anthraquinone part of AFEs. This is supported by the presence of the intermediate iodoanthracene (B) (Figure 3.4). However, the exact pathway from (A) to the enediyne core was unknown, and no early-stage alkyne-bearing intermediate had been identified. BGCs do not contain known alkyne-synthesizing enzyme genes but include three unknown genes, E3, E4, and E5, near the PKS and thioesterase genes, forming the enediyne PKS cassette.

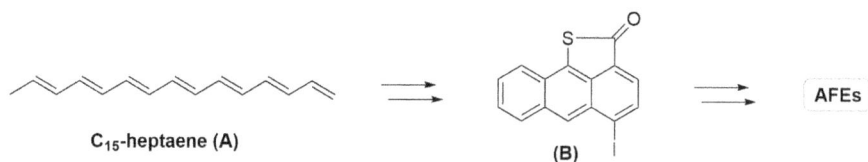

C_{15}-heptaene (A)   (B)

FIGURE 3.4   Intermediates in the biosynthesis of anthraquinone-fused enediynes (AFEs).

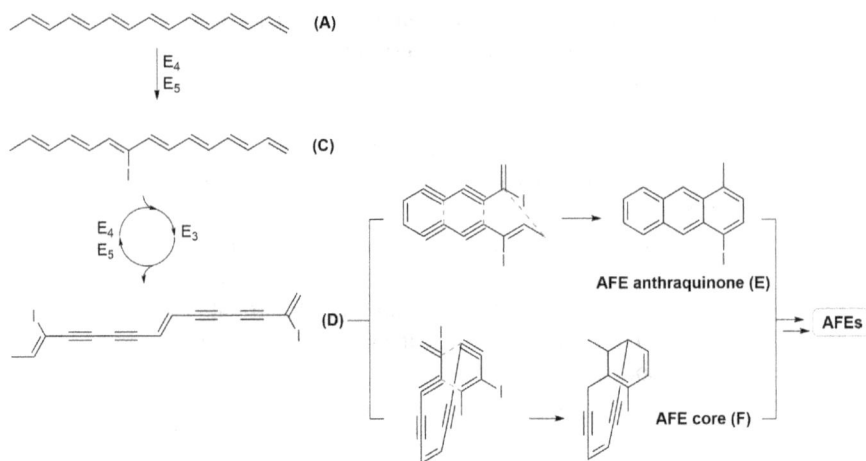

FIGURE 3.5   Biosynthesis of AFEs: a schematic presentation.

A recent study (Gui et al. 2024) has revealed that diiodotetrayne (D) is a direct product of the enediyne PKS cassette (Figure 3.5). PKSE and thioesterase produce compound (A), and E3, E4, and E5 convert (A) into (D), demonstrating how diio-dotetrayne (D) can transform into AFE anthraquinone (E), as well as the enediyne core (F). Different sources of enediyne PKS cassettes show that they follow a similar biosynthesis pathway. They share compound (D) as a common intermediate, which is created through an iodinated heptaene intermediate (C).

3.3   NATURAL OCCURRENCE

Deoxydynemicin A (3.1), along with DYN A (3.2), was produced in the culture broth of *Micromonospora globose* MG331-hF6 (Shiomi et al. 1990). In comparison with DYN A, Deoxydynemicin A lacks the hydroxy group at position 15. Treating the spore suspension of *M. chersina* M956-1 (ATCC-53710) with Methylnitronitrosoguanidine (MNNG) produced the mutant strain F1085 (Saitoh et al. 1992a, 1992b). Cultivating this strain produced DYN C (3.3). During the purification of DYN A, produced from *M. chersina* sp. M956-1, three more components were isolated (Konishi et al. 1991). These compounds, named DYNs L (3.5), M (3.6), and N (3.7), are structurally similar to DYN A. However, they lack the enediyne (replaced by a phenyl ring) and epoxide (replaced by a vicinal diol) moieties. The culture broth of *M. chersina* sp. M956-1 also produced DYNs O (3.8), P (3.9), and Q (3.10) (Miyoshi-Saitoh et al. 1991). Their structures were determined by Nuclear Magnetic Resonance (NMR) and UV comparison with the data of DYNs A and M. DYNs O, P, and Q contain the same 1,2,4,5,8-pentasubstituted anthraquinone present in DYN M. Fermentation broth and mycelia of *Micromonospora strain* C5308 resulted in the isolation of DYN H (3.4) (Qi-wei et al. 2002). As another non-enediyne DYN, its structure was confirmed by comparison with the synthetic version introduced earlier by Miyoshi et al. (1991). In the early 2000s, sequencing the genomes of *Streptomyces* bacteria revealed a valuable set of undiscovered natural compounds. This discovery initiated a profound shift in the natural product discovery field. Genome mining, a method relying on the analysis of genomic data, has surfaced as a dominant approach for identifying genes responsible for generating novel compounds. This marked a departure from the traditional method, which centred on bioactivity. Genome mining has significantly enhanced our capacity to explore and employ the diverse array of natural products available (Medema et al. 2021; Bauman et al. 2021). A survey of thousands of *actinomycetes* genomes revealed numerous strains containing genes responsible for enediyne polyketide synthesis (Yan et al. 2016). These genes were grouped into distinct categories based on evolutionary analysis. The researchers confirmed the presence of unique enediyne BGCs in each category through genome sequencing. Employing a genome neighbourhood network, they predicted novel structural elements that could facilitate enediyne discovery. The study introduced a new category of enediyne natural products, referred to as TNMs. TNM A (3.17) was isolated from the fermentation culture of *Streptomyces* sp. CB03234 wild-type strain (Liu et al. 2018), while TNM C (3.19) was isolated from the fermentation culture of the $\Delta tnmH$ mutant strain SB20002. The researchers were able to produce TNMs in large quantities through microbial fermentation and manipulate their biosynthesis to engineer new analogues. TnmH is an *O*-methyltransferase enzyme of particular interest in the biosynthesis of TNMs due to its ability to catalyse regiospecific methylation at the C-7 hydroxyl group (Adhikari et al. 2020; Annaval et al. 2021). In the course of conducting comparative analyses on the BGCs associated with anthraquinone-fused enediynes, four distinct genes, namely *tnmE6*, *tnmH*, *tnmL*, and *tnmQ*, were discerned as exclusive components of the *tnm* gene cluster (Yan et al. 2018). Subsequently, through the execution of larger-scale fermentation procedures involving both the wild-type *Streptomyces* CB03234 strain and the mutant strains lacking either *tnmH* ($\Delta tnmH$) or *tnmL* ($\Delta tnmL$), a comprehensive investigation was conducted, leading to the identification and characterization of a total of several novel TNM congeners (3.24–3.40). Six new TNM congeners, in addition to TNM A, were isolated from the CB03234 WT. Eight new TNM congeners were isolated from the $\Delta tnmL$ mutant strain SB20020, and additional new TNM congeners were isolated from the $\Delta tnmH$ mutant strain SB20002. This study encompasses the revelation of five enediynes among these congeners. The incorporation of these outcomes has paved the way for the formulation of a theoretical framework detailing the latter stages of TNM biosynthesis, with a notable emphasis on a potential intermediate compound that might be shared across the entire spectrum of AFEs. Gui et al. (2022) have identified and conducted a comprehensive biochemical and structural analysis of TnmK1, an integral component of the α/β-hydrolase fold superfamily. TnmK1 plays a fundamental role in facilitating the formation of the C−C bond, which functions to connect the anthraquinone moiety with the enediyne core in the complex process of TNM biosynthesis. This discovery marks the first instance of such chemistry within the α/β-hydrolases, shedding light on an exceptional enzymatic capability in natural product biosynthesis. Large-scale fermentation of the *Streptomyces* sp. CB03234 wild-type strain enabled the isolation and characterization of TNM D (3.20) as well as five Bergman cyclization products/congeners (Zhuang et al. 2019). Using a similar approach, YPM A (3.43) was extracted from *Micromonospora yangpuensis* DSM 45577; the process was aided by the identification of enediyne BGCs in publicly available actinobacterial genome databases (Yan et al. 2017). The selection of potential drug candidates leveraged analysis of the enediyne genome neighbourhood network to aid in the discovery process. Four cycloaromatization compounds (Bergman cyclization congeners) of YPM A, i.e., YPM B (3.44), YPM C (3.45), YPM D (3.46), and YPM E (3.47), were also identified from the fermentation broth of *M. yangpuensis* (Yan et al. 2017). Wang et al. (2019) reported the isolation and characterization of two additional AFEs, YPM F (3.48) and YPM G (3.49), along with five known YPMs above, from *M.yangpuensis* DSM 45577 by adding microporous resins to the fermentation medium. Unciaphenol (3.42) and UCM (3.41) were isolated from cultures of the actinomycete *Streptomyces uncialis* (Williams et al. 2015). Unciaphenol is a type of compound that comes from the Bergman cyclization product of an enediyne called UCM (Davies et al. 2005). It is proposed to form when a nucleophilic oxygen attacks the p-benzyne diradical intermediate. Another type of bacteria, *Nonomuraea* sp. strain MM565M-173N2, was found in deep-sea sediment off the Sanriku coast in Japan. From this bacterium, researchers discovered four new compounds called sealutomicins A–D, which contain the THQN-dione structure. Sealutomicin A (3.13) was identified as a new enediyne antibiotic, and sealutomicins B–D (3.14–3.16) were found to be aromatized derivatives of sealutomicin A. Marmycins A and B (3.11 and 3.12) were isolated from the culture broth of a marine sediment-derived *Streptomyces*-related actinomycete, strain [plate number 1] (Martin et al. 2007), with Marmycin B being the halogenated (chlorinated) congener of Marmycin A. Halogenated natural products from the marine-derived actinobacteria are well documented (Tan et al. 2013).

All THQN-dione alkaloids are found in bacterial cultures, particularly in the genus *Streptomyces* (Streptomycetaceae), *Micromonospora* (Micromonosporaceae), and *Nonomuraea* (Streptosporangiaceae). Table 3.2 shows the natural occurrence of THQN-dione alkaloids in these genera. In summary, DYNs, TNMs, YPMs, and Sealutomicins were isolated from the species *M. chersina*, *Streptomyces* sp. strain CB03234, *Micromonospora yangpuensis* DSM 45577, and *Nonomuraea* sp. strain MM565M-173N2, respectively.

3.4   BIOLOGICAL ACTIVITY

The DNA damage activity of DYNs was initially investigated in the early 90s (Sugiura et al. 1990; Miyoshi-Saitoh et al. 1991; Shimazawa et al. 1994). It was found that both enediyne-containing DYNs (such as DYN A and deoxydynemicin A) and non-enediyne DYNs (like DYNs H, M, O, P, and Q) cause damage to DNA functions. It was suggested that DNA strand cleavage by enediyne DYNs may not be the sole mechanism responsible for this bioactivity and that the aza-anthraquinone core (THNQ) can interact with DNA through intercalation. DYN A exhibited strong antibacterial activities, particularly against Gram-positive bacteria (Konishi et al. 1989). Similarly, deoxydynemicin A and DYN C demonstrated remarkable antimicrobial activity against Gram-positive bacteria at low doses. DYN A also displayed significant cytotoxic and anti-tumour activity against B16 melanoma, Moser human carcinoma, HCT-116 human carcinoma, as well as normal and vincristine-resistant P388 leukaemia cells. Additionally, DYN A extended the lifespan of mice inoculated

TABLE 3.2
Natural Occurrence of Tetrahydronaphtoquinoline-dione (THQN-dione) Alkaloids

Kingdom	Family	Genus	Species	Compound	References
Bacteria	Micromonosporaceae	*Micromonospora*	globosa MG331-hF6	3.1	Shiomi et al. (1990)
Bacteria	Micromonosporaceae	*Micromonospora*	globosa MG331-hF6	3.2	Shiomi et al. (1990)
Bacteria	Micromonosporaceae	*Micromonospora*	chersina sp. M956-1	3.2	Konishi et al. (1989)
Bacteria	Micromonosporaceae	*Micromonospora*	chersina sp. M956-1 mutant F1085 (ATCC-55077)	3.3	Saitoh et al. (1992)
Bacteria	Micromonosporaceae	*Micromonospora*	strain C5308	3.4	Kangping et al. (2000)
Bacteria	Micromonosporaceae	*Micromonospora*	chersina sp. nov. M956-1	3.5	Konishi et al. (1991)
Bacteria	Micromonosporaceae	*Micromonospora*	chersina sp. nov. M956-1	3.6	Konishi et al. (1991)
Bacteria	Micromonosporaceae	*Micromonospora*	chersina sp. nov. M956-1	3.7	Konishi et al. (1991)
Bacteria	Micromonosporaceae	*Micromonospora*	chersina	3.8	Miyoshi-Saitoh et al. (1991)
Bacteria	Micromonosporaceae	*Micromonospora*	chersina	3.9	Miyoshi-Saitoh et al. (1991)
Bacteria	Micromonosporaceae	*Micromonospora*	chersina	3.10	Miyoshi-Saitoh et al. (1991)
Bacteria	Streptomycetaceae	*Streptomyces*	strain CNH990	3.11	Martin et al. (2007)
Bacteria	Streptomycetaceae	*Streptomyces*	strain CNH990	3.12	Martin et al. (2007)
Bacteria	Streptosporangiaceae	*Nonomuraea*	sp. strain MM565M-173N2	3.13	Igarashi et al. (2021)
Bacteria	Streptosporangiaceae	*Nonomuraea*	sp. strain MM565M-173N2	3.14	Igarashi et al. (2021)
Bacteria	Streptosporangiaceae	*Nonomuraea*	sp. strain MM565M-173N2	3.15	Igarashi et al. (2021)
Bacteria	Streptosporangiaceae	*Nonomuraea*	sp. strain MM565M-173N2	3.16	Igarashi et al. (2021)
Bacteria	Streptomycetaceae	*Streptomyces*	sp. strain CB03234	3.17	Yan et al. (2016)

(Continued)

TABLE 3.2 (*Continued*)

Natural Occurrence of Tetrahydronaphtoquinoline-dione (THQN-dione) Alkaloids

Kingdom	Family	Genus	Species	Compound	References
Bacteria	Streptomycetaceae	*Streptomyces*	sp. strain CB03234	3.18	Yan et al. (2018)
Bacteria	Streptomycetaceae	*Streptomyces*	ΔtnmH mutant strain SB20002	3.19	Yan et al. (2016)
Bacteria	Streptomycetaceae	*Streptomyces*	sp. strain CB03234	3.20	Yan et al. (2018)
Bacteria	Streptomycetaceae	*Streptomyces*	sp. strain CB03234	3.21	Yan et al. (2018)
Bacteria	Streptomycetaceae	*Streptomyces*	sp. strain CB03234	3.22	Yan et al. (2018)
Bacteria	Streptomycetaceae	*Streptomyces*	sp. strain CB03234	3.23	Yan et al. (2018)
Bacteria	Streptomycetaceae	*Streptomyces*	sp. strain CB03234	3.24	Yan et al. (2018)
Bacteria	Streptomycetaceae	*Streptomyces*	sp. strain CB03234	3.25	Yan et al. (2018)
Bacteria	Streptomycetaceae	*Streptomyces*	sp. strain CB03234	3.26	Yan et al. (2018)
Bacteria	Streptomycetaceae	*Streptomyces*	sp. strain CB03234	3.27	Yan et al. (2018)
Bacteria	Streptomycetaceae	*Streptomyces*	sp. strain CB03234	3.28	Yan et al. (2018)
Bacteria	Streptomycetaceae	*Streptomyces*	sp. strain CB03234	3.29	Yan et al. (2018)
Bacteria	Streptomycetaceae	*Streptomyces*	sp. strain CB03234	3.30	Yan et al. (2018)
Bacteria	Streptomycetaceae	*Streptomyces*	sp. strain CB03234	3.31	Yan et al. (2018)
Bacteria	Streptomycetaceae	*Streptomyces*	sp. strain CB03234	3.32	Yan et al. (2018)
Bacteria	Streptomycetaceae	*Streptomyces*	sp. strain CB03234	3.33	Yan et al. (2018)
Bacteria	Streptomycetaceae	*Streptomyces*	sp. strain CB03234	3.34	Yan et al. (2018)
Bacteria	Streptomycetaceae	*Streptomyces*	sp. strain CB03234	3.35	Yan et al. (2018)
Bacteria	Streptomycetaceae	*Streptomyces*	sp. strain CB03234	3.36	Yan et al. (2018)
Bacteria	Streptomycetaceae	*Streptomyces*	sp. strain CB03234	3.37	Yan et al. (2018)
Bacteria	Streptomycetaceae	*Streptomyces*	sp. strain CB03234	3.38	Yan et al. (2018)

(Continued)

TABLE 3.2 (*Continued*)
Natural Occurrence of Tetrahydronaphtoquinoline-dione (THQN-dione) Alkaloids

Kingdom	Family	Genus	Species	Compound	References
Bacteria	Streptomycetaceae	*Streptomyces*	sp. strain CB03234	3.39	Yan et al. (2018)
Bacteria	Streptomycetaceae	*Streptomyces*	sp. strain CB03234	3.40	Yan et al. (2018)
Bacteria	Streptomycetaceae	*Streptomyces*	Strain C42	3.41	Davies et al. (2005)
Bacteria	Streptomycetaceae	*Streptomyces*	*uncialis*	3.42	Williams et al. (2015)
Bacteria	Micromonosporaceae	*Micromonospora*	*yangpuensis* DSM 45577	3.43	Yan et al. (2017)
Bacteria	Micromonosporaceae	*Micromonospora*	*yangpuensis* DSM 45577	3.44	Yan et al. (2017)
Bacteria	Micromonosporaceae	*Micromonospora*	*yangpuensis* DSM 45577	3.45	Yan et al. (2017)
Bacteria	Micromonosporaceae	*Micromonospora*	*yangpuensis* DSM 45577	3.46	Yan et al. (2017)
Bacteria	Micromonosporaceae	*Micromonospora*	*yangpuensis* DSM 45577	3.47	Yan et al. (2017)
Bacteria	Micromonosporaceae	*Micromonospora*	*yangpuensis* DSM 45577	3.48	Wang et al. (2019)
Bacteria	Micromonosporaceae	*Micromonospora*	*yangpuensis* DSM 45577	3.49	Wang et al. (2019)

with B16 melanoma and P388 leukaemia. DYN C has strong antifungal properties; it was more effective than ketoconazole in combating an amphotericin-resistant strain of *Candida albicans*. Additionally, DYN C outperformed both amphotericin B and ketoconazole in fighting *Cryptococcus neoformans* spp. D49 and IAM4514. DYNs O (3.8) and Q (3.10) demonstrated potent activity against Gram-positive bacteria, though not as much as DYN A. The five AFE TNMs TNM C (3.19), TNM D (3.20), TNM B (3.18), TNM E (3.21), and TNM F (3.22) were tested for cytotoxicity against various human cancer cell lines. TNM A exhibited the most potency, followed by TNM D, TNM E, TNM B, TNM C, and TNM F sequentially. Initial evaluations showed that TNMs are highly effective against a wide range of cancer cell lines, with sub-nanomolar IC_{50} values. For instance, TNM A is more potent than UCM, especially against breast cancer cell lines. Both yangpumicins F (YPM F) (3.48) and G (YPM G) (3.49) exhibited potent cytotoxicity against several human cancer cell lines (CaCo2, Jurkat, A549, and SKBR-3) and strong inhibitory effects against the bacteria tested (MRSA, *S. aureus* ATCC 29213, and *E. coli*). Unciaphenol (3.42) inhibited *in vitro* HIV-1 replication against viral strains resistant to clinically utilized antiretroviral therapies. Sealutomicin A (3.13) showed a strong *in vitro* antibacterial activity against multidrug-resistant (MDR) Gram-negative bacteria, including New Delhi metallo-beta-lactamase- and *Klebsiella pneumoniae* carbapenemase-producing strains. 3.13 also showed potent antimicrobial activities against susceptible and MDR Gram-positive bacteria, including methicillin-resistant *Staphylococcus aureus* (MRSA) and vancomycin-resistant *Enterococci* (VRE) strains. Sealutomicins B-D (3.14-3.16) displayed strong or moderate antibacterial activity against Gram-positive bacteria such as *S. aureus* and *E. faecalis/faecium*, but hardly showed any activity against Gram-negative bacteria. Marmycins A (3.11) and B (3.12) were tested for their cytotoxicity against the human colon tumour cell line (HCT-116). Marmycin A proved much more potent (18 times) than its halogenated congener Marmycin B. They were further evaluated for their *in vitro* cytotoxicity against a panel of 12 human tumour cell lines (breast, prostate, colon, lung, and leukaemia). Marmycin A showed a modest induction of apoptosis in a human ovarian tumour A2780 cell line after 24 h of drug exposure, with a corresponding loss of cells from G2 and an arrest of cells in the G1 phase.

Table 3.3 provides a comprehensive overview of the diverse biological activities displayed by THQN-dione alkaloids (also see Khadem and Marles 2025b). Cytotoxicity, predominantly against cancer cell lines, is the most common bioactivity observed among THQN-diones and was found in 32 compounds (65.3%). All the potent cytotoxic compounds such as DYN A, TNM A and B, UCM, YPM F and G belong to the AFE family. Their IC_{50} values against cancerous cell lines were as low as 3 *p*M (*pico*-molar $= 10^{-12}$ M). DYN A and deoxy-DYN A possess anti-phytopathogenic properties against the bacterial genus *Xanthomonas*, many of which cause plant diseases (An et al. 2020). The antifungal properties, specifically against the genus *Candida*, are found in DYN A, deoxy-DYN A, and DYN C, with a MIC value of less than 1 µg/mL. The antibacterial effect of THQN-diones was observed in 10 compounds (over 20%). Four compounds, DYN A, deoxy-DYN A, Sealutomicin A, and UCM showed antibacterial activity at the ng/mL level. Unciaphenol has anti-viral activity against HIV-1 at the micromolar level.

Of the 49 compounds, 32 (65.3%) were cytotoxic, which strongly suggests a potential for anticancer applications, subject to toxicological evaluation. Antibacterial was shown in 11 compounds (22.4%), and antifungal was shown in three compounds (6.1%). Four compounds (7.1%) exhibited anti-phytopathogen activity, and one (2.0%) was antiviral. These results (Figure 3.6) show a diversity of bioactivities with a major focus on cytotoxicity.

3.5 DISCUSSION

The synthesis of THQN-diones (specifically AFEs) presents challenges due to their structural complexity and reactive nature. Controlling reactive intermediates, selective functionalization, sensitive ring closure reactions, and low yields and purity are some of the key challenges. Moreover, enhancing synthesis while ensuring functional group compatibility and addressing safety concerns complicates the process. To overcome these challenges, researchers utilize strategies such as total synthesis approaches, biosynthetic insights, precursor-directed synthesis, meticulous optimization of reaction conditions, and strategic use of protective groups. These efforts aim

TABLE 3.3

Biological Activity of Tetrahydronaphthoquinoline-dione (THQN-dione) Alkaloids

Biological Activity	Target	Assay/Method	Results	Compound	References
Antibacterial	Bacteria (Gram-positive): *Staphylococcus aureus,* *S. aureus Smith,* *Staphylococcus epidermidis 109,* *Micrococcus luteus FDA 16,* *M. luteus PCI 1001,* *Bacillus anthracis,* *Bacillus subtilis PCI 219,* *B. subtilis NRRL B-558,* *Bacillus cereus ATCC 10702,* *Corynebacterium bovis 1810,* Bacteria (Gram-negative): *Escherichia coli NIHJ,* *E. coli K-12,* *E. coli K-12 ML1629,* *Klebsiella pneumoniae PCI 602,* *Shigella dysenteriae JS11910,* *Salmonella typhi T-63,* *Proteus rettgeri GN466,* *Serratia marcescens,* *Pseudomonas aeruginosa A3,* *Aeromonas punctata IAM 1646,* *Vibrio anguillarum NCMB6,* *Pseudomonas fluorescens,* *Erwinia aroideae*	Minimum inhibitory concentration (MIC) assay, Mueller–Hinton agar	MIC = 0.06 to 30 ng/ml	3.1	Shiomi et al. (1990)

(Continued)

TABLE 3.3 (Continued)
Biological Activity of Tetrahydronaphtoquinoline-dione (THQN-dione) Alkaloids

Biological Activity	Target	Assay/Method	Results	Compound	References
Antibacterial	Bacteria (Gram-positive):	MIC assay, Mueller–Hinton agar	MIC = 0.06 to 60 ng/ml	3.2	Shiomi et al. (1990)
	S. aureus,				
	S. aureus Smith,				
	S. epidermidis 109,				
	M. luteus FDA 16,				
	M. luteus PCI 1001,				
	B. anthracis,				
	B. subtilis PCI 219,				
	B. subtilis NRRL B-558,				
	B. cereus ATCC 10702,				
	Corynebacterium bovis 1810,				
	Bacteria (Gram-negative):				
	Escherichia coli NIHJ,				
	E. coli K-12,				
	E. coli K-12 MLJ629,				
	Klebsiella pneumoniae PCI 602,				
	Shigella dysenteriae JS11910,				
	Salmonella typhi T-63,				
	Proteus rettgeri GN466,				
	Serratia marcescens,				
	P. aeruginosa A3,				
	Aeromonas punctata IAM 1646,				
	Vibrio anguillarum NCMB6,				
	P. fluorescens,				
	Erwinia aroideae				

(Continued)

TABLE 3.3 (Continued)
Biological Activity of Tetrahydronaphthoquinoline-dione (THQN-dione) Alkaloids

Biological Activity	Target	Assay/Method	Results	Compound	References
Antibacterial	Gram-positive bacteria	—	MIC < 0.8 µg/kg	3.3	Saitoh et al. (1992)
Antibacterial	S. aureus FDA209P,	—	MIC = 2.5 to 63 ng/mL	3.8	Miyoshi-Saitoh et al. (1991)
	S. aureus Smith,				
	S. epidermidis 11-1168,				
	S. epidermidis 11-1230,				
	Enterococcus feacalis A9808,				
	E. faecium A24817, M. luteus PCI,				
	B. subtilis PCI 219,				
	Comamonas terrigena IFO 12685				
Antibacterial	Staphylococcus aureus FDA209P,	—	MIC = 0.01 to 10 µg/mL	3.10	Miyoshi-Saitoh et al. (1991)
	S. aureus Smith,				
	S. epidermidis 11-1168,				
	S. epidermidis 11-1230,				
	Enterococcus feacalis A9808,				
	E. faecium A24817, M. luteus PCI,				
	B. subtilis PCI 219,				
	Comamonas terrigena IFO 12685				

(Continued)

TABLE 3.3 (Continued)
Biological Activity of Tetrahydronaphtoquinoline-dione (THQN-dione) Alkaloids

Biological Activity	Target	Assay/Method	Results	Compound	References
Antibacterial	*S. aureus* FDA 209P, *S. aureus* MRSA No.5, *S. aureus* MRSA No. 17, *S. aureus* Mu50, *Enterococcus faecalis* JCM 5803, *E. faecalis* NCTC12201, *E. faecium* JCM 5804, *E. faecium* NCTC12202, *Escherichia coli* K-12, *E. coli* DH5α/pUC57-Kan-MCR-1, *E. coli* DH5α/pUC57-Kan-MCR-2, *E. coli* NDM-1 Dok01, *Enterobacter cloacae* ATCC BAA-2468, *Klebsiella pneumoniae* PCI 602, *K. pneumoniae* ATCC BAA-1705, *K. pneumoniae* NDM-1 Sai01, *P. aeruginosa* A3, *P. aeruginosa* K-Ps102, *Acinetobacter baumannii* ATCC 19606, *A. baumannii* ATCC BAA-1605, *A. baumannii* NCGM237	Standard agar dilution method, Mueller–Hinton agar (Becton Dickinson)	MIC = 0.0063 to 0.4 μg/mL	3.13	Igarashi et al. (2021)

(Continued)

TABLE 3.3 (Continued)
Biological Activity of Tetrahydronaphthoquinoline-dione (THQN-dione) Alkaloids

Biological Activity	Target	Assay/Method	Results	Compound	References
Antibacterial	*Staphylococcus aureus* FDA 209P, *S. aureus* MRSA No.5, *S. aureus* MRSA No. 17, *S. aureus* Mu50, *Enterococcus faecalis* JCM 5803, *E. faecalis* NCTC12201, *E. faecium* JCM 5804, *E. faecium* NCTC12202, *Escherichia coli* K-12, *E. coli* DH5a/pUC57-Kan-MCR-1, *E. coli* DH5a/pUC57-Kan-MCR-2, *E. coli* NDM-1 Dok01, *Enterobacter cloacae* ATCC BAA-2468, *Klebsiella pneumoniae* PCI 602, *K. pneumoniae* ATCC BAA-1705, *K. pneumoniae* NDM-1 Sai01, *P. aeruginosa* A3, *P. aeruginosa* K-Ps102, *Acinetobacter baumannii* ATCC 19606, *Acinetobacter baumannii* ATCC BAA-1605, *A. baumannii* NCGM237	Standard agar dilution method, Mueller–Hinton agar (Becton Dickinson)	MIC = 0.2 to 6.4 μg/mL	3.14	Igarashi et al. (2021)

(Continued)

TABLE 3.3 (Continued)
Biological Activity of Tetrahydronaphtoquinoline-dione (THQN-dione) Alkaloids

Biological Activity	Target	Assay/Method	Results	Compound	References
Antibacterial	*Staphylococcus aureus* FDA 209P,	Standard agar dilution method,	MIC = 0.2 to	3.15	Igarashi et al. (2021)
	S. aureus MRSA No.5,	Mueller–Hinton agar (Becton	6.4 µg/mL		
	S. aureus MRSA No. 17,	Dickinson)			
	S. aureus Mu50,				
	Enterococcus faecalis JCM 5803,				
	E. faecalis NCTC12201,				
	E. faecium JCM 5804,				
	E. faecium NCTC12202,				
	Escherichia coli K-12,				
	E. coli DH5a/pUC57-Kan-MCR-1,				
	E. coli DH5a/pUC57-Kan-MCR-2,				
	E. coli NDM-1 Dok01,				
	Enterobacter cloacae ATCC BAA-2468,				
	Klebsiella pneumoniae PCI 602,				
	K. pneumoniae ATCC BAA-1705,				
	K. pneumoniae NDM-1 Sai01,				
	P. aeruginosa A3,				
	P. aeruginosa K-Ps102,				
	Acinetobacter baumannii ATCC 19606,				
	Acinetobacter baumannii ATCC BAA-1605,				
	A. baumannii NCGM237				

(Continued)

TABLE 3.3 (Continued)
Biological Activity of Tetrahydronaphtoquinoline-dione (THQN-dione) Alkaloids

Biological Activity	Target	Assay/Method	Results	Compound	References
Antibacterial	*Staphylococcus aureus* FDA 209P, *S. aureus* MRSA No.5, *S. aureus* MRSA No. 17, *S. aureus* Mu50. *Enterococcus faecalis* JCM 5803, *E. faecalis* NCTC12201, *E. faecium* JCM 5804, *E. faecium* NCTC12202, *Escherichia coli* K-12, *E. coli* DH5α/pUC57-Kan-MCR-1, *E. coli* DH5α/pUC57-Kan-MCR-2, *E. coli* NDM-1 Dok01, *Enterobacter cloacae* ATCC BAA-2468, *Klebsiella pneumoniae* PCI 602, *K. pneumoniae* ATCC BAA-1705, *K. pneumoniae* NDM-1 Sai01, *P. aeruginosa* A3, *P. aeruginosa* K-Ps102, *Acinetobacter baumannii* ATCC 19606, *A. baumannii* ATCC BAA-1605, *A. baumannii* NCGM237	Standard agar dilution method, Mueller–Hinton agar (Becton Dickinson)	MIC = 0.8 to 6.4 µg/mL	3.16	Igarashi et al. (2021)

(Continued)

TABLE 3.3 (Continued)
Biological Activity of Tetrahydronaphtoquinoline-dione (THQN-dione) Alkaloids

Biological Activity	Target	Assay/Method	Results	Compound	References
Antibacterial	S. aureus, E. coli, Burkholderia cepacian	–	MIC = 0.0064, 2, and 1 ng/mL, respectively	3.41	Davies et al. (2005)
Antibacterial	Methicillin-resistant S. aureus (MRSA), S. aureus, S. epidermidis, Bacillus cereus, Lysteria monocytogenes, vancomycin-resistant Enterococcus faecalis (VRE), Streptococcus pneumoniae, E. coli, B. cepacian, Salmonella typhimurium, P. aeruginosa	–	MIC = 0.2, 0.2, 0.09, 0.3, 1, 2, 0.4, 6, 0.4, 9, and 20ng/mL, respectively	3.41	Nicolaou et al. (2008)
Antifungal	Candida pseudotropicalis F-2	MIC assay, nutrient agar + glucose 1%	MIC = 1.0 µg/ml	3.1	Shiomi et al. (1990)
Antifungal	C. pseudotropicalis F-2	MIC assay, nutrient agar + glucose 1%	MIC = 1.0 µg/ml	3.2	Shiomi et al. (1990)
Antifungal	Amphotericin-resistant strain of Candida albicans (ATCC 38247), Cryptococcus neoformans spp. D49 and IAM4514	–	MIC = 0.8 µg/ml; better than controls for C. neoformans spp.	3.3	Saitoh et al. (1992)
Anti-phytopathogen	Xanthomonas citri, Xanthomonas oryzae	MIC assay, nutrient agar + glucose 1%	MIC = 0.001 µg/ml	3.1	Shiomi et al. (1990)
Anti-phytopathogen	X. citri, X.f oryzae	MIC assay, nutrient agar + glucose 1%	MIC = 4.0 and 0.5 ng/ml, respectively	3.2	Shiomi et al. (1990)
Antiviral	HIV-1 replication (virus strain NL4.3)	–	$EC_{50} = 9.9\ \mu M$	3.42	Williams et al. (2015)

(Continued)

TABLE 3.3 (Continued)
Biological Activity of Tetrahydronaphtoquinoline-dione (THQN-dione) Alkaloids

Biological Activity	Target	Assay/Method	Results	Compound	References
Cytotoxicity	B16-F10 (murine melanoma), HCT-116 (human colon carcinoma), P388 (murine leukaemia), vincristine-resistant subline of P388 (P388/VCR) and doxorubicin-resistant subline (P388/ADM), K562 (human myelogenous leukaemia)	—	$IC_{50} = 0.0027$ to 4.1 ng/mL	3.2	Kamei et al. (1991)
Cytotoxicity	B16-F10 (murine melanoma), HCT-116 (human colon carcinoma)	—	$IC_{50} = 2.8$ and 8.6 µg/kg, respectively	3.3	Saitoh et al. (1992)
Cytotoxicity	HL-60 and K-562 cell lines	—	showed cytotoxic activity	3.4	Shirai et al. (1995)
Cytotoxicity	B16F10: Mouse melanoma, HCT116: human colon carcinoma, P388: mouse leukaemia, K562/S: human chronic myelogenous leukaemia	—	$IC_{50} = 0.48$, 2.8, 0.04, and 0.13 µg/mL, respectively	3.8	Miyoshi-Saitoh et al. (1991)
Cytotoxicity	B16F10: Mouse melanoma, HCT116: human colon carcinoma, P388: mouse leukaemia, K562/S: human chronic myelogenous leukaemia	—	$IC_{50} = 3.6$, 5.3, 0.18, and 0.45 µg/mL, respectively	3.9	Miyoshi-Saitoh et al. (1991)
Cytotoxicity	B16F10: Mouse melanoma, HCT116: human colon carcinoma, P388: mouse leukaemia, K562/S: human chronic myelogenous leukaemia	—	$IC_{50} = 8.3$, 9.3, 0.54, and 1.71 µg/mL, respectively	3.10	Miyoshi-Saitoh et al. (1991)

(Continued)

TABLE 3.3 (*Continued*)

Biological Activity of Tetrahydronaphtoquinoline-dione (THQN-dione) Alkaloids

Biological Activity	Target	Assay/Method	Results	Compound	References
Cytotoxicity	A panel of 12 human tumour cell lines (breast, prostate, colon, lung, leukaemia)	—	A mean IC_{50} of 22 nM (range 7–58 nM), $IC_{50} = 60.5$ nM against human colon tumour cell line (HCT-116)	3.11	Martin et al. (2007)
Cytotoxicity	A panel of 12 human tumour cell lines (breast, prostate, colon, lung, leukaemia)	—	Less potent than 3.11: a mean IC_{50} of 3.5 μM (range 1.0 μM to 4.4 μM); $IC_{50} = 1.09$ μM against human colon tumour cell line (HCT-116)	3.12	Martin et al. (2007)
Cytotoxicity	Breast cancer: MDA-MB-468, MDA-MB-231, SKBR-3, KPL-4, BT474, and DYT2 Melanoma: M14 and SK-MEL-5 Lung cancer: NCI-H226 Central nervous system (CNS): SF-295 and SF-539	—	IC_{50} (nM) Breast cancer: 0.31, 2.3, 2.3, 0.33, 2.2, and 2.3 Melanoma: 0.84 and 0.33 Lung cancer: 9.2 CNS: 1.1 and 0.91	3.17	Yan et al. (2016)

(Continued)

TABLE 3.3 (*Continued*)
Biological Activity of Tetrahydronaphthoquinoline-dione (THQN-dione) Alkaloids

Biological Activity	Target	Assay/Method	Results	Compound	References
Cytotoxicity	Human cancer cell lines: melanoma (SKMEL-5), breast (MDA-MB-231 and SKBR-3), central nervous system (SF-295), and non-small cell lung (NCI-H226)	–	$IC_{50} = 0.1, 0.26, 0.50,$ 0.19, and 4.1 nM, respectively	3.17	Yan et al. (2017)
Cytotoxicity	Human cervical cancer cells HeLa, mouse bladder cancer cell MB49, mouse melanoma cell B16, human bladder cancer cell BIU87	MTT assay	$IC_{50} = 4.48, 0.38, 0.48,$ and 1.16 nM, respectively	3.17	Lu et al. (2022)
Cytotoxicity	Human embryonic kidney (HEK 293T), multidrug-resistant uterus sarcoma (MES SA/DX), multidrug-resistant uterine sarcoma cell line with PGP inhibitor elacridar (MES SA/DXE), breast cancer cell line (SK-BR-3), ovarian cancer cell line (SKOV3), and cervical cancer cell line (HeLa)	–	$IC_{50} = 0.30, 0.064,$ 0.037, 0.0253, 0.104, and 1.679 nM, respectively	3.17	Nicolaou et al. (2020)
Cytotoxicity	Human cancer cell lines: SF-295 (glioblastoma), SKMEL-5 (melanoma), MDA-MB-231 (breast), NCI-H226 (ncn-small cell lung), SKBR-3 (breast)	–	$IC_{50} = 0.11, 0.16, 1.2,$ 1.7, and 2.6 nM, respectively	3.17	Yang et al. (2022)
Cytotoxicity	Four cancer cell lines: Jurkat (human T lymphocytic leukaemia cells), A549 (lung adenocarcinoma cells), CaCo-2 (human intestinal epithelium cells), and KPL-4 (HER2-positive human cancer cells); and two normal human cell lines: human umbilical vein endothelial cells (HUVECs) ard NCM460 (colon mucosal epithelial cells)	–	$IC_{50} = 0.05, 0.04, 04,$ 0.02, 0.2, and 0.2 nM, respectively	3.17	Wen et al. (2024)

(Continued)

TABLE 3.3 (*Continued*)

Biological Activity of Tetrahydronaphtoquinoline-dione (THQN-dione) Alkaloids

Biological Activity	Target	Assay/Method	Results	Compound	References
Cytotoxicity	HEK 293T, MES SA/DX, multidrug-resistant uterine sarcoma cell line with PGP inhibitor elacridar (MES SA/DXE), breast cancer cell line (SK–BR-3), ovarian cancer cell line (SKOV3), and cervical cancer cell line (HeLa)	–	$IC_{50} = 0.404, 0.661, 0.599, 4.473, 1.818,$ and $24.08\,nM,$ respectively	3.18	Nicolaou et al. (2020)
Cytotoxicity	Human cervical cancer cells HeLa, mouse bladder cancer cell MB49, mouse melanoma cell B16, human bladder cancer cell BIU87	MTT assay	$IC_{50} = 16.6, 2.80, 9.18,$ and $1.10\,nM,$ respectively	3.20	Lu et al. (2022)
Cytotoxicity	Human cancer cell lines: SF-295 (glioblastoma), SKMEL-5 (melanoma), MDA-MB-231 (breast), NCI-H226 (non-small cell lung), SKBR-3 (breast)	–	$IC_{50} = 0.89, 0.93, 32, 20,$ and $21\,nM,$ respectively	3.20	Yang et al. (2022)
Cytotoxicity	Four cancer cell lines: Jurkat (human T lymphocytic leukaemia cells), A549 (lung adenocarcinoma cells), CaCo-2 (human intestinal epithelium cells), and KPL-4 (HER2-positive human cancer cells); and two normal human cell lines: HUVECs and NCM460 (colon mucosal epithelial cells)	–	$IC_{50} = 0.4, 0.1, 0.4, 0.04, 0.4,$ and $0.4\,nM,$ respectively	3.20	Wen et al. (2024)
Cytotoxicity	Four cancer cell lines: Jurkat (human T lymphocytic leukaemia cells), A549 (lung adenocarcinoma cells), CaCo-2 (human intestinal epithelium cells), and KPL-4 (HER2-positive human cancer cells); and two normal human cell lines: HUVECs and NCM460 (colon mucosal epithelial cells)	–	$IC_{50} = 0.1, 2.8, 6.8, 0.5, 2.8,$ and $15.5\,nM,$ respectively	3.23	Wen et al. (2024)

(Continued)

TABLE 3.3 (Continued)
Biological Activity of Tetrahydronaphthoquinoline-dione (THQN-dione) Alkaloids

Biological Activity	Target	Assay/Method	Results	Compound	References
Cytotoxicity	Breast cancer: MDA-MB-468, MDA-MB-231, SKBR-3, KPL-4, BT474, and DYT2 Melanoma: M14 and SK-MEL-5 Lung cancer: NCI-H226 Central nervous system (CNS): SF-295 and SF-539	—	IC_{50} (nM) = Breast cancer: 0.35, 12, 19, 5.8, 33, 8.2; Melanoma: 0.41, 2.9; Lung cancer: 4.3; CNS: 0.97, 1.0	3.41	Yan et al. (2016)
Cytotoxicity	HEK 293T, MES SA/DX, multidrug-resistant uterine sarcoma cell line with PGP inhibitor elacridar (MES SA/DXE), breast cancer cell line (SK-BR-3), ovarian cancer cell line (SKOV3), and cervical cancer cell line (HeLa)	—	IC_{50} = 0.086, 0.153, 0.121, 0.550, 0.329, and 41.24 nM, respectively	3.41	Nicolaou et al. (2020)
Cytotoxicity	Tumour cell lines: A-2780, CCRF-CEM, A549, H-2087, HCT-116, and MDA-MB231	—	IC_{50} = 0.003, 0.014, 0.076, 0.037, 0.007, and 0.085 nM, respectively	3.41	Chowdari et al. (2019)
Cytotoxicity	Ovarian tumour cell lines: 1A9, 1A9/PTX10, 1A9/PTX22, 1A9/A8	—	IC_{50} = 0.01, 0.06, 0.03, and 0.009 nM, respectively	3.41	Nicolaou et al. (2008)
Cytotoxicity	Cancer cell lines: SK-MEL-5 (melanoma), MDA-MB-468 (breast), NCI-H226 (non-small cell lung), SF-295 and SF-539 (central nervous system)	—	LC_{50} = 9, 20, 8, 20, 6, and 9 nM, respectively	3.41	Nicolaou et al. (2008)

(Continued)

TABLE 3.3 (Continued)

Biological Activity of Tetrahydronaphtoquinoline-dione (THQN-dione) Alkaloids

Biological Activity	Target	Assay/Method	Results	Compound	References
Cytotoxicity	Human cancer cell lines: melanoma (SKMEL-5), breast (MDA-MB-231 and SKBR-3), central nervous system (SF-295), and non-small cell lung (NCI-H226)	—	$IC_{50} = 0.87$, 1.2, 3.2, 0.26, and 1.9 nM, respectively	3.41	Yan et al. (2017)
Cytotoxicity	HEK 293T, MES SA/DX, multidrug-resistant uterine sarcoma cell line with PGP inhibitor elacridar (MES SA/DXE), breast cancer cell line (SK-BR-3), ovarian cancer cell line (SKOV3), and cervical cancer cell line (HeLa)	—	$IC_{50} = 0.013$, 0.0207, 0.013, 0.042, 0.014, and 0.244 nM, respectively	3.43	Nicolaou et al. (2020)
Cytotoxicity	Human cancer cell lines: CaCo2 (colon), Jurkat (lymphoma), A549 (lung), SKBR-3 (breast)	—	$IC_{50} = 3.40$, 0.24, 0.02, and 0.86 nM, respectively	3.43	Wang et al. (2019)
Cytotoxicity	Human cancer cell lines: melanoma (SKMEL-5), breast (MDA-MB-231 and SKBR-3), central nervous system (SF-295), and non-small cell lung (NCI-H226)	—	$IC_{50} = 0.40$, 0.57, 2.9, 0.26, and 2.8 nM, respectively	3.43	Yan et al. (2017)
Cytotoxicity	Four cancer cell lines: Jurkat (human T lymphocytic leukaemia cells), A549 (lung adenocarcinoma cells), CaCo-2 (human intestinal epithelium cells), and KPL-4 (HER2-positive human cancer cells); and two normal human cell lines: HUVECs and NCM460 (colon mucosal epithelial cells)	—	$IC_{50} = 0.009$, 0.06, 1.4, 0.006, 0.07, and 0.2 nM, respectively	3.43	Wen et al. (2024)

(Continued)

TABLE 3.3 (Continued)
Biological Activity of Tetrahydronaphthoquinoline-dione (THQN-dione) Alkaloids

Biological Activity	Target	Assay/Method	Results	Compound	References
Cytotoxicity	Human cancer cell lines: CaCo2 (colon), Jurkat (lymphoma), A549 (lung), SKBR-3 (breast)	–	$IC_{50} = 3.16, 0.17, 0.24,$ and 2.56 nM, respectively	3.48	Wang et al. (2019)
Cytotoxicity	Four cancer cell lines: Jurkat (human T lymphocytic leukaemia cells), A549 (lung adenocarcinoma cells), CaCo-2 (human intestinal epithelium cells), and KPL-4 (HER2-positive human cancer cells); and two normal human cell lines: HUVECs and NCM460 (colon mucosal epithelial cells)	–	$IC_{50} = 0.02, 0.2, 13.5,$ 0.01, 0.2, and 0.8 nM, respectively	3.48	Wen et al. (2024)
Cytotoxicity	Human cancer cell lines: CaCo2 (colon), Jurkat (lymphoma), A549 (lung), SKBR-3 (breast)	–	$IC_{50} = 55.20, 0.62,$ 3.41, and 24.20 nM, respectively	3.49	Wang et al. (2019)
Cytotoxicity	Four cancer cell lines: Jurkat (human T lymphocytic leukaemia cells), A549 (lung adenocarcinoma cells), CaCo-2 (human intestinal epithelium cells), and KPL-4 (HER2-positive human cancer cells); and two normal human cell lines: HUVECs and NCM460 (colon mucosal epithelial cells)	–	$IC_{50} = 0.01, 0.2, 5.4,$ 0.004, 0.1, and 0.3 nM, respectively	3.49	Wen et al. (2024)

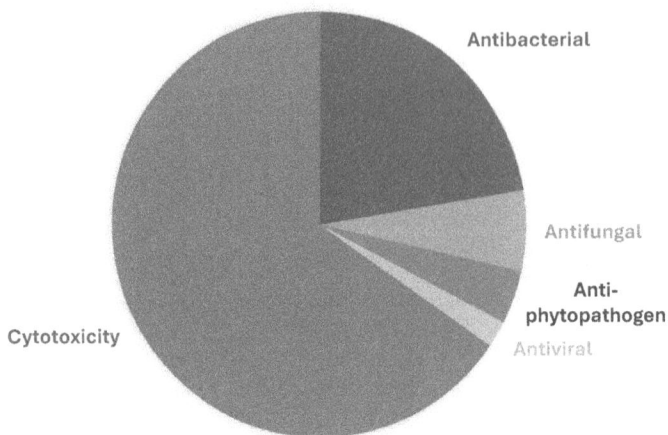

FIGURE 3.6 Biological activity of THNQ-dione alkaloids.

to improve the efficiency and scalability of THQN-diones' synthesis, which makes these complex molecules more accessible for further study and potential therapeutic applications.

The biosynthesis study of THQN-diones also raises numerous challenges, including the identification and characterization of complex gene clusters responsible for their production within microbial genomes. Understanding the specific roles and interactions of biosynthetic enzymes is crucial, as these enzymes often exhibit unique activities and require detailed biochemical and structural characterization. Synthesizing THQN-dione is complicated by several factors. First, it is important to control the biosynthetic pathway to ensure there are enough raw materials and to balance the flow of chemicals needed for production. Additionally, managing toxicity is crucial. Selecting the right host organisms for producing THQN-dione adds further challenges. Recovering and purifying substances from fermentation broth is also a challenge. Scientists face difficulties because of complex regulations and natural differences in the ability of organisms to produce these substances. Addressing these challenges requires combining knowledge from genomics, enzymology, metabolic engineering, and synthetic biology.

The presence of THQN-diones exclusively in bacteria suggests that bacteria have specialized metabolic pathways, possibly involving unique enzymes and gene clusters. This could indicate a separate evolutionary history for these alkaloids, potentially linked to bacterial adaptation and survival strategies. These alkaloids could be valuable for developing new antibiotics, as they may have evolved to provide competitive advantages against other microorganisms. They offer unique chemical structures and biological activities, which could lead to new pharmaceutical applications. Important to the ecosystem, they may help bacteria communicate, defend against threats, or interact with other organisms in their environment.

REFERENCES

Adhikari, A., Teijaro, C. N., Yan, X., Chang, C., Gui, C., Liu, Y., Crnovcic, I., Yang, D., Annaval, T., & Rader, C. (2020). Characterization of TnmH as an *O*-methyltransferase revealing insights into tiancimycin biosynthesis and enabling a biocatalytic strategy to prepare antibody–tiancimycin conjugates. *Journal of Medicinal Chemistry, 63*(15), 8432–8441.

Albooyeh, F., & Aghapour, G. (2023). Zinc-mediated efficient and selective reductive aza-claisen rearrangement of *N*-allyland *N*-propargylaminoanthraquinones in ionic liquid. *Turkish Journal of Chemistry, 47*(3), 514–526.

An, S., Potnis, N., Dow, M., Vorhölter, F., He, Y., Becker, A., Teper, D., Li, Y., Wang, N., & Bleris, L. (2020). Mechanistic insights into host adaptation, virulence and epidemiology of the phytopathogen *Xanthomonas*. *FEMS Microbiology Reviews, 44*(1), 1–32.

Annaval, T., Teijaro, C. N., Adhikari, A., Yan, X., Chen, J., Crnovcic, I., Yang, D., & Shen, B. (2021). Cytochrome P450 hydroxylase TnmL catalyzing sequential hydroxylation with an additional proofreading activity in tiancimycin biosynthesis. *ACS Chemical Biology, 16*(7), 1172–1178.

Bauman, K. D., Butler, K. S., Moore, B. S., & Chekan, J. R. (2021). Genome mining methods to discover bioactive natural products. *Natural Product Reports, 38*(11), 2100–2129. 10.1039/d1np00032b

Belecki, K., & Townsend, C. A. (2012). Environmental control of the calicheamicin polyketide synthase leads to detection of a new octaketide and proposal for enediyne biosynthesis. *Angewandte Chemie, 51*(45), 11316.

Belecki, K., & Townsend, C. A. (2013). Biochemical determination of enzyme-bound metabolites: Preferential accumulation of a programmed octaketide on the enediyne polyketide synthase CalE8. *Journal of the American Chemical Society, 135*(38), 14339–14348.

Bhardwaj, M., Cui, Z., Daniel Hankore, E., Moonschi, F. H., Saghaeiannejad Esfahani, H., Kalkreuter, E., Gui, C., Yang, D., Phillips Jr, G. N., & Thorson, J. S. (2023). A discrete intermediate for the biosynthesis of both the enediyne core and the anthraquinone moiety of enediyne natural products. *Proceedings of the National Academy of Sciences of the United States of America, 120*(9), e2220468120.

Chowdari, N. S., Pan, C., Rao, C., Langley, D. R., Sivaprakasam, P., Sufi, B., Derwin, D., Wang, Y., Kwok, E., & Passmore, D. (2019). Uncialamycin as a novel payload for antibody drug conjugate (ADC) based targeted cancer therapy. *Bioorganic & Medicinal Chemistry Letters*, 29(3), 466–470.

Cohen, D. R., & Townsend, C. A. (2018a). Characterization of an anthracene intermediate in dynemicin biosynthesis. *Angewandte Chemie International Edition, 57*(20), 5650–5654.

Cohen, D. R., & Townsend, C. A. (2018b). A dual role for a polyketide synthase in dynemicin enediyne and anthraquinone biosynthesis. *Nature Chemistry, 10*(2), 231–236.

Davies, J., Wang, H., Taylor, T., Warabi, K., Huang, X., & Andersen, R. J. (2005). Uncialamycin, a new enediyne antibiotic. *Organic Letters, 7*(23), 5233–5236.

Diaz-Munoz, G., Miranda, I. L., Sartori, S. K., de Rezende, D. C., & Diaz, M. A. (2018). Anthraquinones: An overview. *Studies in Natural Products Chemistry, 58*, 313–338.

Flemming, A. (2014). Fine-tuning antibody–drug conjugates. *Nature Reviews Drug Discovery, 13*(3), 178.

Gao, Q., & Thorson, J. S. (2008). The biosynthetic genes encoding for the production of the dynemicin enediyne core in *Micromonospora chersina* ATCC53710. *FEMS Microbiology Letters, 282*(1), 105–114.

Gartman, J. A., & Tambar, U. K. (2022). Recent total syntheses of anthraquinone-based natural products. *Tetrahedron, 105*, 132501.

Gui, C., Kalkreuter, E., Lauterbach, L., Yang, D., & Shen, B. (2024). Enediyne natural product biosynthesis unified by a diiodotetrayne intermediate. *Nature Chemical Biology, 20*, 1–10.

Gui, C., Kalkreuter, E., Liu, Y., Adhikari, A., Teijaro, C. N., Yang, D., Chang, C., & Shen, B. (2022). Intramolecular C–C bond formation links anthraquinone and enediyne scaffolds in tiancimycin biosynthesis. *Journal of the American Chemical Society, 144*(44), 20452–20462.

Hindra, Y. D., Luo, J., Huang, T., Yan, X., Adhikari, A., Teijaro, C. N., Ge, H., & Shen, B. (2021). Submerged fermentation of streptomyces uncialis providing a biotechnology platform for uncialamycin biosynthesis, engineering, and production. *Journal of Industrial Microbiology and Biotechnology, 48*(3–4), kuab025.

Igarashi, M., Sawa, R., Umekita, M., Hatano, M., Arisaka, R., Hayashi, C., Ishizaki, Y., Suzuki, M., & Kato, C. (2021). Sealutomicins, new enediyne antibiotics from the deep-sea *Actinomycete nonomuraea* sp. MM565M-173N2. *The Journal of Antibiotics, 74*(5), 291–299.

Kamei, H., Nishiyama, Y., Takahashi, A., Obi, Y., & Oki, T. (1991). Dynemicins, new antibiotics with the 1,5-diyn-3-ene and anthraquinone subunit II. Antitumor activity of dynemicin A and its triacetyl derivative. *The Journal of Antibiotics, 44*(12), 1306–1311.

Kangping, D., Jing, W., Rong, J., Baoyi, L., & Jianbo, W. (2000). Isolation and structure identification of antibiotics from a *Micromonospora* strain C5308. *Zhongguo Kang Sheng Su Za Zhi = Chinese Journal of Antibiotics, 25*(5), 333–335,385.

Khadem, S., & Marles, R. J. (2025a). Tetrahydroquinoline-containing natural products discovered within the last decade: Occurrence and bioactivity. *Natural Product Research, 39*(1), 182–194.

Khadem, S., & Marles, R. J. (2025b). The occurrence and bioactivity of tetrahydronaphthoquinoline-diones (THNQ-dione). *Natural Product Research, 39*(6), 1622–1635.

Konishi, M., Ohkuma, H., Matsumoto, K., Saitoh, K., Miyaki, T., Oki, T., & Kawaguchi, H. (1991). Dynemicins, new antibiotics with the 1, 5-diyn-3-ene and anthraquinone subunit i. production, isolation and physico-chemical properties. *The Journal of Antibiotics, 44*(12), 1300–1305.

Konishi, M., Ohkuma, H., Matsumoto, K., Tsuno, T., Kamei, H., Miyaki, T., Oki, T., & Kawaguchi, H. (1989). Dynemicin A, a novel antibiotic with the anthraquinone and 1,5-diyn-3-ene subunit. *The Journal of Antibiotics, 42*(9), 1449–1452.

Konishi, M., Ohkuma, H., Tsuno, T., Oki, T., VanDuyne, G. D., & Clardy, J. (1990). Crystal and molecular structure of dynemicin A: A novel 1,5-diyn-3-ene antitumor antibiotic. *Journal of the American Chemical Society, 112*(9), 3715–3716.

Liu, L., Pan, J., Wang, Z., Yan, X., Yang, D., Zhu, X., Shen, B., Duan, Y., & Huang, Y. (2018). Ribosome engineering and fermentation optimization leads to overproduction of tiancimycin A, a new enediyne natural product from *Streptomyces* sp. CB03234. *Journal of Industrial Microbiology and Biotechnology, 45*(3), 141–151.

Low, Z. J., Ma, G., Tran, H. T., Zou, Y., Xiong, J., Pang, L., Nuryyeva, S., Ye, H., Hu, J., & Houk, K. N. (2020). Sungeidines from a non-canonical enediyne biosynthetic pathway. *Journal of the American Chemical Society, 142*(4), 1673–1679.

Lu, X., Lihua, Z., Chengyu, Z., Xin, Z., Weifan, D., Zhaoxin, W., Chunhua, W., Tongchuan, S., & Xiaohui, Y. (2022). Discovery of tiancimycin congeners from *Streptomyces* sp. CB03234-S. *Chinese Journal of Organic Chemistry, 42*(4), 1241.

Malik, E. M., & Muller, C. E. (2016). Anthraquinones as pharmacological tools and drugs. *Medicinal Research Reviews, 36*(4), 705–748. 10.1002/med.21391

Martin, G. D., Tan, L. T., Jensen, P. R., Dimayuga, R. E., Fairchild, C. R., Raventos-Suarez, C., & Fenical, W. (2007). Marmycins A and B, cytotoxic pentacyclic C-glycosides from a marine sediment-derived actinomycete related to the genus *Streptomyces*. *Journal of Natural Products, 70*(9), 1406–1409.

Medema, M. H., de Rond, T., & Moore, B. S. (2021). Mining genomes to illuminate the specialized chemistry of life. *Nature Reviews Genetics, 22*(9), 553–571.

Mellor, J. M., Merriman, G. D., & Riviere, P. (1991). Synthesis of tetrahydroquinolines from aromatic amines, formaldehyde and electron rich alkenes: Evidence for non-concertedness. *Tetrahedron Letters, 32*(48), 7103–7106.

Miyoshi, M., Morisaki, N., Tokiwa, Y., Kobayashi, H., Iwasaki, S., Konishi, M., & Oki, T. (1991). Facile reductive rearrangement of dynemicin A to dynemicin H: The direct evidence for the *p*-phenylene diradical intermediate. *Tetrahedron Letters, 32*(42), 6007–6010.

Miyoshi-Saitoh, M., Morisaki, N., Tokiwa, Y., Iwasaki, S., Konishi, M., Saitoh, K., & Oki, T. (1991). Dynemicins O, P and Q: Novel antibiotics related to dynemicin A isolation, characterization and biological activity. *The Journal of Antibiotics, 44*(10), 1037–1044.

Muthukrishnan, I., Sridharan, V., & Menendez, J. C. (2019). Progress in the chemistry of tetrahydroquinolines. *Chemical Reviews, 119*(8), 5057–5191. doi:10.1021/acs.chemrev.8b00567

Myers, A. G., Fraley, M. E., Tom, N. J., Cohen, S. B., & Madar, D. J. (1995). Synthesis of (+)-dynemicin A and analogs of wide structural variability: Establishment of the absolute configuration of natural dynemicin A. *Chemistry & Biology, 2*(1), 33–43.

Nicolaou, K. C., Das, D., Lu, Y., Rout, S., Pitsinos, E. N., Lyssikatos, J., Schammel, A., Sandoval, J., Hammond, M., & Aujay, M. (2020). Total synthesis and biological evaluation of tiancimycins A and B, yangpumicin A, and related anthraquinone-fused enediyne antitumor antibiotics. *Journal of the American Chemical Society, 142*(5), 2549–2561.

Nicolaou, K. C., Wang, Y., Lu, M., Mandal, D., Pattanayak, M. R., Yu, R., Shah, A. A., Chen, J. S., Zhang, H., & Crawford, J. J. (2016). Streamlined total synthesis of uncialamycin and its application to the synthesis of designed analogues for biological investigations. *Journal of the American Chemical Society, 138*(26), 8235–8246.

Nicolaou, K. e., Chen, J. S., Zhang, H., & Montero, A. (2008). Asymmetric synthesis and biological properties of uncialamycin and 26-*epi*-uncialamycin. *Angewandte Chemie, 120*(1), 191–195.

Nicolaou, K. e., Zhang, H., Chen, J. S., Crawford, J. J., & Pasunoori, L. (2007). Total synthesis and stereochemistry of uncialamycin. *Angewandte Chemie International Edition, 46*(25), 4704–4707.

Qi-wei, Y., Cui-min, F., & Ji-lan, H. (2002). A micromonospora strain C5308 producing antitumor antibiotic dynemicin H. *Chinese Journal of Antibiotics, 27*(11), 647–650.

Saitoh, K., Miyaki, T., Tsunashima, K., Yamamoto, H., & Oda, N. (1992a). Dynemicin C antitumor antibiotic. *Patent* No. EP 0484856 A2, 1–21.

Saitoh, K., Miyaki, T., Yamamoto, H., & Oda, N. (1992b). Dynemicin c antibiotic, its triacetyl derivative and pharmaceutical composition containing same. *Patent* No. US5162330, 1–11.

Shen, B. (2003). Polyketide biosynthesis beyond the type I, II and III polyketide synthase paradigms. *Current Opinion in Chemical Biology, 7*(2), 285–295.

Shimazawa, R., Shirai, R., Hashimoto, Y., & Iwasaki, S. (1994). DNA-binding ability of non-diynene class of dynemicins and aza-anthraquinones. *Bioorganic & Medicinal Chemistry Letters, 4*(20), 2377–2382.

Shiomi, K., Iinuma, H., Naganawa, H., Hamada, M., Hattori, S., Nakamura, H., & Takeuchi, T. (1990). New antibiotic produced by *Micromonospora globosa*. *The Journal of Antibiotics, 43*(8), 1000–1005.

Shirai, R., Shimazawa, R., Shichita, M., Takahashi, M., Hashimoto, Y., & Iwasaki, S. (1995). Cytotoxicity and DNA-binding property of non-diynene class of dynemicins and aza-anthraquinones. *Paper presented at the Nucleic Acids Symposium Series,* (34) 151–152.

Sugiura, Y., Shiraki, T., Konishi, M., & Oki, T. (1990). DNA intercalation and cleavage of an antitumor antibiotic dynemicin that contains anthracycline and enediyne cores. *Proceedings of the National Academy of Sciences of the United States of America, 87*(10), 3831–3835.

Tan, Y., Zhou, H. X., Wang, Y. G., Gan, M. L., & Yang, Z. Y. (2013). Halogenated natural products from the marine-derived actinobacteria and their halogenation mechanism. *Yao Xue Xue Bao = Acta Pharmaceutica Sinica, 48*(9), 1369–1375.

Tokiwa, Y., Miyoshi-Saitoh, M., Kobayashi, H., Sunaga, R., Konishi, M., Oki, T., & Iwasaki, S. (1992). Biosynthesis of dynemicin A, a 3-ene-1,5-diyne antitumor antibiotic. *Journal of the American Chemical Society, 114*(11), 4107–4110.

Unno, R., Michishita, H., Inagaki, H., Suzuki, Y., Baba, Y., Jomori, T., Nishikawa, T., & Isobe, M. (1997). Synthesis and antitumor activity of water-soluble enediyne compounds related to dynemicin A. *Bioorganic & Medicinal Chemistry, 5*(5), 987–999.

Wang, Z., Wen, Z., Liu, L., Zhu, X., Shen, B., Yan, X., Duan, Y., & Huang, Y. (2019). Yangpumicins F and G, enediyne congeners from *Micromonospora yangpuensis* DSM 45577. *Journal of Natural Products, 82*(9), 2483–2488.

Wen, Z., Zhuang, Z., Liu, H., Wang, Z., Feng, X., Zhu, X., Yan, X., Duan, Y., & Huang, Y. (2024). DNA interaction and cleavage modes of anthraquinone-fused enediynes: A study on tiancimycins, yangpumicins, and their semisynthetic analogues. *Journal of Medicinal Chemistry, 67*(6), 4624–4640.

Williams, D. E., Bottriell, H., Davies, J., Tietjen, I., Brockman, M. A., & Andersen, R. J. (2015). Unciaphenol, an oxygenated analogue of the Bergman cyclization product of uncialamycin exhibits anti-HIV activity. *Organic Letters, 17*(21), 5304–5307.

Yan, S., Zeng, M., Wang, H., & Zhang, H. (2022). *Micromonospora*: A prolific source of bioactive secondary metabolites with therapeutic potential. *Journal of Medicinal Chemistry, 65*(13), 8735–8771.

Yan, X. (2022). Anthraquinone-fused enediynes: Discovery, biosynthesis and development. *Natural Product Reports, 39*(3), 703–728.

Yan, X., Chen, J., Adhikari, A., Teijaro, C. N., Ge, H., Crnovcic, I., Chang, C., Annaval, T., Yang, D., & Rader, C. (2018). Comparative studies of the biosynthetic gene clusters for anthraquinone-fused enediynes shedding light into the tailoring steps of tiancimycin biosynthesis. *Organic Letters, 20*(18), 5918–5921.

Yan, X., Chen, J., Adhikari, A., Yang, D., Crnovcic, I., Wang, N., Chang, C., Rader, C., & Shen, B. (2017). Genome mining of *micromonospora yangpuensis* DSM 45577 as a producer of an anthraquinone-fused enediyne. *Organic Letters, 19*(22), 6192–6195.

Yan, X., Ge, H., Huang, T., Hindra, Yang, D., Teng, Q., Crnovčić, I., Li, X., Rudolf, J. D., & Lohman, J. R. (2016). Strain prioritization and genome mining for enediyne natural products. *MBio, 7*(6), 10.1128/mbio.02104-16.

Yang, D., Ye, F., Teijaro, C. N., Hwang, D., Annaval, T., Adhikari, A., Li, G., Yan, X., Gui, C., & Rader, C. (2022). Functional characterization of cytochrome P450 hydroxylase YpmL in yangpumicin A biosynthesis and its application for anthraquinone-fused enediyne structural diversification. *Organic Letters, 24*(5), 1219–1223.

Zhuang, Z., Jiang, C., Zhang, F., Huang, R., Yi, L., Huang, Y., Yan, X., Duan, Y., & Zhu, X. (2019). Streptomycin-induced ribosome engineering complemented with fermentation optimization for enhanced production of 10-membered enediynes tiancimycin-A and tiancimycin-D. *Biotechnology and Bioengineering, 116*(6), 1304–1314.

4 3,4-Dihydro-2(1*H*)-quinolinones (2O-THQ)

4.1 INTRODUCTION

The 2-quinolinone skeleton, shown in Figure 4.1, is a versatile structural component found in natural products and biologically active substances. It is widely used in drug design and discovery because it can act as a useful building block, scaffold, fragment, and pharmacophore with various substituent groups (Hong et al. 2020). 2-Quinolinone can interact with pharmaceutical target proteins through hydrogen bonding or π-π stacking. The skeleton is metabolically stable in a live organism, and its ring structure cannot be opened, particularly at the amide bond. The internal amide within 2-quinolinone and its derivatives, such as 3,4-dihydro-2(1*H*)-quinolinone (4.1), largely exists in the fixed *cis* conformation of the lactam amide group. This configuration allows it to form hydrogen bonds with peptides, nucleic acids, or water molecules. The internal amide within 2-quinolinones interacts with proteases, the ubiquitin-proteasome system, or other metabolic pathways. Nucleophilic addition to the carbonyl is feasible following the Bürgi–Dunitz trajectory, but steric hindrance prevents large molecules like proteases from accessing this carbon atom. Additionally, the amide bond present in 2-quinolinone derivatives, distinct from external amide bonds in other compounds, exhibits resistance to cleavage by cysteine proteases, serine proteases, metalloproteases, and aspartic acid proteases. This characteristic significantly contributes to the favourable absorption, distribution, metabolism, excretion, and toxicity properties observed in 2-quinolinone derivatives (Tashima 2015).

3,4-Dihydro-2(1*H*)-quinolinone (4.1), also known as 2-oxo-tetrahydroquinoline (2O-THQ), is a chemical compound with a bicyclic structure consisting of a δ-lactam ring fused to a phenyl ring. The amide group in this compound can act as both a hydrogen donor and acceptor when interacting with protein residues. Additionally, this amide group is resistant to acid- and base-catalyzed hydrolysis reactions. The presence of two sp³-hybridized carbons in its structure contributes to the non-planar geometry of the 2O-THQ moiety (Meiring et al. 2018).

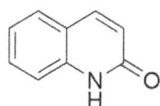

2-Quinolinone   3,4-Dihydro-2(1*H*)-quinolinone
(4.1)

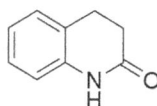

FIGURE 4.1 Chemical structures of 2-quinolinone and 3,4-qihydro-2(1*H*)-quinolinone [2O-THQ].

DOI: 10.1201/9781003534600-4

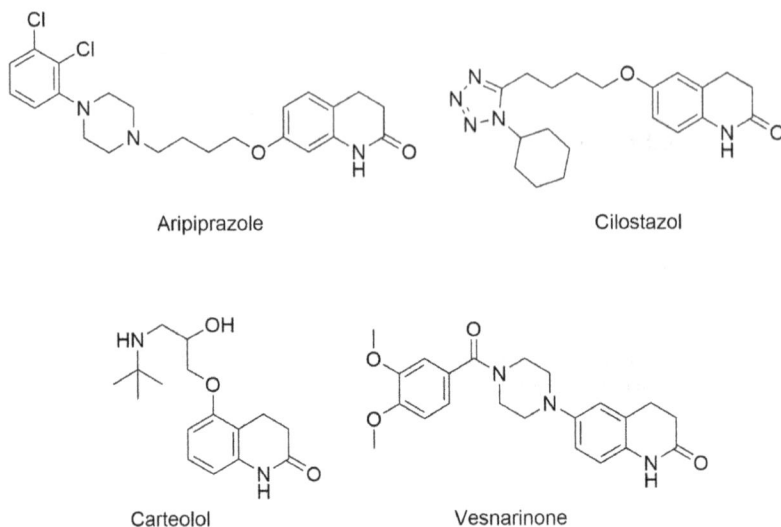

FIGURE 4.2 Chemical structures of aripiprazole, cilostazol, carteolol, and vesnarinone.

Compounds containing the structural motif of 2O-THQ have various effects on both peripheral and central tissues. These effects include blocking certain enzymes, stopping the action of specific receptors, and influencing serotonin and dopamine levels. They can also block vasopressin receptors and interact with sigma receptors in different ways. These substances act as helpers for muscarinic acetylcholine receptors and can prevent the activation of N-methyl-D-aspartate (NMDA) receptors. They inhibit the activity of monoamine oxidase, block oxytocin receptors, show properties that can help control seizures, function as contraceptives, and improve heart strength (Meiring et al. 2018; Wu and Piao 2013).

Notable synthetic instances of 2O-THQ, depicted in Figure 4.2, include aripiprazole (antipsychotic), carteolol (non-selective beta blocker), vesnarinone (cardiotonic, investigation for the treatment of HIV infections and Kaposi sarcoma, but not marketed at this time), and cilostazol (phosphodiesterase-3 inhibitor). These substances have received approval from the US Food and Drug Administration for their clinical application in human therapeutics.

Despite the undeniable importance of 2O-THQ alkaloids in natural product chemistry and pharmacology, current review literature has not provided a comprehensive and systematic exploration of their prevalence in nature and their diverse biological functions. This chapter aims to fill this knowledge gap by providing a detailed analysis of the synthesis, occurrence, and biological activities associated with 2O-THQ alkaloids.

4.1.1　CHEMICAL IDENTITY

The chemical identity of 2O-THQ alkaloids in this chapter is shown in Table 4.1. The structures can be classified into several subclasses: (a) simple bicyclic 2O-THQs with one or more substitutions within rings; (b) Melodinus alkaloids, such as Scandine

TABLE 4.1
The Chemical Identity Of 2-Oxo-Tetrahydroquinoline (2O-Thq) Alkaloids

Structure (Number)	Systematic Name	Common Name	CAS
(4.1)	3,4-Dihydroquinolin-2(1*H*)-one [2-Oxo-1,2,3,4-tetrahydroquinoline]	2O-THQ (Hydro-carbostyril)	553-03-7
(4.2)	3,4-Dihydro-6,7-dihydroxy-2(1*H*)-quinolinone	Alternamide B	1890808-73-7
(4.3)	Methyl 1,2,3,4-tetrahydro-3,6-dihydroxy-2-oxo-3-quinolinecarboxylate	–	103142-43-4
(4.4)	Methyl 1,2,3,4-tetrahydro-6-hydroxy-2-oxo-3-quinolinecarboxylate	–	2409943-10-6
(4.5)	10H-Indolizino[1′,8′:2,3,4]cyclopenta[1,2-c]quinoline-6a(7H)-carboxylic acid, 7a-ethenyl-7a,11a,12,13-tetrahydro-, methyl ester, (6aR,7aS,11aS,13aS)-	Scandine	24314-59-8
(4.6)	10*H*-Indolizino[1′,8′:2,3,4]cyclopenta[1,2-c]quinoline-6a(7H)-carboxylic acid, 7a-ethenyl-5,6,7a,11a,12,13-hexahydro-2-hydroxy-6-oxo-, methyl ester, (6aR,7aS,11aS,13aS)-	10-Hydro-xyscandine	119188-47-5
(4.7)	Meloscine, 3-(methoxycarbonyl)-, 9-oxide	Scandine N-oxide	140701-69-5
(4.8)	methyl (6aS,7aS,7a1S,13aS)-6,10-dioxo-7a-vinyl-5,6,7a,7a1,12,13-hexahydro-10*H*-indolizino[1′,8′:2,3,4]cyclopenta[1,2-c]quinoline-6a(7H)-carboxylate	3-Oxo-scandine	–

(Continued)

TABLE 4.1 (Continued)
The Chemical Identity Of 2-Oxo-Tetrahydroquinoline (2O-Thq) Alkaloids

Structure (Number)	Systematic Name	Common Name	CAS
(4.9)	Meloscine, 3-(methoxycarbonyl)-, (3α)-	3-Episcandine	73951-55-0
(4.10)	(6aS,7aS,11aS,13aR)-7a-Ethenyl-6a,7,7a,11a,12,13-hexahydro-10H-indolizino [1′,8′:2,3,4]cyclopenta [1,2-c]quinolin-6(5H)-one	Meloscine	24314-51-0
(4.11)	10H-Indolizino[1′,8′:2,3,4] cyclopenta[1,2-c]quinolin-6 (5H)-one, 7a-ethenyl-6a,7,7a, 11a,12,13-hexahydro-, (6aR,7aS,11aS,13aR)-	Epimeloseine	24314-58-7
(4.12)	Meloscine, 14-hydroxy-, (3α)-	14-Hydroxy-3-epimeloscine	169700-70-3
(4.13)	Meloscine, 14-methoxy-, (3α)-	14-Methoxy-3-epimeloscine	169700-71-4
(4.14)	(6aS,7aS,7a1S,13aR)-6-oxo-7a-vinyl-5,6,6a,7,7a,10,12,13-octahydroindolizino[1′,8′:2,3,4] cyclopenta[1,2-c]quinoline 11 (7a1H)-oxide	Meloscine N-oxide	1802540-44-8

(Continued)

TABLE 4.1 (*Continued*)
The Chemical Identity Of 2-Oxo-Tetrahydroquinoline (2O-Thq) Alkaloids

Structure (Number)	Systematic Name	Common Name	CAS
(**4.15**)	Meloscine, 15-methoxy-3-(methoxycarbonyl)-	10-Methoxyscandine	119188-48-6
(**4.16**)	(6aR,8R,8aR,14aS,14bS)-13,14-Dihydro-8-methyl-6a,8a-methano-11H,12aH-benzo[k]pyrrolo[3,2,1-mn][1,8]phenanthroline-6,7(5H,8H)-dione	Meloscandonine	28645-27-4
(**4.17**)	6a,8a-Methano-11H,12aH-benzo[k]pyrrolo[3,2,1-mn][1,8]phenanthroline-6,7(5H,8H)-dione, 13,14-dihydro-8-methyl-, (6aR,8S,8aR,14aS,14bS)-	19-epi-Meloscandonine	151636-21-4
(**4.18**)	Meloscine, 6,7-epoxy-6,7-dihydro-3-(methoxycarbonyl)-	14,15-Epoxyscandine	151563-68-7
(**4.19**)	Methyl (6bS,6b1R,10aR,11aS,11bR,12aR)-1-oxo-11b-vinyl-1,2,7,8,10a,11a,11b,12-octahydro-10H-oxireno[2²,3²:6′,7′]indolizino[1′,8′:2,3,4]cyclopenta[1,2-c]quinoline-12a(6b1H)-carboxylate	14,15β-Epoxyscandine	1802540-43-7
(**4.20**)	10-Hydroxy-14,15-β-epoxysandine	–	1801530-00-6

(*Continued*)

TABLE 4.1 (*Continued*)
The Chemical Identity Of 2-Oxo-Tetrahydroquinoline (2O-Thq) Alkaloids

Structure (Number)	Systematic Name	Common Name	CAS
 (4.21)	Methyl (6aR,7aS,7a1S,13aS)-6-oxo-7a-vinyl-5,6,7a,7a1,9,10,12,13-octahydro-8H-indolizino[1',8':2,3,4]cyclopenta[1,2-c]quinoline-6a(7H)-carboxylate	14,15-Dihydroscandine	–
 (4.22)	15β-hydroxy-14,15-dihydroscandine	–	–
 (4.23)	(6aR,8R,8aR,9S,12aR,14aS)-11,12,13,14-Tetrahydro-9-hydroxy-8-methyl-6a,8a-methano-9H,12aH-benzo[k]pyrrolo[3,2,1-mn][1,8]phenanthroline-6,7(5H,8H)-dione	15α-Hydroxymeloscandonine	1801530-01-7
 (4.24)	10H-Indolizino[1',8':2,3,4]cyclopenta[1,2-c]quinolin-6(5H)-one, 7a-ethenyl-6a,7,7a,11a,12,13-hexahydro-, 11-oxide, (6aR,7aS,11aS,13aR)-	9-Oxide 3-epimeloscine	34210-64-5
 (4.25)	Methyl (3aS,4aS,10bS,13aR)-3a-acetyl-3a,5,6,11,12,13a-hexahydro-5-oxo-1H-indolizino[1',8':2,3,4]cyclopenta[1,2-c]quinoline-4a(4H)-carboxylate	Melodinine T	1356156-01-8
 (4.26)	1H-Indolizino[1',8':2,3,4]cyclopenta[1,2-c]quinoline-4a(4H)-carboxylic acid, 3a-ethenyl-3a,5,6,11,12,13a-hexahydro-9-hydroxy-5-oxo-, methyl ester, 13-oxide, (3aS,4aS,10bS,13aS)-	Melodinine U	1356156-03-0

(Continued)

TABLE 4.1 (*Continued*)
The Chemical Identity Of 2-Oxo-Tetrahydroquinoline (2O-Thq) Alkaloids

Structure (Number)	Systematic Name	Common Name	CAS
(4.27)	Methyl (αR,3aR,4aR,10bR, 13aS)-4,4a,5,6,11,12-hexahydro-α-methyl-5-oxo-1H-indolizino[1′,8′:2,3,4]cyclopenta[1,2-c]quinoline-3a (13aH)-acetate	Melaxilline A (Melokhasine A)	2241558-61-0
(4.28)	Methyl (2S,3aR,4aS,10bS, 13aS)-3a-ethenyl-2,3,3a,5, 6,11,12,13a-octahydro-2-hydroxy-1,5-dioxo-1H-indolizino[1′,8′:2,3,4]cyclopenta[1,2-c]quinoline-4a(4H)-carboxylate	Melaxilline B	2241558-62-1
(4.29)	19-epi-Meloyine III	Meloyine II	2139305-25-0
(4.30)	Methyl (2β,12R,19α,20R)-6,7-didehydro-3-[(6aR,8S, 8aR,12aS,14aS)-5,6,7,8,13,14-hexahydro-8-methyl-6,7-dioxo-6a,8a-methano-11H,12aH-benzo[k]pyrrolo[3,2,1-mn][1,8]phenanthrolin-2-yl]-2,20-cycloaspidospermidine-2-carboxylate	Melokhanine K	2413295-99-3
(4.31)	19-epi-Meloyine II	Meloyine III	2139305-26-1

(*Continued*)

TABLE 4.1 (Continued)

The Chemical Identity Of 2-Oxo-Tetrahydroquinoline (2O-Thq) Alkaloids

Structure (Number)	Systematic Name	Common Name	CAS
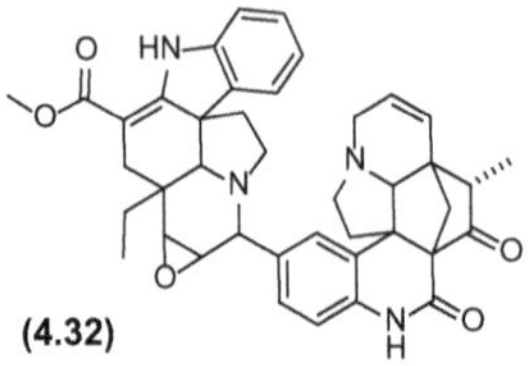 (4.32)	Aspidospermidine-3-carboxylic acid, 2,3-didehydro-6,7-epoxy-8-[(6aR,8S,8aR,14aS,14bS)-5,6,7,8,13,14-hexahydro-8-methyl-6,7-dioxo-6a,8a-methano-11H,12aH-benzo[k]pyrrolo[3,2,1-mn][1,8]phenanthrolin-15-yl]-, methyl ester, (5α,6β,7β,8α,12R,19α)-	Episcandomelonine	59830-06-7
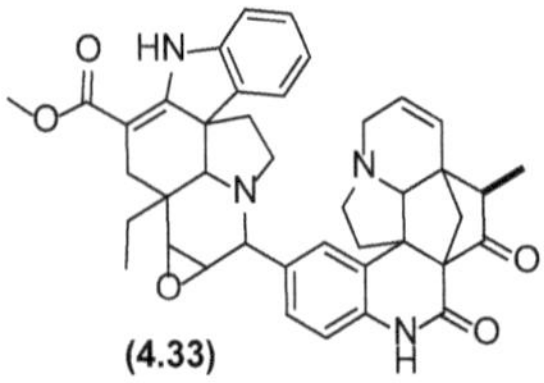 (4.33)	Aspidospermidine-3-carboxylic acid, 2,3-didehydro-6,7-epoxy-8-[(6aR,8R,8aR,14aS,14bS)-5,6,7,8,13,14-hexahydro-8-methyl-6,7-dioxo-6a,8a-methano-11H,12aH-benzo[k]pyrrolo[3,2,1-mn][1,8]phenanthrolin-15-yl]-, methyl ester, (5α,6β,7β,8α,12R,19α)-	Scandomelonine	59813-31-9
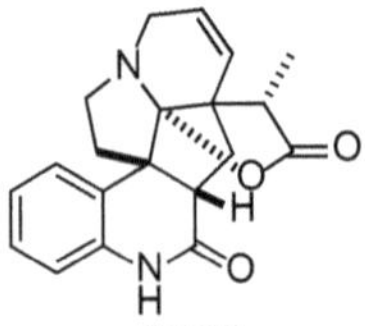 (4.34)	(6aS,7a^{1}R,13aR,16S)-16-methyl-6a,7,12,13-tetrahydro-10H-7a^1,7a-(epoxyethano)indolizino[1′,8′:2,3,4]cyclopenta[1,2-c]quinoline-6,15(5H)-dione	Melohemsine J	–
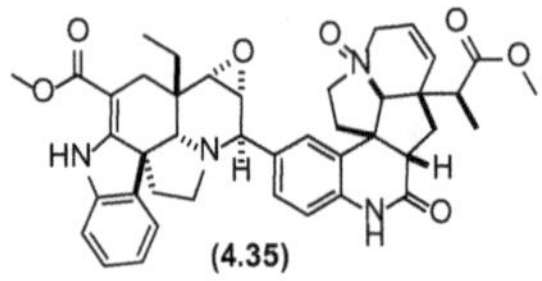 (4.35)	(6aS,7aS,7a^{1}S,11S,13aR)-2-((1R,1aR,2aS,2bS,2b1R,9bR)-2b-ethyl-4-(methoxycarbonyl)-1a,2,2b,2b1,3,5,10,11-octahydro-1H-oxireno[2′,3′:6,7]indolizino[8,1-cd]carbazol-1-yl)-7a-(1-methoxy-1-oxo-2λ^3-propan-2-yl)-6-oxo-5,6,6a,7,7a,10,12,13-octahydroindolizino[1′,8′:2,3,4]cyclopenta[1,2-c]quinoline 11(7a^{1}H)-oxide	Melofusinine C	–

(*Continued*)

TABLE 4.1 (*Continued*)

The Chemical Identity Of 2-Oxo-Tetrahydroquinoline (2O-Thq) Alkaloids

Structure (Number)	Systematic Name	Common Name	CAS
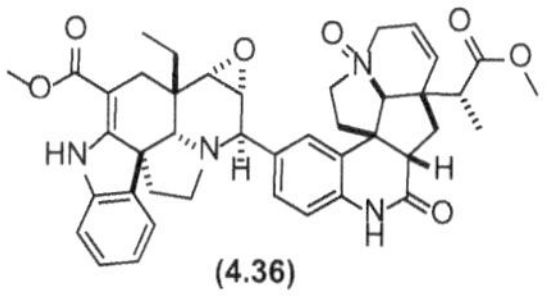 **(4.36)**	(6aS,7aS,7a¹S,11S,13aR)-2-((1R,1aR,2aS,2bS,2b1R,9bR)-2b-ethyl-4-(methoxycarbonyl)-1a,2a,2b,2b1,3,5,10,11-octahydro-1H-oxireno[2′,3′:6,7]indolizino[8,1-cd]carbazol-1-yl)-7a-(1-methoxy-1-oxo-2λ³-propan-2-yl)-6-oxo-5,6,6a,7,7a,10,12,13-octahydroindolizino[1′,8′:2,3,4]cyclopenta[1,2-c]quinoline 11(7a¹H)-oxide	Melofusinine D	–
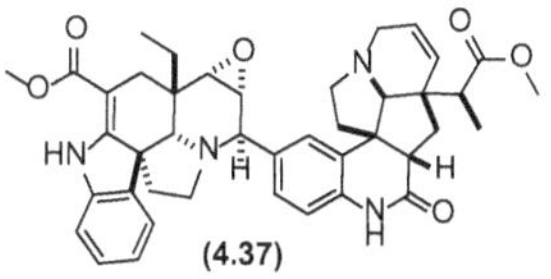 **(4.37)**	methyl (1R,1aR,2aS,2bS,2b¹R,9bR)-2b-ethyl-1-((6aS,7aR,7a¹S,13aR)-7a-((S)-1-methoxy-1-oxopropan-2-yl)-6-oxo-5,6,6a,7,7a,7a1,12,13-octahydro-10H-indolizino[1′,8′:2,3,4]cyclopenta[1,2-c]quinolin-2-yl)-1a,2a,2b,2b1,3,5,10,11-octahydro-1H-oxireno[2′,3′:6,7]indolizino[8,1-cd]carbazole-4-carboxylate	Melofusinine E	–
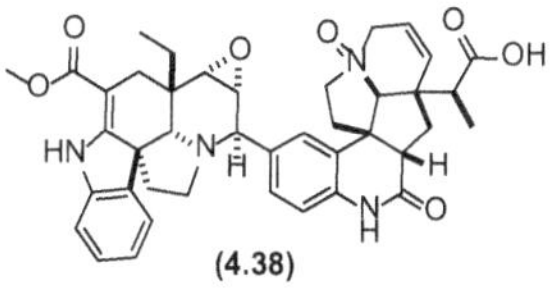 **(4.38)**	(6aS,7aR,7a¹S,11S,13aR)-7a-((S)-1-carboxyethyl)-2-((1R,1aR,2aS,2bS,2b¹R,9bR)-2b-ethyl-4-(methoxycarbonyl)-1a,2a,2b,2b1,3,5,10,11-octahydro-1H-oxireno[2′,3′:6,7]indolizino[8,1-cd]carbazol-1-yl)-6-oxo-5,6,6a,7,7a,10,12,13-octahydroindolizino[1′,8′:2,3,4]cyclopenta[1,2-c]quinoline 11(7a¹H)-oxide	Melofusinine F	–
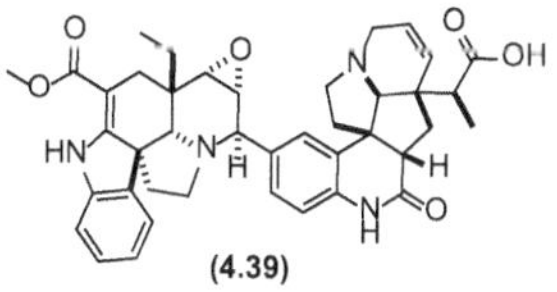 **(4.39)**	(S)-2-((6aS,7aR,7a¹S,13aR)-2-((1R,1aR,2aS,2bS,2b¹R,9bR)-2b-ethyl-4-(methoxycarbonyl)-1a,2a,2b,2b1,3,5,10,11-octahydro-1H-oxireno[2′,3′:6,7]indolizino[8,1-cd]carbazol-1-yl)-6-oxo-5,6,6a,7,12,13-hexahydro-10H-indolizino[1′,8′:2,3,4]cyclopenta[1,2-c]quinolin-7a(7a¹H)-yl)propanoic acid	Melofusinine G	–

(Continued)

TABLE 4.1 (*Continued*)
The Chemical Identity Of 2-Oxo-Tetrahydroquinoline (2O-Thq) Alkaloids

Structure (Number)	Systematic Name	Common Name	CAS
(4.40)	methyl (3aR,3a¹S,10bR)-3a-ethyl-8-hydroxy-9-((6aS,7aR,7a¹S,12R,13aR)-7a-((S)-1-methoxy-1-oxopropan-2-yl)-6-oxo-5,6,6a,7,7a,7a¹,12,13-octahydro-10H-indolizino[1′,8′:2,3,4]cyclopenta[1,2-c]quinolin-12-yl)-3a,3a¹,4,6,11,12-hexahydro-1H-indolizino[8,1-cd]carbazole-5-carboxylate	Melofusinine H	–
(4.41)	(6aS,7aS,7a¹S,11S,13aR)-7a-(1-methoxy-1-oxo-2λ³-propan-2-yl)-6-oxo-5,6,6a,7,7a,10,12,13-octahydroindolizino[1′,8′:2,3,4]cyclopenta[1,2-c]quinoline 11(7a¹H)-oxide	Melofusinine I	–
(4.42)	(6aS,7aS,7a¹S,11S,13aR)-7a-(1-methoxy-1-oxo-2λ³-propan-2-yl)-6-oxo-5,6,6a,7,7a,10,12,13-octahydroindolizino[1′,8′:2,3,4]cyclopenta[1,2-c]quinoline 11(7a¹H)-oxide	Melofusinine J	–
(4.43)	Methyl 2-((6aS,7aS,7a¹S,13aR)-6-oxo-5,6,6a,7,12,13-hexahydro-10H-indolizino[1′,8′:2,3,4]cyclopenta[1,2-c]quinolin-7a(7a¹H)-yl)-2λ³-propanoate	Melodinusine D	–
(4.44)	Methyl 2-((6aS,7aS,7a¹S,13aR)-6-oxo-5,6,6a,7,12,13-hexahydro-10H-indolizino[1′,8′:2,3,4]cyclopenta[1,2-c]quinolin-7a(7a¹H)-yl)-2λ³-propanoate	Melofusinine K	–

(Continued)

TABLE 4.1 (*Continued*)
The Chemical Identity Of 2-Oxo-Tetrahydroquinoline (2O-Thq) Alkaloids

Structure (Number)	Systematic Name	Common Name	CAS
(4.45)	Methyl (2S,2aR,3S,3aS, 3a¹S,4aR,10bS)-2-hydroxy-3-methyl-5-oxo-1,2,2a,3,5,6, 11,12-octahydro-3a¹H-cyclopropa[7′,8′]indolizino [1′,8′:2,3,4]cyclopenta [1,2-c]quinoline-4a(4H)-carboxylate	Melohemsine A	2239301-87-0
(4.46)	Methyl (3aR,4aR,10bS,13aS)-3a,5,6,11,12,13a-hexahydro-5-oxo-1H-indolizino[1′,8′: 2,3,4]cyclopenta[1,2-c] quinoline-4a(4H)-carboxylate	Melohemsine B	2239301-88-1
(4.47)	Methyl (3R,3aS,4aR, 10bS,13aS)-2,3,3a,5,6, 11,12,13a-octahydro-3-hydroxy-5-oxo-1H-indolizino[1′,8′:2,3,4] cyclopenta[1,2-c] quinoline-4a(4H)-carboxylate	Melohemsine C	2239301-89-2
(4.48)	Methyl (6aR,7aR,7a¹R, 8S,13aS)-8-hydroxy-6-oxo-7 a-vinyl-5,6,7a,7a¹,9,10,12,13-octahydro-8H-indolizino [1′,8′:2,3,4]cyclopenta [1,2-c]quinoline-6a(7H)-carboxylate	Melohemsine D	2239301-90-5
(4.49)	methyl (6aR,7aR,7a¹R,8R, 13aS)-8-hydroxy-6-oxo-7a-vinyl-5,6,7a,7a¹,9,10,12,13-octahydro-8H-indolizino [1′,8′:2,3,4]cyclopenta [1,2-c]quinoline-6a(7H)-carboxylate	Melohemsine E	2239301-91-6
(4.50)	Methyl (3S,3aR,4aR, 10bS,13aR)-3a-ethenyl-2,3,3a,5,6,11,12,13a-octahydro-3-hydroxy-1,5-dioxo-1H-indolizino [1′,8′:2,3,4]cyclopenta [1,2-c]quinoline-4a(4H)-carboxylate	Melohemsine F	2239301-92-7

(Continued)

TABLE 4.1 (*Continued*)

The Chemical Identity Of 2-Oxo-Tetrahydroquinoline (2O-Thq) Alkaloids

Structure (Number)	Systematic Name	Common Name	CAS
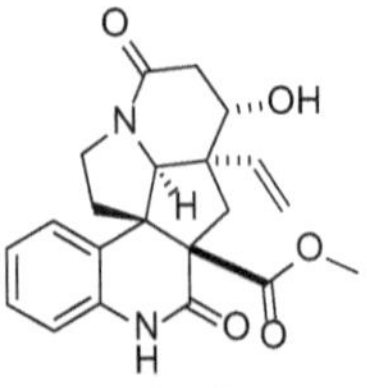 (4.51)	Methyl (3R,3aR,4aR,10bS,13aR)-3a-ethenyl-2,3,3a,5,6,11,12,13a-octahydro-3-hydroxy-1,5-dioxo-1H-indolizino[1′,8′:2,3,4]cyclopenta[1,2-c]quinoline-4a(4H)-carboxylate	Melohemsine G	2239301-93-8
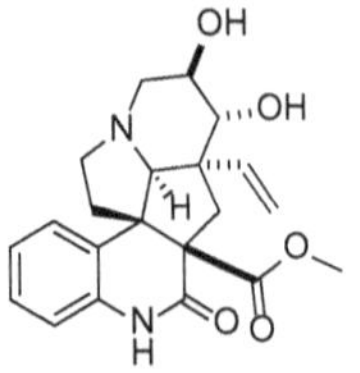 (4.52)	methyl (6aR,7aR,7a¹R,8R,9R,13aS)-8,9-dihydroxy-6-oxo-7a-vinyl-5,6,7a,7a¹,9,10,12,13-octahydro-8H-indolizino[1′,8′:2,3,4]cyclopenta[1,2-c]quinoline-6a(7H)-carboxylate	Melohemsine H	2239301-94-9
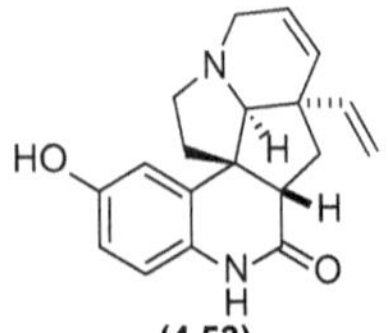 (4.53)	(6aS,7aS,7a¹S,13aR)-2-hydroxy-7a-vinyl-6a,7,7a,7a¹,12,13-hexahydro-10H-indolizino[1′,8′:2,3,4]cyclopenta[1,2-c]quinolin-6(5H)-one	Melohemsine I	2239301-95-0
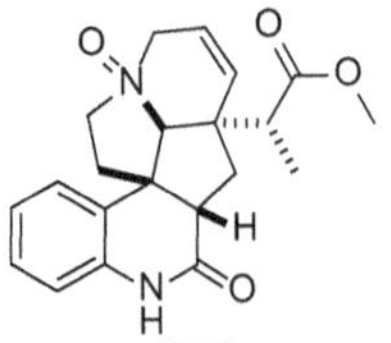 (4.54)	(6aS,7aR,7a¹S,11S,13aR)-7a-((R)-1-methoxy-1-oxopropan-2-yl)-6-oxo-5,6,6a,7,7a,10,12,13-octahydroindolizino[1′,8′:2,3,4]cyclopenta[1,2-c]quinoline 11(7a¹H)-oxide	Melohenryine A	–
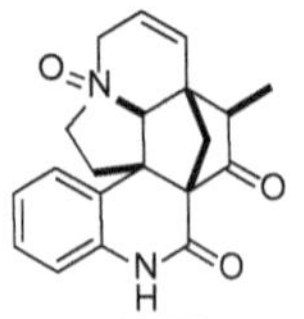 (4.55)	(6bS,6b¹S,9S,12aR,13R,14aR)-13-methyl-1,14-dioxo-1,2,7,8,13,14-hexahydro-10H-12a,14a-methanobenzo[k]pyrrolo[3,2,1-mn][1,8]phenanthroline 9(6b¹H)-oxide	Melohenryine B	–
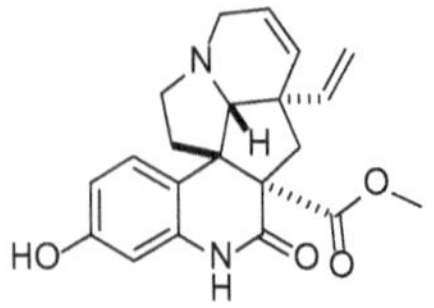 (4.56)	Methyl (3aS,4aS,10bS,13aR)-3a-ethenyl-3a,5,6,11,12,13a-hexahydro-8-hydroxy-5-oxo-1H-indolizino[1′,8′:2,3,4]cyclopenta[1,2-c]quinoline-4a(4H)-carboxylate	Melodinhenine C	1841154-46-8

(Continued)

TABLE 4.1 (*Continued*)
The Chemical Identity Of 2-Oxo-Tetrahydroquinoline (2O-Thq) Alkaloids

Structure (Number)	Systematic Name	Common Name	CAS
(4.57)	Methyl (3aS,4aS,10bS,13aR)-3a-ethenyl-3a,5,6,11,12,13a-hexahydro-5-oxo-1H-indolizino[1′,8′:2,3,4]cyclopenta[1,2-c]quinoline-4a(4H)-carboxylate	Melodinhenine D	1841154-47-9
(4.58)	(6bS,6b¹R,12aR,13S,14aR)-13-methyl-7,8-dihydro-6b¹H,10H-12a,14a-methanobenzo[k]pyrrolo[3,2,1-mn][1,8]phenanthroline-1,14(2H,13H)-dione	Melodinhenine E	1841154-48-0
(4.59)	(6bS,6b¹R,12aR,13R,14aR)-13-methyl-7,8-dihydro-6b¹H,10H-12a,14a-methanobenzo[k]pyrrolo[3,2,1-mn][1,8]phenanthroline-1,14(2H,13H)-dione	Melodinhenine F	1841154-49-1
(4.60)	Methyl (6aR,7aR,7a¹S,13aS)-2-hydroxy-6-oxo-5,6,7a,7a¹,12,13-hexahydro-10H-indolizino[1′,8′:2,3,4]cyclopenta[1,2-c]quinoline-6a(7H)-carboxylate	Melomorine A	–
(4.61)	Methyl (6aR,7aS,7a¹S,8R,13aS)-8-hydroxy-2-methoxy-6-oxo-5,6,7a,7a¹,9,10,12,13-octahydro-8H-indolizino[1′,8′:2,3,4]cyclopenta[1,2-c]quinoline-6a(7H)-carboxylate	Melomorine B	–
(4.62)	Dimethyl (6aR,6′aR,7aS,7a¹S,7′aS,7′a¹S,10′R,13aS,13′aS)-6,6′-dioxo-7a,7′a-divinyl-5,5′,6,6′,7a,7a¹,7′a,7′a¹,12,12′,13,13′-dodecahydro-10H,10′H-[9,10′-biindolizino[1′,8′:2,3,4]cyclopenta[1,2-c]quinoline]-6a,6′a(7H,7′H)-dicarboxylate	Suadimin A	2560588-87-4

(*Continued*)

TABLE 4.1 (*Continued*)
The Chemical Identity Of 2-Oxo-Tetrahydroquinoline (2O-Thq) Alkaloids

Structure (Number)	Systematic Name	Common Name	CAS
(4.63)	Dimethyl (6aR,6'aR,7aS,7a^{1}S, 7'aS,7'a^{1}S,10'S,13aS,13'aS)-6, 6'-dioxo-7a,7'a-divinyl-5,5',6,6', 7a,7a^1,7'a,7'a^1,12,12',13,13'- dodecahydro-10H,10'H-[9,10'- biindolizino[1',8':2,3,4] cyclopenta[1,2-c]quinoline]- 6a,6'a(7H,7'H)-dicarboxylate	Suadimin B	2560588-89-6
(4.64)	–	Suadimin C	2560588-92-1
(4.65)	methyl (6aR,7aS,7a^{1}S,13aS)-10- ((3aR,3a^{1}S,10bR)-3a-ethyl-8- methoxy-5-(methoxycarbonyl)- 3a,3a^1,4,6,11,12-hexahydro-1H- indolizino[8,1-cd]carbazol-2-yl)- 6-oxo-7a-vinyl-5,6,7a,7a^1,12,13- hexahydro-10H-10λ^3-indolizino [1',8':2,3,4]cyclopenta[1,2-c] quinoline-6a(7H)-carboxylate	Suadimin D	–

(Continued)

TABLE 4.1 (*Continued*)

The Chemical Identity Of 2-Oxo-Tetrahydroquinoline (2O-Thq) Alkaloids

Structure (Number)	Systematic Name	Common Name	CAS
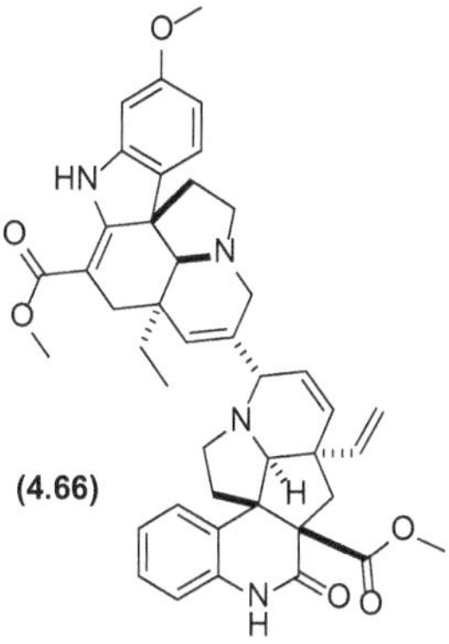 (4.66)	–	Suadimin E	–
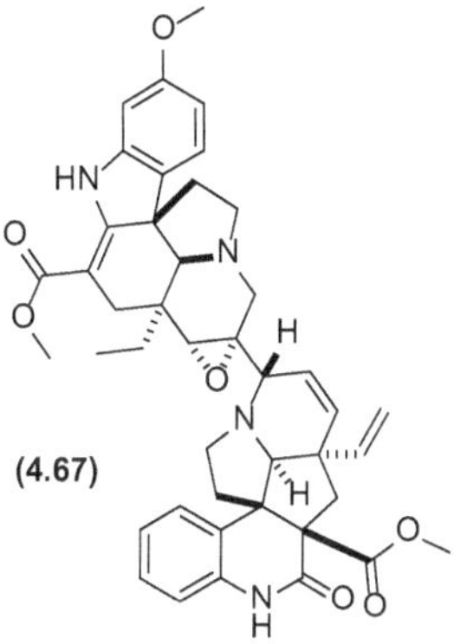 (4.67)	Methyl (1aR,2aR,2bS,2b¹R,9bR)-2b-ethyl-7-methoxy-1a-((6aR,7aS,7a¹S,10S,13aS)-6a-(methoxycarbonyl)-6-oxo-7a-vinyl-5,6,6a,7,7a,7a¹,12,13-octahydro-10H-indolizino[1′,8′:2,3,4]cyclopenta[1,2-c]quinolin-10-yl)-1a,2a,2b,2b¹,3,5,10,11-octahydro-1H-oxireno[2′,3′:6,7]indolizino[8,1-cd]carbazole-4-carboxylate	Suadimin F	–
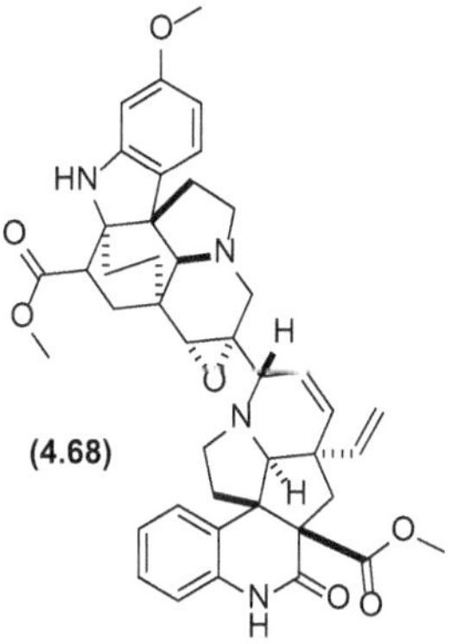 (4.68)	Methyl (1aR,2aR,2bS,2b¹R,4aR,9bR)-7-methoxy-1a-((6aR,7aS,7a¹S,10S,13aS)-6a-(methoxycarbonyl)-6-oxo-7a-vinyl-5,6,6a,7,7a,7a¹,12,13-octahydro-10H-indolizino[1′,8′:2,3,4]cyclopenta[1,2-c]quinolin-10-yl)-1a,2a,3,4,10,11-hexahydro-1H,2b1H,5H-2b,4a-ethanooxireno[2′,3′:6,7]indolizino[8,1-cd]carbazole-4-carboxylate	Suadimin G	–

(Continued)

TABLE 4.1 (*Continued*)

The Chemical Identity Of 2-Oxo-Tetrahydroquinoline (2O-Thq) Alkaloids

Structure (Number)	Systematic Name	Common Name	CAS
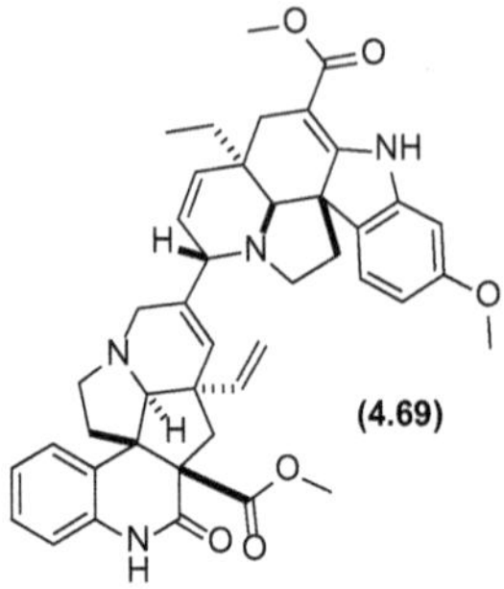 (4.69)	Methyl (6aR,7aS,7a^{1}S,13aS)-9-((1S,3aR,3a^{1}S,10bR)-3a-ethyl-8-methoxy-5-(methoxycarbonyl)-3a,3a^1,4,6,11,12-hexahydro-1H-indolizino[8,1-cd]carbazol-1-yl)-6-oxo-7a-vinyl-5,6,7a,7a^1,12,13-hexahydro-10H-indolizino[1′,8′:2,3,4]cyclopenta[1,2-c]quinoline-6a(7H)-carboxylate	Suadimin H	–
(4.70) 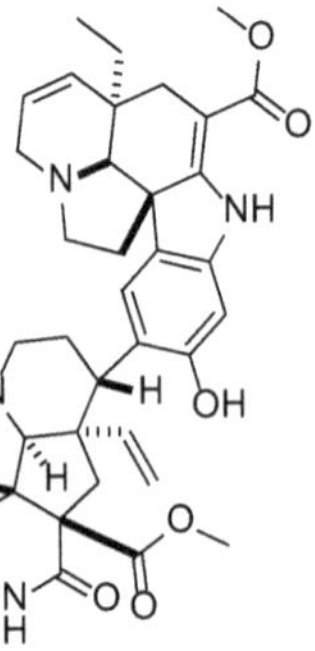	Methyl (6aR,7aS,7a^{1}S,8S,13aS)-8-((3aR,3a^{1}S,10bR)-3a-ethyl-8-hydroxy-5-(methoxycarbonyl)-3a,3a^1,4,6,11,12-hexahydro-1H-indolizino[8,1-cd]carbazol-9-yl)-6-oxo-7a-vinyl-5,6,7a,7a^1,9,10,12,13-octahydro-8H-indolizino[1′,8′:2,3,4]cyclopenta[1,2-c]quinoline-6a(7H)-carboxylate	Suadimin I	–
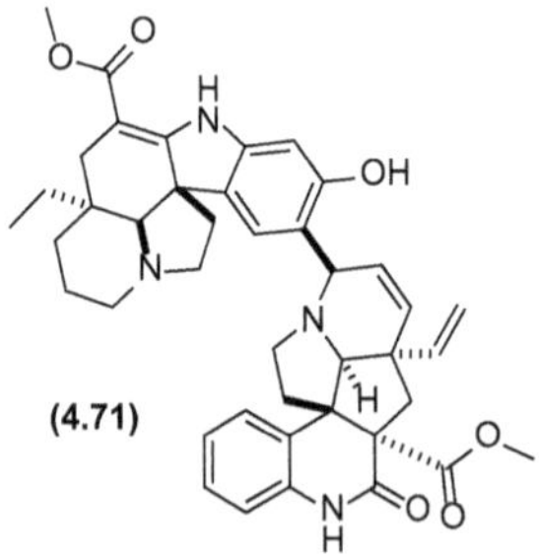 (4.71)	Methyl (5α,12R,19α)-2,3-didehydro-15-[(6aS,7aS,10R,13aS,13bS)-7a-ethenyl-5,6,6a,7,7a,12,13,13a-octahydro-6a-(methoxycarbonyl)-6-oxo-10H-indolizino[1′,8′:2,3,4]cyclopenta[1,2-c]quinolin-10-yl]-16-hydroxyaspidospermidine-3-carboxylate	Melosuavine A	1499181-64-4
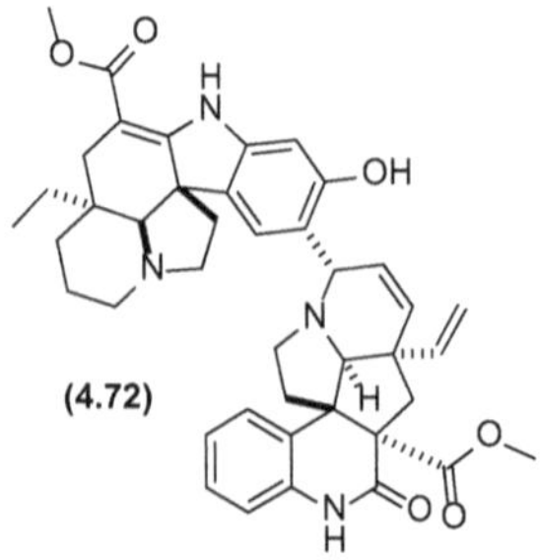 (4.72)	Methyl (5α,12R,19α)-2,3-didehydro-15-[(6aS,7aS,10S,13aS,13bS)-7a-ethenyl-5,6,6a,7,7a,12,13,13a-octahydro-6a-(methoxycarbonyl)-6-oxo-10H-indolizino[1′,8′:2,3,4]cyclopenta[1,2-c]quinolin-10-yl]-16-hydroxyaspidospermidine-3-carboxylate	Melosuavine B	1499181-65-5

(*Continued*)

TABLE 4.1 (*Continued*)

The Chemical Identity Of 2-Oxo-Tetrahydroquinoline (2O-Thq) Alkaloids

Structure (Number)	Systematic Name	Common Name	CAS
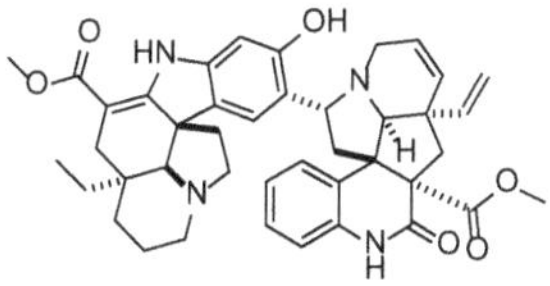 (4.73)	Methyl (5α,12R,19α)-2,3-didehydro-15-[(6aS,7aS,12R,13aS,13bS)-7a-ethenyl-5,6,6a,7,7a,12,13,13a-octahydro-6a-(methoxycarbonyl)-6-oxo-10H-indolizino[1′,8′:2,3,4]cyclopenta[1,2-c]quinolin-12-yl]-16-hydroxyaspidospermidine-3-carboxylate	Melosuavine C	1499181-66-6
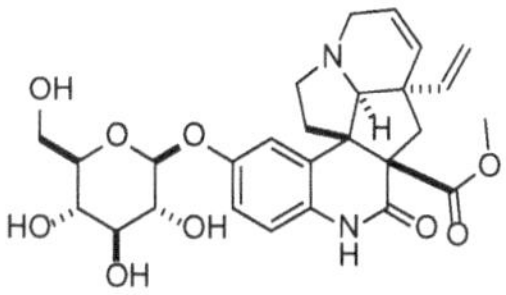 (4.74)	10-*O*-Glucosylscandine	–	2138868-99-0
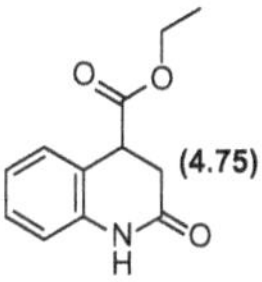 (4.75)	Ethyl 2-oxo-1,2,3,4-tetrahydroquinoline-4-carboxylate	–	67752-52-7
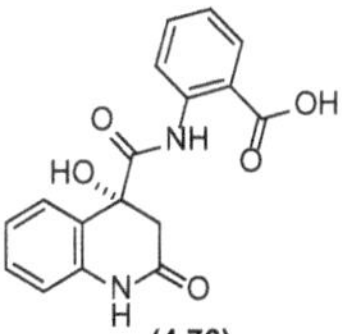 (4.76)	(R)-2-(4-hydroxy-2-oxo-1,2,3,4-tetrahydroquinoline-4-carboxamido)benzoic acid	Isatindigoticoic acid A	–
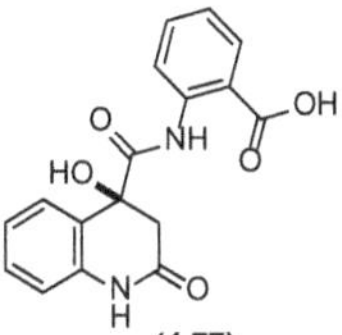 (4.77)	(S)-2-(4-hydroxy-2-oxo-1,2,3,4-tetrahydroquinoline-4-carboxamido)benzoic acid	Epi-isatindigoticoic acid A	–
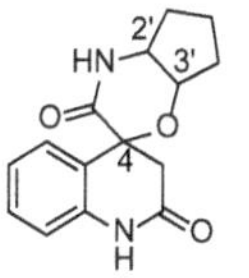 (4.78) (4S,2′R,3′R)	4,4a,5,6,7,7a-hexahydro-1′H,3H-spiro[cyclopenta[b][1,4]oxazine-2,4′-quinoline]-2′,3(3′H)-dione	(4S,2′R,3′R)-Isatisindigoti-canine D	–

(*Continued*)

TABLE 4.1 (*Continued*)

The Chemical Identity Of 2-Oxo-Tetrahydroquinoline (2O-Thq) Alkaloids

Structure (Number)	Systematic Name	Common Name	CAS
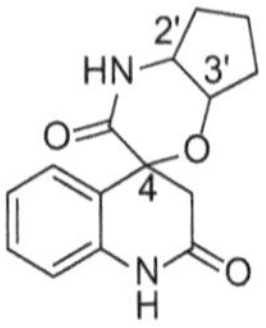 **(4.79)** (4*R*,2'*S*,3'*S*)	–	(4R,2'S,3'S)-isatisindi-goticanine D	–
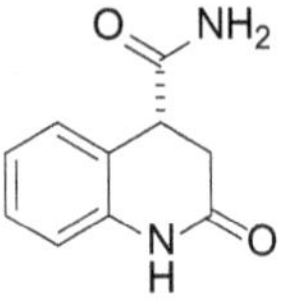 **(4.80)**	(4R)-1,2,3,4-Tetrahydro-2-oxo-4-quinolinecarboxamide	–	1380540-74-8
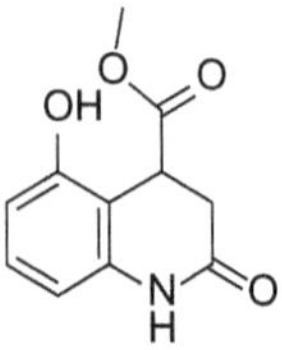 **(4.81)**	Methyl 1,2,3,4-tetrahydro-5-hydroxy-2-oxo-4-quinolinecarboxylate	–	2090463-17-3
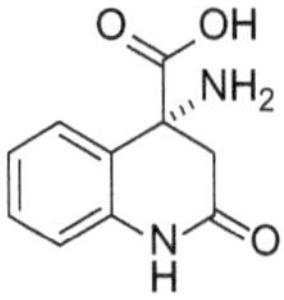 **(4.82)**	(4R)-4-Amino-1,2,3,4-tetrahydro-2-oxo-4-quinolinecarboxylic acid	–	1529789-60-3
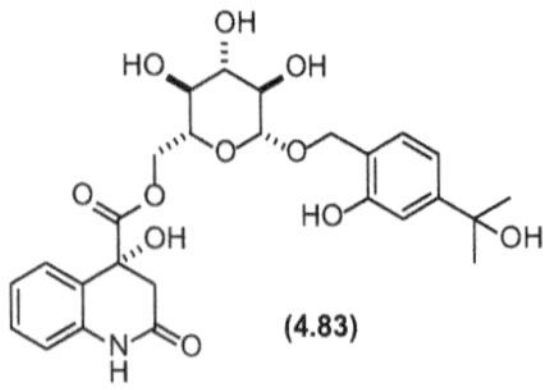 **(4.83)**	β-D-Glucopyranoside, [2-hydroxy-4-(1-hydroxy-1-methylethyl)phenyl]methyl, 6-[(4R)-1,2,3,4-tetrahydro-4-hydroxy-2-oxo-4-quinolinecarboxylate]	Superbusine A	2306896-60-4
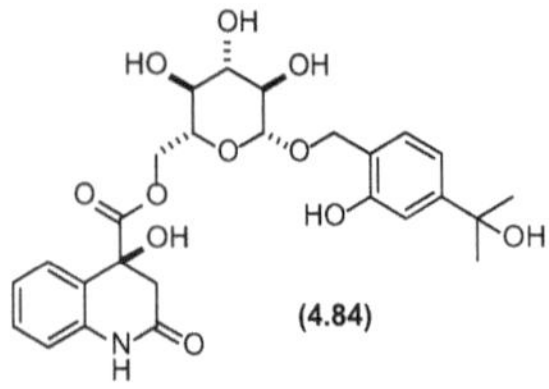 **(4.84)**	β-D-Glucopyranoside, [2-hydroxy-4-(1-hydroxy-1-methylethyl)phenyl]methyl, 6-[(4S)-1,2,3,4-tetrahydro-4-hydroxy-2-oxo-4-quinolinecarboxylate]	Superbusine B	2306896-61-5

(*Continued*)

TABLE 4.1 (*Continued*)
The Chemical Identity Of 2-Oxo-Tetrahydroquinoline (2O-Thq) Alkaloids

Structure (Number)	Systematic Name	Common Name	CAS
(4.85)	rel-(-)-(3R,4R,5R)-4,5-Dihydro-7'-hydroxy-5-(hydroxymethyl)-4-(1H-indol-3-yl)spiro[furan-3(2H),4'(1'H)-quinoline]-2,2'(3'H)-dione	Trigolutesin A	1426413-85-5
(4.86)	rel-(+)-(3R,4R,5R)-4,5-Dihydro-7'-hydroxy-5-(hydroxymethyl)-4-[1-(methoxymethyl)-1H-indol-3-yl]spiro[furan-3(2H),4'(1'H)-quinoline]-2,2'(3'H)-dione	Trigolutesin B	1426413-86-6
(4.87)	4-[[6-O-[(2E,4E,8R)-9-Carboxy-8-hydroxy-2,7-dimethyl-1-oxo-2,4-nonadien-1-yl]-β-D-glucopyranosyl]oxy]-1,2,3,4-tetrahydro-2-oxo-4-quinolinecarboxylic acid	Glansreginin A	860267-30-7
(4.88)	–	Glansreginin C	2459626-49-2
(4.89)	3,4-Dihydro-3,6,7-trihydroxy-2(1H)-quinolinone	–	1512853-36-9
(4.90)	4-[Acetyl[(3R)-1,2,3,4-tetrahydro-3,7-dihydroxy-2-oxo-6-quinolinyl]amino]butanamide	Donglingine	955938-88-2

(Continued)

TABLE 4.1 (*Continued*)
The Chemical Identity Of 2-Oxo-Tetrahydroquinoline (2O-Thq) Alkaloids

Structure (Number)	Systematic Name	Common Name	CAS
(4.91)	Methyl 1,2,3,4-tetrahydro-4-hydroxy-2-oxo-4-quinolinecarboxylate	Excavatine B	1449380-78-2
(4.92)	Methyl 1,2,3,4-tetrahydro-2-oxo-4-quinolinecarboxylate	–	78941-89-6
(4.93)	(+)-Methyl 1,2,3,4-tetrahydro-2-oxo-3-quinolinecarboxylate	Coixlactam	1060659-87-1
(4.94)	Methyl 1,2,3,4-tetrahydro-6-hydroxy-2-oxo-4-quinolinecarboxylate	–	2092157-58-7
(4.95)	4-Quinolinecarboxylic acid, 1,2,3,4-tetrahydro-4,6-dihydroxy-2-oxo-	–	2922238-92-2
(4.96)	3,4-Dihydro-6-hydroxy-2(1H)-quinolinone	–	54197-66-9

(Continued)

TABLE 4.1 (*Continued*)
The Chemical Identity Of 2-Oxo-Tetrahydroquinoline (2O-Thq) Alkaloids

Structure (Number)	Systematic Name	Common Name	CAS
(4.97)	1,2,3,4-Tetrahydro-4-hydroxy-2-oxo-4-quinolinecarboxylic acid	–	2089723-47-5
(4.98)	(3S,4R)-3,4-Dihydro-3,4-dihydroxy-1-methyl-3-(3-methyl-2-buten-1-yl)-2(1H)-quinolinone	Pinolinone	299422-02-9
(4.99)	(4R)-1,4-Dihydro-4-methoxy-1,4-dimethyl-3-(3-methyl-2-buten-1-yl)-2,7-quinolinediol	–	1588847-76-0
(4.100)	rel-(4bR,4cR,11bR,11cS)-4c,11b-Dihydro-11-methoxy-13,16,16-trimethyl-4b,11c-(epoxy[2]propeno)-5H-furo[3'',2''':6',7'][1]benzopyrano[3',4':3,4]cyclobuta[1,2-c]quinoline-5,12(13H)-dione	Melicodenine G	1363039-75-1
(4.101)	3,4-Dihydro-8-hydroxy-2(1H)-quinolinone	–	52749-50-5
(4.102)	Methyl (S)-8-hydroxy-2-oxo-1,2,3,4-tetrahydroquinoline-4-carboxylate	–	–

(*Continued*)

TABLE 4.1 (*Continued*)

The Chemical Identity Of 2-Oxo-Tetrahydroquinoline (2O-Thq) Alkaloids

Structure (Number)	Systematic Name	Common Name	CAS
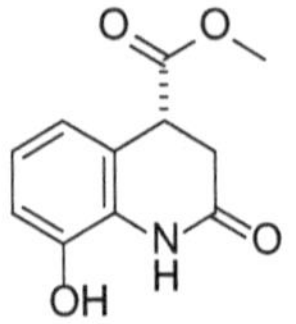 (4.103)	Methyl (R)-8-hydroxy-2-oxo-1,2,3,4-tetrahydroquinoline-4-carboxylate	–	–
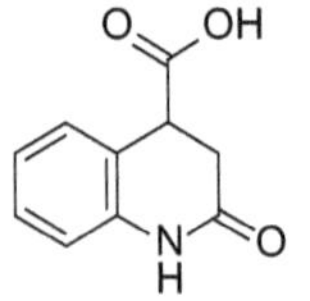 (4.104)	1,2,3,4-Tetrahydro-2-oxo-4-quinolinecarboxylic acid	–	14179-84-1
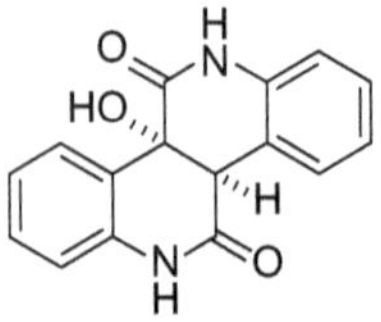 (4.105)	(4bR,10bR)-4b,6,10b,12-Tetrahydro-4b-hydroxydibenzo[c,h][2,6]naphthyridine-5,11-dione	–	1615197-94-8
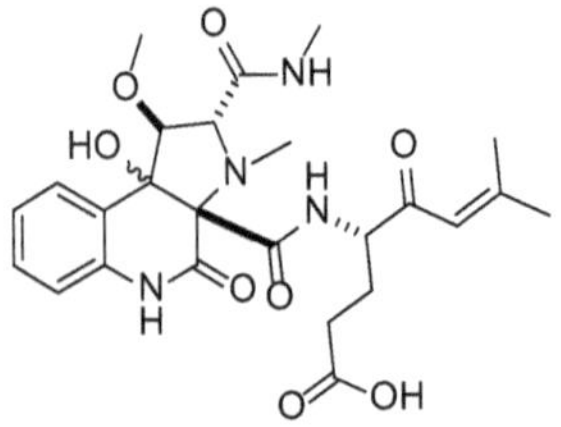 (4.106)	6-Octenoic acid, 4-[[[(1S,2R,3aR)-1,2,3,4,5,9b-hexahydro-9b-hydroxy-1-methoxy-3-methyl-2-[(methylamino)carbonyl]-4-oxo-3aH-pyrrolo[2,3-c]quinolin-3a-yl]carbonyl]amino]-7-methyl-5-oxo-, (4S)-	Albogrisin C	2629222-11-1
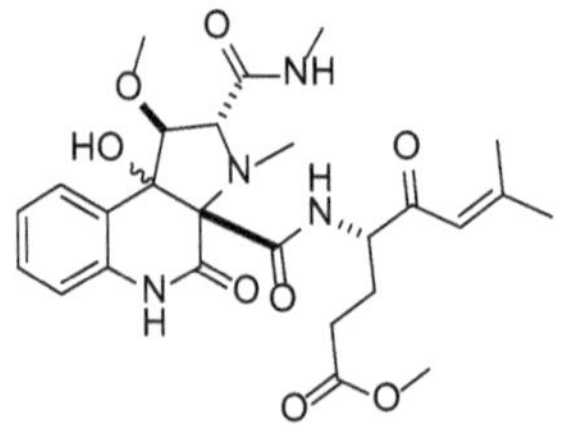 (4.107)	Methyl (4S)-4-[[[(1S,2R,3aR)-1,2,3,4,5,9b-hexahydro-9b-hydroxy-1-methoxy-3-methyl-2-[(methylamino)carbonyl]-4-oxo-3aH-pyrrolo[2,3-c]quinolin-3a-yl]carbonyl]amino]-7-methyl-5-oxo-6-octenoate	Albogrisin D	2629227-20-7

(*Continued*)

TABLE 4.1 (*Continued*)
The Chemical Identity Of 2-Oxo-Tetrahydroquinoline (2O-Thq) Alkaloids

Structure (Number)	Systematic Name	Common Name	CAS
(4.108)	6-Octenoic acid, 4-[[[(1R,2R,3aR)-1,2,3,4,5,9b-hexahydro-9b-hydroxy-1-methoxy-3-methyl-2-[(methylamino)carbonyl]-4-oxo-3aH-pyrrolo[2,3-c]quinolin-3a-yl]carbonyl]amino]-7-methyl-5-oxo-, (4S)-	Albogrisin C'	2629222-12-2
(4.109)	Methyl (4S)-4-[[[(1R,2R,3aR)-1,2,3,4,5,9b-hexahydro-9b-hydroxy-1-methoxy-3-methyl-2-[(methylamino)carbonyl]-4-oxo-3aH-pyrrolo[2,3-c]quinolin-3a-yl]carbonyl]amino]-7-methyl-5-oxo-6-octenoate	Albogrisin D'	2629227-22-9
(4.110)	3,4-Dihydro-8-[(4-*O*-methyl-α-L-ribopyranosyl)oxy]-2(1H)-quinolinone	–	1588764-74-2
(4.111)	8-[(6-Deoxy-4-*O*-methyl-α-L-mannopyranosyl)oxy]-3,4-dihydro-2(1H)-quinolinone	–	1588764-70-8
(4.112)	3,4-Dihydro-8-methoxy-2(1H)-quinolinone	–	53899-19-7
(4.113)	(3S,4S)-3,4-Dihydro-4,5-dihydroxy-6-[(1S,6R)-6-(1-hydroxy-1-methylethyl)-3-methyl-2-cyclohexen-1-yl]-3-methoxy-4-(4-methoxyphenyl)-2(1H)-quinolinone	Pesimquinolone A	2522921-34-0

(Continued)

TABLE 4.1 (*Continued*)

The Chemical Identity Of 2-Oxo-Tetrahydroquinoline (2O-Thq) Alkaloids

Structure (Number)	Systematic Name	Common Name	CAS
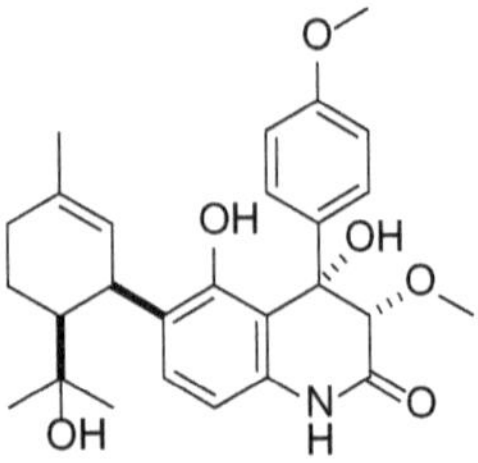 (4.114)	(3S,4S)-3,4-Dihydro-4,5-dihydroxy-6-[(1R,6S)-6-(1-hydroxy-1-methylethyl)-3-methyl-2-cyclohexen-1-yl]-3-methoxy-4-(4-methoxyphenyl)-2(1H)-quinolinone	Pesimquinolone B	2522921-35-1
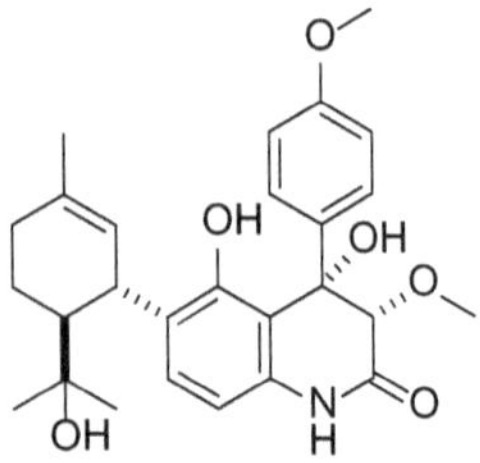 (4.115)	(3S,4S)-3,4-Dihydro-4,5-dihydroxy-6-[(1S,6S)-6-(1-hydroxy-1-methylethyl)-3-methyl-2-cyclohexen-1-yl]-3-methoxy-4-(4-methoxyphenyl)-2(1H)-quinolinone	Pesimquinolone C	2522926-29-8
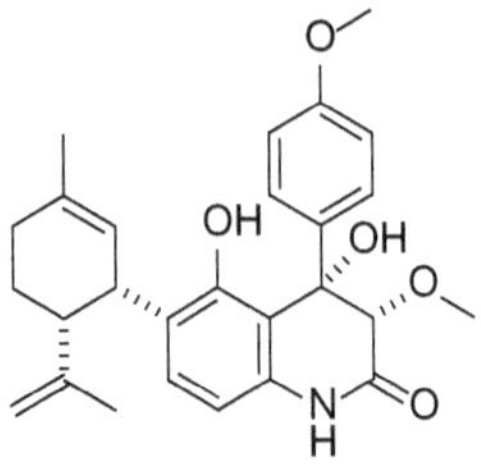 (4.116)	(3S,4S)-3,4-Dihydro-4,5-dihydroxy-3-methoxy-4-(4-methoxyphenyl)-6-[(1S,6R)-3-methyl-6-(1-methylethenyl)-2-cyclohexen-1-yl]-2(1H)-quinolinone	Pesimquinolone D	2522922-16-1
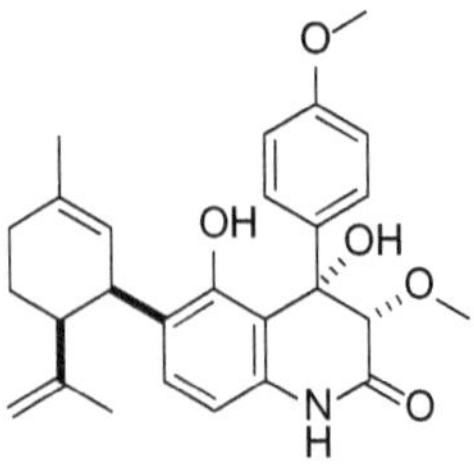 (4.117)	(3S,4S)-3,4-Dihydro-4,5-dihydroxy-3-methoxy-4-(4-methoxyphenyl)-6-[(1R,6S)-3-methyl-6-(1-methylethenyl)-2-cyclohexen-1-yl]-2(1H)-quinolinone	Pesimquinolone E	2522921-37-3
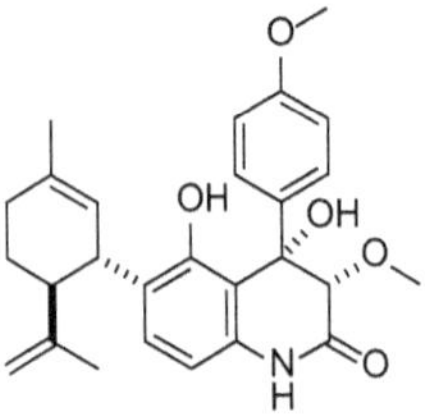 (4.118)	(3S,4S)-3,4-Dihydro-4,5-dihydroxy-3-methoxy-4-(4-methoxyphenyl)-6-[(1S,6S)-3-methyl-6-(1-methylethenyl)-2-cyclohexen-1-yl]-2(1H)-quinolinone	Pesimquinolone F	2522920-50-7

(Continued)

TABLE 4.1 (*Continued*)
The Chemical Identity Of 2-Oxo-Tetrahydroquinoline (2O-Thq) Alkaloids

Structure (Number)	Systematic Name	Common Name	CAS
(4.119)	(3S,4S)-3,4-Dihydro-4,5-dihydroxy-3-methoxy-4-(4-methoxyphenyl)-6-[(1R,6R)-3-methyl-6-(1-methylethenyl)-2-cyclohexen-1-yl]-2(1*H*)-quinolinone	Pesimquinolone G	2575750-06-8
(4.120)	(3S,4S)-3,4-Dihydro-4,5-dihydroxy-3-methoxy-4-(4-methoxyphenyl)-6-[(1E)-2-[(2R,5R)-tetrahydro-2-methyl-5-(1-methylethenyl)-2-furanyl]ethenyl]-2(1*H*)-quinolinone	Pesimquinolone H	2575766-22-0
(4.121)	3,4-Dihydro-4,5-dihydroxy-3-methoxy-4-(4-methoxyphenyl)-6-[(1E)-2-[tetrahydro-2-methyl-5-(1-methylethenyl)-2-furanyl]ethenyl]-2(1*H*)-quinolinone	Yaequinolone F	889659-59-0
(4.122)	(2R,9S,10S)-2,7,9,10-Tetrahydro-10-hydroxy-9-methoxy-10-(4-methoxyphenyl)-2-methyl-2-(4-methyl-3-penten-1-yl)-8H-pyrano[2,3-f]quinolin-8-one	Yaequinolone J_1	872452-60-3
(4.123)	(2S,9S,10S)-2,7,9,10-Tetrahydro-10-hydroxy-9-methoxy-10-(4-methoxyphenyl)-2-methyl-2-(4-methyl-3-penten-1-yl)-8H-pyrano[2,3-f]quinolin-8-one	Yaequinolone J_2	872452-61-4

(*Continued*)

TABLE 4.1 (*Continued*)
The Chemical Identity Of 2-Oxo-Tetrahydroquinoline (2O-Thq) Alkaloids

Structure (Number)	Systematic Name	Common Name	CAS
(4.124)	(3R,4R)-3,4-Dihydro-4,5-dihydroxy-3-methoxy-4-(4-methoxyphenyl)-6-[(1E)-2-[(2S)-tetrahydro-2,5,5-trimethyl-2H-pyran-2-yl]ethenyl]-2(1*H*)-quinolinone	Penigequinolone A	180045-91-4
(4.125)	(3R,4R)-3,4-Dihydro-4,5-dihydroxy-3-methoxy-4-(4-methoxyphenyl)-6-[(1E)-2-[(2R)-tetrahydro-2,5,5-trimethyl-2H-pyran-2-yl]ethenyl]-2(1*H*)-quinolinone	Penigequinolone B	180185-69-7
(4.126)	(3R,4R)-3,4-Dihydro-4,5-dihydroxy-3-methoxy-4-(4-methoxyphenyl)-6-(3-methyl-2-buten-1-yl)-2(1H)-quinolinone	Peniprequinolone	328556-06-5
(4.127)	(3S,4S)-4-hydroxy-3-methoxy-4-(4-methoxyphenyl)-6,6,9-trimethyl-1,3,4,6,6a,7,8,10a-octahydro-2H-isochromeno[3,4-f]quinolin-2-one	Pesimquinolone I	2760451-36-1
(4.128)	(3S,4S,6aS,10aS)-4-hydroxy-3-methoxy-4-(4-methoxyphenyl)-6,6,9-trimethyl-1,3,4,6,6a,7,8,10a-octahydro-2H-isochromeno[3,4-f]quinolin-2-one	Pesimquinolone J	2760451-38-3
(4.129)	(3S,4S)-1,3,4,6-Tetrahydro-4-hydroxy-3-methoxy-4-(4-methoxyphenyl)-6,6,9-trimethyl-2H-[2]benzopyrano[3,4-f]quinolin-2-one	Pesimquinolone K	2760451-41-8

(*Continued*)

TABLE 4.1 (*Continued*)
The Chemical Identity Of 2-Oxo-Tetrahydroquinoline (2O-Thq) Alkaloids

Structure (Number)	Systematic Name	Common Name	CAS
(4.130)	(2R,9S,10S)-2,7,9,10-Tetrahydro-10-hydroxy-2-(4-hydroxy-3,3-dimethylbutyl)-9-methoxy-10-(4-methoxyphenyl)-2-methyl-8H-pyrano[2,3-f]quinolin-8-one	Pesimquinolone L	2760451-55-4
(4.131)	(3S,4S)-3,4-Dihydro-4-hydroxy-3-methoxy-4-(4-methoxyphenyl)-6-[(1E)-2-[tetrahydro-5-(1-hydroxy-1-methylethyl)-2-methyl-2-furanyl]ethenyl]-2(1H)-quinolinone	Pesimquinolone M	2760451-66-7
(4.132)	(3S,4S)-3,4-Dihydro-4-hydroxy-3-methoxy-4-(4-methoxyphenyl)-6-[(1E)-2-[(2R,5R)-tetrahydro-5-(1-hydroxy-1-methylethyl)-2-methyl-2-furanyl]ethenyl]-2(1H)-quinolinone	Pesimquinolone N	2760451-81-6
(4.133)	(3S,4S)-3,4-Dihydro-4-hydroxy-3-methoxy-4-(4-methoxyphenyl)-6-[(1E)-2-[(2S,5S)-tetrahydro-5-(1-hydroxy-1-methylethyl)-2-methyl-2-furanyl]ethenyl]-2(1H)-quinolinone	Pesimquinolone O	2760451-93-0
(4.134)	(8S,9S)-3,6,8,9-Tetrahydro-9-hydroxy-2-(1-hydroxy-1-methylethyl)-8-methoxy-9-(4-methoxyphenyl)furo[2,3-f]quinolin-7(2H)-one	Pesimquinolone P	2760451-96-3
(4.135)	(3S,9S,10S)-2,3,4,7,9,10-Hexahydro-3,10-dihydroxy-9-methoxy-10-(4-methoxyphenyl)-2,2-dimethyl-8H-pyrano[2,3-f]quinolin-8-one	Pesimquinolone Q	2760452-00-2

(Continued)

TABLE 4.1 (*Continued*)
The Chemical Identity Of 2-Oxo-Tetrahydroquinoline (2O-Thq) Alkaloids

Structure (Number)	Systematic Name	Common Name	CAS
(4.136)	(3R,4R)-3,4-Dihydro-4,5-dihydroxy-3-methoxy-4-(4-methoxyphenyl)-6-[(1E)-2-[tetrahydro-5-(1-hydroxy-1-methylethyl)-2-methyl-2-furanyl]ethenyl]-2(1H)-quinolinone	Yaequinolone C	889659-56-7
(4.137)	rel-(-)-(3S,4R)-3,4-Dihydro-3,4-dihydroxy-4-(4-methoxyphenyl)-2(1H)-quinolinone	Yaequinolone A₁	866487-88-9
(4.138)	(3R,4R)-3,4-Dihydro-3,4-dihydroxy-4-(4-methoxyphenyl)-2(1H)-quinolinone	Yaequinolone A₂ (Peneciraistin E)	866487-89-0
(4.139)	(3R,4R)-3,4-Dihydro-4-hydroxy-3-methoxy-4-(4-methoxyphenyl)-2(1H)-quinolinone	Peneciraistin F	183854-01-5
(4.140)	rel-(+)-(3R,4R)-3,4-Dihydro-4,5-dihydroxy-3-methoxy-4-(4-methoxyphenyl)-2(1H)-quinolinone	–	328530-34-3

(Continued)

TABLE 4.1 (*Continued*)
The Chemical Identity Of 2-Oxo-Tetrahydroquinoline (2O-Thq) Alkaloids

Structure (Number)	Systematic Name	Common Name	CAS
(4.141)	(3S,4S)-4,5-dihydroxy-6-((E)-2-((1R,3S)-3-hydroxy-1,3-dimethyl-4-oxocyclohexyl)vinyl)-3-methoxy-4-phenyl-3,4-dihydroquinolin-2(1H)-one	Aflaquinolone H	–
(4.142)	(3S,4S)-3,4-Dihydro-4,5-dihydroxy-3-methoxy-4-phenyl-6-[(1E)-2-[(2R,5R)-tetrahydro-2-methyl-5-(1-methylethenyl)-2-furanyl]ethenyl]-2(1H)-quinolinone	Aniduquinolone A	1465233-97-9
(4.143)	(3S,4S)-3,4-Dihydro-4,5-dihydroxy-3-methoxy-4-phenyl-6-[(1E)-2-[(2R,5S)-tetrahydro-2-methyl-5-(1-methylethenyl)-2-furanyl]ethenyl]-2(1H)-quinolinone	Aniduquinolone D	2088364-43-4
(4.144)	(3R,4R)-3,4-Dihydro-3,4-dihydroxy-4-(3-hydroxyphenyl)-2(1H)-quinolinone	Aflaquinolone I	2376068-30-1
(4.145)	cis-(-)-3,4-dihydro-4,5-dihydroxy-3-methoxy-4-(4-methoxyphenyl)-2(1H)-Quinolinone	–	184046-65-9
(4.146) (*= R and S)	rel-(1R,3S,4S,5'E)-3,4-Dihydro-5'-(1H-imidazol-5-ylmethylene)-3-(3-methyl-2-buten-1-yl)spiro[2H-1,4-methanoquinoline-9,2'-imidazolidine]-2,4'-dione	Penispirolloid A	1356613-04-1

(Continued)

TABLE 4.1 (Continued)
The Chemical Identity Of 2-Oxo-Tetrahydroquinoline (2O-Thq) Alkaloids

Structure (Number)	Systematic Name	Common Name	CAS
(4.147)	(3R,4S)-3,4-Dihydro-3,4-dihydroxy-4-phenyl-2(1H)-quinolinone	Aflaquinolone G	1356937-47-7
(4.148)	Benzoic acid, 2-amino-4-[(3-methyl-2-buten-1-yl)oxy]-, (3R,4S)-1,2,3,4-tetrahydro-4-hydroxy-2-oxo-4-phenyl-3-quinolinyl ester	Asperalin A	2924565-49-9
(4.149)	Benzoic acid, 4-hydroxy-2-[(3-methyl-2-buten-1-yl)amino]-, (3R,4S)-1,2,3,4-tetrahydro-4-hydroxy-2-oxo-4-phenyl-3-quinolinyl ester	Asperalin B	2924565-50-2
(4.150)	Benzoic acid, 5-chloro-4-hydroxy-2-[(3-methyl-2-buten-1-yl)amino]-, (3R,4S)-1,2,3,4-tetrahydro-4-hydroxy-2-oxo-4-phenyl-3-quinolinyl ester	Asperalin C	2924565-51-3
(4.151)	Benzoic acid, 2-amino-5-chloro-4-[(3-methyl-2-buten-1-yl)oxy]-, (3R,4S)-1,2,3,4-tetrahydro-4-hydroxy-2-oxo-4-phenyl-3-quinolinyl ester	Asperalin D	2924565-52-4
(4.152)	Benzoic acid, 5-chloro-4-[(3-methyl-2-buten-1-yl)oxy]-2-[(3-oxobutyl)amino]-, (3R,4S)-1,2,3,4-tetrahydro-4-hydroxy-2-oxo-4-phenyl-3-quinolinyl ester	Asperalin E	2924565-53-5

(Continued)

TABLE 4.1 (*Continued*)
The Chemical Identity Of 2-Oxo-Tetrahydroquinoline (2O-Thq) Alkaloids

Structure (Number)	Systematic Name	Common Name	CAS
(4.153)	Benzoic acid, 2-amino-3,5-dichloro-4-hydroxy-, (3R,4S)-1,2,3,4-tetrahydro-4-hydroxy-2-oxo-4-phenyl-3-quinolinyl ester	Asperalin F	2924565-54-6
(4.154)	(3S,4S)-3,4-Dihydro-4,5-dihydroxy-3-methoxy-4-phenyl-6-[(1E)-2-[(1S,2R,5R)-2,4,4-trimethyl-3-oxabicyclo[3.1.0]hex-2-yl]ethenyl]-2(1*H*)-quinolinone	Aspoquinolone E	2688836-09-9
(4.155)	(3S,4S)-3,4-Dihydro-4,5-dihydroxy-3-methoxy-4-phenyl-6-[(1E)-2-[(1R,2R,5S)-2,4,4-trimethyl-3-oxabicyclo[3.1.0]hex-2-yl]ethenyl]-2(1H)-quinolinone	Aspoquinolone F	2688836-17-9
(4.156)	(3S,4S)-3,4-Dihydro-4-hydroxy-3-methoxy-4-phenyl-2(1*H*)-quinolinone	5-Deoxyafla-quinolone E	1465234-03-0
(4.157)	(3S,4S)-3,4-Dihydro-4,5-dihydroxy-3-methoxy-4-phenyl-2(1H)-quinolinone	Aflaquinolone E	1356937-45-5
(4.158)	(3S,4S)-3,4-Dihydro-3,4-dihydroxy-4-phenyl-2(1*H*)-quinolinone	Aflaquinolone F	1356937-46-6

(*Continued*)

TABLE 4.1 (*Continued*)
The Chemical Identity Of 2-Oxo-Tetrahydroquinoline (2O-Thq) Alkaloids

Structure (Number)	Systematic Name	Common Name	CAS
(4.159)	2(1H)-Quinolinone, 3,4-dihydro-4,5-dihydroxy-6-[(1E)-2-[(1R,3S,4S)-4-hydroxy-1,3-dimethylcyclohexyl]ethenyl]-3-methoxy-4-phenyl-, (3S,4S)-	Scopuquinolone B	2088364-44-5
(4.160)	2(1H)-Quinolinone, 3,4-dihydro-4,5-dihydroxy-3-methoxy-4-(4-methoxyphenyl)-	NTC 47B	168010-08-0
(4.161)	12H-Oxecino[2,3-f]quinolin-12-one, 2,3,4,5,6,11,13,14-octahydro-6,14-dihydroxy-13-methoxy-14-(4-methoxyphenyl)-3,3,6-trimethyl-	NTC 47A	168010-07-9
(4.162)	(3S,4S)-6-[(1E)-2-[(1R,3S)-1,3-Dimethyl-4-oxocyclohexyl]ethenyl]-3,4-dihydro-4,5-dihydroxy-3-methoxy-4-phenyl-2(1H)-quinolinone	Aflaquinolone A	1356937-41-1
(4.163)	(3S,4S)-3,4-Dihydro-4,5-dihydroxy-6-[(1E)-2-[(1R,3S,4R)-4-hydroxy-1,3-dimethylcyclohexyl]ethenyl]-3-methoxy-4-phenyl-2(1H)-quinolinone	Aflaquinolone B	1356937-42-2
(4.164)	(3S,4S)-6-[(1E)-2-[(1S,3R)-1,3-Dimethyl-4-oxocyclohexyl]ethenyl]-3,4-dihydro-4,5-dihydroxy-3-methoxy-4-phenyl-2(1H)-quinolinone	Aflaquinolone C	1356937-43-3

(Continued)

TABLE 4.1 (*Continued*)
The Chemical Identity Of 2-Oxo-Tetrahydroquinoline (2O-Thq) Alkaloids

Structure (Number)	Systematic Name	Common Name	CAS
(4.165)	(3S,4S)-6-[(1E)-2-[(1S,3S)-1,3-Dimethyl-4-oxocyclohexyl]ethenyl]-3,4-dihydro-4,5-dihydroxy-3-methoxy-4-phenyl-2(1*H*)-quinolinone	Aflaquinolone D	1356937-44-4
(4.166)	rel-(3R,4R)-3,4-Dihydro-4,5-dihydroxy-3-methoxy-4-(4-methoxyphenyl)-6-[(1E)-2-[(1R,2S,5S)-2,4,4-trimethyl-3-oxabicyclo[3.1.0]hex-2-yl]ethenyl]-2(1*H*)-quinolinone	Aspoquinolone A	913256-32-3
(4.167)	rel-(3R,4R)-3,4-Dihydro-4,5-dihydroxy-3-methoxy-4-(4-methoxyphenyl)-6-[(1E)-2-[(1R,2R,5S)-2,4,4-trimethyl-3-oxabicyclo[3.1.0]hex-2-yl]ethenyl]-2(1*H*)-quinolinone	Aspoquinolone B	913253-59-5
(4.168)	rel-(3R,4R)-3,4-Dihydro-4,5-dihydroxy-3-methoxy-4-(4-methoxyphenyl)-6-[(1E)-2-(tetrahydro-5-hydroxy-2,6,6-trimethyl-2H-pyran-2-yl)ethenyl]-2(1*H*)-quinolinone	Aspoquinolone C	913256-33-4
(4.169)	rel-(3R,4R)-3,4-Dihydro-4,5-dihydroxy-3-methoxy-4-(4-methoxyphenyl)-6-[(1E)-2-(tetrahydro-5-hydroxy-2,6,6-trimethyl-2H-pyran-2-yl)ethenyl]-2(1*H*)-quinolinone	Aspoquinolone D	913256-34-5
(4.170)	(3R,4R)-3,4-Dihydro-4,5-dihydroxy-3-methoxy-4-(4-methoxyphenyl)-2(1*H*)-quinolinone	–	858357-96-7

(Continued)

TABLE 4.1 (*Continued*)
The Chemical Identity Of 2-Oxo-Tetrahydroquinoline (2O-Thq) Alkaloids

Structure (Number)	Systematic Name	Common Name	CAS
(4.171)	3,4-Dihydro-4,5-dihydroxy-3-methoxy-4-(4-methoxyphenyl)-6-[(1E)-3-methyl-1,3-butadien-1-yl]-2(1*H*)-quinolinone	Yaequinolone E	889659-58-9
(4.172)	3,4-Dihydro-4,5-dihydroxy-3-methoxy-4-(4-methoxyphenyl)-6-[(1E)-3-oxo-1-buten-1-yl]-2(1H)-quinolinone	Yaequinolone B	889659-55-6
(4.173)	(3R,4R)-3,4-Dihydro-4,5-dihydroxy-3-methoxy-4-(4-methoxyphenyl)-6-[(1E)-2-(tetrahydro-6-hydroxy-2,5,5-trimethyl-2H-pyran-2-yl)ethenyl]-2(1H)-quinolinone	Yaequinolone D	889659-57-8
(4.174)	(3S,4S)-6-((E)-2-((1R,3S)-1,3-dimethyl-4-oxocyclohexyl)vinyl)-4,5-dihydroxy-3-methoxy-4-(4-methoxyphenyl)-3,4-dihydroquinolin-2(1*H*)-one	Aflaquinolone H'	–
(4.175)	(3S,4S)-4,5-dihydroxy-6-((E)-2-((1R,3S,4S)-4-hydroxy-1,3-dimethylcyclohexyl)vinyl)-3-methoxy-4-(4-methoxyphenyl)-3,4-dihydroquinolin-2(1H)-one	Aflaquinolone I'	–
(4.176)	(3S,4S)-4,5-dihydroxy-6-((E)-2-((1R,3S,4S)-4-hydroxy-1,3-dimethylcyclohexyl)vinyl)-3-methoxy-4-phenyl-3,4-dihydroquinolin-2(1*H*)-one	22-epi-aflaquinolone B	–

(*Continued*)

TABLE 4.1 (*Continued*)
The Chemical Identity Of 2-Oxo-Tetrahydroquinoline (2O-Thq) Alkaloids

Structure (Number)	Systematic Name	Common Name	CAS
(4.177)	N-Methyl-L-valine (1R,2S,4R)-2,4-dimethyl-4-[(1E)-2-[(3S,4S)-1,2,3,4-tetrahydro-4,5-dihydroxy-3-methoxy-2-oxo-4-phenyl-6-quinolinyl]ethenyl]cyclohexyl ester	22-O-(N-Me-L-valyl) aflaquinolone B	1638761-72-4
(4.178)	N-Methyl-L-valine (1R,2R,4R)-2,4-dimethyl-4-[(1E)-2-[(3S,4S)-1,2,3,4-tetrahydro-4,5-dihydroxy-3-methoxy-2-oxo-4-phenyl-6-quinolinyl]ethenyl]cyclohexyl ester	22-O-(N-Me-L-valyl)-21-epi-aflaquinolone B	1638761-73-5
(4.179)	(3S,4S)-3,4-Dihydro-4,5-dihydroxy-3-methoxy-4-phenyl-6-[(1E)-2-[(2R,5R)-tetrahydro-5-(1-hydroxy-1-methylethyl)-2-methyl-2-furanyl]ethenyl]-2(1*H*)-quinolinone	Aniduquinolone B	1465233-98-0
(4.180)	(3S,4S)-3,4-Dihydro-4,5-dihydroxy-3-methoxy-6-(3-methyl-2-buten-1-yl)-4-phenyl-2(1*H*)-quinolinone	Aniduquinolone C	1465234-00-7
(4.181)	(3S,4S)-3,4-Dihydro-4,6-dihydroxy-3-methoxy-4-phenyl-2(1H)-quinolinone	Isoaflaquinolone E	1465234-05-2
(4.182)	(3S,4S)-3,4-Dihydro-3,4-dihydroxy-4-(4-hydroxyphenyl)-2(1*H*)-quinolinone	14-Hydroxyfla-quinolone F	1465234-06-3

(*Continued*)

TABLE 4.1 (*Continued*)
The Chemical Identity Of 2-Oxo-Tetrahydroquinoline (2O-Thq) Alkaloids

Structure (Number)	Systematic Name	Common Name	CAS
(4.183)	(1R,2aR,8bS,8b¹R, 12aS,13aR,13bR)-1-methyl-9,10,12a, 13a-tetrahydro-8b¹H,12H-2a,13b-methanobenzo[k]oxireno[2,3-c]pyrrolo[3,2,1-mn][1,8]phenanthroline-2,3(1H,4H)-dione	Melotenucadine A	–
(4.184)	Methyl (6aR,7aR,7a¹R,8R, 13aS)-3,8-dihydroxy-6,10-dioxo-7a-vinyl-5,6,7a,7a¹,9,10,12,13-octahydro-8H-indolizino[1′,8′:2,3,4]cyclopenta[1,2-c]quinoline-6a(7H)-carboxylate	Melotenucadine B	–

(4.5), with the 6/6/5/6/5 pentacyclic core structure (the 6/6 refers to THQ system); (c) bis-melodinus alkaloids, such as meloyine II (4.29); (d) 3,4-dioxygenated-2O-THQs, such as pesimquinolones (4.113–4.120); and (e) others, such as spiro- (e.g. 4.78) and gluco- (e.g. 4.83) compounds.

4.1.2 SYNTHESIS

Initially called dihydrocarbostyril, 2O-THQ has been known for almost 150 years! Its first synthesis is believed to have been an intramolecular N-acylation of aniline derivatives (Gabriel and Zimmermann 1879, 1880; Baeyer and Jackson 1880; Friedländer 1883). Since then, numerous synthetic methods have been presented for constructing the 2O-THQ skeleton. Here, we will briefly mention a few synthetic approaches.

4.1.2.1 Retrosynthesis

Mieriņa and co-workers (2016) comprehensively explained the retrosynthetic approach to construct 4-aryl-3,4-dihydroquinolin-2(1H)-ones (Figure 4.3). The retrosynthetic analysis illustrates routes (a)–(e) for bond disconnection. Route (a) C_4–C_{4a} involves an internal Friedel–Crafts alkylation, which is sometimes performed using free radicals. Route (b) C_3–C_4 is an aldol reaction. A transition metal-catalysed coupling reaction mostly involves route (c) C_2–C_3. Route (d) N_1–C_2 is an intramolecular N-acylation of anilines (the early method explained above). The Schmidt rearrangement or internal cyclization of 3-aryl-N-methoxypropanamides is used for the route (e) N_1–C_{8a}.

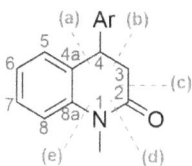

FIGURE 4.3   Retrosynthesis analysis for constructing 4-aryl-3,4-dihydroquinolin-2(1*H*)-ones. (Adopted from Mieriņa et al. 2016.)

4.1.2.2   Catalytic Processes

Several synthetic strategies for making 2O-THQs focus on catalytic processes. One key strategy is the Pd-catalysed Heck reduction–cyclization. This method forms carbon–carbon bonds and is known for its versatility in creating complex molecules (Felpin et al. 2009). Another strategy is the Pd-catalysed cyclopropane ring expansion, which involves the expansion of a cyclopropane ring into a larger ring structure using palladium catalysis. This approach is useful for forming larger and more complex cyclic structures from simpler three-membered rings (Tsuritani et al. 2009). The Mn-mediated intramolecular cyclization strategy uses manganese (Mn) as a mediator to facilitate the intramolecular cyclization of a substrate. Manganese is often favoured for its ability to promote reactions difficult to achieve with other metals (Tsubusaki et al. 2009). The Rh-mediated Michael-addition strategy utilizes rhodium (Rh) to mediate the Michael addition of boronic acid to an enone, followed by cyclization. Rhodium catalysis is highly efficient and selective, making it valuable for forming complex structures (Horn et al. 2009). Another strategy is the Ru-catalysed (ruthenium) cyclization of 1,4,2-dioxazol-5-ones to form 2O-THQ structures (Sun et al. 2021). The photoredox strategy uses light (photons) to drive a redox reaction, typically involving anilines, oxalyl chloride, and electron-deficient alkenes. Photoredox catalysis offers high selectivity and the ability to perform reactions under mild conditions (He et al. 2022; Correia et al. 2020). Hydrogenation (reduction) of the corresponding 2-quinolinone using Pd, Ru, Rh, and samarium diiodide (SmI$_2$) is a general and practical method (Xie and Zhang 2022; Hu et al. 2023; Horn et al. 2009; Xie and Zhang 2022). A recent review by Niu et al. (2023) summarized different strategies for the synthesis of 2O-THQs via the catalytic annulation of α,β-unsaturated *N*-arylamides, including electrophilic cyclization, radical-initiated cyclization, and photochemical cyclization reactions. The review also discussed the substrate scope and mechanistic details of these reactions.

4.1.2.3   3,4-Dioxygenated 4-aryl-2O-THQ

There are approximately 70 compounds listed in Table 4.1, having C–O bonds at both C3 and C4, and an aryl group at C4. These substances are named 3,4-dioxygenated 4-aryl-quinolin-2(1*H*)-one alkaloids. The synthesis of 3,4-dioxygenated 4-aryl-quinolin-2(1*H*)-ones involves several key steps and approaches, ranging from biotransformations to total synthesis (Simonetti et al. 2016). Biotransformations utilize enzymatic or microbial systems to perform selective transformations on precursor molecules. It offers a potentially more environmentally friendly and selective approach to certain steps in the synthesis (Boyd et al. 2002). The primary synthesis challenge lies in building the 2O-THQ core structure with the correct substitution pattern. This typically involves forming the aromatic ring system and attaching the lactam functionality. The 3,4-dioxygenated pattern and the 4-aryl substituent are key features that need to be incorporated. These steps often require careful control of reaction conditions and the use of protecting groups to achieve the desired selectivity. A significant milestone was the total synthesis of yaequinolone A2 (4.138). This work demonstrated the feasibility of constructing these complex natural products from simple, commercially available starting materials (Lie et al. 2009). The total synthesis not only confirms the proposed structure of the natural product but also provides a route for producing larger quantities of the compound for further study. The synthesis of key intermediates is another important aspect. For example, an important intermediate towards peniprequinolone and related compounds was synthesized (Benakki et al. 2008). This approach was valuable because it allowed for more efficient synthesis of multiple target molecules that share common structural features. By creating structurally similar compounds for structure–activity relationship studies, researchers can investigate which parts of the molecule are essential for biological activity and potentially develop improved compounds for biological testing or drug development. Stereochemistry is another important consideration in these syntheses. Many of the natural products in this family have chiral centres, particularly at C3 and C4. Stereoselective synthesis methods are employed to control the configuration at these centres. For compounds with complex side chains, such as the isoprenoid chains in penigequinolones, additional synthetic steps are required to construct and attach these moieties. This adds another layer of complexity to the total synthesis of these molecules (Jia et al. 2021).

4.1.2.4   Melodinus Alkaloids

Melodinus is a genus of plants in the family Apocynaceae. These plants have demonstrated significant medicinal properties, particularly in traditional medicine for treating ailments such as meningitis, rheumatic heart disease, and dyspepsia. The genus is rich in bioactive compounds, especially alkaloids (Lu et al. 2014; Jiang et al. 2015; Teng et al. 2023). One specific group of Melodinus alkaloids, known as monoterpenoid quinoline alkaloids, contains the Melodinus skeleton (Figure 4.4) (Buckingham 2015). These alkaloids have a 6/6/5/6/5 (ABCDE) pentacyclic core structure. In this chapter's list of 2O-THQ (Table 4.1), there are approximately 70 substances belonging to this class. Synthesizing these alkaloids often involves

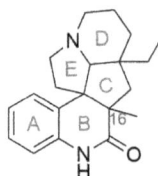

Melodinus skeleton

FIGURE 4.4   Melodinus skeleton (ABCDE system).

generating complex pentacyclic structures, establishing multiple stereogenic centres, and carefully manipulating functional groups. The choice of synthetic route
depends on the specific Melodinus alkaloid being targeted and can vary in terms of
the number of steps, overall yield, and stereoselectivity. We are sharing only a few
examples here.

Hugel and Levy (1986) described the first biomimetic synthesis of two Melodinus
alkaloids: scandine and meloscine. Starting with Δ^{18}-tabersonine, it is converted
to an unstable 16-chloroindolenine intermediate, which is then reduced to form an
aziridine compound. A key flow thermolysis step of this aziridine leads to an imine
intermediate. Selective oxidation of this imine yields scandine. Scandine can be
further transformed into meloscine through decarbomethoxylation. The synthetic
route is proposed to mimic the natural biosynthetic pathway, involving a rearrangement with complete inversion at the C-16 position (position 3 in 2O-THQ). Goldberg
and Stoltz (2011) performed a synthetic approach to the core structure of Melodinus
alkaloids. The key step in the synthesis was a palladium-catalyzed [3+2] cycloaddition between a vinylcyclopropane and a β-nitrostyrene, which rapidly assembled the
cyclopentane core of these alkaloids. This reaction is followed by a zinc reduction and
lactamization to form the ABC ring system (6/6/5). The D ring is then added through
a series of steps, including reductive amination, acylation, and ring-closing metathesis. Importantly, this approach allows installing a quaternary stereocentre at C16, a
challenging feature of many Melodinus alkaloids. The synthesis achieved the ABCD
ring system of the target molecules in six steps from commercially available materials, which demonstrated its efficiency. An interesting synthetic route to produce the
alkaloids epimeloscine and meloscine was described by Zhang and Curran (2011).
The key innovation was a cascade radical annulation of a divinyl-cyclopropane,
which allowed the researchers to construct two rings (B and C) of the molecule in a
single step. This approach significantly shortened the total synthesis to 13 steps, with
the longest linear sequence being 10 steps. The core part of the synthesis took 5 steps
and proceeded with almost 20% overall yield. The method produced epimeloscine
stereoselectively, which can then be converted to meloscine. In a study by Xiao et al.
(2016), a synthetic method for constructing tricyclic dihydroquinolinones was introduced. Dihydroquinolinones are key structural components of certain alkaloids like
scandine. The researchers developed an intramolecular [3+2] annulation reaction
using cyclopropane 1,1-diester derivatives with an amide linker. This reaction, promoted by titanium tetrachloride, allowed for the rapid and highly diastereoselective
formation of complex tricyclic structures containing multiple stereocentres, including all-carbon quaternary centres. The method is notable for its efficiency, forming
three rings in a single step with excellent stereoselectivity. The researchers explored
various substrate modifications and found that the amide linker played a crucial role
in facilitating the ring-opening process. Recently, Mani et al. (2023) synthesized the
new fused polycyclic compounds containing pyrroloquinoline structures (ABE rings)
using an environmentally friendly solid-state melt reaction. The synthesis involved
a tandem multicomponent reaction sequence, including a [3+2]-cycloaddition followed by two annulation steps. The authors used Baylis–Hillman products as dipo-
larophiles, synthesized from various substituted aryl/heteroaryl aldehydes, DABCO
(1,4-diazabicyclo[2.2.2]octane), and methyl acrylate. The 1,3-dipole component was

derived in situ from indoline-2,3-dione and either *N*-methylglycine or L-proline. This method resulted in the formation of three new rings, five new bonds, and three adjoining stereocentres with complete diastereomeric control.

4.2 NATURAL OCCURRENCE

4.2.1 ANIMAL SOURCES

The presence of alkaloids in animals shows the diverse strategies that organisms utilize for survival and defense. For example, poison dart frogs (family Dendrobatidae) produce alkaloids like batrachotoxins and pumiliotoxins as defence mechanism (Caty et al. 2019). Some ant species produce alkaloids in their venom or defensive secretions, which can be neurotoxic or antimicrobial (Xu and Chen 2023). Alkaloids have also been discovered in marine organisms such as sponges, bryozoans, and ascidians, where they serve for protection and competition (Elissawy et al. 2021).

Coral reefs are considered the most fragile, biologically diverse, and economically important ecosystems on Earth. They offer essential ecosystem services and unique biochemical compounds. Within these reefs, organisms such as fungi find vital shelter. Recent research has emphasized the varied fungal communities closely associated with coral ecosystems. These findings indicate that fungi linked to marine invertebrates actively contribute to the formation of biofilms and serve as the main sources of chemical defence mechanisms for host organisms. The discovery of marine natural products, including those produced by coral symbiotic microorganisms, has led to the development of bioactive compounds with potential clinical uses (Chen et al. 2022; Liu et al. 2019). However, the 2O-THQ alkaloids isolated from coral-assisted fungi will be discussed in Section 4.2.3.

The American cockroach, *Periplaneta americana* L. (Blattidae), contains several alkaloids (Liang et al. 2022). These alkaloids have been studied for their potential interactions with nicotinic acetylcholine receptors of insects. For instance, quinolizidine alkaloids have been found to bind with these receptors, with aloperine acting as an antagonist (Liu et al. 2008). In addition, a study discovered the steroid alkaloid conessine, which inhibited moulting, in the American cockroach (Richter et al. 1989). Several 2O-THQs have also been isolated from *Periplaneta americana*, as listed in Table 4.2 (Khadem and Marles 2025b).

4.2.2 BACTERIAL SOURCES

Bacteria can serve as important microorganisms in the production of bioactive substances within cellular factories. Both Gram-negative bacteria (such as *Escherichia coli*) and Gram-positive bacteria (such as *Lactococcus lactis*, *Streptomyces* spp., and *Bacillus* spp.) can be used as host systems for the production of natural products (Pham et al. 2019). For example, quinoline alkaloids have been isolated from various bacterial cultures (Shang et al. 2018a, 2018b). The Actinomycetales (actinomycetes) and Streptomycetales (streptomycetes) orders of bacteria are particularly interesting in microbiology due to their ability to produce a wide range of natural products, including alkaloids (Selim et al. 2021; Salwan and Sharma 2020).

A crude extract of the actinomycete *Actinomadura* BCC27169 (Thermomono-sporaceae) was investigated (Intaraudom et al. 2014). This led to the isolation of several compounds, including two 2O-THQs, namely 3,4-dihydro-8-[(4-O-methyl-α-L-ribopyranosyl)oxy]-2(1H)-quinolinone (4.110) and 8-[(6-deoxy-4-O-methyl-α-L-m annopyranosyl)oxy]-3,4-dihydro-2(1H)-quinolinone (4.111).

The study conducted by Mahmoud et al. (2018) revealed the chemical composition of an organic extract from *Streptomyces* sp. LGE21 (Streptomycetaceae) was obtained from *Lemna gibba* L. (duckweed) leaves. This investigation identified three 2O-THQs: 3,4-dihydro-2(1H)-quinolinone (4.1), 3,4-dihydro-8-hydroxy-2(1H)-quinolinone (4.101), and 3,4-dihydro-8-methoxy-2(1H)-quinolinone (4.112). Albogrisins C–C′ and D–D′ (4.106–4.109) were isolated from mangrove-derived *Streptomyces albogriseolus* (Gao et al. 2019).

4.2.3 Fungal Sources

Among a subset of over 60,000 natural products reported in the collection of open natural products database, 23% originated from fungi (Capecchi and Raymond 2021). The potential therapeutic applications of these fungal natural products are vast and hold great promise for developing novel drugs (Bhattarai et al. 2021). In particular, the fungal family Apergillaceae, composed of various fungal genera such as *Aspergillus* and *Penicillium*, is a promising source of diverse natural products, including alkaloids, with significant pharmaceutical potential (Youssef et al. 2021; Zhang et al. 2022; Dai et al. 2020b).

In Table 4.2, various 2O-THQ alkaloids were found in fungal families, with Aspergillaceae and Microascaceae being the primary sources (Khadem and Marles 2025b). The table also includes 2O-THQ sources from plants (+ Rhizophoraceae), corals (+ Acanthogorgiidae), and bacterial co-culture (+ Bacillaceae). Penicillium and Aspergillus species are the main producers, with 37 and 32 compounds, respectively. The most common alkaloids found in fungal sources are peniprequinolone (4.126), peneciraistin F (4.139), and penigequinolone A (4.124).

4.2.4 Plant Sources

Various plant species have been discovered as significant sources of alkaloids. These compounds are often produced in specialized plant tissues, such as roots, stems, leaves, or seeds. Some well-known plant families that are rich in alkaloids include Apocynaceae, Solanaceae, Ranunculaceae, Papaveraceae, and Rubiaceae. In these families, different genera and species have been found to contain high concentrations of alkaloids. For example, the Apocynaceae family includes plants like *Catharanthus roseus* (L.) G. Don (Madagascar periwinkle), which is known for its abundance of alkaloids such as vincristine and vinblastine. These alkaloids have been widely used in the treatment of various cancers. Similarly, plants from the Solanaceae family, like *Atropa belladonna* L. (belladonna, deadly nightshade) and *Datura stramonium* L. (Jimson weed) contain alkaloids with well-known medicinal properties like atropine (used as a muscarinic antagonist for ophthal-mic examination pupil dilation and treatment of organophosphate poisoning) and

TABLE 4.2

Natural Occurrence of 2O-THQ Alkaloids

Kingdom	Family	Genus	Species	Compound(s)	Reference(s)
Animalia	Blattidae	Periplaneta	americana	4.1	Block and McChesney (1974)
"	"	"	"	4.101	Li et al. (2015)
"	"	"	"	4.96, 4.101	Lv et al. (2021)
"	"	"	"	4.102–4.104	Zhang et al. (2021)
Bacteria	Actinomycetes	Actinomudura	BCC27169	4.110, 4.111	Intaraudom et al. (2014)
"	Streptomycetaceae	Streptomyces	Albogriseolus/MGR072	4.106–4.109	Gao et al. (2019)
"	"	"	LGE21	4.1, 4.101, 4.112	Mahmoud et al. (2018)
Fungi	Agaricaceae	Agaricales	sp. 3034/F12	4.126	dos Santos et al. (2019)
"	Trichocomaceae	Aspergillus	Versicolor/KU258497	4.141	Ebada et al. (2018)
"	"	"	Versicolor /Eich.5.2.2	4.142, 4.143	Ebada and Ebrahim (2020)
"	"	"	Alabamensis/ SYSU-6778	4.147–4.153	Hu et al. (2023)
"	"	"	nidulans	4.154, 4.155	Li et al. (2020)
"	"	"	creber EN-602	4.156–4.158	Li et al. (2021)
"	"	"	SF-5044/MYC-2048 (NRRL 58570)	4.147, 4.157, 4.158, 4.162–4.165	Neff et al. (2012)
"	"	"	nidulans	4.166–4.169	Scherlach and Hertweck (2006)
"	"	"	CMB-MRF324	4.174, 4.175	Wu et al. (2022)
Fungi /bacteria co-exist	Trichocomaceae /Bacillaceae	Aspergillus /Bacillus	Versicolor/subtilis	4.147, 4.158, 4.162, 4.176	Abdel-Wahab et al. (2019)

(Continued)

TABLE 4.2 (Continued)
Natural Occurrence of 2O-THQ Alkaloids

Kingdom	Family	Genus	Species	Compound(s)	Reference(s)
Fungi /coral-based	Trichocomaceae /Acanthogorgiidae	Aspergillus /Muricella	XS-20090B15/ abnormalis	4.162, 4.165, 4.177, 4.178	Chen et al. (2014)
Fungi /plant-based	Trichocomaceae /Rhizophoraceae	Aspergillus /Rhizophora	Nidulans/MA-143/ stylosa	4.142, 4.156, 4.162, 4.179–4.182	An et al. (2013)
Fungi	Trichocomaceae	Metarhizium / Paecilomyces	marquandii	4.144	El-Kashef et al. (2019)
"	"	Penicillium	simplicissimum	4.113–4.126	Dai et al. (2020)
"	"	"	"	4.122–4.139	Dai et al. (2021)
"	"	"	strain NTC-47	4.124, 4.125, 4.139, 4.145	Hayashi et al. (1996)
"	"	"	"	4.139, 4.145	Hayashi et al. (1997)
"	"	"	Janczewskii /H-TW5/869	4.126, 4.137–4.139	He et al. (2005)
"	"	"	OUCMDZ-776	4.146	He et al. (2012)
"	"	"	No. 410	4.124, 4.125	Kimura et al. (1996)
"	"	"	simplicissimum	4.124–4.126, 4.139, 4.140	Kusano et al. (2000)
"	"	"	scabrosum	4.124, 4.125	Larsen et al. (1999)
"	"	"	strain NTC-47	4.160, 4.161	Nakaya (1995)
"	"	"	janczewskii	4.126	Schmeda-Hirsch. et al. (2005)
"	"	"	FKI-2140	4.122, 4.123	Uchida et al. (2005)
"	"	"	"	4.121–4.126, 4.136–4.139, 4.170–4.173	Uchida et al. (2006)
"	"	"	MCCC 3A00228	4.144	Wang et al. (2021)

(Continued)

TABLE 4.2 (Continued)
Natural Occurrence of 2O-THQ Alkaloids

Kingdom	Family	Genus	Species	Compound(s)	Reference(s)
"	"	"	namyslowskii	4.126	Wubshet et al. (2013)
"	Microascaceae	Scopulariopsis	TAO1-33	4.159	Mou et al. (2018)
"	"	"	"	4.143	Shao et al. (2016)
"	"	"	2468	4.142, 4.147, 4.156, 4.158, 4.162, 4.165	Shao et al. (2015)
"	"	"	(EU821474)	4.1	Kovalenko et al. (2002); Neuwoehner et al. (2009); Reineke et al. (2007)
Geologic	oil/tar-contaminated sites	—	—	4.1	Belyagoubi-Benhammou et al. (2019)
Plantae	Amaranthaceae	Anabasis	articulata	4.2	Koolen et al. (2017)
"	"	Alternanthera	littoralis	4.3, 4.4	Lin et al. (2023)
"	"	Chenopodium	formosanum	4.6, 4.7, 4.9, 4.10	Fu et al. (2014)
"	Apocynaceae	Bousigonia	angustifolia	4.5	Du et al. (2023)
"	"	Melodinus	axillaris	4.5, 4.16, 4.17, 4.25, 4.27, 4.28	Fang et al. (2016)
"	"	"	"	4.5, 4.29, 4.30, 4.32	Li et al. (2021)
"	"	"	cochinchinensis	4.5, 4.6, 4.16, 4.27	Yang et al. (2021)
"	"	"	"	4.5, 4.6, 4.16, 4.17	Cai et al. (2011)
"	"	"	fusiformis	4.5, 4.6, 4.7, 4.16	He et al. (1992)
"	"	"	"	4.16, 4.17, 4.35–4.44	Liu et al. (2023)
"	"	"	hemsleyanus	4.5, 4.6, 4.10, 4.16–4.18	Guo and Zhou (1993)
"	"	"	"	4.5, 4.16, 4.19, 4.25	Wang et al. (2019)
"	"	"	"	4.5, 4.6	Yan and Feng (1998)

(Continued)

TABLE 4.2 (Continued)
Natural Occurrence of 2O-THQ Alkaloids

Kingdom	Family	Genus	Species	Compound(s)	Reference(s)
"	"	"		4.5, 4.6, 4.17, 4.20, 4.23	Zhang et al. (2016)
"	"	"		4.5, 4.6, 4.7, 4.10, 4.14, 4.19, 4.20, 4.26, 4.45–4.53	Zhang J. et al. (2016)
"	"	"		4.16, 4.17, 4.34, 4.43	Zhang J. et al. (2020)
"	"	"	*henryi*	4.5–4.7, 4.10, 4.16, 4.17, 4.54, 4.55	Guo et al. (2017)
"	"	"		4.5	Li et al. (1987)
"	"	"		4.56–4.59	Ma et al. (2015)
"	"	"		4.5, 4.6, 4.16, 4.27	Yang et al. (2020)
"	"	"		4.5, 4.8, 4.17	Yu et al. (2019b)
"	"	"	*khasianus*	4.5, 4.6, 4.10, 4.16	Hui-Lan and Han-Dong (1994)
"	"	"	*morsei*	4.46, 4.60, 4.61	Yin et al. (2018)
"	"	"	*oblongus*	4.5, 4.6, 4.10, 4.16, 4.17	Lien et al. (2002a); Lien et al. (2002b)
"	"	"	*scandens*	4.5, 4.10, 4.11	Bernauer et al. (1969)
"	"	"		4.5, 4.10, 4.11, 4.16	Chazelet et al. (1986); Daudon et al. (1975)
"	"	"		4.32, 4.33	Daudon et al. (1976)
"	"	"		4.5, 4.11, 4.16, 4.24	Mehri et al. (1971)
"	"	"		4.11–4.13	Mehri et al. (1995)
"	"	"		4.16	Plat et al. (1970)
"	"	"	*suaveolens*	4.62–4.64	Gao et al. (2019)
"	"	"		4.5, 4.51, 4.65–4.70	Gao et al. (2023)
"	"	"		4.5, 4.6, 4.25, 4.26	Liu et al. (2012)

(Continued)

TABLE 4.2 (Continued)
Natural Occurrence of 2O-THQ Alkaloids

Kingdom	Family	Genus	Species	Compound(s)	Reference(s)
"	"	"	"	4.71–4.73	Liu et al. (2013)
"	"	"	"	4.5, 4.6, 4.10, 4.19, 4.21, 4.22	Lu et al. (2014)
"	"	"	"	4.5, 4.7, 4.10, 4.14, 4.16, 4.19	Zhang et al. (2013)
"	"	"	tenuicaudatus	4.5, 4.15, 4.16	Feng et al. (2010)
"	"	"	"	4.5, 4.6, 4.43	Yi et al. (2018)
"	"	"	"	4.5, 4.6, 4.15	Zhou et al. (1988)
"	"	"	"	4.6, 4.9, 4.15, 4.53, 4.183, 4.184	Wu. et al. (2022)
"	"	"	yunnanensis	4.5, 4.16	Wang et al. (2011)
"	"	"	"	4.5, 4.6, 4.7	Wu et al. (2020)
"	"	"	"	4.5, 4.6, 4.11, 4.16, 4.17, 4.23, 4.29, 4.31, 4.74	Zhang et al. (2017)
"	Brassicaceae	Brassica	campestris	4.75	Jing et al. (2019)
"	"	Isatis	indigotica (tinctoria)	4.80	Chen et al. (2012)
"	"	"	"	4.76, 4.77	Liu et al. (2016)
"	"	"	"	4.81	Xu et al. (2020)
"	"	"	"	4.78, 4.79	Zhang et al. (2019)
"	Capparaceae	Capparis	spinosa	4.82	Zhang et al. (2014)
"	Caryophyllaceae	Dianthus	superbus	4.83, 4.84	Sun et al. (2019)
"	Euphorbiaceae	Trigonostemon	lutescens	4.85, 4.86	Ma et al. (2013)
"	Fabaceae	Prosopis	juliflora	4.86	Choudhari et al. (2023)
"	"	Spartium	junceum	4.1	Nanni et al. (2018)
"	Juglandaceae	Juglans	regia	4.87, 4.88	Haramiishi et al. (2020)
"	"	"	"	4.87	Ito et al. (2007)

(Continued)

TABLE 4.2 (Continued)
Natural Occurrence of 2O-THQ Alkaloids

Kingdom	Family	Genus	Species	Compound(s)	Reference(s)
"	Lamiaceae	Rabdosia	rubescens	4.90	Feng et al. (2007)
"	"	Mentha	canadensis	4.89	Li et al. (2013)
"	Onagraceae	Chamaenerion (Epilobium)	angustifolium	4.91, 4.92	Deng et al. (2019)
"	Ranunculaceae	Aconitum	ferox	4.96	Hanuman and Katz (1993)
"	Rosaceae	Rubus	chingii	4.97	Chai (2008); Yu et al. (2019b)
"	Poaceae	Coix	lachryma-jobi/vari. Ma-yuen	4.93	Chung et al. (2011); Lee et al. (2008)
"	"	Oryza	sativa	4.94	Feng et al. (2021)
"	"	"	"	4.95	Jia et al. (2023)
"	Rutaceae	Boronia	pinnata	4.98	Ito et al. (2000)
"	"	Clausena	excavata	4.91	Peng et al. (2013)
"	"	Melicope	denhamii	4.100	Nakashima et al. (2012)
"	"	Ruta	graveolens	4.99	Salib et al. (2014)

scopolamine (an anticholinergic used for the treatment of nausea and vomiting) (Bhambhani et al. 2021).

Quinoline alkaloids have a wide range of biological activities, including antimalarial, antibacterial, antifungal, and anticancer effects (Shang et al. 2018a, 2018b). Several plant families are known to produce quinoline alkaloids, with some prominent examples being Rubiaceae, Rutaceae, and Annonaceae. The Rubiaceae family is particularly rich in quinoline alkaloids. Notable plants from this family include *Cinchona* spp., which is the natural source of quinine, a well-known antimalarial alkaloid. Additionally, plants like *Uncaria tomentosa* (cat's claw) and *Psychotria* spp. also belong to the Rubiaceae family and contain quinoline alkaloids with potential therapeutic applications. Furthermore, the Rutaceae family is another significant source of quinoline alkaloids. Species like *Ruta graveolens* (rue) and *Dictamnus albus* (white dittany, burning bush) have been found to contain quinoline alkaloids that possess antibacterial and antifungal properties (Matada et al. 2021).

According to Table 4.2, numerous plant families are known to produce 2O-THQ alkaloids (Khadem and Marles 2025a). Prominent among these families are Apocynaceae, Brassicaceae, Amaranthaceae, and Rutaceae. The Apocynaceae family, specifically the genera *Melodinus* and *Bousigonia*, is responsible for producing at least seventy 2O-THQs. The phytochemistry of *Melodinus* has been extensively reviewed by Lu et al. (2014); Jiang et al. (2015); and Teng et al. (2023). The Brassicaceae family includes plants like those of the genus *Isatis*, which contain several 2O-THQs. Interestingly, 3,4-dihydro-2(1*H*)-quinolinone (4.1) is found in *Anabasis articulata* (Amaranthaceae) and *Spartium junceum* L. (Spanish broom, Fabaceae) (Belyagoubi-Benhammou et al. 2019, Nanni et al. 2018).

Table 4.2 highlights that 2O-THQ alkaloids are present in at least 13 families in 11 eudicot orders: Brassicales (Brassicaceae, Capparaceae), Caryophyllales (Amaranthaceae, Caryophyllaceae), Fabales (Fabaceae), Fagales (Juglandaceae), Gentianales (Apocynaceae), Lamiales (Lamiaceae), Malpighiales (Euphorbiaceae), Myrtales (Onagraceae), Ranunculales (Ranunculaceae), Rosales (Rosaceae), and Sapindales (Rutaceae); and in one family of monocots, the Poales (Poaceae). At a higher taxonomic level, the Fabid clade includes the Fabales, Fagales, Malpighiales, and Rosales, while the Malvid clade includes the Brassicales, Myrtales, and Sapindales. Fabids and Malvids are genetically related as part of the larger Rosid clade. The Lamiid clade comprises the Lamiales and Gentianales, which are distantly related to the Caryophyllales at the Superastrid clade level; Lamiids are even further separated from the Fabids and Malvids. The Ranunculales and the Poales (as monocots) are not closely related to any of these other eudicot groups (Angiosperm Phylogeny Group et al. 2016). From a chemotaxonomic perspective, the lack of close relationships between many of the plant families containing 2O-THQs suggests that this class of phytochemicals may have evolved independently multiple times, rather than from a common ancestor. Among the 2O-THQs group, scandine (4.5), 10-hydroxyscandine (4.6), meloscandonine (4.16), 19-epi-meloscandonine (4.17), meloscine (4.10), and 3-epimeloscine (4.11) are the most commonly found compounds across various plant families. Table 4.2 provides details of the natural occurrence of 2O-THQ derivatives in different plant families, including specific information about the corresponding genera and species.

4.2.5 OTHER SOURCES

3,4-Dihydro-2(1H)-quinolinone (4.1) has been found in plants, animals, fungi, and bacteria, as well as in oil, petroleum, and tar-contaminated environments (Kovalenko et al. 2002; Reineke et al. 2007; Neuwoehner et al. 2009; Khadem and Marles 2025b). However, these geological sources are generally not considered as conventional natural sources within the scientific communities, and the presence of these alkaloids may be attributed to the presence of microbes in such environments.

4.3 BIOLOGICAL ACTIVITY

4.3.1 ANTIBACTERIAL

Researchers have studied several 2O-THQs to determine their ability to inhibit both Gram-positive and Gram-negative bacteria. Among them, some have shown strong antibacterial effects. For example, a combination of aniduquinolone A and its 22-epimer aniduquinolone D (4.142 and 4.143) displayed significant activity against the *Staphylococcus aureus* bacterium (ATCC700699) with a minimum inhibitory concentration of 0.4 mg/mL, without causing any cytotoxic effects. This demonstrates the specificity of the antibacterial effect. 5-Deoxyaflaquinolone E (4.156) showed extensive antibacterial activity against several bacteria, especially *Staphylococcus aureus* (ATCC 27154), *Bacillus cereus* (ATCC 11077), *Vibrio parahaemolyticus* (ATCC 17802), *Nocardia brasiliensis* (ATCC 19019), and *Pseudomonas putida* (ATCC 17485), with MIC values ranging from 0.78 to 6.25 μM. Finally, suadimin A (4.62) exhibited significant *in vitro* antimycobacterial activity against *Mycobacterium tuberculosis* strain H37Rv, with an MIC_{90} value of 6.76 μM and a selectivity index of 21.2 concerning general cytotoxicity.

4.3.2 CYTOTOXICITY/ANTICANCER

The 2O-THQ alkaloids have been shown to possess anti-proliferative and cytotoxic activities against both human and murine cancer cell lines. These effects include growth inhibition, induction of apoptosis, and reduced viability (Olofinsan et al. 2023). For example, scandomelonine (4.33) and episcandomelonine (4.32) demonstrated strong efficacy against human T-ALL cell lines (MOLT-4), comparable to the effects of cisplatin. Additionally, Melosuavine A and B (4.71 and 4.72) exhibited stronger inhibitory effects on one or more of the five human cancer cell lines, with lower IC_{50} values than those observed with cisplatin.

4.3.3 ANTIFOULING

Researchers conducted a recent study to test the effectiveness of certain compounds (2O-THQs) in preventing barnacle larvae from settling. They identified three promising compounds, such as aniduquinolone A (4.142), aflaquinolone A (4.162), and scopuquinolone B (4.159), as potential non-toxic antifouling agents. These compounds showed very high therapeutic ratios and were present in pico/nanomolar concentrations. Penispirolloid A (4.146) demonstrated significant antifouling activity against the larvae of a sessile marine animal known as *Bugula neritina*, with an EC_{50} of 2.40 µg/mL. For a compound to be considered a promising non-toxic antifouling candidate, it must have a therapeutic ratio of $LC_{50}/EC_{50} > 50$ and an $EC_{50} < 5.0$ µg/mL.

4.3.4 ANTI-INFLAMMATORY

Several studies have found that different 2O-THQs can inhibit the production of nitric oxide in RAW 264.7 cells. These cells are a type of murine macrophage cell line commonly used in immunopharmacology. For instance, pesimquinolones A, E, G, and H (4.113, 4.117, 4.119, and 4.120) have demonstrated promising results in suppressing nitric oxide production, with IC_{50} values below 2 µM (Dai et al. 2020a, 2020b; Dai et al. 2021).

Analysing the inhibitory ratios of β-glucuronidase release in rat polymorphonuclear leukocytes is crucial in understanding how cells respond to inflammation caused by platelet-activating factor *in vitro* (Fang et al. 2016). This research offers insights into the modulation of cellular degranulation and contributes to our understanding of inflammatory activities at the cellular level. Melaxillines A and B (4.27 and 4.28), melodinine T (4.25), and scandine (4.5) have exhibited significant anti-inflammatory activities, with IC_{50} values ranging from 1.51 to 21.56 µM, respectively.

4.3.5 ANTIOXIDANT

The DPPH (1,1-diphenyl-2-picrylhydrazyl) free radical scavenging assay is a widely used method for assessing the antioxidant properties of compounds. This method involves reducing the stable 2,2-diphenyl-1-picrylhydrazyl (DPPH) free radical, which leads to a noticeable colour change. The test provides quantitative information about whether a substance can neutralize free radicals and act as an antioxidant (Macáková et al. 2019). Table 4.2 demonstrates that a few 2O-THQ compounds exhibited antioxidant activity.

4.3.6 BRINE SHRIMP LETHALITY

The brine shrimp lethality assay is a commonly used method to assess the toxicity of different compounds, including natural products. This assay involves exposing brine shrimp nauplii (Artemia salina L.) to the compounds and observing whether they cause lethality. According to a study by Ntungwe et al. (2020), among the 2O-THQs tested, yaequinolone F, penigequinones A and B (4.121, 4.124, 4.125) were found to be the most potent, with a minimum inhibitory concentration value of 0.19 µg/mL.

4.3.7 NEMATICIDAL

The control of *Pratylenchus penetrans* Cobb, a plant parasite of horticultural importance, involves the assessment of various compounds, including 2O-THQs, for their ability to combat this nematode. For instance, penigequinones A and B (4.124 and

4.125) showed nematicidal activity within 3 days of incubation. Interestingly, the selectivity of these compounds towards *P. penetrans* was observed through their low activity against the free-living nematode, *Caenorhabditis elegans* Maupas (Kusano et al. 2000).

4.3.8 Neuroprotectivity

MPP⁺ (1-methyl-4-phenylpyridinium) is a type of neurotoxin commonly used in experimental studies to cause damage or injury to primary cerebral cortical neurons. Melofusinines E and H (4.37 and 4.40) showed a significant neuroprotective effect on primary cortical neurons that have been injured by MPP⁺. Even at low concentrations of 6.25 µM, these compounds have demonstrated effectiveness.

4.3.9 Other Bioactivities

In Table 4.3, it is shown that 2O-THQ alkaloids have a range of biological activities (also Khadem and Marles 2025c). These include the inhibition of acetylcholinesterase, adipogenesis, lipase, and pollen growth. Additionally, they exhibit neurotropic and antiviral activities, as well as ecotoxicity. Furthermore, the alkaloids can inhibit the protein binding of indoleamine 2,3-dioxygenase 1 and serve as metabolic markers for gut microbiota.

The data shows that out of 100 compounds, there are various biological activities. Cytotoxic (anticancer) activity was found in 25 compounds, accounting for 25% of the total, indicating significant potential for cancer treatment. Anti-inflammatory activity and brine shrimp lethality were noted in 17 compounds each, representing 17%. This highlights potential applications in anti-inflammatory therapies and general toxicity screening. Antibacterial activity was identified in 11 compounds (11%). It suggests the potential for developing antibacterial agents. Additionally, antifouling activity was seen in 7 compounds (7%), indicating usefulness in preventing the accumulation of organisms on surfaces. Nematicidal and neuroprotective activities were observed in 4 compounds (4%). It predicts the potential roles in protecting against nerve damage and targeting nematodes. Furthermore, antioxidant activity was found in three compounds (3%), indicating a minor potential for neutralizing free radicals. Finally, in the "Others" category, several activities were noted, including acetylcholinesterase inhibition, antiviral activity, lipase inhibition, metabolic marker (gut microbiota), neurotrophic activity, ecotoxicity, adipogenesis inhibition, IDO1 inhibition, and pollen-growth inhibition, collectively accounting for 12 compounds (12%). This distribution (Figure 4.5) underscores the diverse applications of these compounds, particularly in cytotoxicity and anti-inflammatory properties, warranting further research to optimize their use and ensure safety and efficacy.

4.4 DISCUSSION

The structural variation in 2O-THQ alkaloids is important because it directly affects their chemical and biological properties. These differences affect how strong, specific, and toxic the compounds are. This, in turn, determines how they work in the

TABLE 4.3

Biological Activity of 2O-THQ Alkaloids

Biological Activity	Target	Assay, Method	Results	Compound	References
Acetylcholinesterase inhibition	Acetylcholinesterase (AChE)	in vitro AChE inhibition assay	14.56% inhibition at 50 μg/mL	4.85	Ma et al. (2013)
Adipogenesis inhibition	3T3-L1 cells	3T3-L1 cells treated with compounds (100 mg/mL)	36% decrease compared to oxidized LDL group	4.3	Lin et al. (2023)
"	"		31.2% decrease compared to oxidized LDL group	4.4	Lin et al. (2023)
Antibacterial	Human-pathogenic bacterium Staphylococcus aureus (ATCC700699)	Broth microdilution assay	MIC = 0.4 mg/mL using moxifloxacin as a reference antibiotic (MIC = 4 mg/mL) with no cytotoxic activity	4.142 + 4.143	Ebada and Ebrahim (2020)
"	S. iniae, S. parauberis, B. subtilis, S. aureus, E. ictalurid	"	MIC values (moderate to potent against S. iniae and S. parauberis)	4.148	Hu et al. (2023)
"	"	"	MIC values (moderate to potent against S. iniae and S. parauberis)	4.149	Hu et al. (2023)
"	"	"	MIC = 10.1 μM against S. aureus, 5.0 μM against S. iniae and S. parauberis	4.150, 4.151	Hu et al. (2023)
"	"	"	MIC = 2.2 μM against S. iniae	4.152	Hu et al. (2023)
"	"	"	MIC = 10.9 μM against E. ictalurid, moderate to potent against other strains	4.153	Hu et al. (2023)
"	S. aureus, B. cereus, V. parahaemolyticus, N. brasiliensis, and P. putida	—	MIC = 0.78, 1.56,6.25, 0.78,1.56 μM, respectively	4.156	Shao et al. (2015)

(Continued)

TABLE 4.3 (Continued)
Biological Activity of 2O-THQ Alkaloids

Biological Activity	Target	Assay, Method	Results	Compound	References
"	H$_{37}$Rv strain of *M. tuberculosis*	Microplate Alamar Blue Assay (MABA), using general cytotoxicity assays against VERO cells	MIC$_{90}$ = 6.76 µM, selectivity index: 21.2	4.62	Gao et al. (2019)
"	"	"	MIC$_{90}$ = 33.47 µM	4.63	Gao et al. (2019)
Anticancer (Anti-proliferative/ Cytotoxicity)	Liver cancer ⁻HEPG2 cell line	Sulforhodamine-B (SRB) assay	0.45/12.5 µg/mL (displaying 0.23 alive cell lines at a concentration of 100 µg/mL)	4.101	Mahmoud et al. (2018)
"	"	"	0.68/12.5 µg/mL (displaying 0.23 alive cell lines at a concentration of 100 µg/mL)	4.112	Mahmoud et al. (2018)
"	"	"	0.96/12.5 µg/mL (displaying 0.23 alive cell lines at a concentration of 100 µg/mL)	4.1	Mahmoud et al. (2018)
"	"	—	EC$_{50}$ = 11.2 µg/mL	4.101	Mahmoud et al. (2018)
"	Human lung cancer cell A549, human colorectal carcinoma cell HT-29, COLO 205 cell	MTT assay	IC$_{50}$ = 28.57 µg/mL (A549), 32.70 µg/mL (HT-29), 41.29 µg/mL (COLO 205)	4.93	Lee et al. (2008)
"	Human leukaemia cell line K-562	—	GI$_{50}$ = 17.8 µg/mL	4.166	Scherlach and Hertweck (2006)
"	"	—	GI$_{50}$ = 21.2 µg/mL	4.167	Scherlach and Hertweck (2006)

(Continued)

TABLE 4.3 (*Continued*)

Biological Activity of 2O-THQ Alkaloids

Biological Activity	Target	Assay, Method	Results	Compound	References
"	Human breast cancer cell lines MCF-7 and T-47D	MTT assay	32.4% inhibition at 50 μM 28.9% inhibition at 50 μM	4.93	Chung et al. (2011)
"	DLD-1 human colon cancer cell	Trypan blue dye-exclusion assay	$IC_{50} = 9.40$ mM (induction of apoptosis)	4.100	Nakashima et al. (2012)
"	"	—	$EC_{50} = 33$ μg/Ml	4.112	Mahmoud et al. (2018)
"	"	—	$EC_{50} = 46.6$ μg/mL	4.1	Mahmoud et al. (2018)
"	Mouse lymphoma L5178Y cell line	*in vitro* cytotoxicity (MTT) assay	95% growth inhibition at 10 μg/mL $IC_{50} = 10.3$ μM	4.141	Ebada et al. (2018)
"	SKOV-3 cells (human ovary adenocarcinoma)	Sulforhodamine B (SRB) staining assay	8.1% viability at 10 μg/mL	4.138	He et al. (2005)
"	HT-29 (human colon carcinoma)	"	4.0 % viability at 10 μg/mL	4.126	He et al. (2005)
"	Human cancer cell lines HL-60, SMMC-7721, A-549, MCF-7, SW480	MTT assay	$IC_{50} = 2.6, 5.9, 11.0, 11.6, 9.8$ μM (respectively)	4.71	Liu et al. (2013)
"	"	"	$IC_{50} = 2.8, 3.1, 8.9, 6.1, 2.9$ μM (respectively)	4.72	Liu et al. (2013)
"	MOLT-4 (human T-ALL cell)	"	$IC_{50} = 5.2$ μM	4.32	Li et al. (2021)
"	"	"	$IC_{50} = 1.5$ μM	4.33	Li et al. (2021)
"	L-929 mouse fibroblast cell lines	—	$GI_{50} = 10.6$ μg/mL	4.166	Scherlach and Hertweck (2006)

(*Continued*)

TABLE 4.3 (Continued)
Biological Activity of 2O-THQ Alkaloids

Biological Activity	Target	Assay, Method	Results	Compound	References
"	"	—	GI_{50}= 11.4 µg/mL	4.167	Scherlach and Hertweck (2006)
"	A-549 cells (lung cancer), BGC-823 cells (human gastric carcinoma), HepG2 cells (human hepatocellular carcinoma), HL-60 (human myeloid leukemia), and MCF-7 cells (human breast cancer)	MTT method	IC_{50}= 11.3, 9.7, 10.1, 11.1, 10.4 µM, respectively	4.60	Yin et al. (2018)
"	"	"	IC_{50}= 12.2, 11.7, 12.7, 12.3, 15.9 µM, respectively	4.61	Yin et al. (2018)
"	"	"	IC_{50}= 17.0, 15.1, 14.7, 17.1, 18.3 µM, respectively	4.46	Yin et al. (2018)
"	Permanent fibroblast cell line derived from human lung (MRC-5); human gastric adenocarcinoma cells (AGS)	—	IC_{50}= 116 µM (fibroblasts); IC_{50}= 89 µM (AGS)	4.126	Schmeda-Hirschmann et al. (2005)
"	Human cancer cell lines (MCF-7)	MTT assay	IC50= 27.3 µM	4.184	Wu et al. (2022)
Antifouling	*Bugula neritina* larvae	Larval settlement bioassay	EC_{50}= 2.40 µg/mL	4.146	He et al. (2012)
"	Larval settlement of the barnacle *Balanus amphitrite*	—	EC_{50}= 0.103 µM and a high therapeutic ratio (LC_{50}/ EC_{50}= 222)	4.159	Mou et al. (2018)

(Continued)

TABLE 4.3 (Continued)
Biological Activity of 2O-THQ Alkaloids

Biological Activity	Target	Assay, Method	Results	Compound	References
"	"	—	$EC_{50} = 17.5$ pM $LC_{50}/EC_{50} = 1200$	4.142	Shao et al. (2015)
"	"	—	$EC_{50} = 28$ nM $LC_{50}/EC_{50} = 205$	4.162	Shao et al. (2015)
"	"	—	$EC_{50} = 2.8$ nM $LC_{50}/EC_{50} = 57$	4.165	Shao et al. (2015)
"	"	—	$EC_{50} = 1.04$ μM $LC_{50}/EC_{50} = 89$	4.156	Shao et al. (2015)
"	"	—	$EC_{50} = 0.86$ μM $LC_{50}/EC_{50} = 91$	4.158	Shao et al. (2015)
Anti-inflammatory	Inhibition of NO secretion induced by LPS	RAW 264.7 cells	$IC_{50} = 10.13$ μM	4.127	Dai et al. (2021)
"	"	"	$IC_{50} = 8.10$ μM	4.128	Dai et al. (2021)
"	"	"	$IC_{50} = 1.94$ μM	4.113	Dai et al. (2020)
"	"	"	$IC_{50} = 1.29$ μM	4.117	Dai et al. (2020)
"	"	"	$IC_{50} = 1.20$ μM	4.119	Dai et al. (2020)
"	"	"	$IC_{50} = 1.23$ μM	4.120	Dai et al. (2020)
"	"	"	$IC_{50} = 39.05$ μM	4.116	Dai et al. (2021)
"	"	"	$IC_{50} = 9.10$ μM	4.118	Dai et al. (2020)

(Continued)

TABLE 4.3 (Continued)
Biological Activity of 2O-THQ Alkaloids

Biological Activity	Target	Assay, Method	Results	Compound	References
"	Inhibitory ratios of β-glucuronidase release in rat polymorphonuclear leukocytes (PMNs) induced by platelet-activating factor (PAF) in vitro	—	IC_{50} = 1.51 μM	4.27	Fang et al. (2016)
"	"	—	IC_{50} = 2.62 μM	4.28	Fang et al. (2016)
"	"	—	IC_{50} = 11.68 μM	4.25	Fang et al. (2016)
"	"	—	IC_{50} = 21.56 μM	4.5	Fang et al. (2016)
"	LPS-induced systemic inflammation mouse model/ Microglia in hippocampus of LPS-induced mouse	Open field test/ Immunohistochemistry with Iba1 antibody	Suppressed LPS-induced higher locomotive activity; Suppressed LPS-induced microglial activation to ameboid form; Significantly decreased (p<0.05) LPS-induced increase in Iba1 immune-positive signals	4.87	Haramiishi et al. (2020)
"	Inflammatory mediators (NO, TNF-α, IL-6) in RAW 264.7 macrophages	Inhibition of NO production/ IL-6 expression/ TNF-α expression	Inhibited LPS-induced NO production at 30 μM; Concentration-dependently decreased LPS-induced elevation of IL-6 at 10–20 μM; Concentration-dependently decreased LPS-induced elevation of TNF-α at 10–20 μM	4.29, 4.30	Li et al. (2021)
"	Lipopolysaccharide (LPS)-induced murine macrophages (RAW264.7)	enzyme-linked immunosorbent assay (ELISA)	significantly inhibited NO, IL-6 and IL-8 generation in LPS-stimulated macrophages	4.21, 4.22	Lu et al. (2014)

(Continued)

TABLE 4.3 (*Continued*)
Biological Activity of 2O-THQ Alkaloids

Biological Activity	Target	Assay, Method	Results	Compound	References
Antioxidant	DPPH free radical scavenging	–	IC_{50} = 16.7 mg/mL	4.105	Bhattarai et al. (2013)
"	Free-radical scavenging effectiveness	$ORAC_{FL}$ assay	Active: 1.10 relative Trolox equivalent (RTE)	4.2	Koolen et al. (2017)
"	Radical scavenging activity	microplate-based $ABTS^+$ reduction assay	Trolox equivalent antioxidant capacity (TEAC)= 0.62 mM (at 0.25 mM concentration)	4.126	Wubshet et al. (2013)
Antiviral	Human respiratory syncytial virus (RSV)-induced cytopathogenicity in human laryngeal carcinoma (Hep-2) cells	Cytopathic effect (CPE) assay	IC_{50} = 42 nM Therapeutic ratio: TC_{50}/IC_{50} = 520	4.178	Chen et al. (2014)
Brine shrimp lethality	Brine shrimp (*Artemia salina*)	–	LC_{50} = 1.16 μM	4.165	Shao et al. (2015)
"	"	–	LC_{50} = 3.79 μM	4.156	Shao et al. (2015)
"	"	–	LC_{50} = 7.85 μM	4.158	Shao et al. (2015)
"	"	Brine shrimp lethality assay (BSLA)	LD_{50} = 7.1 μM	4.179	An et al. (2013)
"	"	"	LD_{50} = 4.5 μM	4.180	An et al. (2013)
"	"	"	LD_{50} = 5.5 μM	4.162	An et al. (2013)
"	"	Meyer bioassay	LC_{50} = 20 μg/mL	4.145	Hayashi et al. (1997)
"	"	–	LD_{50} = 0.9 ppm	4.161	Nakaya et al. (1995)
"	"	–	LD_{50} = 19.6 ppm	4.160	Nakaya et al. (1995)
"	"	microplate assay	MIC = 6.25 μg/mL	4.122	Uchida et al. (2005)

(Continued)

TABLE 4.3 (Continued)
Biological Activity of 2O-THQ Alkaloids

Biological Activity	Target	Assay, Method	Results	Compound	References
"	"	microtiter-plate method	MIC= 6.25 µg/mL	4.123	Uchida et al. (2005)
"	"	"	MIC= 0.19 µg/mL	4.121	Uchida et al. (2006)
"	"	"	MIC= 6.25 µg/mL	4.173	Uchida et al. (2006)
"	"	"	MIC= 6.25 µg/mL	4.171	Uchida et al. (2006)
"	"	"	MIC= 0.78 µg/mL	4.126	Uchida et al. (2006)
"	"	"	MIC= 0.19 µg/mL	4.124	Uchida et al. (2006)
"	"	"	MIC= 0.19 µg/mL	4.125	Uchida et al. (2006)
IDO1 inhibition	Indoleamine 2,3-dioxygenase 1 (IDO1) protein binding	surface plasmon resonance (SPR)- Kinetic parameters	K_D (equilibrium dissociation constant)= 5.978×10^{-6} M; K_i= 40.7 µM (reversible); IC_{50}= 12.2 µM	4.107	Gao et al. (2019)
"	"	"	K_D (equilibrium dissociation constant)= 5.084×10^{-6} M; K_i= 15 µM (reversible); IC_{50}= 10.8 µM	4.109	Gao et al. (2019)
Lipase inhibition	Porcine pancreatic lipase		IC_{50}= 187.1 µM	4.95	Jia et al. (2023)
Metabolic marker (gut microbiota)	Nymphs and adults of *P. americana*	principal component analysis, orthogonal partial least squares discriminant analysis	The distinguished metabolic marker between nymphs and adults	4.101	Lv et al. (2021)
Nematicidal	*Pratylenchus penetrans*	—	82.4% activity (at 1 g/L)	4.126	Kusano et al. (2000)

(Continued)

TABLE 4.3 (Continued)
Biological Activity of 2O-THQ Alkaloids

Biological Activity	Target	Assay, Method	Results	Compound	References
"	P. penetrans; Caenorhabditis elegans	–	69.2% activity (at 1 g/L), LD_{50}= 100 mg/L; No effect on C. elegans (selective nematicide)	4.124, 4.125	Kusano et al. (2000)
"	Pratylenchus penetrans	–	57.7% activity (at 1 g/L)	4.140	Kusano et al. (2000)
Neuroprotectivity	Protection on neuron SH-SY5Y cells against amyloid-β(Aβ)-induced injury	Cell viability	Mild neuroprotective effect at 10–40 μM	4.83, 4.84	Sun et al. (2019)
"	MPP-injured primary cerebral cortical neurons	Cell viability by the MTT method	Neuroprotective effect at 6.25–25 μM	4.37	Liu et al. (2023)
"	"	"	Neuroprotective effect at 6.25–25 μM	4.40	Liu et al. (2023)
Neurotrophic activity	Mouse cortical neurons	Mouse primary cultured cortical neurons	Promotes neurite growth at 10 μM: The longest neurites of treated neurons were two times longer than control	4.7	Wu et al. (2020)
Pollen-growth inhibition	Growth of tea pollen tubes	-	40% Pollen-growth inhibition at 10 mg/L, and 100% at 100 mg/L	4.124+4.125	Kimura et al. (1996)
Ecotoxicity	Algae growth inhibition, Daphnia magna immobilization, Vibrio fischeri luminescence inhibition, V. fischeri growth inhibition, Pseudomonas putida growth inhibition	-	EC_{50}= >176.7, 53.1, 4.4, >110.5, >110.5 mg/L, respectively	4.1	Neuwoehner et al. (2009)

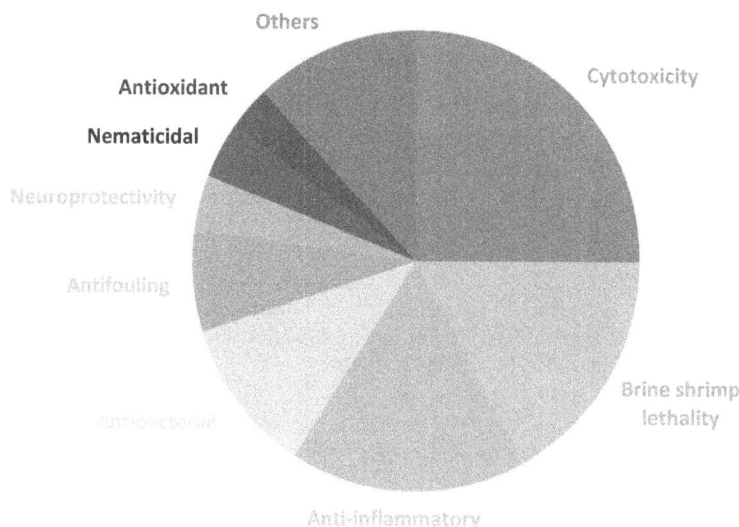

FIGURE 4.5 Biological activity of 2O-THQ alkaloids.

body and how they can be used in medicine. This knowledge allows researchers to design more effective and safer medications based on alkaloid structures. The diverse bioactivities within this class of alkaloids are significant because they can interact with a wide range of biological targets. This diversity is crucial ecologically, enabling plants to employ multiple defense mechanisms against various predators and pathogens. In pharmacology, the broad spectrum of 2O-THQ activities has helped in discovering numerous therapeutic agents, highlighting their importance in medicine. Future directions may include exploring the variety of 2O-THQs, studying how they interact biologically, improving biotechnology for sustainable production, and promoting research across different fields. Personalized medicine may also benefit from tailored 2O-THQ-based treatments, enhancing therapeutic efficacy and safety.

REFERENCES

Abdel-Wahab, N. M., Scharf, S., Özkaya, F. C., Kurtán, T., Mándi, A., Fouad, M. A., Kamel, M. S., Müller, W. E. G., Kalscheuer, R., Lin, W., Daletos, G., Ebrahim, W., Liu, Z., & Proksch, P. (2019). Induction of secondary metabolites from the marine-derived fungus *Aspergillus versicolor* through co-cultivation with bacillus subtilis. *Planta Medica, 85*(6), 503–512. doi: 10.1055/a-0835-2332

An, C., Li, X., Luo, H., Li, C., Wang, M., Xu, G., & Wang, B. (2013). 4-Phenyl-3,4-dihydroquinolone derivatives from *Aspergillus nidulans* MA-143, an endophytic fungus isolated from the mangrove plant *Rhizophora stylosa. Journal of Natural Products, 76*(10), 1896–1901. doi: 10.1021/np4004646

Angiosperm Phylogeny Group, Chase, M. W., Christenhusz, M. J., Fay, M. F., Byng, J. W., Judd, W. S., Soltis, D. E., Mabberley, D. J., Sennikov, A. N., & Soltis, P. S. (2016). An update of the angiosperm phylogeny group classification for the orders and families of flowering plants: APG IV. *Botanical Journal of the Linnean Society, 181*(1), 1–20.

Baeyer, A., & Jackson, O. R. (1880). Ueber die synthese der homologen des hydrocarbostyrils und des chinolins. *Berichte Der Deutschen Chemischen Gesellschaft, 13*(1), 115–123.

Belyagoubi-Benhammou, N., Belyagoubi, L., Gismondi, A., Di Marco, G., Canini, A., & Atik Bekkara, F. (2019). GC/MS analysis, and antioxidant and antimicrobial activities of alkaloids extracted by polar and apolar solvents from the stems of *Anabasis articulata. Medicinal Chemistry Research, 28*(5), 754–767. doi: 10.1007/s00044-019-02332-6

Benakki, H., Colacino, E., Andre, C., Guenoun, F., Martinez, J., & Lamaty, F. (2008). Microwave-assisted multi-step synthesis of novel pyrrolo-[3, 2-c] quinoline derivatives. *Tetrahedron, 64*(25), 5949–5955.

Bernauer, K., Englert, G., Vetter, W., & Weiss, E. (1969). Die konstitution der Melodinus-Alkaloide (+)-Meloscin,(+)-Epimeloscin und (+)-Scandin. 1. Mitteilung über alkaloide aus *melodinus scandens* FORST. *Helvetica Chimica Acta, 52*(7), 1886–1905.

Bhambhani, S., Kondhare, K. R., & Giri, A. P. (2021). Diversity in chemical structures and biological properties of plant alkaloids. *Molecules, 26*(11), 3374.

Bhattarai, H. D., Paudel, B., Chan, K. I., Oh, H., & Yim, J. H. (2013). A new fused tetracyclic heterocyclic antioxidant from *Serratia* sp. PAMC 25557. *Phytochemistry Letters, 6*(4), 536–538. doi: 10.1016/j.phytol.2013.07.006

Bhattarai, K., Kabir, M. E., Bastola, R., & Baral, B. (2021). Fungal natural products galaxy: biochemistry and molecular genetics toward blockbuster drugs discovery. *Advances in Genetics, 107*, 193–284.

Block, E. F., & McChesney, J. D. (1974). Two new tryptophane metabolites of the American cockroach. *Journal of Insect Physiology, 20*(9), 1683–1686. doi: 10.1016/0022-1910(74)90197-8

Buckingham, J. (2015). *Natural products desk reference.* Boca Raton, FL: CRC Press.

Boyd, D.R., Sharma, N.D., & Allen, C.C.R. (2002). Biotransformations in organic synthesis. In S. Terashima & I. Ojima (Eds.), *Comprehensive asymmetric synthesis II* (Vol. 1, pp. 1–32). Elsevier.

Cai, X., Jiang, H., Li, Y., Cheng, G., Liu, Y., Feng, T., & Luo, X. (2011). Cytotoxic indole alkaloids from *Melodinus fusiformis* and *M. morsei. Chinese Journal of Natural Medicines, 9*(4), 259–263. doi: 10.1016/S1875-5364(11)60061-7

Capecchi, A., & Reymond, J. (2021). Classifying natural products from plants, fungi or bacteria using the COCONUT database and machine learning. *Journal of Cheminformatics, 13*, 1–11.

Caty, S. N., Alvarez-Buylla, A., Byrd, G. D., Vidoudez, C., Roland, A. B., Tapia, E. E., Budnik, B., Trauger, S. A., Coloma, L. A., & O'Connell, L. A. (2019). Molecular physiology of chemical defenses in a poison frog. *Journal of Experimental Biology, 222*(12), jeb204149.

Chai, W. (2008). Studies on the chemical constituents and quality standard of *Rubus chingii* Hu. (Doctoral dissertation, PhD Thesis). China Academy of Chinese Medical Science, Beijing, China.

Chazelet, I., Batchily, F., Mehri, H., Plat, M., & Rabaron, A. (1986). Alcaloïdes des fruits de *Melodinus scandens* forster. Paper presented at the *Annales Pharmaceutiques Françaises, 44*(5) 355–362.

Chen, M., Shao, C., Meng, H., She, Z., & Wang, C. (2014). Anti-respiratory syncytial virus prenylated dihydroquinolone derivatives from the gorgonian-derived fungus *Aspergillus* sp. XS-20090B15. *Journal of Natural Products, 77*(12), 2720–2724. doi: 10.1021/np500650t

Chen, M., Gan, L., Lin, S., Wang, X., Li, L., Li, Y., Zhu, C., Wang, Y., Jiang, B., Jiang, J., Yang, Y., & Shi, J. (2012). Alkaloids from the root of *Isatis indigotica. Journal of Natural Products, 75*(6), 1167–1176. doi: 10.1021/np3002833

Chen, Y., Pang, X., He, Y., Lin, X., Zhou, X., Liu, Y., & Yang, B. (2022). Secondary metabolites from coral-associated fungi: Source, chemistry and bioactivities. *Journal of Fungi, 8*(10), 1043.

Choudhari, J., Nimma, R., Nimal, S. K., Totakura Venkata, S. K., Kundu, G. C., & Gacche, R. N. (2023). *Prosopis juliflora* (sw.) DC phytochemicals induce apoptosis and inhibit cell proliferation signaling pathways, EMT, migration, invasion, angiogenesis and stem cell markers in melanoma cell lines. *Journal of Ethnopharmacology, 312*, 116472. doi: 10.1016/j.jep.2023.116472

Chung, C., Hsu, C., Lin, J., Kuo, Y., Chiang, W., & Lin, Y. (2011). Antiproliferative lactams and spiroenone from adlay bran in human breast cancer cell lines. *Journal of Agricultural and Food Chemistry, 59*(4), 1185–1194. doi: 10.1021/jf104088x

Correia, J. T. M., Piva da Silva, G., Kisukuri, C. M., André, E., Pires, B., Carneiro, P. S., & Paixão, M. W. (2020). Metal-free photoinduced hydroalkylation cascade enabled by an electron-donor–acceptor complex. *The Journal of Organic Chemistry, 85*(15), 9820–9834.

Dai, C., Chen, C., Guan, D., Chen, H., Wang, F., Wang, W., Zang, Y., Li, Q., Wei, M., Li, X., Zhang, X., Wang, J., Zhou, Q., Zhu, H., & Zhang, Y. (2020a). Pesimquinolones produced by penicillium simplicissimum and their inhibitory activity on nitric oxide production. *Phytochemistry, 174*, 112327. doi: 10.1016/j.phytochem.2020.112327

Dai, C., Li, X., Zhang, K., Li, X., Wang, W., Zang, Y., Chen, X., Li, Q., Wei, M., & Chen, C. (2021). Pesimquinolones I–S, eleven new quinolone alkaloids produced by *Penicillium simplicissimum* and their inhibitory activity on NO production. *Bioorganic Chemistry, 108*, 104635.

Dai, Z., Wang, X., & Li, G. (2020b). Secondary metabolites and their bioactivities produced by paecilomyces. *Molecules, 25*(21), 5077.

Daudon, M., Mehri, H., Plat, M. M., Hagaman, E. W., Schell, F. M., & Wenkert, E. (1975). Carbon-13 nuclear magnetic resonance spectroscopy of naturally occurring substances. XXXIV. Monomeric quinolinic melodinus alkaloids. *Journal of Organic Chemistry, 40*(19), 2838–2839. doi: 10.1021/jo00907a031

Daudon, M., Mehri, M. H., Plat, M. M., Hagaman, E. W., & Wenkert, E. (1976). Carbon-13 nuclear magnetic resonance spectroscopy of naturally occurring substances. 48. Dimeric quinolinic melodinus alkaloids. *Journal of Organic Chemistry, 41*(20), 3275–3278. doi: 10.1021/jo00882a014

Deng, L., Zong, W., Tao, X., Liu, S., Feng, Z., Lin, Y., Liao, Z., & Chen, M. (2019). Evaluation of the therapeutic effect against benign prostatic hyperplasia and the active constituents from *Epilobium angustifolium* L. *Journal of Ethnopharmacology, 232*, 1–10. doi: 10.1016/j.jep.2018.11.045

dos Santos, D. L., Vieira, J. D. G., de Sousa Fiuza, T., & de Paula, J. R. (2019). Isolation and identification of endophytic microorganisms of *Campomanesia adamantium* (cambess) O. berg (myrtaceae) and their metabolites. *Revista De Biotecnologia & Ciência, 8*(2), 25–42.

Du, K., Li, X., Zheng, C., Lai, L., Shen, M., Wang, Y., & Meng, D. (2023). Monoterpenoid indole alkaloid dimers from the *melodinus axillaris* induce G2/M phase arrest and apoptosis via p38 MAPK activation in HCT116 cells. *Bioorganic Chemistry, 140*, 106841. doi: 10.1016/j.bioorg.2023.106841

Ebada, S. S., & Ebrahim, W. (2020). A new antibacterial quinolone derivative from the endophytic fungus *Aspergillus versicolor* strain eich.5.2.2. *South African Journal of Botany, 134*, 151–155. doi: 10.1016/j.sajb.2019.12.004

Ebada, S. S., El-Neketi, M., Ebrahim, W., Mándi, A., Kurtán, T., Kalscheuer, R., Müller, W. E. G., & Proksch, P. (2018). Cytotoxic secondary metabolites from the endophytic fungus *Aspergillus versicolor* KU258497. *Phytochemistry Letters, 24*, 88–93. doi: 10.1016/j.phytol.2018.01.010

Elissawy, A. M., Soleiman Dehkordi, E., Mehdinezhad, N., Ashour, M. L., & Mohammadi Pour, P. (2021). Cytotoxic alkaloids derived from marine sponges: A comprehensive review. *Biomolecules, 11*(2), 258.

El-Kashef, D. H., Daletos, G., Plenker, M., Hartmann, R., Mandi, A., Kurtan, T., Weber, H., Lin, W., Ancheeva, E., & Proksch, P. (2019). Polyketides and a dihydroquinolone alkaloid from a marine-derived strain of the fungus *Metarhizium marquandii. Journal of Natural Products, 82*(9), 2460–2469.

Fang, L., Tian, S., Zhou, J., Lin, Y., Wang, Z., & Wang, X. (2016). Melaxillines A and B, monoterpenoid indole alkaloids from *Melodinus axillaris. Fitoterapia, 115*, 173–176. doi: 10.1016/j.fitote.2016.10.012

Felpin, F. X., Coste, J., Zakri, C., & Fouquet, E. (2009). Preparation of 2-quinolones by sequential Heck reduction–cyclization (HRC) reactions by using a multitask palladium catalyst. *Chemistry–A European Journal, 15*(29), 7238–7245.

Feng, T., Li, Y., Wang, Y., Cai, X., Liu, Y., & Luo, X. (2010). Cytotoxic indole alkaloids from *Melodinus tenuicaudatus. Journal of Natural Products, 73*(6), 1075–1079. doi: 10.1021/np100086x

Feng, W., Li, Q., Zheng, X., & Kuang, H. (2007). A new alkaloid from the aerial part of *Rabdosia rubescens. Chinese Journal of Natural Medicines, 5*(2), 92–95.

Feng, Z., Zhang, M., Zhang, R., Liu, L., Chi, J., Dong, L., & Jia, X. (2021). Isolation, structural characterization and antioxidant activity of brown rice bound phenolics. *Modern Food Science & Technology, 37*(7), 83–89.

Friedländer, P. (1883). *Untersuchungen über die innern anhydride der orthoamidozimmtsäure und der orthoamidohydrozimmtsäure: Inaug.-diss. d. U. münchen.* Druck von F. Straub.

Fu, Y., Di, Y., He, H., Li, S., Zhang, Y., & Hao, X. (2014). Angustifonines A and B, cytotoxic bisindole alkaloids from *Bousigonia angustifolia. Journal of Natural Products, 77*(1), 57–62. doi: 10.1021/np4005823

Gabriel, S., & Zimmermann, J. (1879). Dinitrohydrozimmtsäure und derivate. *Berichte Der Deutschen Chemischen Gesellschaft, 12*(1), 600–603.

Gabriel, S., & Zimmermann, J. (1880). Ueber dinitrohydrozimmtsäure und derivate derselben. *Berichte Der Deutschen Chemischen Gesellschaft, 13*(2), 1680–1684.

Gao, D., Zhou, T., Da, L., Bruhn, T., Guo, L., Chen, Y., Xu, J., & Xu, M. (2019). Characterization and nonenzymatic transformation of three types of alkaloids from *Streptomyces albogriseolus* MGR072 and discovery of inhibitors of indoleamine 2,3–dioxygenase. *Organic Letters, 21*(21), 8577–8581. doi: 10.1021/acs.orglett.9b03149

Gao, X., Fan, Y., Liu, Q., Cho, S., Pauli, G. F., Chen, S., & Yue, J. (2019). Suadimins A–C, unprecedented dimeric quinoline alkaloids with antimycobacterial activity from *Melodinus suaveolens. Organic Letters, 21*(17), 7065–7068. doi: 10.1021/acs.orglett.9b02630

Gao, X., Wu, P., Fan, Y., Zhou, B., & Yue, J. (2023). Suadimins D—J, monoterpenoid indole-quinoline and bisindole alkaloids from *Melodinus suaveolens. Chinese Journal of Chemistry, 41*(18), 2296–2304. doi: 10.1002/cjoc.202300209

Ge, H. X., & Wang, L. C. (2005). Pharmacological activities and structure-activity relationship of 2 (1h)-quinolinone compounds. *Progress in Pharmaceutical Sciences, 29*, 309–316.

Goldberg, A. F., & Stoltz, B. M. (2011). A palladium-catalyzed vinylcyclopropane (3 + 2) cycloaddition approach to the Melodinus alkaloids. *Organic Letters, 13*(16), 4474–4476.

Guo, L., Yuan, Y., He, H., Li, S., Zhang, Y., & Hao, X. (2017). Melohenryines A and B, two new indole alkaloids from *Melodinus henryi. Phytochemistry Letters, 21*, 179–182. doi: 10.1016/j.phytol.2017.07.001

Guo, L., & Zhou, Y. (1993). Alkaloids from *Melodinus hemsleyanus. Phytochemistry, 34*(2), 563–566.

Hanuman, J. B., & Katz, A. (1993). Isolation of quinolinones from ayurvedic processed root tubers of *Aconitum ferox. Natural Product Letters, 3*(3), 227–231.

Haramiishi, R., Okuyama, S., Yoshimura, M., Nakajima, M., Furukawa, Y., Ito, H., & Amakura, Y. (2020). Identification of the characteristic components in walnut and anti-inflammatory effect of glansreginin A as an indicator for quality evaluation. *Bioscience, Biotechnology, and Biochemistry, 84*(1), 187–197. doi: 10.1080/09168451.2019.1670046

Hayashi, H., Nakatani, T., Inoue, Y., Teraguchi, S., Nakayama, M., & Nozaki, H. (1996). New fungal metabolites toxic to brine shrimps. *Tennen Yuki Kagobutsu Toronkai Koen Yoshishu, 38*, 271–276.

Hayashi, H., Nakatani, T., Inoue, Y., Nakayama, M., & Nozaki, H. (1997). New dihydro-quinolinone toxic to artemia salina produced by *Penicillium* sp. NTC-47. *Bioscience, Biotechnology, and Biochemistry, 61*(5), 914–916. doi: 10.1271/bbb.61.914

He, F., Liu, Z., Yang, J., Fu, P., Peng, J., Zhu, W., & Qi, S. (2012). A novel antifouling alkaloid from halotolerant fungus *Penicillium* sp. OUCMDZ-776. *Tetrahedron Letters, 53*(18), 2280–2283. doi: 10.1016/j.tetlet.2012.02.063

He, J., Lion, U., Sattler, I., Gollmick, F.A., Grabley, S., Cai, J., Meiners, M., Schünke, H., Schaumann, K., Dechert, U., & Krohn, M. (2005). Diastereomeric quinolinone alkaloids from the marine-derived fungus *Penicillium janczewskii*. *Journal of Natural Products, 68*(9), 1397–1399. doi: 10.1021/np058018g

He, J. Y., Bai, Q. F., Li, X., Shou, J., & Feng, G. (2022). Photoredox one-pot synthesis of 3, 4-dihydroquinolin-2 (1H)-ones. *Synlett, 33*(07), 679–683.

He, X., Zhou, Y., & Huang, Z. (1992). Study on the alkaloids of *Melodinus fusiformis*. *Acta Chimica Sinica, 50*(1), 96.

Hong, W. P., Shin, I., & Lim, H. N. (2020). Recent advances in one-pot modular synthesis of 2-quinolones. *Molecules, 25*(22), 5450.

Horn, J., Li, H. Y., Marsden, S. P., Nelson, A., Shearer, R. J., Campbell, A. J., … & Weingarten, G. G. (2009). Convergent synthesis of dihydroquinolones from o-aminoarylboro-nates. *Tetrahedron, 65*(44), 9002–9007.

Hu, Z., Zhu, Y., Chen, J., Chen, J., Li, C., Gao, Z., Li, J., & Liu, L. (2023). Discovery of novel bactericides from *Aspergillus alabamensis* and their antibacterial activity against fish pathogens. *Journal of Agricultural and Food Chemistry, 71*(10), 4298–4305. doi: 10.1021/acs.jafc.2c09141

Hugel, G., & Levy, J. (1986). Methyleneindolines, indolenines, and indoleniniums. 20. The first biomimetic synthesis of scandine and meloscine. *The Journal of Organic Chemistry, 51*(9), 1594–1595.

Hui-Lan, L. C. Z., & Han-Dong, W. S. S. (1994). Quinolinic melodinus alkaloids from stem bark *Melodinus khasianus*. *Plant Diversity, 16*(03), 1.

Intaraudom, C., Dramae, A., Supothina, S., Komwijit, S., & Pittayakhajonwut, P. (2014). 3-oxyanthranilic acid derivatives from *Actinomadura* sp. BCC27169. *Tetrahedron, 70*(17), 2711–2716. doi: 10.1016/j.tet.2014.03.005

Ito, C., Itoigawa, M., Otsuka, T., Tokuda, H., Nishino, H., & Furukawa, H. (2000). Constituents of *Boronia pinnata*. *Journal of Natural Products, 63*(10), 1344–1348. doi: 10.1021/np0000318

Ito, H., Okuda, T., Fukuda, T., Hatano, T., & Yoshida, T. (2007). Two novel dicarboxylic acid derivatives and a new dimeric hydrolyzable tannin from walnuts. *Journal of Agricultural and Food Chemistry, 55*(3), 672–679. doi: 10.1021/jf062872b

Jia, W., Ces, S. V., & Fernandez-Ibanez, M. A. (2021). Divergent total syntheses of yaequi-nolone-related natural products by late-stage C-H olefination. *The Journal of Organic Chemistry, 86*(9), 6259–6277. doi: 10.1021/acs.joc.1c00042

Jia, X., Zhang, M., Zhang, R., Chi, J., Huang, F., Dong, L., Ma, Q., & Zhao, D. (2023). Green preparation of rice bran alkaloid with lipase inhibitory activity, pharmaceutical compo-sition and its application. *Assignee Sericulture & Agri-Food Research Institute, GAAS, China Source China, 202211312127.8* (CN115894358 A), 1–16.

Jiang, J., Zhang, W., & Chen, Y. (2015). Phytochemical and pharmacological properties of the genus melodinus – A review. *Tropical Journal of Pharmaceutical Research, 14*(12), 2325–2344. doi: 10.4314/tjpr.v14i12.25

Jing, W., Sun, X., Lan, Q., & Liu, A. (2019). Nitrogenous chemical constituents from *Brassica campestris* seeds. *Chinese Journal of Experimental Traditional Medical Formulae, 10*, 127–133.

Khadem, S., & Marles, R. J. (2025a). Natural 3, 4-dihydro-2 (1H)-quinolinones-Part I: Plant sources. *Natural Product Research, 39*(3), 593–608.

Khadem, S., & Marles, R. J. (2025b). Natural 3, 4-dihydro-2 (1H)-quinolinones-Part II: Animal, bacterial, and fungal sources. *Natural Product Research, 39*(2), 374–387.

Khadem, S., & Marles, R. J. (2025c). Natural 3, 4-Dihydro-2 (1H)-quinolinones–Part III: Biological activities. *Natural Product Research, 39*(8), 2252–2259.

Kimura, Y., Kusano, M., Koshino, H., Uzawa, J., Fujioka, S., & Tani, K. (1996). Penigequinolones A and B, pollen-growth inhibitors produced by *Penicilium* sp., no. 410. *Tetrahedron Letters, 37*(28), 4961–4964.

Koolen, H., Elizabeth, M. F., Silvia, C., Jane, V. N., Alessandra, F., Alvaro, J., Nathalia, L., & Marcos, J. (2017). Antiprotozoal and antioxidant alkaloids from *Alternanthera littoralis*. *Phytochemistry, 134*, 106–113.

Kovalenko, E., Sagachenko, T., & Kadychagov, P. (2002). Composition of nitrogen-containing bases in petroleums from the middle-jurassic deposits of western Siberia. *Khimiya V Interesakh Ustoichivogo Razvitiya, 10*(3), 313–319.

Kusano, M. (Tottori Univ. (Japan). Faculty of Agriculture), Koshino, H., Uzawa, J., Fujioka, S., Kawano, T., & Kimura, Y. (2000). Nematicidal alkaloids and related compounds produced by the fungus *Penicillium* cf. *simplicissimum. Bioscience, Biotechnology, and Biochemistry, 64*(12), 2559–2568. doi: 10.1271/bbb.64.2559

Larsen, T. O., Smedsgaard, J., Frisvad, J. C., Anthoni, U., & Christophersen, C. (1999). Consistent production of penigequinolone A and B by *Penicillium scabrosum. Biochemical Systematics and Ecology, 27*(3), 329–332. doi: 10.1016/S0305-1978(98)00078-7

Lee, M., Lin, H., Cheng, F., Chiang, W., & Kuo, Y. (2008). Isolation and characterization of new lactam compounds that inhibit lung and colon cancer cells from adlay (*Coix lachryma-jobi* L. var. *ma-yuen* Stapf) bran. *Food and Chemical Toxicology, 46*(6), 1933–1939. doi: 10.1016/j.fct.2008.01.033

Li, C. M., Wu, S. G., Tao, G. D., Zhong, J. Y., You, C., Zhou, Y. L., & Huang, L. (1987). Chemical constituents of Simao shanchen (*Melodinus henryi*). *Zhongcaoyao, 18*, 52–53.

Li, F., Liu, L., Liu, Y., Wang, J., Yang, M., Khan, A., Qin, X., Wang, Y., & Cheng, G. (2021). HRESIMS-guided isolation of aspidosperma-scandine type bisindole alkaloids from *Melodinus cochinchinensis* and their anti-inflammatory and cytotoxic activities. *Phytochemistry, 184*, 112673.

Li, H., Yang, S., Li, X., Li, X., & Wang, B. (2021). Structurally diverse alkaloids produced by *Aspergillus creber* EN-602, an endophytic fungus obtained from the marine red alga *Rhodomela confervoides. Bioorganic Chemistry, 110*, 104822.

Li, M., Xu, L., Li, Z., Qian, S., & Qin, M. (2013). Chemical constituents from *Mentha canadensis. Biochemical Systematics and Ecology, 49*, 144–147. doi: 10.1016/j.bse.2013.03.021

Li, Q., Chen, C., He, Y., Wei, M., Cheng, L., Kang, X., Wang, J., Hao, X., Zhu, H., & Zhang, Y. (2020). Prenylated quinolinone alkaloids and prenylated isoindolinone alkaloids from the fungus *Aspergillus nidulans. Phytochemistry, 169*, 112177.

Li, X., Huo, X., Li, J., She, X., & Pan, X. (2009). A concise synthesis of (±)-Yaequinolone A2. *Chinese Journal of Chemistry, 27*(7), 1379–1381.

Li, Y., Wang, F., Zhang, P. Z., & Yang, M. (2015). Chemical constituents from *Periplaneta americana. Zhong Yao Cai= Zhongyaocai= Journal of Chinese Medicinal Materials, 38*(10), 2038–2041.

Liang, S., Zhang, Y., Li, J., & Yao, S. (2022). Phytochemical profiling, isolation, and pharmacological applications of bioactive compounds from insects of the family blattidae together with related drug development. *Molecules, 27*(24), 8882. doi: 10.3390/molecules27248882

Lien, L. N., Dien, P. G., & Pais, M. (2002a). The alkaloids from the leaves of *Melodinus oblongus* (Apocynacea). *Tap Chi Hoa Hoc, 40*(3), 47–51.

Lien, L. N., Dien, P. G., & Pais, M. (2002b). Study on the alkaloids from the stem barks of *Melodinus oblongus* (Apocynaceae). *Tap Chi Hoa Hoc, 40*(1), 79–85.

Lin, Y., Lin, Y., Lin, Y., & Chiang, C. (2023). Novel compounds of Djulis (*Chenopodium formosanum Koidz*) increases collagen, antioxidants, inhibits adipogenesis. Natural Product Research, 1–10. doi: 10.1080/14786419.2023.2235064

Liu, L., Alam, M. S., Hirata, K., Matsuda, K., & Ozoe, Y. (2008). Actions of quinolizidine alkaloids on *Periplaneta americana* nicotinic acetylcholine receptors. *Pest Management Science: Formerly Pesticide Science, 64*(12), 1222–1228.

Liu, L., Zheng, Y., Shao, C., & Wang, C. (2019). Metabolites from marine invertebrates and their symbiotic microorganisms: Molecular diversity discovery, mining, and application. *Marine Life Science & Technology, 1*(1), 60–94. doi: 10.1007/s42995-019-00021-2

Liu, Y., Li, Y., Cai, X., Li, X., Kong, L., Cheng, G., & Luo, X. (2012). Melodinines M–U, cytotoxic alkaloids from *Melodinus suaveolens. Journal of Natural Products, 75*(2), 220–224.

Liu, Y., Zhao, Y., Feng, T., Cheng, G., Zhang, B., Li, Y., Cai, X., & Luo, X. (2013). Melosuavines A–H, cytotoxic bisindole alkaloid derivatives from *Melodinus suaveolens. Journal of Natural Products, 76*(12), 2322–2329.

Liu, Y., Wang, X., Chen, M., Lin, S., Li, L., & Shi, J. (2016). Three pairs of alkaloid enantiomers from the root of *Isatis indigotica. Acta Pharmaceutica Sinica B, 6*(2), 141–147.

Liu, Z., Song, M., Wang, J., Wang, D., Sun, B., Shi, L., Jiang, R., Ma, M., & Zhang, X. (2023). Monoterpenoid indole alkaloid adducts and dimers from *melodinus fusiformis. Phytochemistry, 211*, 113678.

Lu, F., Liu, P., Yin, S., Zhang, J., & Wang, J. (2014). Two new alkaloids from *Melodinus suaveolens. Natural Product Communications, 9*(10), 1934578X1400901011.

Lu, Y., Khoo, T. J., & Wiart, C. (2014). The genus *Melodinus* (Apocynaceae): Chemical and pharmacological perspectives. *Pharmacology & Pharmacy, 5*(05), 540–550.

Lv, W., Cui, Y., Xue, G., Wang, Z., Niu, L., Chai, X., & Wang, Y. (2021). Potential value and chemical characterization of gut microbiota derived nitrogen containing metabolites in feces from *Periplaneta americana* (L.) at different growth stages. *Scientific Reports, 11*(1), 21191.

Ma, K., Wang, J., Luo, J., & Kong, L. (2015). Six new alkaloids from *Melodinus henryi. Fitoterapia, 100*, 133–138.

Ma, S., Mei, W., Guo, Z., Liu, S., Zhao, Y., Yang, D., Zeng, Y., Jiang, B., & Dai, H. (2013). Two new types of bisindole alkaloid from *Trigonostemon lutescens. Organic Letters, 15*(7), 1492–1495.

Macáková, K., Afonso, R., Saso, L., & Mladěnka, P. (2019). The influence of alkaloids on oxidative stress and related signaling pathways. *Free Radical Biology and Medicine, 134*, 429–444.

Mahmoud, M. M., Abdel-Razek, A. S., Frese, M., Soliman, H. S., Sewald, N., & Shaaban, M. (2018). 3, 4-Dihydro-quinolin-2-one derivatives from extremophilic *Streptomyces* sp. LGE21. *Medicinal Chemistry Research, 27*, 1834–1842.

Matada, B. S., Pattanashettar, R., & Yernale, N. G. (2021). A comprehensive review on the biological interest of quinoline and its derivatives. *Bioorganic & Medicinal Chemistry, 32*, 115973.

Mehri, H., Plat, M., & Potier, P. (1971). Plantes de nouvelle-calédonie. V. *Melodinus scandens* forst. Isolement de dix alcaloïdes monomères. Description de deuz alcaloïdes nouveaux: N-oxy-épiméloscine et méloscandonine. *Annales Pharmaceutiques Françaises, 29*(4), 291–296.

Mehri, H., Diallo, A. O., & Plat, M. (1995). An alkaloid from leaves of *Melodinus scandens*. *Phytochemistry, 40*(3), 1005–1006.

Mieriņa, I., Jure, M., & Stikute, A. (2016). Synthetic approaches to 4-(het) aryl-3, 4-dihydro-quinolin-2 (1 H)-ones. *Chemistry of Heterocyclic Compounds, 52*(8), 509–523.

Meiring, L., Petzer, J. P., & Petzer, A. (2018). A review of the pharmacological properties of 3, 4-dihydro-2(1*H*)-quinolinones. *Mini Reviews in Medicinal Chemistry, 18*(10), 828–836.

Mou, X., Liu, X., Xu, R., Wei, M., Fang, Y., & Shao, C. (2018). Scopuquinolone B, a new mono-terpenoid dihydroquinolin-2(1*H*)-one isolated from the coral-derived *Scopulariopsis* sp. fungus. *Natural Product Research, 32*(7), 773–776.

Nakashima, K., Oyama, M., Ito, T., Akao, Y., Witono, J. R., Darnaedi, D., Tanaka, T., Murata, J., & Iinuma, M. (2012). Novel quinolinone alkaloids bearing a lignoid moiety and related constituents in the leaves of *Melicope denhamii*. *Tetrahedron, 68*(10), 2421–2428.

Nakaya, T. (1995). Anti-brine shrimp substances produced by *Penicillium* sp. NTC-47. *SEIKATSU EISEI (Journal of Urban Living and Health Association), 39*(3), 141–143.

Nanni, V., Canuti, L., Gismondi, A., & Canini, A. (2018). Hydroalcoholic extract of *Spartium junceum* L. flowers inhibits growth and melanogenesis in B16-F10 cells by inducing senescence. *Phytomedicine, 46*, 1–10.

Neff, S. A., Lee, S. U., Asami, Y., Ahn, J. S., Oh, H., Baltrusaitis, J., Gloer, J. B., & Wicklow, D. T. (2012). Aflaquinolones A–G: Secondary metabolites from marine and fungicolous isolates of *Aspergillus* spp. *Journal of Natural Products, 75*(3), 464–472.

Neuwoehner, J., Reineke, A., Hollender, J., & Eisentraeger, A. (2009). Ecotoxicity of quinoline and hydroxylated derivatives and their occurrence in groundwater of a tar-contaminated field site. *Ecotoxicology and Environmental Safety, 72*(3), 819–827.

Niu, Y., Tian, L., Lv, H., & Li, P. (2023). Recent advances for the synthesis of dihydroquino-lin-2(1*H*)-ones via catalytic annulation of α,β-unsaturated *N*-arylamides. *Catalysts, 13*(7), 1105. doi: 10.3390/catal13071105

Ntungwe N, E., Domínguez-Martín, E. M., Roberto, A., Tavares, J., Isca, V., Pereira, P., Cebola, M., & Rijo, P. (2020). Artemia species: An important tool to screen general toxicity samples. *Current Pharmaceutical Design, 26*(24), 2892–2908.

Olofinsan, K., Abrahamse, H., & George, B. P. (2023). Therapeutic role of alkaloids and alka-loid derivatives in cancer management. *Molecules, 28*(14), 5578.

Peng, W., Zeng, G., Song, W., & Tan, N. (2013). A new cytotoxic carbazole alkaloid and two new other alkaloids from *Clausena excavata*. *Chemistry & Biodiversity, 10*(7), 1317–1321.

Pham, J. V., Yilma, M. A., Feliz, A., Majid, M. T., Maffetone, N., Walker, J. R., Kim, E., Cho, H. J., Reynolds, J. M., & Song, M. C. (2019). A review of the microbial production of bioactive natural products and biologics. *Frontiers in Microbiology, 10*, 1404.

Plat, M., Hachem-Mehri, M., Koch, M., Scheidegger, U., & Potier, P. (1970). Structure et stereochimie de la meloscandonine, alcaloide du *Melodinus scandens* forst. *Tetrahedron Letters, 11*(39), 3395–3398.

Reineke, A., Göen, T., Preiss, A., & Hollender, J. (2007). Quinoline and derivatives at a tar oil contaminated site: Hydroxylated products as indicator for natural attenuation? *Environmental Science & Technology, 41*(15), 5314–5322.

Richter, K., Birkenbeil, H., & Adam, G. (1989). Moult-inhibiting effect of the steroid alkaloid conessine in the cockroach *Periplaneta americana*. *Naturwissenschaften, 76*(7), 333–334.

Salib, J. Y., El-Toumy, S. A., Hassan, E. M., Shafik, N. H., Abdel-Latif, S. M., & Brouard, I. (2014). New quinoline alkaloid from *ruta graveolens* aerial parts and evaluation of the antifertility activity. *Natural Product Research, 28*(17), 1335–1342.

Salwan, R., & Sharma, V. (2020). Bioactive compounds of streptomyces: Biosynthesis to applications. *Studies in Natural Products Chemistry, 64*, 467–491.

Scherlach, K., & Hertweck, C. (2006). Discovery of aspoquinolones A–D, prenylated quino-line-2-one alkaloids from *aspergillus nidulans*, motivated by genome mining. *Organic & Biomolecular Chemistry, 4*(18), 3517–3520.

Schmeda-Hirschmann, G., Hormazabal, E., Astudillo, L., Rodriguez, J., & Theoduloz, C. (2005). Secondary metabolites from endophytic fungi isolated from the Chilean gymnosperm *Prumnopitys andina* (lleuque). *World Journal of Microbiology and Biotechnology, 21*, 27–32.

Selim, M. S. M., Abdelhamid, S. A., & Mohamed, S. S. (2021). Secondary metabolites and biodiversity of actinomycetes. *Journal of Genetic Engineering and Biotechnology, 19*(1), 72.

Shang, X., Morris-Natschke, S. L., Liu, Y., Guo, X., Xu, X., Goto, M., Li, J., Yang, G., & Lee, K. (2018a). Biologically active quinoline and quinazoline alkaloids part I. *Medicinal Research Reviews, 38*(3), 775–828.

Shang, X., Morris-Natschke, S. L., Yang, G., Liu, Y., Guo, X., Xu, X., Goto, M., Li, J., Zhang, J., & Lee, K. (2018b). Biologically active quinoline and quinazoline alkaloids part II. *Medicinal Research Reviews, 38*(5), 1614–1660.

Shao, C. L., Chao, R., Xu, R. F., Cao, F., & Wei, M. Y. (2016). Scopuquinolone A, a new terpenoids dihydroquinolone alkaloid from a gorgonian coral-derived *Scopulariopsis* sp. fungus. *Chinese Journal of Marine Drugs, 35*, 1–5.

Shao, C., Xu, R., Wang, C., Qian, P., Wang, K., & Wei, M. (2015). Potent antifouling marine dihydroquinolin-2(1H)-one-containing alkaloids from the gorgonian coral-derived fungus *Scopulariopsis* sp. Marine Biotechnology, *17*(4), 408–415. doi: 10.1007/s10126-015-9628-x

Simonetti, S. O., Larghi, E. L., & Kaufman, T. S. (2016). The 3,4-dioxygenated 5-hydroxy-4-aryl-quinolin-2(1*H*)-one alkaloids. Results of 20 years of research, uncovering a new family of natural products. *Natural Product Reports, 33*(12), 1425–1446. doi: 10.1039/c6np00064a

Sun, J., Yu, J., Song, J., Jiang, C., & Zhang, H. (2019). Two new quinolone alkaloids from *dianthus superbus* var. superbus. *Tetrahedron Letters, 60*(2), 161–163.

Tashima, T. (2015). The structural use of carbostyril in physiologically active substances. *Bioorganic & Medicinal Chemistry Letters, 25*(17), 3415–3419. doi: 10.1016/j.bmcl.2015.06.027

Teng, S., Li, F., Cui, Q., Khan, A., He, T., Luo, X., Liu, Y., & Cheng, G. (2023). A review on the genus *Melodinus*: Traditional uses, phytochemical diversity and pharmacological activities of indole alkaloids. *Phytochemistry Reviews*, 1–54. doi: 10.1007/s11101-023-09871-2

Tsuritani, T., Yamamoto, Y., Kawasaki, M., & Mase, T. (2009). Novel approach to 3, 4-dihydro-2 (1H)-quinolinone derivatives via cyclopropane ring expansion. *Organic Letters, 11*(5), 1043–1045.

Uchida, R., Imasato, R., Shioml, K., Tomoda, H., & Ōmura, S. (2005). Yaequinolones J$_1$ and J$_2$, novel insecticidal antibiotics from *Penicillium* sp. FKI-2140. *Organic Letters, 7*(25), 5701–5704.

Uchida, R., Imasato, R., Tomoda, H., & Ōmura, S. (2006). Yaequinolones, new insecticidal antibiotics produced by *Penicillium* sp. FKI-2140. *The Journal of Antibiotics, 59*(10), 652–658.

Wang, C., Huang, X., Xiao, H., Hao, Y., Xu, L., Yan, Q., Zou, Z., Xie, C., Xu, Y., & Yang, X. (2021). Chemical constituents of the marine fungus *Penicillium* sp. MCCC 3A00228. *Chemistry & Biodiversity, 18*(10), e2100697.

Wang, J., Liang, L., Li, F., Khan, A., Liu, L., Cao, J., Cheng, G., & Zhao, T. (2019). Chemical constituents of *Melodinus hemsleyanus* diels. *Biochemical Systematics and Ecology, 84*, 71–74.

Wang, S., Li, L., Wu, Y., & Wu, J. (2011). Alkaloids from *Melodinus yunnanensis*. *Shizhen Guoyi Guoyao, 22*(10), 2356–2357. doi: 10.3969/j.issn.1008-0805.2011.10.017

Wu, F., Chen, Y., Yin, J., Zhu, C., Zhang, Y., Liu, J., & He, J. (2022). Monoterpenoid quinoline alkaloids from stems and leaves of *Melodinus tenuicaudatus*. *Chemistry & Biodiversity, 19*(7), e202200209.

Wu, J., Zhao, S., Shi, B., Bao, M., Schinnerl, J., & Cai, X. (2020). Cage-monoterpenoid quinoline alkaloids with neurite growth promoting effects from the fruits of *Melodinus yunnanensis*. *Organic Letters, 22*(19), 7676–7680.

Wu, T., Salim, A. A., Bernhardt, P. V., & Capon, R. J. (2022). Molecular networking and cultivation profiling reveals diverse natural product classes from an Australian soil-derived fungus *Aspergillus* sp. CMB-MRF324. *Molecules, 27*(24), 9066.

Wu, Y., & Piao, H. (2013). The progress towards the development of DHQO derivatives and related analogues with inotropic effects. *Mini Reviews in Medicinal Chemistry, 13*(12), 1801–1811.

Wubshet, S. G., Nyberg, N. T., Tejesvi, M. V., Pirttilä, A. M., Kajula, M., Mattila, S., & Staerk, D. (2013). Targeting high-performance liquid chromatography–high-resolution mass spectrometry–solid-phase extraction–nuclear magnetic resonance analysis with high-resolution radical scavenging profiles – Bioactive secondary metabolites from the endophytic fungus *Penicillium namyslowskii*. *Journal of Chromatography A, 1302*, 34–39.

Xie, D., & Zhang, S. (2022). Selective reduction of quinolinones promoted by a SmI2/H2O/MeOH system. *The Journal of Organic Chemistry, 87*(13), 8757–8763.

Xu, G., & Chen, L. (2023). Biological activities and ecological significance of fire ant venom alkaloids. *Toxins, 15*(7), 439.

Xu, Z., Xi, Y., Zhou, W., Lou, L., Wang, X., Huang, X., & Song, S. (2020). Alkaloids and monoterpenes from the leaves of *Isatis tinctoria* Linnaeus and their chemotaxonomic significance. *Biochemical Systematics and Ecology, 92*, 104089.

Yan, K., & Feng, X. (1998). Chemical constituents of *Melodinus hemsleyanus*. *Zhongcaoyao, 29*(12), 793–795.

Yang, M., Wang, Y., Fan, Z., Xue, Q., Njateng, G. S. S., Liu, Y., Cao, J., Khan, A., & Cheng, G. (2021). Chemical constituents and anti-inflammatory activity of the total alkaloid extract from *Melodinus cochinchinensis* (Lour.) Merr. and its inhibition of the NF-κB and MAPK signaling pathways. *Phytomedicine, 91*, 153684.

Yang, M., Wang, Y., Fan, Z., Xue, Q., Njateng, G. S. S., Liu, Y., Cao, J., Zhao, T., & Cheng, G. (2020). Acute and sub-acute toxicological evaluations of bioactive alkaloidal extract from *Melodinus henryi* and their main chemical constituents. *Natural Products and Bioprospecting, 10*, 227–241.

Yi, W., Chen, D., Ding, X., Li, X., Li, S., Di, Y., Zhang, Y., & Hao, X. (2018). Cytotoxic indole alkaloids from *Melodinus khasianus* and *Melodinus tenuicaudatus*. *Fitoterapia, 128*, 162–168.

Yin, J., Mao, X., Hu, J., & Song, Y. (2018). Cytotoxic monoterpenoid-type alkaloids from the aerial parts of *Melodinus morsei*. *Journal of Asian Natural Products Research, 20*(6), 525–530.

Youssef, F. S., Alshammari, E., & Ashour, M. L. (2021). Bioactive alkaloids from genus *Aspergillus*: Mechanistic interpretation of their antimicrobial and potential SARS-CoV-2 inhibitory activity using molecular modelling. *International Journal of Molecular Sciences, 22*(4), 1866.

Yu, G., Luo, Z., Wang, W., Li, Y., Zhou, Y., & Shi, Y. (2019a). *Rubus chingii* Hu: A review of the phytochemistry and pharmacology. *Frontiers in Pharmacology, 10*(799), 1–22. doi: 10.3389/fphar.2019.00799

Yu, J., Sun, X., Wang, Z., Fang, L., & Wang, X. (2019b). Alkaloids from *Melodinus henryi* with anti-inflammatory activity. *Journal of Asian Natural Products Research, 21*(8), 820–825.

Zhang, B., Liu, C., Bao, M., Zhong, X., Ni, L., Wu, J., & Cai, X. (2017). Novel monoterpenoid indole alkaloids from *Melodinus yunnanensis*. *Tetrahedron, 73*(40), 5821–5826.

Zhang, D., Shi, Y., Xu, R., Du, K., Guo, F., Chen, K., Li, Y., & Wang, R. (2019). Alkaloid enantiomers from the roots of *Isatis indigotica*. *Molecules, 24*(17), 3140.

Zhang, H., & Curran, D. P. (2011). A short total synthesis of (±)-epimeloscine and (±)-meloscine enabled by a cascade radical annulation of a divinylcyclopropane. *Journal of the American Chemical Society*, 133(27), 10376–10378.

Zhang, H., Yan, Y., Wang, D., Lv, Q., Cheng, Y., & Wang, S. (2021). Small molecule constituents of *Periplaneta americana* and their IL-6 inhibitory activities. *Natural Product Communications, 16*(9), 1934578X211033180.

Zhang, J., Huang, X., Jiang, R., Wang, Y., Sun, P., Fan, R., Zhang, X., & Ye, W. (2016). Melohemsines AI, melodinus-type alkaloids from *Melodinus hemsleyanus*. *RSC Advances, 6*(95), 92218–92224.

Zhang, J., Liu, Z., Li, Y., Wei, C., Xie, J., Yuan, M., Zhang, D., Ye, W., & Zhang, X. (2020). Structurally diverse indole alkaloids with vasorelaxant activity from *Melodinus hemsleyanus*. *Journal of Natural Products, 83*(8), 2313–2319.

Zhang, P., Zhang, Y., Gu, J., & Zhang, G. (2016). Two new alkaloids from *Melodinus hemsleyanus* Diels. *Natural Product Research, 30*(2), 162–167.

Zhang, S., Hu, D., He, J., Guan, K., & Zhu, H. (2014). A novel tetrahydroquinoline acid and a new racemic benzofuranone from *Capparis spinosa* L., a case study of absolute configuration determination using quantum methods. *Tetrahedron, 70*(4), 869–873.

Zhang, T., Liu, Z., Wang, W., Tong, Y., Xu, F., Yuan, J., Liu, B., Zhang, X., & Ye, W. (2013). Alkaloids from *Melodinus suaveolens*. *Heterocycles, 87*(10), 2047–2052.

Zhang, X., Yin, Q., Li, X., Liu, X., Lei, H., & Wu, B. (2022). Structures and bioactivities of secondary metabolites from Penicillium genus since 2010. *Fitoterapia, 163*, 105349.

Zhou, Y. L., Ye, J. H., Li, Z. M., & Huang, Z. H. (1988). Study on the alkaloids of *Melodinus tenuicaudatus*. *Planta Medica, 54*(04), 315–317.

5 2-Quinolinones

5.1 INTRODUCTION

The compound 2-quinolinone [also known as 2-quinolone, carbostyril, and 2(1*H*)-quinolinone] (Figure 5.1) is valuable in natural and synthetic substances, including drugs and fluorescent materials. Its unique structure allows for the design of diverse 2-quinolinone-derived compounds with adaptable steric structures (Silva et al. 2022). This makes it suitable for interacting with the active sites of pharmaceutical target molecules. The presence of the 2-quinolinone motif in drugs suggests favourable pharmacokinetic features for its derivatives. Additionally, the benzene ring in 2-quinolinone can interact with aromatic rings found in peptides or nucleic acids. These interactions create non-covalent bonds through π-π stacking. The internal amide, which is in a fixed *cis* (Z) form, can also form hydrogen bonds with peptides, nucleic acids, or water molecules (Tashima 2005). As part of the α,β-unsaturated carbonyls, 2-quinolinones are inherent Michael acceptors. Although the unsaturated amide functionality in 2-quinolinone is not as electrophilically active as unsaturated aldehydes and ketones, it can interact with biological nucleophilic residues, such as proteins, in living organisms (Liang et al. 2022; Khadem et al. 2004). This interaction can lead to the formation of complexes capable of modulating protein pathways to fulfill specific physiological functions.

2-Quinolinone can exist in equilibrium with 2-hydroxyquinoline as tautomers (Figure 5.2). However, the equilibrium between the amide (lactam) and the hydroxyimine (lactim) forms favours the amide form. Various spectroscopic analyses, calculations of dissociation constants and dipole moments, and computational methods have all confirmed that the dominant form is 2-quinolinone over 2-hydroxyquinoline (Mirek and Sygula 1982; Pan et al. 2014; Volle et al. 2008; Hong et al. 2020; Li et al. 2020).

Synthetic 2-quinolinones and their derivatives exhibit several biological activities (Aly et al. 2021). They have significant anticancer activity, enhancing natural killer (NK) lymphocyte activity, macrophage cytotoxicity, and inhibiting

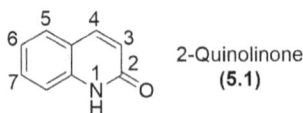

FIGURE 5.1 Chemical structure and numbering system for 2-quinolinone.

FIGURE 5.2 Tautomeric forms of 2-quinolinone.

DOI: 10.1201/9781003534600-5

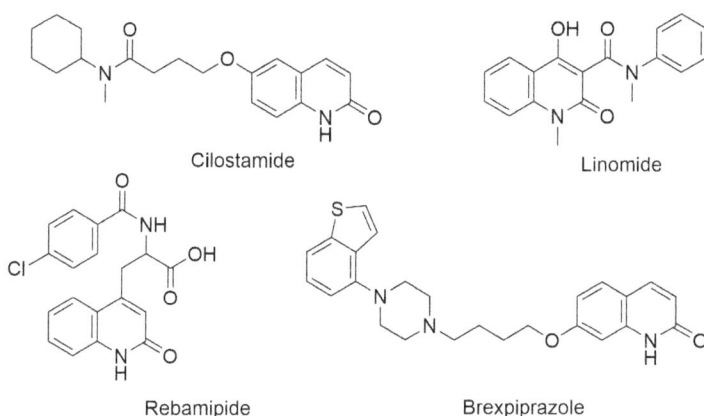

FIGURE 5.3 Chemical structures of Cilostamide, Linomide, Rebamipide, and Brexpiprazole.

angiogenesis. Regarding antibacterial activity, 2-quinolinones are effective against both Gram-positive and Gram-negative bacteria, inhibiting DNA gyrase, a crucial enzyme in bacterial replication. Their antifungal activity includes the inhibition of chitin synthase. Additionally, 2-quinolinone derivatives possess antiviral activity, inhibiting the replication of hepatitis C virus and HIV-1 integrase and reverse transcriptase. They also demonstrate anticonvulsant activity, acting as NMDA receptor antagonists with sedative effects comparable to diazepam. Moreover, the antioxidant activity of 2-quinolinones includes radical scavenging and metal chelating. Several drugs contain a 2-quinolinone skeleton, such as Cilostamide, Linomide, Rebamipide, and Brexpiprazole, each with its unique therapeutic application (Figure 5.3). Cilostamide is a selective phosphodiesterase 3 inhibitor, which increases cyclic adenosine monophosphate levels, leading to vasodilation and inhibition of platelet aggregation. Linomide, also known as roquinimex, is an immunomodulatory drug investigated for treating autoimmune diseases such as multiple sclerosis, but was discontinued due to severe side effects. Rebamipide is a gastroprotective agent that promotes mucosal healing and increases mucus production, and it is commonly used to treat gastric ulcers and gastritis. Brexpiprazole is an atypical antipsychotic that acts as a partial agonist at serotonin 5-HT1A and dopamine D2 receptors and is used to treat schizophrenia and as an adjunct in major depressive disorder.

In our investigation aimed at determining the natural presence and bioactivity of 2-quinolinones, we identified approximately 170 compounds with a simple (bicyclic) 2-quinolinone framework.

5.1.1 CHEMICAL IDENTITY

The chemical identity (structure, name, and CAS number) of 2-quinolinone alkaloids in this chapter is shown in Table 5.1. Some of the reliable literature sources were reviewed to verify the accuracy of this data (e.g. Buckingham et al. 2010; Buckingham 2023).

TABLE 5.1
The Chemical Identity of 2-quinolinone Alkaloids

Structure (Number)	Systematic Name	Common Name	CAS #
(5.1)	2(1H)-Quinolinone	2-Quinolinone (Carbostyril)	59-31-4
(5.2)	3-[(2S)-2,3-Dihydroxy-3-methylbutyl]-4-methoxy-1-methyl-2(1H)-quinolinone	(S)-Edulinine	53585-47-0
(5.3)	(+)-4-(2,3-Dihydroxy-3-methylbutoxy)-1-methyl-2(1H)-quinolinone	—	80357-89-7
(5.4)	1,2-Dihydro-1-mercapto-2-oxo-4-quinolinecarboxaldehyde	—	402934-24-1

(Continued)

TABLE 5.1 (Continued)
The Chemical Identity of 2-quinolinone Alkaloids

Structure (Number)	Systematic Name	Common Name	CAS #
(5.5)	1,2-Dihydro-2-oxo-4-quinolinecarboxylic acid	–	15733-89-8
(5.6)	Methyl 1,2-dihydro-4-methoxy-1-methyl-2-oxo-3-quinolinecarboxylate	–	79966-24-8
(5.7)	1,2-Dihydro-8-methoxy-2-oxo-4-quinolinecarboxylic acid	–	37749-17-0
(5.8)	1-Methyl-2(1H)-quinolinone	–	606-43-9

(Continued)

TABLE 5.1 (Continued)
The Chemical Identity of 2-quinolinone Alkaloids

Structure (Number)	Systematic Name	Common Name	CAS #
(5.9)	4-Quinolinecarboxylic acid, 1,2-dihydro-2-oxo-, methyl ester	—	39497-01-3
(5.10)	3-(1,2-Dihydroxyethyl)-4-methoxy-2(1H)-quinolinone	—	1845748-12-0
(5.11)	3-(2,3-Dihydroxy-2-methylbutyl)-4,8-dimethoxy-1-methyl-2(1H)-quinolinone	—	1194691-62-7
(5.12)	3-(2,3-Dihydroxy-3-methylbutyl)-4,7-dimethoxy-1-methyl-2(1H)-quinolinone	—	894351-36-1

(Continued)

TABLE 5.1 (*Continued*)
The Chemical Identity of 2-quinolinone Alkaloids

Structure (Number)	Systematic Name	Common Name	CAS #
(5.13)	3-(2,3-Dihydroxy-3-methylbutyl)-4,8-dimethoxy-1-methyl-2(1H)-quinolinone	—	101599-43-3
(5.14)	3-(2-Hydroxyethyl)-4,8-dimethoxy-1-methyl-2(1H)-quinolinone	—	107522-48-5
(5.15)	3-(3-Methyl-2-buten-1-yl)-4-[(3-methyl-2-buten-1-yl)oxy]-2(1H)-quinolinone	—	18118-29-1

(Continued)

TABLE 5.1 (Continued)
The Chemical Identity of 2-quinolinone Alkaloids

Structure (Number)	Systematic Name	Common Name	CAS #
(5.16)	3,4,8-Trimethoxy-2(1H)-quinolinone	—	164163-88-6
(5.17)	3,4-Dimethyl-2(1H)-quinolinone	—	17336-90-2
(5.18)	3,6-Dimethoxy-4-phenyl-2(1H)-quinolinone	—	1899094-63-3
(5.19)	3-[(1R)-2-(β-D-Glucopyranosyloxy)-1-hydroxyethyl]-4-meth oxy- 2(1H)-quinolinone	—	1393580-72-7

(Continued)

TABLE 5.1 (*Continued*)

The Chemical Identity of 2-quinolinone Alkaloids

Structure (Number)	Systematic Name	Common Name	CAS #
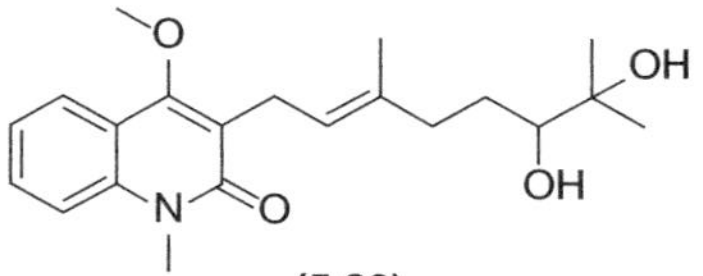 **(5.20)**	3-[(2E)-6,7-Dihydroxy-3,7-dimethyl-2-octen-1-yl]-4-methoxy-1-methyl-2(1H)-quinolinone	—	1174716-86-9
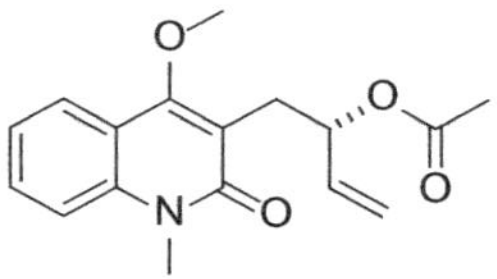 **(5.21)**	3-[(2S)-2-(Acetyloxy)-3-buten-1-yl]-4-methoxy-1-methyl-2(1H)-quinolinone	—	953079-52-2
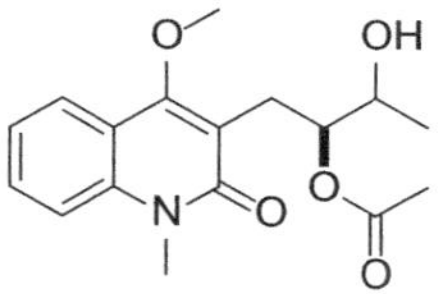 **(5.22)**	3-[(2S)-2-(Acetyloxy)-3-hydroxybutyl]-4-methoxy-1-methyl-2(1H)-quinolinone	—	953079-54-4
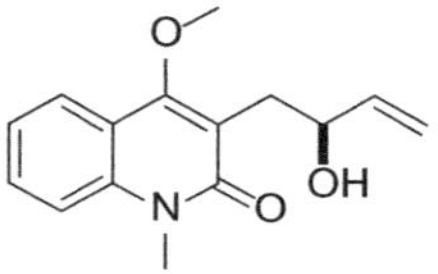 **(5.23)**	3-[(2S)-2-Hydroxy-3-buten-1-yl]-4-methoxy-1-methyl-2(1H)-quinolinone	—	953079-56-6

(*Continued*)

TABLE 5.1 (Continued)
The Chemical Identity of 2-quinolinone Alkaloids

Structure (Number)	Systematic Name	Common Name	CAS #
(5.24)	3-[(3,3-Dimethyl-2-oxiranyl)methyl]-4-hydroxy-1-methyl-2(1H)-quinolinone	—	316148-46-6
(5.25)	3-[(3,3-Dimethyl-2-oxiranyl) methyl]-4-hydroxy-7-methoxy-1-methyl- 2(1H)-quinolinone	—	894351-35-0
(5.26)	3-[[(2R)-3,3-Dimethyl-2-oxiranyl] methyl]-4-methoxy-1-methyl- 2(1H)-quinolinone	—	1384841-46-6
(5.27)	3-Ethyl-4-hydroxy-7-methoxy-2(1H)-quinolinone	—	22048-12-0

(Continued)

TABLE 5.1 (Continued)
The Chemical Identity of 2-quinolinone Alkaloids

Structure (Number)	Systematic Name	Common Name	CAS #
(5.28)	3-Hydroxy-2(1H)-quinolinone	—	26386-86-7
(5.29)	3-Hydroxy-4-(3-methoxyphenyl)-2(1H)-quinolinone	—	94298-63-2
(5.30)	3-Hydroxy-4-methyl-2(1H)-quinolinone	—	24186-98-9
(5.31)	3-Methoxy-4-phenyl-2(1H)-quinolinone	3-O-Methylviridicatin	6152-57-4

(Continued)

TABLE 5.1 (*Continued*)
The Chemical Identity of 2-quinolinone Alkaloids

Structure (Number)	Systematic Name	Common Name	CAS #
(5.32)	4-(3-Hydroxyphenyl)-3-methoxy-2(1H)-quinolinone	3-*O*-Methylviridicatol	1620963-31-6
(5.33)	4-(4-Hydroxyphenyl)-2(1H)-quinolinone	Isatisindigoticanine F	1402932-05-1
(5.34)	10,11-Dihydrocinchonan-2′,9(1′H)-dione	—	916163-80-9

(Continued)

TABLE 5.1 (*Continued*)
The Chemical Identity of 2-quinolinone Alkaloids

Structure (Number)	Systematic Name	Common Name	CAS #
(5.35)	4-Quinolinecarboxylic acid, 1,2-dihydro-2-oxo-, ethyl ester	—	5466-27-3
(5.36)	4-(β-D-Glucopyranosyloxy)-3-hydroxy-2(1H)-quinolinone	—	1144037-45-5
(5.37)	4,6,8-Trimethoxy-1-methyl-3-(3-methyl-2-buten-1-yl)-2(1H)-quinolinone	—	57499-55-5

(Continued)

TABLE 5.1 (Continued)
The Chemical Identity of 2-quinolinone Alkaloids

Structure (Number)	Systematic Name	Common Name	CAS #
(5.38)	4,6-Dimethoxy-1-methyl-2(1H)-quinolinone	—	52345-94-5
(5.39)	4,7,8-Trimethoxy-1-methyl-2(1H)-quinolinone	N-Methylhaplobungine	43215-43-6
(5.40)	3-Hydroxy-4-(4-methoxyphenyl)-2(1H)-quinolinone	4'-Methoxyviridicatin	99557-34-3

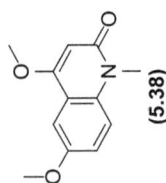

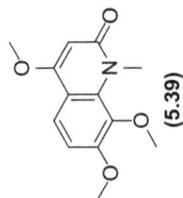

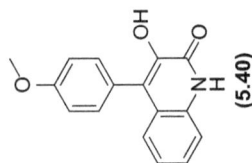

(Continued)

TABLE 5.1 (Continued)
The Chemical Identity of 2-quinolinone Alkaloids

Structure (Number)	Systematic Name	Common Name	CAS #
(5.41)	4-Acetyl-6-methyl-7-(1-methylethyl)-2(1H)-quinolinone	—	2676925-32-7
(5.42)	4-Acetyl-7-isopropyl-5-methylquinolin-2(1H)-one	—	—
(5.43)	Ethyl 1,2-dihydro-6-hydroxy-2-oxo-4-quinolinecarboxylate	—	872267-31-7

(Continued)

TABLE 5.1 (Continued)
The Chemical Identity of 2-quinolinone Alkaloids

Structure (Number)	Systematic Name	Common Name	CAS #
(5.44)	4-Hydroxy-1-methyl-2(1H)-quinolinone	—	1677-46-9
(5.45)	4-Hydroxy-1-methyl-3-(3-methyl-2-buten-1-yl)-2(1H)-quinolinone	—	15954-07-1
(5.46)	4-Hydroxy-2(1H)-quinolinone	—	86-95-3
(5.47)	4-Hydroxy-3-(2-hydroxy-3-methylbutyl)-8-methoxy-1-methyl-2(1H)-quinolinone	—	94621-58-6

(Continued)

TABLE 5.1 (Continued)
The Chemical Identity of 2-quinolinone Alkaloids

Structure (Number)	Systematic Name	Common Name	CAS #
(5.48)	4-Hydroxy-3-methoxy-1-methyl-2(1H)-quinolinone	—	90061-39-5
(5.49)	4-Hydroxy-3-methoxy-2(1H)-quinolinone	—	15151-56-1
(5.50)	4-Hydroxy-3-methyl-2(1H)-quinolinone	—	1873-59-2
(5.51)	4-Hydroxy-6-methoxy-2(1H)-quinolinone	—	14300-45-9

(Continued)

TABLE 5.1 (*Continued*)

The Chemical Identity of 2-quinolinone Alkaloids

Structure (Number)	Systematic Name	Common Name	CAS #
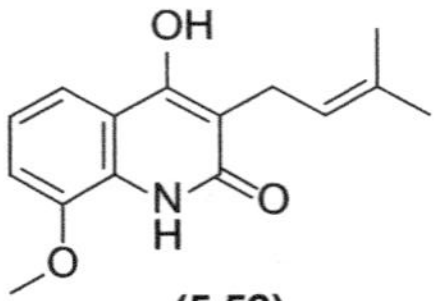 (5.52)	4-Hydroxy-8-methoxy-3-(3-methyl-2-buten-1-yl)-2(1H)-quinolinone	—	7691-03-4
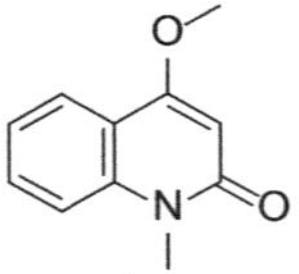 (5.53)	4-Methoxy-1-methyl-2(1H)-quinolinone	—	32262-18-3
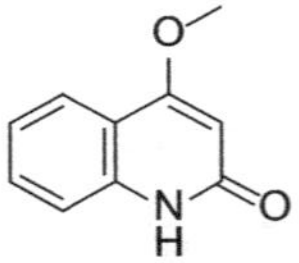 (5.54)	4-Methoxy-2(1H)-quinolinone	—	27667-34-1
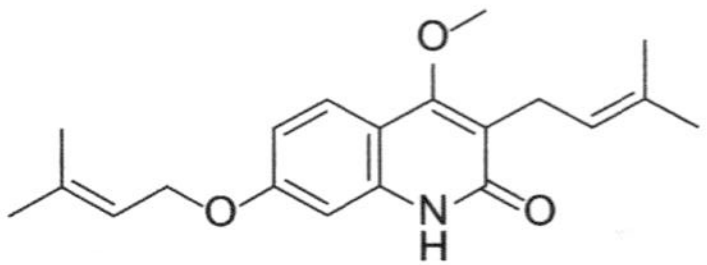 (5.55)	4-Methoxy-3-(3-methyl-2-buten-1-yl)-7-[(3-methyl-2-buten-1-yl) oxy]-2(1H)-quinolinone	—	2583703-09-5

(Continued)

TABLE 5.1 (*Continued*)
The Chemical Identity of 2-quinolinone Alkaloids

Structure (Number)	Systematic Name	Common Name	CAS #
(5.56)	4-Quinolinecarboxylic acid, 1,2-dihydro-2-oxo-, methyl ester	–	39497-01-3
(5.57)	4-Phenyl-2(1H)-quinolinone	–	5855-57-2
(5.58)	5-Hydroxy-4-methyl-2(1H)-quinolinone	–	131195-67-0

(*Continued*)

TABLE 5.1 (Continued)
The Chemical Identity of 2-quinolinone Alkaloids

Structure (Number)	Systematic Name	Common Name	CAS #
(5.59)	5-Methoxy-1-methyl-2(1H)-quinolinone	—	64330-12-7
(5.60)	6,7-Dibromo-4-hydroxy-2(1H)-quinolinone	—	376598-45-7
(5.61)	6-Bromo-4-hydroxy-2(1H)-quinolinone	—	54675-23-9
(5.62)	6-Hydroxy-4-[5-(hydroxymethyl)-2-furanyl]-2(1H)-quinolinone	—	1380540-72-6

(Continued)

TABLE 5.1 (Continued)
The Chemical Identity of 2-quinolinone Alkaloids

Structure (Number)	Systematic Name	Common Name	CAS #
(5.63)	6-Hydroxy-4-methyl-2(1H)-quinolinone	–	34982-01-9
(5.64)	7-[(3,7-Dimethyl-2,6-octadien-1-yl)oxy]-2(1H)-quinolinone	–	1640102-00-6
(5.65)	7-Hydroxy-3-methoxy-4-phenyl-2(1H)-quinolinone	–	1993466-13-9
(5.66)	7-Hydroxy-4-[5-(hydroxymethyl)-2-furanyl]-2(1H)-quinolinone	–	460719-15-7

(Continued)

TABLE 5.1 (Continued)
The Chemical Identity of 2-quinolinone Alkaloids

Structure (Number)	Systematic Name	Common Name	CAS #
(5.67)	7-Hydroxy-6-methoxy-2(1H)-quinolinone	—	2248627-10-1
(5.68)	7-Hydroxy-6-methyl-4-(4-methylfuran-2-yl)quinolin-2(1H)-one	—	—
(5.69)	7-Methoxy-6-methyl-4-(4-methylfuran-2-yl)quinolin-2(1H)-one	—	—

(Continued)

TABLE 5.1 (Continued)
The Chemical Identity of 2-quinolinone Alkaloids

Structure (Number)	Systematic Name	Common Name	CAS #
(5.70)	8-(β-D-Glucopyranosyloxy)-4-methoxy-1-methyl-2(1H)-quinolinone	—	780825-79-8
(5.71)	8-(β-D-Glucopyranosyloxy)-4-methoxy-2(1H)-quinolinone	—	780825-80-1

(*Continued*)

TABLE 5.1 (*Continued*)
The Chemical Identity of 2-quinolinone Alkaloids

Structure (Number)	Systematic Name	Common Name	CAS #
(5.72)	8-Hydroxy-3-methoxy-2(1H)-quinolinone	—	1515871-97-2
(5.73)	8-Hydroxy-4-methoxy-2(1H)-quinolinone	—	82613-19-2
(5.74)	(-)-3-[2-(Acetyloxy)-3-hydroxy-3-methylbutyl]-4-methoxy-1-methyl-2(1H)-quinolinone	—	207603-52-9
(5.75)	(-)-3-[2-(Acetyloxy)-3-methyl-3-buten-1-yl]-4-methoxy-1-methyl-2(1H)-quinolinone	—	207603-51-8

(Continued)

TABLE 5.1 (*Continued*)
The Chemical Identity of 2-quinolinone Alkaloids

Structure (Number)	Systematic Name	Common Name	CAS #
(5.76)	8-(Acetyloxy)-4-methoxy-1-methyl-2(1H)-quinolinone	Acusine	3148-26-3
(5.77)	(+)-3-(2-Hydroxy-3-methyl-3-buten-1-yl)-4,8-dimethoxy-2(1H)-quinolinone	Acutifolidine	145237-08-7
(5.78)	3-[(2S)-2-Hydroxy-3-methyl-3-buten-1-yl]-4,8-dimethoxy-1-methyl-2(1H)-quinolinone	Acutifoline	145237-07-6

(*Continued*)

TABLE 5.1 (Continued)
The Chemical Identity of 2-quinolinone Alkaloids

Structure (Number)	Systematic Name	Common Name	CAS #
(5.79)	(+)-1-[(1,2-Dihydro-4,8-dimethoxy-1-methyl-2-oxo-3-quinolinyl) methyl]-2-methyl-2-propen-1-yl hexadecanoate	Acutifoline palmitate	145204-98-4
(5.80)	2-[(1,2-Dihydro-2-oxo-4-quinolinyl)oxy]acetic acid	Acutinine	887572-73-8
(5.81)	4-Methoxy-3-(3-methyl-2-buten-1-yl)-2(1H)-quinolinone	Atanine	7282-19-1

(Continued)

TABLE 5.1 (*Continued*)
The Chemical Identity of 2-quinolinone Alkaloids

Structure (Number)	Systematic Name	Common Name	CAS #
(5.82)	3-[(2R)-2,3-Dihydroxy-3-methylbutyl]-4,8-dimethoxy-1-methyl-2(1H)-quinolinone	Balfourolone	478-68-2
(5.83)	4-[[(2E)-6,7-Dihydroxy-3,7-dimethyl-2-octen-1-yl]oxy]-2(1H)-quinolinone	Bucharaine	21059-47-2
(5.84)	2(1H)-Quinolinone, 4-hydroxy-3-[1-(tetrahydro-5-hydroxy-2,6,6-trimethyl-2H-pyran-2-yl)ethyl]-	Bucharidine	25865-94-5

(Continued)

TABLE 5.1 (Continued)
The Chemical Identity of 2-quinolinone Alkaloids

Structure (Number)	Systematic Name	Common Name	CAS #
(5.85)	3-((1S)-1-hydroxy-2-(((3S,4R,5R,6S)-3,4,5-trihydroxy-6-(hydroxymethyl)tetrahydro-2H-pyran-2-yl)oxy)ethyl)-4,8-dimethoxyquinolin-2(1H)-one	Chrysanthemumside A	–
(5.86)	4-methoxy-3-(3-methyl-2-oxo-3-(((3S,4R,5R,6S)-3,4,5-trihydroxy-6-(hydroxymethyl)tetrahydro-2H-pyran-2-yl)oxy)butyl)quinolin-2(1H)-one	Chrysanthemumside B	–
(5.87)	4,5,8-Trimethoxy-3-(3-methyl-2-buten-1-yl)-2(1H)-quinolinone	Dasycarine	304909-68-0
(5.88)	2(1H)-Quinolinone, 3-[(R)-1,3-dioxolan-2-ylmethoxymethyl]-4-methoxy-	Dasycarine C	2938167-19-0

(*Continued*)

TABLE 5.1 (*Continued*)
The Chemical Identity of 2-quinolinone Alkaloids

Structure (Number)	Systematic Name	Common Name	CAS #
(5.89)	3-[(2S)-2-Hydroxy-3-methyl-3-buten-1-yl]-4,7,8-trimethoxy-2(1H)-quinolinone	Dasycarine D	2811623-28-4
(5.90)	4,7,8-Trimethoxy-3-(3-methyl-2-oxo-3-buten-1-yl)-2(1H)-quinolinone	Dasycarine E	2811625-31-5
(5.91)	4-Methoxy-1-methyl-8-[(3-methyl-2-buten-1-yl)oxy]-2(1H)-quinolinone	Daurine	54357-79-8

(*Continued*)

TABLE 5.1 (*Continued*)
The Chemical Identity of 2-quinolinone Alkaloids

Structure (Number)	Systematic Name	Common Name	CAS #
(5.92)	3-[(2R)-2,3-Dihydroxy-3-methylbutyl]-4-methoxy-1-methyl-2(1H)-quinolinone	(R)-Edulinine	27495-36-9
(5.93)	4,8-Dimethoxy-2(1H)-quinolinone	Edulitine (Robustinine)	15272-24-9
(5.94)	Ethyl 1,2-dihydro-6-methoxy-2-oxo-4-quinolinecarboxylate	—	93257-70-6
(5.95)	(-)-3-(1-Hydroxy-3-oxobutyl)-4,7,8-trimethoxy-2(1H)-quinolinone	Evomeliaefoline	168146-20-1

(Continued)

TABLE 5.1 (Continued)
The Chemical Identity of 2-quinolinone Alkaloids

Structure (Number)	Systematic Name	Common Name	CAS #
(5.96)	2(1H)-Quinolinone,7-bromo-4-[2-(dimethylamino)ethyl]-3-(2-methyl-1-oxopropyl)-	Flustramine Q	2986537-68-0
(5.97)	4-Methoxy-1-methyl-8-(3-methyl-2-oxobutoxy)-2(1H)-quinolinone	Folidine	102719-91-5
(5.98)	8-Hydroxy-4-methoxy-1-methyl-2(1H)-quinolinone	Folifidine	3148-23-0
(5.99)	4,8-Dimethoxy-1-methyl-2(1H)-quinolinone	Folimine	3148-24-1

(Continued)

TABLE 5.1 (Continued)
The Chemical Identity of 2-quinolinone Alkaloids

Structure (Number)	Systematic Name	Common Name	CAS #
(5.100)	8-(2,3-Dihydroxy-3-methylbutoxy)-4-methoxy-1-methyl-2(1H)-quinolinone	Foliosidine	2520-38-9
(5.101)	2(1H)-Quinolinone,8-[2-(acetyloxy)-3-hydroxy-3-methylbutoxy]-4-methoxy-1-methyl-	Foliphorin	42907-17-5
(5.102)	3-[(1E)-3-Hydroxy-3-methyl-1-buten-1-yl]-4-methoxy-1-methyl-2(1H)-quinolinone	Glycocitlone A	152628-33-6

(Continued)

TABLE 5.1 (*Continued*)
The Chemical Identity of 2-quinolinone Alkaloids

Structure (Number)	Systematic Name	Common Name	CAS #
(5.103)	8-Hydroxy-3-[(1E)-3-hydroxy-3-methyl-1-buten-1-yl]-4-methoxy-1-methyl-2(1H)-quinolinone	Glycocitlone B	262359-08-0
(5.104)	3-[(1E)-3-Hydroxy-3-methyl-1-buten-1-yl]-4,8-dimethoxy-1-methyl-2(1H)-quinolinone	Glycocitlone C	262359-09-1
(5.105)	1,2-Dihydro-4,7,8-trimethoxy-2-oxo-3-quinolinecarboxaldehyde	Glycocitridine	167504-57-6
(5.106)	4,8-Dimethoxy-3-(3-methyl-2-buten-1-yl)-2(1H)-quinolinone	Glycolone	41303-26-8

(*Continued*)

TABLE 5.1 (*Continued*)
The Chemical Identity of 2-quinolinone Alkaloids

Structure (Number)	Systematic Name	Common Name	CAS #
(5.107)	4-[[(2E)-3,7-Dimethyl-2,6-octadien-1-yl]oxy]-1-methyl-2(1H)-quinolinone	Glycopentanolone A	2366185-99-9
(5.108)	4-[[(2E,5E)-7-Hydroperoxy-3,7-dimethyl-2,5-octadien-1-yl] oxy]-1-methyl- 2(1H)-quinolinone	Glycopentanolone B	2366186-00-5

(*Continued*)

TABLE 5.1 (*Continued*)

The Chemical Identity of 2-quinolinone Alkaloids

Structure (Number)	Systematic Name	Common Name	CAS #
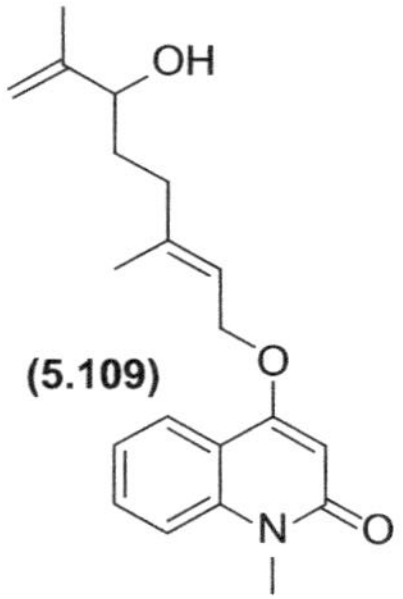 (5.109)	4-[[(2E)-6-Hydroxy-3,7-dimethyl-2,7-octadien-1-yl] oxy]-1-methyl-2(1H)- quinolinone	Glycopentanolone C	2366186-01-6
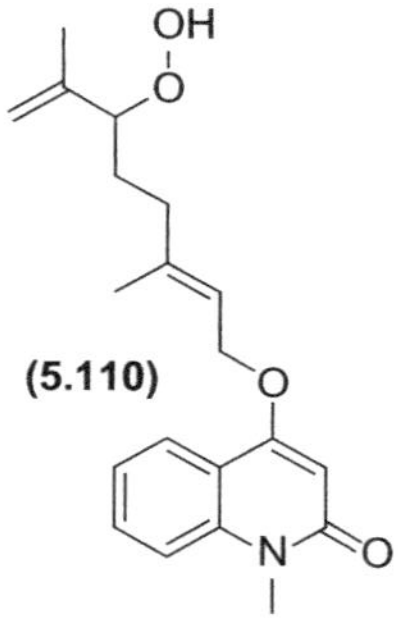 (5.110)	4-[[(2E)-6-Hydroperoxy-3,7-dimethyl-2,7-octadien-1-yl] oxy]-1-methyl-2(1H)- quinolinone	Glycopentanolone D	2366186-02-7

(*Continued*)

TABLE 5.1 (*Continued*)
The Chemical Identity of 2-quinolinone Alkaloids

Structure (Number)	Systematic Name	Common Name	CAS #
(5.111)	4-Hydroxy-8-methoxy-1-methyl-3-(3-methyl-2-buten-1-yl)-2(1H)-quinolinone	Glycophylone	41303-24-6
(5.112)	8-Hydroxy-4-methoxy-1-methyl-3-(3-methyl-2-buten-1-yl)-2(1H)-quinolinone	Glycosolone	67879-81-6
(5.113)	4,6,8-Trimethoxy-2(1H)-quinolinone	Halfordamine	21144-33-2
(5.114)	4-[(3-Methyl-2-buten-1-yl)oxy]-2(1H)-quinolinone	Haplafine	54357-78-7

(*Continued*)

TABLE 5.1 (Continued)
The Chemical Identity of 2-quinolinone Alkaloids

Structure (Number)	Systematic Name	Common Name	CAS #
(5.115)	4,7,8-Trimethoxy-2(1H)-quinolinone	Haplobungine	121949-99-3
(5.116)	3-Ethyl-7-hydroxy-4,8-dimethoxy-2(1H)-quinolinone	Haplosine	70617-27-5
(5.117)	1,4-Dimethoxy-2(1H)-quinolinone	Haplotusine	27667-33-0
(5.118)	4-Ethoxy-8-hydroxy-1-methyl-3-(3-methyl-2-buten-1-yl)-2(1H)-quinolinone	Homoglycosolone	107030-41-1

(Continued)

TABLE 5.1 (Continued)
The Chemical Identity of 2-quinolinone Alkaloids

Structure (Number)	Systematic Name	Common Name	CAS #
(5.119)	6-Hydroxy-4-methoxy-1-methyl-2(1H)-quinolinone	Integriquinolone	81943-13-7
(5.120)	4-Methoxy-3-(3-methyl-2-oxo-3-buten-1-yl)-2(1H)-quinolinone	Limonellone	1446446-75-8
(5.121)	3-[(2R)-2-Hydroxy-3-methylbutyl]-4,8-dimethoxy-1-methyl-2(1H)-quinolinone	Lunacridine	83-58-9 (103303-14-6)
(5.122)	1,2-Dihydro-4,6,7-trimethoxy-2-oxo-3-quinolinecarboxaldehyde	Melisemine	144101-97-3

(Continued)

TABLE 5.1 (Continued)
The Chemical Identity of 2-quinolinone Alkaloids

Structure (Number)	Systematic Name	Common Name	CAS #
(5.123)	Methyl 1,2-dihydro-6-hydroxy-2-oxo-4-quinolinecarboxylate	—	66416-75-9
(5.124)	N-(1,2-Dihydro-4-hydroxy-2-oxo-3-quinolinyl)acetamide	—	65383-48-4
(5.125)	N-[(1,2-Dihydro-2-oxo-4-quinolinyl)carbonyl]-L-aspartic acid	—	1450894-54-8

(Continued)

TABLE 5.1 (Continued)
The Chemical Identity of 2-quinolinone Alkaloids

Structure (Number)	Systematic Name	Common Name	CAS #
(5.126)	4-Methoxy-8-hydroxy-3-(3-methyl-2-buten-1-yl)-2(1H)-quinolinone	Nitidumalkaloid F	—
(5.127)	4-Methoxy-1-methyl-3-(3-methyl-2-buten-1-yl)-2(1H)-quinolinone	N-Methylatanine	22826-43-3
(5.128)	6-hydroxy-4-(4-hydroxyphenyl)quinolin-2(1H)-one	—	—
(5.129)	4,7,8-Trimethoxy-1-methyl-3-(3-methyl-2-buten-1-yl)-2(1H)-quinolinone	N-Methylpreskimmianine	76525-28-5

(Continued)

TABLE 5.1 (Continued)
The Chemical Identity of 2-quinolinone Alkaloids

Structure (Number)	Systematic Name	Common Name	CAS #
(5.130)	2(1H)-Quinolinone, 4-methoxy-1-methyl-3-(3-methyl-1,3-butadienyl)-, (E)-	N-Methylschimifoline	149998-43-6
(5.131)	3,4-Dimethoxy-1-methyl-2(1H)-quinolinone	N-Methylswietenidine B	123497-83-6
(5.132)	4,8-Dimethoxy-1-methyl-3-(3-methyl-2-buten-1-yl)-2(1H)-quinolinone	O-Methylglycosolone	41303-25-7
(5.133)	(-)-4,6,8-Trimethoxy-3-(2-methoxy-3-methyl-3-buten-1-yl)-1-methyl-2(1H)-quinolinone	O-Methylptelefoline	36017-60-4

(Continued)

TABLE 5.1 (Continued)
The Chemical Identity of 2-quinolinone Alkaloids

Structure (Number)	Systematic Name	Common Name	CAS #
(5.134)	(-)-4,8-Dimethoxy-3-(2-methoxy-3-methyl-3-buten-1-yl)-1-methyl-2(1H)-quinolinone	O-Methylacutifoline	145204-97-3
(5.135)	4,8-Dimethoxy-1-methyl-3-(3-methyl-2-oxobutyl)-2(1H)-quinolinone	Orixalone A	797788-72-8
(5.136)	4,8-Dimethoxy-1-methyl-3-(3-methyl-2-oxo-3-buten-1-yl)-2(1H)-quinolinone	Orixalone B	797788-73-9
(5.137)	4,8-Dimethoxy-3-(3-methyl-2-oxo-3-buten-1-yl)-2(1H)-quinolinone	Orixalone C	797788-74-0

(Continued)

TABLE 5.1 (*Continued*)
The Chemical Identity of 2-quinolinone Alkaloids

Structure (Number)	Systematic Name	Common Name	CAS #
(5.138)	4-Methoxy-1-methyl-3-(3-methyl-2-oxobutyl)-2(1H)-quinolinone	Orixiarine (Orijanone)	123613-75-2
(5.139)	(-)-4-Hydroxy-3-(2-hydroxy-1,2-dimethylpropyl)-1-methyl-2(1H)-quinolinone	Paraensine	33172-19-9
(5.140)	Methyl 2-[(1,2-dihydro-2-oxo-4-quinolinyl)oxy]acetate	Pedicine	2387894-91-7
(5.141)	5-Hydroxy-3-methoxy-4-(4-methoxyphenyl)-2(1H)-quinolinone	Pesimquinolone R	2760454-70-2

(*Continued*)

TABLE 5.1 (Continued)
The Chemical Identity of 2-quinolinone Alkaloids

Structure (Number)	Systematic Name	Common Name	CAS #
(5.142)	4-(4-Methoxyphenyl)-2(1H)-quinolinone	Pesimquinolone S	37118-72-2
(5.143)	4,7,8-Trimethoxy-3-(3-methyl-2-buten-1-yl)-2(1H)-quinolinone	Preskimmianine	38695-41-9
(5.144)	3-[(2S)-2-Hydroxy-3-methyl-3-buten-1-yl]-4-methoxy-1-methyl-2(1H)-quinolinone	Ptelefoliarine	207603-50-7
(5.145)	2(1H)-Quinolinone,3-(2-hydroxy-3-methyl-3-butenyl)-4,6,8-trimethoxy-1-methyl-	Ptelefoline	25488-61-3

(Continued)

TABLE 5.1 (Continued)
The Chemical Identity of 2-quinolinone Alkaloids

Structure (Number)	Systematic Name	Common Name	CAS #
(5.146)	2(1H)-Quinolinone,3-(2,3-dihydroxy-3-methylbutyl)-4,6,8-trimethoxy-1-methyl-	Ptelefolinol	59719-25-4
(5.147)	1-Methyl-4-[(3-methyl-2-buten-1-yl)oxy]-2(1H)-quinolinone	Ravenine	20105-22-0
(5.148)	3-(1,2-Dimethyl-2-propen-1-yl)-4-hydroxy-1-methyl-2(1H)-quinolinone	Ravenoline	20105-23-1
(5.149)	4-Methoxy-3-[(1E)-3-methyl-1,3-butadien-1-yl]-2(1H)-quinolinone	Schinifoline	149998-56-1

(Continued)

TABLE 5.1 (Continued)
The Chemical Identity of 2-quinolinone Alkaloids

Structure (Number)	Systematic Name	Common Name	CAS #
(5.150)	4,7-Dimethoxy-3-(methoxymethyl)-2(1H)-quinolinone	Semecarpifoline	366789-86-8
(5.151)	N-[2-[2-[((1,2-Dihydro-4-hydroxy-1-methyl-2-oxo-3-quinolinyl)methyl]-1H-indol-3-yl]ethyl]acetamide	SF 2809 I	271580-72-4
(5.152)	N-[2-[2-[((1,2-Dihydro-4-hydroxy-1-methyl-2-oxo-3- quinolinyl)(4-hydroxyphenyl)methyl]-1H-indol-3-yl]ethyl]acetamide	SF 2809 II	271580-73-5

(Continued)

TABLE 5.1 (Continued)
The Chemical Identity of 2-quinolinones Alkaloids

Structure (Number)	Systematic Name	Common Name	CAS #
(5.153)	4-Hydroxy-3-[[3-(2-hydroxyethyl)-1H-indol-2-yl] methyl]-1-methyl-2(1H)-quinolinone	SF 2809 III	271580-74-6
(5.154)	4-Hydroxy-3-[[3-(2-hydroxyethyl)-1H-indol-2-yl](4-hydroxyphenyl) methyl]-1-methyl-2(1H)-quinolinone	SF 2809 IV	271580-75-7
(5.155)	N-[2-[2-[(1,2-Dihydro-4-hydroxy-1-methyl-2-oxo-3-quinolinyl)phenylmethyl]- 1H-indol-3-yl]ethyl]acetamide	SF 2809 V	271580-76-8

(Continued)

TABLE 5.1 (Continued)
The Chemical Identity of 2-quinolinone Alkaloids

Structure (Number)	Systematic Name	Common Name	CAS #
(5.156)	4-Hydroxy-3-[[3-(2-hydroxyethyl)-1H-indol-2-yl) phenylmethyl]-1-methyl- 2(1H)-quinolinone	SF 2809 VI	271580-77-9
(5.157)	4-Butyl hydrogen N-[(1,2-dihydro-2-oxo-4-quinolinyl)carbonyl]-L-aspartate	Sterculinine I	600166-05-0
(5.158)	4-Methyl hydrogen N-[(1,2-dihydro-2-oxo-4-quinolinyl)carbonyl]-L-aspartate	Sterculinine II	600166-06-1

(Continued)

TABLE 5.1 (*Continued*)
The Chemical Identity of 2-quinolinone Alkaloids

Structure (Number)	Systematic Name	Common Name	CAS #
(5.159)	4-Hydroxy-3,8-dimethoxy-1-methyl-2(1H)-quinolinone	Swietenidine A	64595-74-0
(5.160)	3,4-Dimethoxy-2(1H)-quinolinone	Swietenidine B	2721-56-4
(5.161)	7-[(6-Deoxy-α-L-mannopyranosyl)oxy]-3-ethyl-4,8-dimethoxy- 2(1H)-quinolinone	Tetrahydroglycoperine	61930-00-5
(5.162)	3-Hydroxy-4-phenyl-2(1H)-quinolinone	Viridicatin	129-24-8

(Continued)

TABLE 5.1 (*Continued*)
The Chemical Identity of 2-quinolinone Alkaloids

Structure (Number)	Systematic Name	Common Name	CAS #
(5.163)	3-Hydroxy-4-(3-hydroxyphenyl)-2(1H)-quinolinone	Viridicatol	14484-44-7
(5.164)	3,3′-Methylenebis[4-hydroxy-1-methyl-2(1H)-quinolinone]	Zanthobisquinolone	57147-67-8
(5.165)	4-methoxy-3-((Z)-3-methyl-4-(((2R,3S,4R,5R,6S)-3,4,5-trihydroxy-6-(hydroxymethyl)tetrahydro-2H-pyran-2-yl)oxy)but-2-en- 1-yl) quinolin-2(1H)-one	Zanthonitiside I	—

(*Continued*)

TABLE 5.1 (Continued)
The Chemical Identity of 2-quinolinone Alkaloids

Structure (Number)	Systematic Name	Common Name	CAS #
(5.166)	2(1H)-Quinolinone, 3-[[(2R,3S)-3-(hydroxymethyl)-3-methyloxiranyl]methyl]-4,8-dimethoxy-1-methyl-, rel-	Zascanol epoxide	149998-44-7
(5.167)	1,2-Dihydro-8-hydroxy-2-oxo-4-quinolinecarboxylic acid	Zeanic acid	30536-55-1
(5.168)	8-(β-D-Glucopyranosyloxy)-1,2-dihydro-2-oxo-4-quinolinecarboxylic acid	Zeanoside B	113202-67-8

(Continued)

TABLE 5.1 (Continued)
The Chemical Identity of 2-quinolinone Alkaloids

Structure (Number)	Systematic Name	Common Name	CAS #
(5.169)	3-Methoxy-4-(4-methoxyphenyl)quinolin-2(1H)-one	Raistrimide D	—
(5.170)	6-Hydroxy-4-methoxyquinolin-2(1*H*)-one	Dasycarine F	—
(5.171)	4-Hydroxy-1-((2R)-3-hydroxy-3-methyl-2-(((3R,4S,5S,6R)-3,4,5-trihydroxy-6-(hydroxymethyl)tetrahydro-2H-pyran-2-yl)oxy)butyl)quinolin-2(1H)-one	Dasycarine G	—

5.1.2 SYNTHESIS

The interest in synthesizing 2-quinolinones is obvious because of their abundance in nature and biological activities. Consequently, numerous studies have been conducted to generate a 2-quinolinone core or its derivatives. The synthesis of 2-quinolinones has a rich history dating back to the late 19th and early 20th centuries. The Friedländer quinoline synthesis was one of the earliest methods adapted for 2-quinolinone synthesis (Friedländer 1882). This reaction typically involves the condensation of 2-aminobenzaldehyde or 2-aminoketones with aldehydes or ketones containing an α-methylene group. For 2-quinolinones, a variation using 2-aminophenol derivatives instead of 2-aminobenzaldehyde was employed. The Knorr quinoline synthesis offered another classical approach (Knorr 1886). It involves the condensation of β-ketoesters or β-diketones with aromatic amines, followed by cyclization. These methods were pioneering, but they often needed harsh conditions such as strong acids or high temperatures. This limited their use of sensitive starting materials or even final products. Moreover, yields were often moderate (30%–60%), and the reactions frequently produced mixtures of regioisomers. Consequently, they required challenging purification steps. The introduction of transition metal catalysis revolutionized 2-quinolinone synthesis. Palladium-catalysed reactions, in particular, have been extensively studied (Silva and Silva 2019). For example, the Pd(II)-catalysed oxidative cyclization of N-arylacrylamides was used to synthesize 2-quinolinones with yields often exceeding 80%. This method tolerates a wide range of functional groups and allows for the installation of various substituents at different positions of the quinolinone core. Copper catalysis has emerged as a cost-effective alternative to palladium. Other transition metals have shown unique capabilities. Rhodium catalysts, for instance, have been used in the cyclization of 2-aminophenethyl alcohols with alkynes to form 2-quinolinones, while iridium complexes have facilitated the synthesis of these compounds through C–H activation strategies (Hong et al. 2020). In another study, the Iron-complex-catalysed synthesis of various mono- and di-substituted 2-quinolinones is achieved via the intramolecular dehydrogenative cyclization of amidoalcohols (Bettoni et al. 2024). The synthesis of 2-quinolinones by direct carbonylation of C–H bonds (ortho-alkenylanilines or o-anilines) with CO, triphosgene, and CO_2 under mild conditions was performed by Zhang and Li (2020).

The development of one-pot modular approaches marks a significant advancement in 2-quinolinone synthesis. These methods combine multiple synthetic steps in a single reaction vessel, dramatically improving efficiency (Hong et al. 2020). A notable two-component reaction involves the copper-catalysed coupling of 2-haloanilines with β-ketoesters, followed by in situ cyclization to form 2-quinolinones. Three-component reactions offer even greater versatility. For example, a one-pot, three-component reaction using 2-iodoanilines, terminal alkynes, and carbon monoxide, catalysed by a Pd(0) complex, produces 3-substituted 2-quinolinones in yields ranging from 60% to 80%. This method is particularly valuable for rapidly generating libraries of compounds for medicinal chemistry screening. Four-component reactions, while less common, provide access to highly functionalized 2-quinolinones. One example is the silver-catalysed reaction of

2-aminobenzaldehydes, terminal alkynes, amines, and carbon dioxide. They formed 3,4-disubstituted 2-quinolinones in a single step, often with yields exceeding 70%.

The hypervalent iodine-mediated synthesis of 2-quinolinones is a new approach in the field. This method typically uses PhI(OAc)$_2$ (diacetoxyiodobenzene) as the hypervalent iodine reagent to enable an intramolecular decarboxylative Heck-type reaction of *N*-arylcinnamamides (Fan et al. 2018). The reaction proceeds under mild conditions and can achieve yields of 70%–90% for a wide range of substrates. This metal-free method is particularly valuable for applications where metal contamination must be strictly avoided.

The Diels–Alder approach to 2-quinolinones involves the cycloaddition of exo-diene lactams (typically derived from *N*-arylmaleimides) with various dienophiles (Huang and Chang 2008). This method is useful for creating complex polycycles. For instance, the reaction of *N*-arylmaleimide-derived dienes with maleic anhydride can produce tricyclic 2-quinolinone derivatives in yields of 60%–80% under thermal conditions. The Diels–Alder approach offers excellent control over stereochemistry, often producing single diastereomers.

As indicated, the development of metal-free methodologies for 2-quinolinone synthesis addresses concerns about metal contamination and offers potentially more sustainable approaches. The organoiodine-catalysed electrophilic arene C(sp^2)–H amination method is a good example. This reaction typically uses a combination of an iodine(I) catalyst (e.g., 1,3-diiodo-5,5-dimethylhydantoin) and an oxidant (e.g. *m*-CPBA) to facilitate the intramolecular amination of *N*-arylacrylamides. The reaction proceeds at room temperature in dichloromethane and can achieve yields of 75%–90% for a wide range of substrates (Das et al. 2024).

Another metal-free approach involves the use of organocatalysts. For example, proline-derived organocatalysts have been employed in the asymmetric synthesis of 3-substituted 2-quinolinones through a Michael-addition/cyclization cascade. Photoredox catalysis has also emerged as a powerful tool for metal-free 2-quinolinone synthesis. Researchers have developed visible light-mediated cyclizations of N-arylacrylamides to form 2-quinolinones under mild conditions using organic photocatalysts like Eosin Y, typically achieving yields of 60%–80% (Wu et al. 2023). Modern one-pot methods frequently achieve comparable or superior overall yields to multi-step sequences, while minimizing solvent use and the steps required for purification. For example, the one-pot, four-component synthesis of 3,4-disubstituted 2-quinolinones achieved yields of up to 85% in a single step, compared to a traditional linear synthesis that might require 3–4 steps with an overall yield of 40%–50% (Marcaccini et al. 2004).

Stereoselectivity has been greatly enhanced through careful catalyst and ligand design. For instance, in palladium-catalysed cyclizations, using chiral phosphine ligands has enabled the synthesis of 3-substituted 2-quinolinones with enantiomeric excesses exceeding 95% (Silva and Silva 2019).

The move towards milder reaction conditions includes using room-temperature methods and environmentally friendly solvents. For instance, some recent methods employ water as a solvent or co-solvent, significantly reducing the environmental impact of the synthesis (Atalay et al. 2020).

Microwave-assisted protocols have also been developed. They reduce reaction times from hours to minutes in many cases, while maintaining or improving yields (Jia et al. 2007).

Used as a model, the compound 4-hydroxy-2(1*H*)-quinolinone (5.46) and its derivatives are of particular interest due to the specific synthetic methods, as well as their biological activities (Aly et al. 2020; Aly et al. 2021).

5.1.3   STRUCTURE–ACTIVITY RELATIONSHIP (SAR)

The careful selection and positioning of substituents on 2-quinolinones provide medicinal chemists with a powerful tool to optimize them for different therapeutic purposes. It is crucial to understand these structure–activity relationships (SARs) for the development of more effective quinolinone-based drugs (Moussaoui et al. 2022). A series of *N*-substituted quinolinone-3-aminoamides and their hybrids containing the α-lipoic acid functionality were synthesized as potential bifunctional agents combining antioxidant and anti-inflammatory activity. The new compounds were tested for their antioxidant and anti-inflammatory properties, as well as their ability to inhibit lipoxygenase *in vitro*. The derivatives were found to be potent antioxidant and anti-inflammatory agents (Detsi et al. 2007). In a study conducted by Khamkhenshorngphanuch et al. (2020), a series of 4-hydroxy-2-quinolinones with a long alkyl side chain at C-3 and various substituents at the C-6 and C-7 positions were synthesized. The study aimed to investigate the antibacterial and antifungal activities of these analogues against *Staphylococcus aureus*, *Escherichia coli*, and *Aspergillus flavus*. The results of the SAR study showed that the length of the alkyl chain and the type of substituent significantly influenced the antimicrobial activities. Specifically, the analog with a nonyl side chain and bromine substituents exhibited exceptional antifungal activities against *A. flavus*. Its activity surpassed that of the positive control.

2-Quinolinones, their structural derivatives, and associated mechanistic and SAR studies are also aimed at anticancer applications (Singh et al. 2023). In particular, substitution at the -NH functionality of the quinolone motif provided sufficient space for suitable functionalization and structural modifications. This facilitated the design and discovery of selective anticancer agents binding specifically with various drug targets such as topoisomerase, histone deacetylase, epidermal growth factor receptor, etc. Further substitution at the third and sixth carbons of 2-quinolinones resulted in compounds with target-specific anticancer activity.

Recently, researchers investigated the 4-hydroxy-2-quinolinone scaffold by creating various carboxamides and hybrid derivatives (Kostopoulou et al. 2023; Aly et al. 2020). They tested those compounds to see if they could inhibit soybean lipoxygenase (LOX) and also evaluated their antioxidant properties. The study found that two quinolinone-carboxamides had the best LOX inhibitory activity. Additionally, two other compounds showed promise as multi-target agents, displaying both antioxidant and LOX inhibitory activities. Furthermore, in silico docking studies revealed that several compounds bind to the same alternative binding site of LOX as the reference compound.

5.2   NATURAL OCCURRENCE

5.2.1   ANIMAL SOURCES

Scolopendra multidens is a species of large tropical centipede belonging to the family Scolopendridae. This predatory arthropod is widespread in Asia, in countries such as Vietnam, China, and the Philippines. Like other members of its genus, *S. multidens* is known for its powerful venom and ability to deliver painful bites. Researchers have identified compound 5.72 from this centipede species (Fu et al. 2013). In the marine environment, *Flustra foliacea*, commonly known as hornwrack, is a bryozoan species from the family Flustridae. This colonial animal forms leaf-like structures and is often found in cooler waters. *F. foliacea* has been found to contain compound 5.96 (Di et al. 2020; Kowal et al. 2023). Another marine organism, *Hyrtios erecta*, is a species of sea sponge belonging to the family Thorectidae. This sponge has been a subject of natural product research, with compounds 5.60 and 5.61 isolated from its tissues (Aoki et al. 2001).

5.2.2   BACTERIAL SOURCES

Burkholderia sp. 3Y-MMP is a soil bacterium classified within the family Burkholderiaceae. This genus is recognized for its diverse metabolic capabilities and resilience in various environments. Researchers have identified compound 5.50 from this strain (Li et al. 2020). *Dactylosporangium* strain SF2809, part of the Micromonosporaceae family, is an actinobacterium commonly found in soil. This strain has yielded seven compounds: 5.44 and 5.151–5.156 (Tani et al. 2004a,b). The marine bacterium *Marinomonas* Hel59b, belonging to the Oceanospirillaceae family, is noted for producing compound 5.9 (Mahmoud 2005). *Pseudomonas fluorescens* strain G308, a fluorescent pseudomonad from the Pseudomonadaceae family, has been documented to generate compound 5.4 (Fakhouri et al. 2001). *Streptomyces* sp. TN82, an actinobacterium from the Streptomycetaceae family, is known for its production of compound 5.31 (El Euch et al. 2018). Lastly, *Wautersiella falsenii* YMF 3.00141, a member of the Weeksellaceae family, has been reported to generate compound 5.1 (Chen et al. 2015).

5.2.3   FUNGAL SOURCES

The Aspergillaceae family includes important fungal genera, notably *Aspergillus* and *Penicillium*. *Aspergillus*, with over 300 species, includes *A. oryzae* used in food fermentation and the potentially pathogenic *A. fumigatus*. It produces various alkaloids like fumigaclavines and ergot alkaloids, known for their anti-inflammatory and vasoconstrictive properties (Zhu et al. 2023a). *Penicillium*, comprising over 350 species, is famous for penicillin production and its role in food spoilage. It yields significant alkaloids, particularly indole alkaloids from species like *P. crustosum*, which show potential in treating neurological disorders (Zhang et al. 2020). Both genera also produce compounds with antimicrobial and antitumor effects, which make them crucial for natural product research and biotechnological applications. Various species and strains of these fungi are listed in Table 5.2, demonstrating their capacity

TABLE 5.2

Natural Occurrence of 2-Quinolinone Alkaloids

Kingdom	Family	Genus	Species	Compound(s)	References
Animalia	Scolopendridae	Scolopendra	multidens	5.72	Fu et al. (2013)
"	Flustridae	Flustra	foliacea	5.96	Di et al. (2020); Kowal et al. (2023)
"	Thorectidae	Hyrtios	erecta	5.60, 5.61	Aoki et al. (2001)
Bacteria	Burkholderiaceae	Burkholderia	sp. 3Y-MMP	5.50	Li et al. (2020)
"	Micromonosporaceae	Dactylosporangium	strain SF2809	5.44, 5.151–5.156	Tani (2004)ab
"	Oceanospirillaceae	Marinomonas	Hel59b	5.9	Mahmoud (2005)
"	Pseudomonadaceae	Pseudomonas	fluorescens strain G308	5.4	Fakhouri (2001)
"	Streptomycetaceae	Streptomyces	sp. TN82 strain	5.31	El Euch et al. (2018)
"	Weeksellaceae	Wautersiella	falsenii YMF 3.00141	5.1	Chen et al. (2015)
Fungi	Aspergillaceae	Aspergillus	versicolor	5.31, 5.32	Li et al. (2016)
"	"	"	versicolor D5	5.18, 5.31, 5.32, 5.162	Fu et al. (2020)
"	"	"	versicolor SCSIO 05879	5.18, 5.32	Wang et al. (2016)
"	"	"	versicolor SmT07	5.162, 5.163	Yue et al. (2022)
"	"	Penicillium	citrinum XIA-16	5.30, 5.48	Yu et al. (2023)
"	"	"	Crustosum Strain P11	5.163	Quaglia et al. (2020)
"	"	"	cyclopium	5.162	Bracken et al. (1954)
"	"	"	expansum	5.57, 5.162, 5.163	Li et al. (2023)
"	"	"	OS-F67406	5.31	Heguy et al. (1998)
"	"	"	polonicum AP2T1	5.31, 5.32, 5.162, 5.163	Zhang et al. (2017a)
"	"	"	polonicum H175	5.31, 5.162, 5.163	Liu et al. (2022)
"	"	"	polonicum H92	5.29, 5.31, 5.162, 5.163	Liu et al. (2021a)

(Continued)

TABLE 5.2 (Continued)
Natural Occurrence of 2-Quinolinone Alkaloids

Kingdom	Family	Genus	Species	Compound(s)	References
"	"	"	raistrichii	5.40, 5.141, 5.169	Zhong et al. (2023)
"	"	"	simplicissimum	5.141, 5.142	Dai et al. (2021)
"	"	"	sp. I09F 484	5.125	Gan et al. (2013)
"	"	"	sp. MCCC 3A00228	5.31, 5.163	Wang et al. (2021)
"	"	"	sp. SCSIO 06720	5.48	Guo et al. (2020)
"	"	"	sp. SCSIO41015	5.48	Pang et al. (2019)
"	"	"	sp.DCS82	5.162	Yang et al. (2013)
"	"	"	Spp.	5.31, 5.163	Wei et al. (2011)
"	"	"	steckii P2648	5.48, 5.49	Yao et al. (2021)
"	"	"	verrucosum var. cyclopium	5.31, 5.162	Hodge et al. (1988)
"	"	"	viridicatum	5.162	Cunningham and Freeman (1953)
"	"	"	"	5.162, 5.163	Birkinshaw et al. (1963)
"	Stachybotryaceae	Myrothecium	verrucaria	5.32, 5.65, 5.162, 5.163	Zhang et al. (2017b)
"	Thermoascaceae	Paecilomyces	sp. PDB	5.34	Xu et al. (2009)
"	Pleosporaceae	Alternaria	sp. YD-01	5.36	Yang and Yang (2008)
Plantae	Amaranthaceae	Suaeda	fruticosa	5.28	Al-ghazzawi (2020)
"	Anacardiaceae	Spondias	pinnata	5.67	Ghate (2018)
"	Asteraceae	Echinops	gmelinii	5.70, 5.71	Su et al. (2004)
"	"	Chrysanthemum	indicum	5.85, 5.86	Zhu et al. (2023b)
"	Boraginaceae	Cynoglossum	gansuense	5.7	Jin et al. (2007)
"	Brassicaceae	Isatis	Indigotica (tinctoria)	5.62	Chen et al. (2012)
"	"	"	"	5.33, 5.50, 5.128	Zhang et al. (2019)

(Continued)

TABLE 5.2 (Continued)
Natural Occurrence of 2-Quinolinone Alkaloids

Kingdom	Family	Genus	Species	Compound(s)	References
"	Clusiacea	Garcinia	yunnanensi	5.123	Fang et al. (2018)
"	Ericaceae	Rhododendron	dauricum	5.44, 5.58, 5.63	Sun and Wang (2011)
"	Euphorbiaceae	Sebastiania	corniculata	5.16	Machado et al. (2005)
"	Fabaceae	Cassia (Senna)	auriculata	5.68, 5.69	Zheng et al. (2023)
"	"	Psoralea	corylifolia	5.17	Kumar et al. (2023)
"	"	Ceratonia	siliqua	5.46	El-Neketi et al. (2013)
"	"	Castanea	crenata	5.1	Cho et al. (2015)
"	Gentianaceae	Gentiana	crassicaulis	5.9	He et al. (2011)
"	Irvingiaceae	Irvingia	gabonensis	5.17	Bob-chile and Amali (2020)
"	Lamiaceae	Clerodendrum	paniculatum	5.105	Arba et al. (2023)
"	Malvaceae	Sterculia	lychnophora	5.157, 5.158	Wang et al. (2003)
"	"	"	urens	5.157, 5.158	Darapureddy et al. (2021)
"	Nymphaeaceae	Nuphar	pumila	5.27	Peng et al. (2014)
"	Nyssaceae	Camptotheca	acuminata	5.5	Yin and Hu Li Hong (2005)
"	Papaveraceae	Papaver	somniferum	5.5, 5.9	Schmid and Karrer (1945)
"	Platanaceae	Platanus	acerifolia	5.163	Zhang et al. (2011)
"	Poaceae	Zea	mays	5.168	Tateishi et al. (1987)
"	"	"	"	5.167	Matsushima et al. (1973); Tateishi et al. (1989)
"	"	Oryza	sativa	5.123	Chung and Woo (2001)
"	"	"	sativa	5.43, 5.94	Ryu and Chung (2010)
"	Ranunculaceae	Aquilegia	cv. heugnambyeo ecalcarata	5.66	Chen et al. (2002b)

(Continued)

TABLE 5.2 (Continued)
Natural Occurrence of 2-Quinolinone Alkaloids

Kingdom	Family	Genus	Species	Compound(s)	References
"	"	Aconitum	ferox	5.1	Hanuman and Katz (1993)
"	Schisandraceae	Schisandra	chinensis	5.5	Chen et al. (2011)
"	Simaroubaceae	Brucea	javanica	5.35	Yu and Li (1990)
"	"	Eurycoma	longifolia	5.57	Purwantiningsih and Chan (2011)
"	Solanaceae	Nicotiana	—	5.41	Liu et al. (2021b)
"	"	"	tabacum	5.41, 5.42	Deng et al. (2022)
"	Vitaceae	Parthenocissus	tricuspidata	5.9, 5.35	Wang et al. (1982)
"	Rutaceae	Aegle	marmelos	5.53, 5.119	Yang et al. (1996a, 1996b)
"	"	Agathosma	—	5.38	Campbell et al. (1990)
"	"	Almeidea	guyanensis	5.45, 5.127	Moulis et al. (1983)
"	"	Araliopsis	soyauxii	5.53	Noulala et al. (2020); Mbaveng et al. (2021)
"	"	Balfourodendron	riedelianum	5.82	Rapoport and Holden (1959)
"	"	Boronia	bowmanii	5.143	Ahsan et al. (1994)
"	"	"	pinnata	5.99, 5.143	Curini et al. (2008)
"	"	Casimiroa	edulis	5.92, 5.93	Iriarte et al. (1956)
"	"	Chloroxylon	swietenia	5.159, 5.160	Bhide et al. (1977)
"	"	Clausena	anisata	5.131, 5.160	Ngadjui et al. (1989)
"	"	"	lansium	5.53, 5.54, 5.81, 5.119	Song et al. (2014)
"	"	"	Vestita D. D. Tao	5.53	Shi et al. (2010)
"	"	Conchocarpus	heterophyllus	5.117	Ambrozin (2008)
"	"	"	mastigophorus	5.53, 5.117	Pinto et al. (2022a2022b)
"	"	"	gaudichaudianus	5.20	Ranieri Cortez et al. (2009)

(Continued)

TABLE 5.2 (Continued)
Natural Occurrence of 2-Quinolinone Alkaloids

Kingdom	Family	Genus	Species	Compound(s)	References
"	"	Dictamnus	angustifolius	5.87	Sun et al. (2016)
"	"	"	dasycarpus	5.19, 5.143	Yoon et al. (2012)
"	"	"	"	5.6, 5.53, 5.54, 5.89, 5.127, 5.143	Chang et al. (2021)
"	"	"	"	5.6 5.54, 5.88, 5.90, 5.160, 5.89, 5.53, 5.143, 5.81, 5.127	Gao et al. (2020)
"	"	"	"	5.87	Chen et al. (2000)
"	"	"	"	5.87, 5.106	Yan et al. (2020)
"	"	"	"	5.10, 5.143	Guo et al. (2018)
"	"	"	"	5.19, 5.143, 5.106	Yang et al. (2019)
"	"	"	"	5.170, 5.171	Ye et al. (2024)
"	"	Dictyoloma	vandellianum	5.39	Sartor et al. (2019)
"	"	Drummondita	calida	5.26, 5.127	Yang et al. (2011)
"	"	Eriostemon	gardneri	5.16	Sarker et al. (1995)
"	"	Euodia	roxburghiana	5.15	McCormick et al. (1996)
"	"	Euxylophora	paraensis	5.139	Jurd et al. (1983)
"	"	"	"	5.3	Jurd and Wong (1981)
"	"	Evodia	rutaecarpa	5.81	Perrett and Whitfield (1995)
"	"	Fagara	zanthoxyloides	5.81	Eshiett and Taylor (1968)
"	"	Galipea	officinalis	5.24	Jacquemond-Collet et al. (2000)
"	"	"	"	5.8	Houghton et al. (1999); Jacquemond-Collet et al. (2001)

(Continued)

TABLE 5.2 (Continued)
Natural Occurrence of 2-Quinolinone Alkaloids

Kingdom	Family	Genus	Species	Compound(s)	References
"	"	Geijera	parviflora	5.114	Banbury et al. (2015)
"	"	Glycosmis	citrifolia	5.102, 5.103, 5.104, 5.111, 5.112	Ito et al. (2000)
"	"	"	cochinchinensis	5.119	Sripisut et al. (2013)
"	"	"	craibii var. glabra	5.132, 5.147	Chen et al. (2022)
"	"	"	cyanocarpa	5.149	Wurz et al. (1993)
"	"	"	mauritiana	5.53	Intekhab et al. (2011)
"	"	"	"	5.132	Rastogi et al. (1980)
"	"	"	Pentaphylla	5.118	Kumar et al. (1986)
"	"	"	"	5.107–5.110	Choi et al. (2019)
"	"	"	"	5.111	Bhattacharyya and Chowdhury (1984)
"	"	"	"	5.132	Chakravarty et al. (1999)
"	"	"	"	5.132	Yang et al. (2012a, 2012b)
"	"	"	"	5.106	Bhattacharyya and Chowdhury (1985)
"	"	"	"	5.112	Das and Chowdhury (1978)
"	"	"	"	5.71	Tian (2014)
"	"	Halfordia	scleroxyla	5.113	Crow and Hodgkin (1968)
"	"	Haplophyllum	—	5.76, 5.84, 5.98	Nazrullaev et al. (2001)
"	"	"	acutifolium	5.76, 5.140	Eshonov et al. (2020)
"	"	"	"	5.80	Eshonov and Rasulova (2020)
"	"	"	bucharicum	5.83	Ubaidullaev et al. (1972)
"	"	"	"	5.46, 5.54	Bessonova (2000)

(Continued)

TABLE 5.2 (*Continued*)
Natural Occurrence of 2-Quinolinone Alkaloids

Kingdom	Family	Genus	Species	Compound(s)	References
"	"	"	*bungei*	5.39, 5.54, 5.99, 5.93, 5.115	Bessonova and Yunusov (1989)
"	"	"	*dauricum*	5.53, 5.91, 5.93, 5.99	Bessonova et al. (1984)
"	"	"	*foliosum*	5.99	Razzakova et al. (1972)
"	"	"	*griffithianum*	5.99	Kodirova et al. (2018)
"	"	"	*latifolium*	5.116	Nesmelova et al. (1978)
"	"	"	*myrtifolium*	5.114	Sener et al. (1990)
"	"	"	*obtusifolium*	5.99, 5.117	Razakova et al. (1984)
"	"	"	*patavinum*	5.92	Puricelli et al. (2002)
"	"	"	*pedicellatum*	5.140	Rasulova et al. (2019)
"	"	"	*perforatum*	5.99, 5.100	Razakova et al. (1976)
"	"	"	"	5.114	Bessonova and Yunusov (1986)
"	"	"	"	5.116	Rasulova and Bessonova (1992)
"	"	"	*Spp.*	5.83	Fiot et al. (2006)
"	"	"	*foliosum*	5.93, 5.9	Faizutdinova et al. (1967)
"	"	"	"	5.97	Akhmedzhanova et al. (1985)
"	"	"	"	5.100	Nazrullaev et al. (2001); Pastukhova et al. (1965); Razzakova et al. (1972)
"	"	"	"	5.92, 5.100	Akhmedzhanova et al. (1980)
"	"	"	"	5.101	Akhmedzhanova (1999)
"	"	"	*perforatum*	5.38, 5.51, 5.161	Akhmedzhanova et al. (1976)
"	"	"	*crenulata*	5.53	Nayar et al. (1971)
"	"	*Horatio*	*brasiliana*	5.1	Severino et al. (2014)
"	"	"	*oreadica*	5.1, 5.127	Severino et al. (2014)
"	"	"	*superba*	5.1, 5.53, 5.93, 5.119	Severino et al. (2014)

(*Continued*)

TABLE 5.2 (Continued)
Natural Occurrence of 2-Quinolinone Alkaloids

Kingdom	Family	Genus	Species	Compound(s)	References
"	"	Limonia	acidissima	5.53	Kim et al. (2009)
"	"	"	"	5.119	Wijeratne et al. (1992)
"	"	"	crenulata	5.119	Niu et al. (2001)
"	"	"	amara	5.121	Rüegger and Stauffacher (1963)
"	"	"	var. repanda		
"	"	Lunasia	amara	5.121	Beyerman and Rooda (1959); Prescott et al. (2007)
"	"	"	"	5.11, 5.13, 5.14	Goodwin and Horning (1959)
"	"	"	"	5.47, 5.121	Goodwin et al. (1959)
"	"	"	quercifolia	5.47	Price (1959)
"	"	Melicope	confusa	5.92, 5.105, 5.122	Chen et al. (2002a)
"	"	"	Moluccana	5.55	Tanjung et al. (2017)
"	"	"	ptelefolia	5.54	Li (2017)
"	"	"	semecarpifolia	5.92, 5.105, 5.122	Chen et al. (2003)
"	"	"	"	5.150	Chen et al. (2001)
"	"	Micromelum	falcatum	5.48, 5.131	Luo et al. (2009)
"	"	Murraya	paniculata	5.93	Imai et al. (1989)
"	"	Orixa	japonica	5.138	Noshita et al. (2001)
"	"	"	"	5.92	Feng et al. (2004)
"	"	"	"	5.82, 5.129, 5.135, 5.143	Jian et al. (2023)
"	"	"	"	5.135–5.137	Ito et al. (2004)
"	"	Peltostigma	guatemalense	5.53	Cuca Suarez et al. (2009)
"	"	Phellodendron	amurense	5.53	Min et al. (2007)
"	"	Pilocarpus	grandiflorus	5.53	De Souza et al. (2005)

(Continued)

TABLE 5.2 (Continued)
Natural Occurrence of 2-Quinolinone Alkaloids

Kingdom	Family	Genus	Species	Compound(s)	References
"	"	*Ptelea*	*trifoliata*	5.145	Reisch et al. (1969); Novak et al. (1970)
"	"	"	"	5.133	Reisch et al. (1972)
"	"	"	"	5.37	Reisch et al. (1975)
"	"	"	"	5.145, 5.146	Korosi et al. (1976)
"	"	*Rauia*	*Resinosa*	5.53	Albarici et al. (2010)
"	"	*Ravenia*	*spectabilis*	5.81, 5.147, 5.148	Paul and Bose (1968)
"	"	"	"	5.139	Khan and Waterman (1990)
"	"	"	"	5.81, 5.12, 5.148	Haque et al. (2013)
"	"	*Ruta*	*chalepensis*	5.53	El Sayed et al. (2000)
"	"	"	*graveolens*	5.124	Azalework et al. (2017)
"	"	*Skimmia*	*laureola*	5.74, 5.75, 5.138, 5.144	Atta-ur-Rahman et al. (1998)
"	"	"	"	5.21–5.23, 5.138	Sultana et al. (2007)
"	"	*Spathelia (Sohnreyia)*	*excelsa*	5.99	Carvalho et al. (2012)
"	"	"	"	5.39	Moreira et al. (2009)
"	"	*Teclea*	*simplicifolia*	5.92	Wondimu et al. (1988)
"	"	*Tetradium*	*glabrifolium*	5.53, 5.95	Wu et al. (1995)
"	"	"	*ruticarpum*	5.52	Na et al. (2022)
"	"	*Toddalia*	*asiatica*	5.53	Tsai et al. (1996)
"	"	"	"	5.12, 5.25, 5.53	Zhu et al. (2019)
"	"	*Zanthoxylum*	*acutifolium*	5.77–5.79, 5.134	Arruda et al. (1992)
"	"	"	*ailanthoides*	5.53, 5.93	Cheng et al. (2005)
"	"	"	*asiaticum*	5.12, 5.25	Jain et al. (2006)
"	"	"	*avicennae*	5.2, 5.81, 5.92, 5.149	Ji et al. (2022)

(Continued)

TABLE 5.2 (Continued)
Natural Occurrence of 2-Quinolinone Alkaloids

Kingdom	Family	Genus	Species	Compound(s)	References
"	"	"	Budrunga (rhetsa)	5.121	Ahmad et al. (2003)
"	"	"	"	5.53	Lookpan et al. (2024)
"	"	"	coco	5.10	Garcia et al. (2020)
"	"	"	davyi	5.59	Tarus et al. (2006)
"	"	"	integrifoliolum	5.53, 5.119	Ishii et al. (1982)
"	"	"	limonella	5.120	Tangjitjaroenkun et al. (2012)
"	"	"	monophyllum	5.53	Rodríguez-Guzmán et al. (2010)
"	"	"	nitidium	5.53, 5.54	Deng et al. (2021)
"	"	"	"	5.165	Nguyen et al. (2019)
"	"	"	"	5.2, 5.93	Yang et al. (2012b)
"	"	"	"	5.53	Yang et al. (2009)
"	"	"	"	5.53, 5.92, 5.112, 5.126	Qin et al. (2023)
"	"	"	nitidum var. tomentosum	5.54, 5.73, 5.119	Liao et al. (2021)
"	"	"	rhetsa	5.106, 5.127	Zohora et al. (2022)
"	"	"	rhoifolium	5.64	Santiago-Brugnoli et al. (2013)
"	"	"	scandens	5.106, 5.166	Brader et al. (1992)
"	"	"	schinifolium	5.130, 5.132, 5.149	Brader et al. (1992)
"	"	"	"	5.53	Min (1998); Cheng et al. (2002)
"	"	"	simulans	5.130, 5.149	Brader et al. (1993)
"	"	"	"	5.93	Chen et al. (1994b)
"	"	"	"	5.164	Chen et al. (1994a)
"	"	"	tsihanimposa	5.53	Randrianarivelojosia et al. (2003)
"	"	"	wutaiense	5.53	Huang et al. (2011)
"	"	"	"	5.81	Huang et al. (2008)

to produce a wide range of 2-quinolinone alkaloids. For instance, several strains of *Aspergillus versicolor* are noted to produce compounds such as 5.31, 5.32, 5.18, and 5.162. The *Penicillium* genus is represented by numerous species, including *P. citrinum*, *P. crustosum*, *P. cyclopium*, *P. expansum*, and others, each producing different combinations of 2-quinolinones. The data spans research conducted from as early as 1952 to as recent as 2023, highlighting the ongoing interest in these fungal metabolites. Other fungal species also produce 2-quinolinones, such as *Paecilomyces* sp. YD-01 (Thermoascaceae), *Alternaria* sp. YD-01 (Pleosporaceae), and *Myrothecium verrucaria* (Stachybotryaceae).

5.2.4 PLANT SOURCES

About 80% of 2-quinolinones were isolated from plants, specifically from the Rutaceae family. The genera *Haplophyllum* (Mohammadhosseini et al. 2021), *Zanthoxylum* (Nhiem et al. 2020), and *Glycosmis* alone produce seventy 2-quinolinone alkaloids. Particular attention should be paid to the species *Dictamnus dasycarpus* (Qin et al. 2021), *Glycosmis pentaphylla* (Khandokar et al. 2021), and *Zanthoxylum nitidum* (Lu et al. 2020) due to their abundant alkaloid profiles, including 2-quinolinones. 4-Methoxy-1-methyl-2(1*H*)-quinolinone (5.53) is the most common compound in plants, followed by folimine (5.99), (*R*)-edulinine (5.92), 4-methoxy-2(1*H*)-quinolinone (5.54), and Atanine (5.81). The list of natural sources of 2-quinolinones derived from animals, bacteria, fungi, and plants is presented in Table 5.2.

In Figure 5.4, the distribution of 2-quinolinones across four kingdoms is depicted: Plantae, Bacteria, Animalia, and Fungi. Amongst these, plants are the most common, accounting for 80.5%, followed by fungi at 10.3%. Bacteria account for 6.9%, and animals make up the smallest proportion at 2.3%.

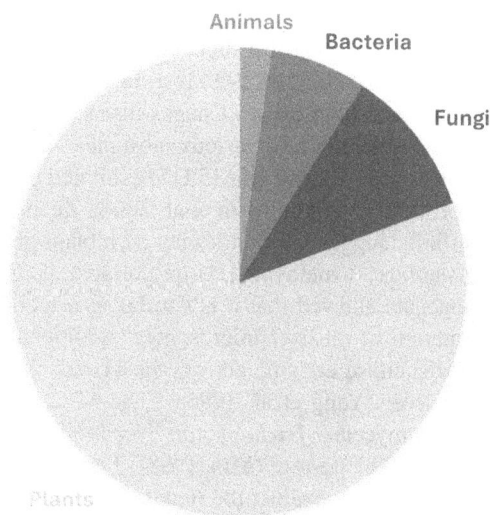

FIGURE 5.4 Distribution of 2-qSuinolinone alkaloids in various life kingdoms.

5.3 BIOLOGICAL ACTIVITY

Several 2-quinolinone alkaloids have a range of biological activities (Khadem and Marles 2025). In a study conducted by Carta et al. (2012), compound 5.1 demonstrated inhibitory activity against specific carbonic anhydrase (CA) isozymes. Although it did not inhibit the off-target isoform CA II, it effectively inhibited the human isoforms hCA III–IX, and XII–XIV with inhibition constant (K_I) values ranging from 0.033 to 0.750 μM. Sulfonamides, the most extensively studied class of inhibitors of the metalloenzyme CA, are associated with many side effects. Based on one study, 2-quinolinone (5.1) has the potential to be a lead molecule for identifying non-sulfonamide CA inhibitors. The enzyme inhibitory analysis of recombinant human chymase was measured for 4-hydroxy-1-methyl-2(1H)-quinolinone (5.44), which inhibited chymase with an IC_{50} value of 300 μM (Tani et al. 2004a, 2000b). Additionally, 4-hydroxy-3-methyl-2(1H)-quinolinone (5.50) exhibited inhibitory activity against *Mycobacterium tuberculosis* H37Ra at MIC 6.8 μM and showed weak cytotoxicity to MRC-5 human lung-derived fibroblasts with a CC_{50} of 84.7 μM (de Macedo et al. 2017). Its antioxidant activity was evaluated using the luminol chemiluminescence extinction assay (Li et al. 2020). The results indicated that compound 5.50, at 10 μM, quenched the hydroxy radical-induced chemiluminescence emitted by luminol by 86%. These data suggest that compound 5.50 has the potential to function as an antioxidant, thereby mitigating the immune response driven by reactive oxygen species in host organisms. The antibacterial activities of 4-hydroxy-3-methoxy-2(1H)-quinolinone (5.50) and 4-hydroxy-3-methoxy-1-methyl-2(1H)-quinolinone (5.48) against various pathogenic bacteria were assessed using the broth microdilution method to determine their minimum inhibitory concentrations (MICs). Compound 5.50 demonstrated a MIC ranging from 16 to 32 μg/mL against all tested pathogens, except for *P. aeruginosa* FRD1. Similarly, compound 5.48 exhibited a MIC of 16 μg/mL against three pathogenic bacteria, *i.e.*, two strains of *E. coli* and one of *E. faecalis*. Interestingly, both compounds displayed significant activity against two drug-resistant bacterial pathogens, namely *E. coli* ATCC 35218 and *E. faecalis* ATCC 51299. Testing the trypomastigote forms of Trypanosoma cruzi, a parasite that causes Chagas disease in humans, allows for a specific and effective assessment of how various substances can inhibit an important stage of the parasite's life cycle. Haplotusine (5.117) exhibited moderate trypanocidal activity with an IC_{50} of 136.9 μM (Ambrozin et al. 2008). Zeanic acid (5.167) stimulated the growth of alfalfa, suggesting that it may be a plant growth regulator. The inhibitory effect of 4-methoxy-1-methyl-2(1H)-quinolinone (5.53) on lipid peroxidation of rat liver homogenate showed that it is similar to α-tocopherol in inhibiting malondialdehyde formation in rat liver microsomes. Additionally, Compound 5.53 increased superoxide dismutase enzyme activity in ferrous-chloride-ascorbic-acid-ADP intoxicated mice liver (Yang et al. 1996a). The MIC for antifungal activity 5.53 against the ascomycete *Trichophyton mentagrophytes* KCTC 6077 (Arthrodermataceae) was 800 μg/mL (Min 1998). Furthermore, compound 5.53 exhibited antiplasmodial activity against the malaria parasite *Plasmodium* with an IC_{50} in the micromolar range (Randrianarivelojosia et al. 2003). Another study on 5.53 for antimalarial properties against two *Plasmodium falciparum* strains showed strong activity (65% inhibition) at a concentration of 35 ppm (Cuca Suarez et al. 2009). The P-glycoprotein-expressing cell lines MES-SA/DX5 and HCT15 displayed inhibition activities with ED_{50} values of 0.335 and 0.017 µg/mL, respectively, for 5.53 concerning multidrug resistance (MDR) reversal activity (Min et al. 2007). Moreover, a study evaluating the inhibition of HIV replication in H9 lymphocyte cells revealed moderate anti-HIV activity for 5.53 with an EC_{50} value of 0.625 µg/mL and a therapeutic index of 36.4 (Cheng et al. 2005). Additionally, Compound 5.53 showed significant activity against methicillin-resistant *Staphylococcus aureus* (MRSA), with an IC_{50} of 1.5 µg/mL vs. 0.06 µg/mL shown by the positive control, ciprofloxacin, and no toxicity up to 4.76 µg /mL (Rodríguez-Guzmán et al. 2010). Kim et al. (2009) found that 5.53 is a potent nitric oxide (NO) production inhibitor in microglia cells (LPS-activated BV-2), suggesting its potential in diseases associated with increased NO production (anti-inflammatory activity). Compound 5.53 showed a strong inhibitory effect on NO production stimulated by LPS in BV-2 microglial cells with an IC_{50} value of 2.5 µM (Gao et al. 2020). Integriquinolone 5.119 increased superoxide dismutase enzyme activity in ferrous-chloride-ascorbic-acid-ADP intoxicated mice liver (Yang et al. 1996a). Compound 5.119 also showed cytotoxic activity against the A549 cancer cell line (human lung adenocarcinoma) with an IC_{50} of 46.3 µM (Song et al. 2014). Fakhouri et al. (2001) investigated the antifungal activity of 1,2-dihydro-1-mercapto-2-oxo-4-quinolinecarboxaldehyde (5.4) against several phytopathogenic fungi *in vitro* and found that at 25 ppm, it inhibited spore germination and germ tube growth of various fungi. Ghate et al. (2018) reported that 7-hydroxy-6-methoxy-2(1*H*)-quinolinone (5.67) has an anti-inflammatory effect by inhibiting NF-κB activation. In the LPS-stimulated murine macrophage RAW 264.7 model, compound 5.67 demonstrated significant inhibition of the excessive production of pro-inflammatory mediators. Deng et al. (2022) tested 4-acetyl-7-isopropyl-5-methylquinolin-2(1*H*)-one (5.42) and 4-acetyl-6-methyl-7-(1-methylethyl)-2(1*H*)-quinolinone (5.41) for their anti-tobacco mosaic virus (anti-TMV) activity. They found that both compounds showed significant anti-TMV activity at a concentration of 20 µM, displaying inhibition rates of 42.2% and 38.5%, respectively. The inhibition rates of compounds 5.41 and 5.42 in inhibiting TMV surpassed that of the positive control, ningnanmycin, indicating their notable effectiveness. Viridicatin (5.162) demonstrated activity against *Mycobacterium tuberculosis* at a dilution of 67 ppm (Cunningham and Freeman 1953) and was also effective in inhibiting UVA-induced ROS generation in HaCaT cells (Teixeira et al. 2021). Additionally, Viridicatin was reported to inhibit matrix metalloproteinases MMP 2 and MMP-9, suggesting an anti-ageing effect (Liang et al. 2019). Viridicatol (5.163) was found to have anti-inflammatory properties, demonstrating a notable mitigating effect in cells exposed to LPS stimulation (Ko et al. 2015). It inhibited the expression of pro-inflammatory mediators and exerted a suppressive influence on the production of iNOS-derived nitric oxide (NO) and COX-2-derived prostaglandin E2 (PGE2) in LPS-stimulated cells. Viridicatol also exhibited a downregulating impact on the *m*RNA expression of key pro-inflammatory cytokines and inhibited the nuclear factor-kappa B (NF-κB) pathway. Atanine (5.81) demonstrated antiparasitic/anthelmintic properties, showing activity against miracidial and cercarial larvae of the human parasite *Schistosoma* *mansoni* Sambon 1907 and adults and larvae of the non-parasite soil nematode *Caenorhabditis elegans* at a concentration of 10 μg/mL (Perrett and Whitfield 1995). Other compounds such as 6-bromo-4-hydroxy-2(1*H*)-quinolinone (5.61), 6,7-dibromo-4-hydroxy-2(1*H*)-quinolinone (5.60), methyl-1,2-dihydro-6-hydroxy-2-oxo-4-quinolinecarboxylate (5.123), methyl 1,2-dihydro-4-methoxy-1-methyl-2-oxo-3-quinolinecarboxylate (5.6) and dasycarine E (5.90) also displayed various activities against specific targets. Compound 5.6 demonstrated potent inhibition of NO with an IC_{50} value of 1.3 μM, while 5.90 showed a moderate inhibitory effect with an IC_{50} of 12.3 μM (Gao et al. 2020). As immune stimulators, ethyl 1,2-dihydro-6-methoxy-2-oxo-4-quinolinecarboxylate (5.94) and ethyl 1,2-dihydro-6-hydroxy-2-oxo-4-quinolinecarboxylate (5.43) significantly inhibited NO generation in the macrophage cell line (murine RAW 264.7) and murine splenocytes at 20 μg/mL, with IC_{50} values of 20.8 and 32.4 μM, respectively (Ryu and Chung 2010). 4-Acetyl-6-methyl-7-(1-methylethyl)-2(1*H*)-quinolinone (5.41) exhibited strong anti-TMV activity with a 38.5% inhibition rate, higher than the reference drug ningnanmycin, which had a relative inhibition rate of 33.0%. Researchers tested two compounds, 7-hydroxy-6-methyl-4-(4-methylfuran-2-yl)quinolin-2(1*H*)-one (5.68) and 7-methoxy-6-methyl-4-(4-methylfuran-2-yl)quinolin-2(1*H*)-one (5.69), for their ability to fight the TMV at a concentration of 20 μM. The results showed that they have good anti-TMV activity, with inhibition rates of 49.2% for 5.68 and 38.4% for 5.69 (Zheng et al. 2023). Raistrimide D (5.169) showed antimicrobial activity against *Staphylococcus aureus* and *Escherichia coli* with MIC values of 12.5 and 50 μg/mL, respectively, evaluated using the micro-broth dilution method. 3-*O*-Methylviridicatin (5.31) was found to block the activation of the human immunodeficiency virus long terminal repeat (HIV LTR) by the cytokine tumour necrosis factor α (TNF-α), with an IC_{50} of 5 μM. It also inhibited virus production in the OM-10.1 cell line, a model of chronic infection responsive to TNF-α induction, with an IC_{50} value of 2.5 μM (Heguy et al. 1998). For its antimicrobial activity, the MICs of the compound 5.31 against the human pathogenic bacteria *Staphylococcus aureus* ATCC 6538, *Listeria monocytogenes* ATCC 19117, and *Salmonella typhimurium* ATCC 14028 were significantly lower than those of kanamycin, at 4.5, 9.0, and 4.5 μg/mL, respectively (El Euch et al. 2018). *O*-Methylviridicatin also displayed inhibitory activity against influenza virus neuraminidase (Liu et al. 2021a, 2021b). Glycocitridine (5.105) demonstrated effective cytotoxicity with an ED_{50} value of 0.52 μg/mL against the HT-29 cell line *in vitro* (Chen et al. 2003). 7-Hydroxy-4-[5-(hydroxymethyl)-2-furanyl]-2(1*H*)-quinolinone (5.66) demonstrated cytotoxicity against GLC-82 and HCT cancer cell lines, with IC_{50} values of 8.8 μM and 10.1 μM, respectively (Chen et al. 2002a). 3-*O*-Methylviridicatol (5.32) showed cytotoxic activity against MCF-7 and SMMC-7721 cell lines, with IC_{50} values of 16.6 μmol/L and 18.2 μmol/L, respectively (Li et al. 2016). For its antimalarial activity, 3-[[(2R)-3,3-dimethyl-2-oxiranyl] methyl]-4-methoxy-1-methyl-2(1*H*)-quinolinone (5.26) was tested against chloroquine-resistant *Plasmodium falciparum* strain Dd2 and displayed 74% inhibition at 80 μM. 3,6-Dimethoxy-4-phenyl-2(1*H*)-quinolinone (5.18) displayed antibacterial activity against *Ralstonia solanacearum* (MIC: 50 μg/mL) and *Xanthomonas campestris* (MIC: 100 μg/mL) (Fu et al. 2020). 3-[(3,3-Dimethyl-2-oxiranyl) methyl]-4-hydroxy-7-methoxy-1-methyl-2(1H)-quinolinone   (5.25)   and   3-(2,3-dihydroxy-3-methylbutyl)-4,7-dimethoxy-1-methyl-2(1H)-quinolinone (5.12) exhibited strong antibacterial activity against *E. coli* (0157:H7), *Bacillus cereus* (M.1.16), and *Lactobacillus lactis* (Jain et al. 2006). 3-(3-methyl-2-buten-1-yl)-4-[(3-methyl-2-buten-1-yl)oxy]-2(1H)-quinolinone (5.15) showed modest anti-HIV activity against HIV-1, with an EC_{50} of 1.64 µM and an IC_{50} of 26.9 µM. Orixalone A (5.135) inhibited nitric oxide (NO) production in murine macrophage-like RAW 264.7 cells. Glycopentanolone A (5.107) and glycopentanolone B (5.108) were tested for their anti-allergic activity; compound 5.107 showed potent inhibitory activity. 4-Methoxy-3-(3-methyl-2-buten-1-yl)-7-[(3-methyl-2-buten-1-yl)oxy]-2(1H)-quinolinone (5.55) was tested for its cytotoxic and antiplasmodial activity (Tanjung et al. 2017). It showed high cytotoxic activity with an IC_{50} value of 0.63 µg/mL against murine lymphocytic leukaemia P-388 cells. In the antiplasmodial assay, 5.55 exhibited a moderate activity with an IC_{50} value of 4.28 µg/mL for parasitaemia decrease. Bucharidine (5.84) was tested for oestrogenic activity in immature rats. Administration of 10 µg/g bucharidine increased uterine mass without liquid by 48.5%, indicating oestrogenic activity. Further administration of 50 µg/g of Bucharidine resulted in a 228.5% increase in uterine mass with liquid, confirming its oestrogenic effect. The antimicrobial activity of chrysanthemumsides A (5.85) and B (5.86) was tested against various bacteria. Chrysanthemumside B showed activity against *Staphylococcus aureus*, *Escherichia coli*, *Pseudomonas aeruginosa*, and *Candida albicans* with MIC values ranging from 3.9 to 7.8 µg/mL. Chrysanthemumside A displayed moderate to low antimicrobial activity against these bacteria with MICs ranging from 15.6 to 31.2 µg/mL. Flustramine Q (5.96) exhibited anti-inflammatory and acetylcholinesterase (AChE) inhibitory activities. It decreased the secretion of the pro-inflammatory cytokine interleukin IL-12p40 by around 55% at a concentration of 10 µg/mL. In AChE inhibitory activity, flustramine Q showed 82% inhibition at a concentration of 100 µM, with an IC_{50} value of 9.6 µM. Finally, six 2-quinolinone compounds (SF 2809 I to SF 2809 VI, 5.151 to 5.156) were assessed for their effects on human chymase, human cathepsin G, and bovine pancreatic chymotrypsin. Compounds SF 2809 II (5.152), IV (5.154), V (5.155), and VI (5.156) significantly inhibited chymase and cathepsin G, while not affecting chymotrypsin at a 20 µM concentration. Compounds SF 2809 I (5.151) and III (5.153) displayed weaker inhibition of chymase. In Table 5.3, the biological activity of 2-quinolinone alkaloids was summarized.

The data shows that out of 169 compounds, there are 49 with various biological activities (Khadem and Marles 2025). The main bioactivities identified were as follows: antibacterial (10 compounds), antiviral (7 compounds), chymase (enzyme) inhibition (7 compounds), cytotoxicity (6 compounds), inhibition of NO production (6 compounds), and antioxidant properties (5 compounds). Additionally, other observed activities include Cathepsin G (enzyme) inhibition, anti-inflammatory effects, and antifungal properties (10 compounds).

TABLE 5.3

Biological Activity of 2-Quinolinone Alkaloids

Biological Activity	Target	Assay, Method	Results	Compound	References
Acetylcholinesterase (AChE) Inhibitor	Acetylcholinesterase (AChE)	In vitro colorimetric assay	82% inhibition at 100 μM; $IC_{50} = 9.6$ μM	5.96	Kowal et al. (2023)
Antiallergic	Inhibition of β-hexosaminidase release from RBL-2H3 cells	Not specified	IC_{50} of 4.3 μM	5.107	Choi et al. (2019)
"	Inhibition of β-hexosaminidase release from RBL-2H3 cells	Not specified	IC_{50} of 85.7 μM	5.108	Choi et al. (2019)
Antibacterial	Mycobacterium tuberculosis H37Ra	IC_{90} assay	Inhibited *M. tuberculosis* H37Ra at IC_{90} 6.8 μM	5.50	de Macedo et al. (2017)
"	Escherichia coli ATCC 25922, E. coli ATCC 35218, Staphylococcus aureus ATCC 2592, Enterococcus faecalis ATCC 51299, Pseudomonas aeruginosa FRD1	Broth microdilution method	Exhibited MIC ranging from 16 to 32 μg/mL against all tested pathogens except *P. aeruginosa* FRD1	5.49	Yao et al. (2021)
"	Escherichia coli ATCC 25922, E. coli ATCC 35218, Enterococcus faecalis ATCC 51299	Broth microdilution method	Compound 5.48 exhibited MIC of 16 μg/mL against two *E. coli* strains and one *E. faecalis* strain	5.48	Yao et al. (2021)
"	Drug-resistant Escherichia coli ATCC 35218 and Enterococcus faecalis ATCC 51299	Broth microdilution method	Both compounds 5.48 and 5.49 showed significant activity against these drug-resistant pathogens	5.48, 5.49	Yao et al. (2021)
"	Mycobacterium tuberculosis	In vitro assay	Active at a dilution of 67 ppm	5.162	Cunningham and Freeman (1953)
"	Ralstonia solanacearum, Xanthomonas campestris	MIC determination	MIC values of 50 μg/mL against *R. solanacearum* and 100 μg/mL against *X. campestris*	5.18	Fu et al. (2020)

(Continued)

TABLE 5.3 (Continued)
Biological Activity of 2-Quinolinone Alkaloids

Biological Activity	Target	Assay, Method	Results	Compound	References
"	E. coli (O157), Bacillus cereus (M.1.16), Lactobacillus lactis (NCC946)	Inhibition at 16 µg/mL	Inhibition ranges: 66%–83% for 5.25 and 70%–81% for 5.12	5.25, 5.12	Jain et al. (2006)
"	Staphylococcus aureus, Escherichia coli	Micro-broth dilution method	MIC values of 12.5 µg/mL against S. aureus and 50 µg/mL against E. coli	5.169	Zhong et al. zhu(2023)
"	Staphylococcus aureus, Listeria monocytogenes, Salmonella typhimurium	Micro-broth dilution method	MIC values of 4.5, 9.0, and 4.5 µg/mL, respectively; lower than kanamycin	5.31	El Euch et al. (2018)
"	Staphylococcus aureus (ATCC 6538), Escherichia coli (8099), and Pseudomonas aeruginosa (ATCC 15442)	MIC	MIC = 3.9, 7.8, and 3.9 µg/mL, respectively	5.86	Zhu et al. (2023b)
Anticancer	P-glycoprotein expressed cell lines (MES-SA/DX5, HCT15)	Cell line inhibition assay; Multidrug resistance (MDR) reversal	ED_{50} values of 0.335 µg/mL (MES-SA/DX5) and 0.017 µg/mL (HCT15)	5.53	Min et al. (2007)
Antifungal	Trichophyton mentagrophytes KCTC 6077	MIC determination	MIC of 800 µg/mL	5.53	Min (1998)
"	Spore germination and germ tube growth of Fusarium oxysporum f. sp. lycopersici, Fusarium culmorum, Cladosporium cucumerinum, Colletotrichum lagenarium	In vitro antifungal assay	Inhibited spore germination and germ tube growth at 25 ppm	5.4	Fakhouri et al. (2001)
"	Germination of teliospores of Ustilago maydis and Ustilago nuda	In vitro antifungal assay	Inhibited germination of teliospores at 100 ppm	5.4	Fakhouri et al. (2001)
"	Candida albicans (ATCC 10231)	MIC	MIC = 7.8 µg/mL	5.86	Zhu et al. (2023)b
Anti-inflammatory	NF-κB activation	Inhibition assay	Inhibited NF-κB activation	5.67	Ghate et al. (2018)

(Continued)

TABLE 5.3 (Continued)

Biological Activity of 2-Quinolinone Alkaloids

Biological Activity	Target	Assay, Method	Results	Compound	References
"	Pro-inflammatory mediators (NO, TNF-α, IL-6, IL-1β, ROS)	LPS-stimulated murine macrophage RAW 264.7 model	Significant inhibition of excessive production of pro-inflammatory mediators induced by LPS	5.67	Ghate et al. (2018)
"	NO-synthesizing enzyme, cyclooxygenase-2, TNF-α, IL-6, IL-1β	Expression level analysis	Expression levels of key enzymes and cytokines returned to baseline levels following treatment	5.67	Ghate et al. (2018)
"	Pro-inflammatory mediators (iNOS, COX-2), NO, PGE2	LPS-stimulated RAW264.7 and BV2 cells	Inhibited expression of iNOS and COX-2; suppressed production of iNOS-derived NO and COX-2-derived PGE2	5.163	Ko et al. (2015)
"	IL-1β, IL-6, TNF-α	mRNA expression analysis	Downregulated mRNA expression of IL-1β, IL-6, and TNF-α	5.163	Ko et al. (2015)
"	Dendritic Cells (DC)	In vitro cytokine secretion	Decreased IL-12p40 by ~55% at 10 μg/mL	5.96	Di et al. (2020)
Antimalarial	Plasmodium falciparum strains	In vitro assay	Strong activity (65% inhibition) at 35 ppm	5.53	Cuca Suarez et al. (2009)
"	Chloroquine-resistant Plasmodium falciparum strain Dd2	In vitro antimalarial assay	74% inhibition at 80 μM	5.26	Yang et al. (2011)
Antimicrobial	Staphylococcus aureus, Escherichia coli, Pseudomonas aeruginosa, Candida albicans	MIC	15.6 to 31.2 μg/mL (moderate to low)	5.85	Zhu et al. (2023b)
Anti-MRSA	Methicillin-resistant Staphylococcus aureus (MRSA)	IC_{50} determination	IC_{50} of 1.5 μg/mL; no toxicity up to 4.76 μg/mL	5.53	Rodríguez-Guzmán et al. (2010)

(Continued)

TABLE 5.3 (Continued)
Biological Activity of 2-Quinolinone Alkaloids

Biological Activity	Target	Assay, Method	Results	Compound	References
Antioxidant	Reactive oxygen species	Luminol chemiluminescence extinction assay	Quenched hydroxy radical-induced chemiluminescence by 86% at 10 μM	5.50	Li et al. (2020)
"	Lipid peroxidation in rat liver homogenate	Lipid peroxidation assay	Similar inhibitory effect as α-tocopherol in inhibiting malondialdehyde formation in rat liver microsomes	5.53	Yang et al. (1996)
"	SOD enzyme activity in mice liver	Enzyme activity assay	Increased SOD activity in ferrous–chloride–ascorbic-acid–ADP intoxicated mice liver	5.53, 5.119	Yang et al. (1996)
"	DPPH free radicals	DPPH free radical scavenging assay	Moderate antioxidant activity	5.123	Chung and Woo (2001)
Antioxidant/ photoprotective	ROS generation in HaCaT cells	In vitro ROS inhibition assay	Effective in inhibiting UVA-induced ROS generation	5.162	Teixeira et al. (2021)
Antiparasitic/ anthelmintic activity	Miracidial and cercarial larvae of Schistosoma mansoni, adults and larvae of Caenorhabditis elegans	In vitro assay against parasites and nematodes	Active against miracidial and cercarial larvae with 50%–80% immobility; effective against adults and larvae at 10 μg/mL	5.81	Perrett and Whitfield (1995)
Antiplasmodial	Plasmodium species	In vitro assay	Exhibited antiplasmodial activity with an IC_{50} in the micromolar range	5.53	Randrianarivelojosia et al. (2003)
"	Plasmodium falciparum	Parasitaemia decrease assay	IC_{50} of 4.28 μg/mL	5.55	Tanjung et al. (2017)

(Continued)

TABLE 5.3 (*Continued*)
Biological Activity of 2-Quinolinone Alkaloids

Biological Activity	Target	Assay, Method	Results	Compound	References
Antiviral	HIV replication in H9 lymphocyte cells	In vitro HIV inhibition assay	Moderate anti-HIV activity with an EC_{50} of 0.625 µg/mL and a therapeutic index of 36.4	5.53	Cheng et al. (2005)
"	Tobacco mosaic virus (TMV)	In vitro anti-TMV activity assay	Inhibition rate of 42.2% at 20 µM	5.42	Deng et al. (2022)
"	Tobacco mosaic virus (TMV)	In vitro anti-TMV activity assay	Inhibition rate of 38.5% at 20 µM	5.41	Deng et al. (2022)
"	Tobacco mosaic virus (TMV)	Inhibition assay	38.5% inhibition rate, higher than reference ningnanmycin (33.0%)	5.41	Liu et al. (2021)b
"	Tobacco mosaic virus (TMV)	Inhibition assay at 20 µM	49.2% inhibition rate	5.68	Zheng et al. (2023)
"	Tobacco mosaic virus (TMV)	Inhibition assay at 20 µM	38.4% inhibition rate	5.69	Zheng et al. (2023)
"	HIV long terminal repeat (LTR)	HeLa-based system	IC_{50} value of 5 µM	5.31	Heguy et al. (1998)
"	HIV virus production (OM-10.1 cell line)	OM-10.1 cell line model	IC_{50} value of 2.5 µM	5.31	Heguy et al. (1998)
"	Influenza virus neuraminidase	Not specified	Inhibitory activity	5.31	Liu et al. (2021a)
"	HIV-1 in CEM-SS cells	XTT-tetrazolium assay	$EC_{50} = 1.64$ µM, $IC_{50} = 26.9$ µM	5.15	McCormic et al. (1996)
"	HIV-1 reverse transcriptase (RT)	RT assay	$IC_{50} = 8$ µM	5.15	McCormic et al. (1996)

(*Continued*)

TABLE 5.3 (Continued)
Biological Activity of 2-Quinolinone Alkaloids

Biological Activity	Target	Assay, Method	Results	Compound	References
Cathepsin G Inhibition	Human Cathepsin G	In vitro enzyme assay	$IC_{50} = 1.0$–$6.3\ \mu M$	5.152, 5.154, 5.155, 5.156	Tani et al. (2004a); Tani et al. (2004b)
Chymase Inhibition	Recombinant human chymase	Enzyme inhibitory analysis	Inhibited chymase with an IC_{50} value of 300 μM	5.44	Tani et al. (2004a); Tani et al. (2004b)
"	Human Chymase	In vitro enzyme assay	$IC_{50} = 0.014$–$0.081\ \mu M$	5.152, 5.154, 5.155, 5.156	Tani et al. (2004a); Tani et al. (2004b)
"	Human Chymase	In vitro enzyme assay	$IC_{50} = 7.3\ \mu M$	5.151	Tani et al. (2004a); Tani et al. (2004b)
"	Human Chymase	In vitro enzyme assay	$IC_{50} = 2.1\ \mu M$	5.153	Tani et al. (2004a); Tani et al. (2004b)
Cytotoxicity	Murine lymphocytic leukemia P-388 cells	Cytotoxic evaluation GI_{50} assay	IC_{50} of 0.63 $\mu g/mL$	5.55	Tanjung et al. (2017)
"	MRC-5 human lung-derived fibroblasts		Weak cytotoxicity to MRC-5 with GI_{50} 84.7 μM	5.50	de Macedo et al. (2017)
"	A549 cancer cell line (human lung adenocarcinoma)	In vitro cytotoxicity assay	IC_{50} of 46.3 μM	5.119	Song et al. (2014)
"	P-388 (murine lymphocytic leukemia), A549 (human lung adenocarcinoma), HT-29 (human colon carcinoma)	In vitro cytotoxicity assay	ED_{50} value of 0.52 $\mu g/mL$ against HT-29 cell line	5.105	Chen et al. (2003)

(Continued)

TABLE 5.3 (Continued)
Biological Activity of 2-Quinolinone Alkaloids

Biological Activity	Target	Assay, Method	Results	Compound	References
''	GLC-82 (human lung cancer), HCT (human colon cancer)	In vitro cytotoxicity assay	IC_{50} values of 8.8 μM against GLC-82 and 10.1 μM against HCT	5.66	Chen et al. (2002a)
''	MCF-7 (human breast carcinoma), SMMC-7721 (human liver cancer)	In vitro cytotoxicity assay	IC_{50} values of 16.6 μmol/L against MCF-7 and 18.2 μmol/L against SMMC-7721	5.32	Li et al. (2016)
Oestrogenic Activity	Immature rats	Administration of compound	10 μg/g: 48.5% increase in uterine mass without liquid; 50 μg/g: 228.5% increase in uterine mass with liquid	5.84	Nazrullaev et al. (2001)
Inhibition of matrix metalloproteinases MMP-2 and MMP-9	Matrix metalloproteinases MMP-2 and MMP-9	Inhibition assay in fibrosarcoma cell line HT 1080	Inhibited MMP-2 and MMP-9, triggering gelatinases degradation; suggested anti-ageing and protective effect	5.162	Liang et al. (2019)
Inhibition of NF-κB pathway	NF-κB, IκB-α	Inhibition of NF-κB pathway in LPS-stimulated RAW264.7 and BV2 cells	Impeded phosphorylation and degradation of IκB-α; hindered translocation of NF-κB p65 and p50 into the nucleus; attenuated NF-κB DNA-binding activity	5.163	Ko et al. (2015)
Inhibition of NO production	Nitric oxide (NO)	Murine RAW 264.7 macrophage cell line and murine splenocytes	Significant inhibition at 20 μg/mL; IC_{50} values of 20.8 μM and 32.4 μM, respectively	5.94, 5.43	Ryu and Chung (2010)

(Continued)

TABLE 5.3 (Continued)

Biological Activity of 2-Quinolinone Alkaloids

Biological Activity	Target	Assay, Method	Results	Compound	References
"	NO production in LPS-activated BV-2 microglia cells	NO production inhibition assay	Strong inhibitory effect on NO production with an IC_{50} of 2.5 μM	5.53	Gao et al. (2020); Kim et al. (2009)
"	NO	LPS-stimulated BV-2 microglial cells	Strong NO inhibition with an IC_{50} value of 1.3 μM	5.6	Gao et al. (2020)
"	NO	LPS-stimulated BV-2 microglial cells	Moderate NO inhibition with an IC_{50} value of 12.3 μM	5.90	Gao et al. (2020)
"	LPS/IFN-γ-induced NO production in RAW 264.7 cells	Not specified	47.3% inhibition at 10 μM; 54.8% inhibition at 50 μM	5.135	Ito et al. (2004)
Inhibitory activity	Isozyme-selective carbonic anhydrase (CA)	Enzymatic inhibition assay	Inhibited human isoforms hCA III-IX, XII-XIV with inhibition constant (K_I) values of 0.033–0.750 μM; Suggested as a potential lead for non-sulfonamide CA inhibitors	5.1	Carta et al. (2012)
Inhibitory activity	Neuronal isozyme of nitric oxide synthase (nNOS)	In vitro inhibition assay	Mild selective inhibitory activity	5.60, 5.61	Aoki et al. (2001)
Stimulation of plant growth	Alfalfa (*Medicago sativa*)	Plant growth assay	Stimulated the growth of alfalfa, indicating potential as a plant growth regulator	5.167	Matsushima et al. (1973)
Trypanocidal	*Trypanosoma cruzi* trypomastigote forms	IC_{50} assay	IC_{50} of 136.9 μM	5.117	Ambrozin et al. (2008)

5.4 DISCUSSION

The ongoing research in 2-quinolinone synthesis is advancing efficiency, selectivity, and sustainability. Current trends suggest that future developments may focus on several key areas. Firstly, the integration of biocatalysis with traditional synthetic methods is an emerging field. Researchers are exploring the use of engineered enzymes to perform selective transformations in synthesizing 2-quinolinones. This potentially offers unparalleled selectivity under mild, aqueous conditions. Secondly, the application of flow chemistry to 2-quinolinone synthesis is gaining attention. Continuous flow reactors are beneficial for large-scale production because they are easy to scale up, safe to use, and allow for better control of processes. These advantages could help in making these compounds more effective in industry. Thirdly, the development of dual catalysis systems, combining transition metal catalysis with organocatalysts or photocatalysis, is opening new paths for complex transformations. For example, merging photoredox catalysis with nickel catalysis has enabled challenging C–H functionalization reactions in the synthesis of 2-quinolinones under exceptionally mild conditions. Lastly, computational chemistry is playing an increasingly important role in reaction design and optimization. High-throughput virtual screening of catalysts and reaction conditions, coupled with machine learning algorithms, accelerates the discovery of novel synthetic routes to 2-quinolinones. Chemists are using advanced computational tools to understand how chemicals react with each other. These tools help predict how a reaction will happen, which products will form, and what side reactions might occur.

This chapter discussed the various biological activities of natural 2-quinolinones, emphasizing their roles in antimicrobial, antiviral, anti-inflammatory, antiparasitic, anticancer, and other activities. It highlighted the importance of studying these compounds for drug discovery. To further explore the biological activities of natural 2-quinolinones, one can utilize techniques that have been recently applied to other NPs, drugs, pesticides, and toxins. This includes high-throughput *in vitro* screening assays (Hubbard et al. 2019), transcriptomics (Harrill et al. 2021), and toxicological new approach methodologies (Stucki et al. 2022). Implementing these approaches to study natural 2-quinolinones would yield comprehensive datasets that could be analysed through quantitative structure–activity relationship (QSAR) analyses (Selvaraj et al. 2021; Soares et al. 2022). Ultimately, this would enable the rational design and optimization of new analogues with improved pharmacological properties for targeted therapeutic uses.

REFERENCES

Ahmad, M. U., Rahman, M. A., Huq, E., & Chowdhury, R. (2003). Alkaloids of *Zanthoxylum budrunga*. *Fitoterapia, 74*(1–2), 191–193.

Ahsan, M., Gray, A. I., Waterman, P. G., & Armstrong, J. A. (1994). 4-quinolinone, 2-quinolinone, 9-acridanone, and furoquinoline alkaloids from the aerial parts of *Boronia bowmanii*. *Journal of Natural Products, 57*(5), 670–672.

Akhmedzhanova, V. I. (1999). New alkaloids from *Haplophyllum*. *Chemistry of Natural Compounds, 35*, 552–553.

Akhmedzhanova, V. I., Bessonova, I. A., & Yunusov, S. Y. (1976). Alkaloids of *Haplophyllum perforatum*. *Chemistry of Natural Compounds, 12*(3), 282–288.

Akhmedzhanova, V. I., Bessonova, I. A., & Yunusov, S. Y. (1980). Alkaloids of *Haplophyllum foliosum*. *Chemistry of Natural Compounds, 16*(6), 574–576.

Akhmedzhanova, V. I., Bessonova, I. A., & Yunusov, S. Y. (1985). Alkaloids of *Haplophyllum foliosum*. III. Structure of folidine. *Chemistry of Natural Compounds, 21*, 782–783.

Albarici, T. R., Vieira, P. C., Fernandes, J. B., Silva, M. F. D. G. F. D., & Pirani, J. R. (2010). Cumarinas e alcaloides de *Rauia resinosa* (Rutaceae). *Química Nova, 33*, 2130–2134.

Al-ghazzawi, A. M., Alshehri, M. A., Deeb, A. A., & Alalmie, A. Y. (2020). Mass spectra analysis of quinoline alkaloids detected in *Sauuda fruticose* leads to novel biosynthesis pathway of quinoline alkaloids. doi: 10.21203/rs.2.24342/v1

Aly, A. A., Ramadan, M., Abuo-Rahma, G. E. A., Elshaier, Y. A., Elbastawesy, M. A., Brown, A. B., & Braese, S. (2021). Chapter three: Quinolones as prospective drugs: Their syntheses and biological applications. *Advances in heterocyclic chemistry* (pp. 147–196). Elsevier.

Ambrozin, A. R. P., Vieira, P. C., Fernandes, J. B., Silva, M. F. D. G. F. D., & Albuquerque, S. D. (2008). New pyranoflavones and trypanocidal activity of compounds isolated from *Conchocarpus heterophyllus*. *Quimica Nova, 31*, 740–743.

Aoki, S., Ye, Y., Higuchi, K., Takashima, A., Tanaka, Y., Kitagawa, I., & Kobayashi, M. (2001). Novel neuronal nitric oxide synthase (nNOS) selective inhibitor, aplysinopsin-type indole alkaloid, from marine sponge *Hyrtios erecta*. *Chemical and Pharmaceutical Bulletin, 49*(10), 1372–1374.

Arba, M., Arfan, A., Yamin, Y., & Zubair, M. S. (2023). The potential of *Clerodendrum paniculatum* leaves fraction as a 3-chymotrypsin-like (3CL) protease inhibitor of SARS-CoV-2. *Indonesian Journal of Chemistry, 23*(3), 770–781.

Arruda, M. S., Fernandes, J. B., da Silva, M. F. d. G., Vieira, P. C., & Pirani, J. R. (1992). Quinolone alkaloids from *Zanthoxylum acutifolium*. *Phytochemistry, 31*(10), 3617–3619.

Atalay, S. S., Assad, M. Y., Amagata, T., & Wu, W. (2020). Mild, efficient, and solvent-free synthesis of 4-hydroxy-2-quinolinones. *Tetrahedron Letters, 61*(16), 151778.

Atta-ur-Rahman, Sultana, N., Choudhary, M. I., Shah, P. M., & Khan, M. R. (1998). Isolation and structural studies on the chemical constituents of *Skimmia laureola*. *Journal of Natural Products, 61*(6), 713–717.

Azalework, H. G., Sahabjada, A. J., Arshad, T. M., & Malik, T. (2017). Phytochemical investigation, GC-MS profile and antimicrobial activity of a medicinal plant *Ruta graveolens* L. from Ethiopia. *International Journal of Pharmacy and Pharmaceutical Sciences, 9*(6), 29–34.

Banbury, L. K., Shou, Q., Renshaw, D. E., Lambley, E. H., Griesser, H. J., Mon, H., & Wohlmuth, H. (2015). Compounds from *Geijera parviflora* with prostaglandin E2 inhibitory activity may explain its traditional use for pain relief. *Journal of Ethnopharmacology, 163*, 251–255.

Bessonova, I. A. (2000). Components of *Haplophyllum bucharicum*. *Chemistry of Natural Compounds, 36*(3), 323–324.

Bessonova, I. A., Batsurén, D., & Yunusov, S. Y. (1984). Alkaloids of *Haplophyllum dauricum*. *Chemistry of Natural Compounds, 20*, 68–70.

Bessonova, I. A., & Yunusov, S. Y. (1986). Haplaphine—A new alkaloid of *Haplophyllum perforatum*. *Chemistry of Natural Compounds, 22*(5), 619–620.

Bessonova, I. A., & Yunusov, S. Y. (1989). Alkaloids of *Haplophyllum bungei*. *Chemistry of Natural Compounds, 25*, 18–20.

Bettoni, L., Joly, N., Mendas, I., Moscogiuri, M. M., Lohier, J., Gaillard, S., Poater, A., & Renaud, J. (2024). Iron-catalyzed synthesis of substituted 3-arylquinolin-2(1H)-ones via an intramolecular dehydrogenative coupling of amido-alcohols. *Organic & Biomolecular Chemistry*. doi: 10.1039/d4ob00649f

Beyerman, H. C., & Rooda, R. W. (1959). Isolation, characterization, and structural studies on six alkaloids from *lunasia amara* blanco. *Proc.Koninkl.Nederl.Akad.Wetenschap. Amsterdam,* 187–199.

Bhattacharyya, P., & Chowdhury, B. K. (1984). Glycophylone: A new quinolone alkaloid from *Glycosmis pentaphylla. Chemistry & Industry, 9,* 352.

Bhattacharyya, P., & Chowdhury, B. K. (1985). Glycolone, a quinolone alkaloid from *Glycosmis pentaphylla. Phytochemistry, 24*(3), 634–635.

Bhide, K. S., Mujumdar, R. B., & Rama Rao A. V. (1977). Phenolic from the bark of *Chloroxylon swietenia* DC. *Indian Journal of Chemistry, Section B: Organic Chemistry Including Medicinal Chemistry, 15B*(5), 440–444.

Birkinshaw, J. H., Luckner, M., Mohammed, Y. S., Mothes, K., & Stickings, C. E. (1963). Studies in the biochemistry of micro-organisms. 114. Viridicatol and cyclopenol, metabolites of *Penicillium viridicatum* westling and *Penicillium cyclopium* westling. *Biochemical Journal, 89*(2), 196.

Bob-Chile, A. A., Amadi, P. U. (2020). Nutritional, phytoconstituents, and free radical scavenging potentials of *Cola lepidota* K. Schum. and *Irvingia gabonensis* (Aubry-Lecomte ex O'Rorke) baill. leaves. *Brazilian Journal of Biological Sciences, 7*(15), 43–57.

Bracken, A., Pocker, A., & Raistrick, H. (1954). Studies in the biochemistry of micro-organisms. 93. Cyclopenin, a nitrogen-containing metabolic product of *Penicillium cyclopium* westling. *Biochemical Journal, 57*(4), 587.

Brader, G., Wurz, G., & Greger, H. (1992). Organ-specific accumulation of novel prenylated quinolin-2-ones from East Asian *Zanthoxylum* species. *Planta Medica,* 58(S1), 692–693.

Brader, G., Wurz, G., Greger, H., & Hofer, O. (1993). Novel prenylated 2-Quinolinones from East Asian *Zanthoxylum* species. *Liebigs Annalen Der Chemie, 4,* 355–358.

Buckingham, J. (2023). *Dictionary of natural products, supplement 2.* Boca Raton, FL: Routledge.

Buckingham, J., Baggaley, K. H., Roberts, A. D., & Szabo, L. F. (2010). *Dictionary of alkaloids with CD-ROM.* Boca Raton, FL: CRC Press.

Campbell, W. E., Davidowitz, B., & Jackson, G. E. (1990). Quinolinone alkaloids from an *Agathosma* species. *Phytochemistry, 29*(4), 1303–1306.

Carta, F., Vullo, D., Maresca, A., Scozzafava, A., & Supuran, C. T. (2012). New chemotypes acting as isozyme-selective carbonic anhydrase inhibitors with low affinity for the off-target cytosolic isoform II. *Bioorganic & Medicinal Chemistry Letters, 22*(6), 2182–2185.

Carvalho, L. E. d., Lima, M. d. P., Máximo, A. d. C., Pereira, E. C. d. S., Moreira, W. A. d. S., Ferreira, A. G., Véras, S. d. M., & Souza, M. G. d. (2012). Estudo em raiz e ráquis foliar de spathelia excelsa: Fitoquímica e atividade frente ao fungo moniliophthora perniciosa associado ao cupuaçuzeiro (*Theobroma grandiflorum*). *Química Nova, 35,* 2237–2240.

Chakravarty, A. K., Sarkar, T., Masuda, K., & Shiojima, K. (1999). Carbazole alkaloids from roots of *Glycosmis arborea. Phytochemistry, 50*(7), 1263–1266.

Chang, K., Gao, P., Lu, Y., Tu, P., Jiang, Y., & Guo, X. (2021). Identification and characterization of quinoline alkaloids from the root bark of *Dictamnus dasycarpus* and their metabolites in rat plasma, urine and feces by UPLC/qtrap-MS and UPLC/Q-TOF-MS. *Journal of Pharmaceutical and Biomedical Analysis, 204,* 114229.

Chen, H., Zhu, S., Tu, P., & Jiang, Y. (2022). Chemical constituents from the stems and leaves of *Glycosmis craibii* var. glabra (Craib) tanaka and their chemotaxonomic significance. *Biochemical Systematics and Ecology, 105,* 104492.

Chen, I., Chen, H., Cheng, M., Chang, Y., Teng, C., Tsutomu, I., Chen, J., & Tsai, I. (2001). Quinoline alkaloids and other constituents of *Melicopes emecarpifolia* with antiplatelet aggregation activity. *Journal of Natural Products, 64*(9), 1143–1147.

Chen, I., Wu, S., Tsai, I., Wu, T., Pezzuto, J. M., Lu, M. C., Chai, H., Suh, N., & Teng, C. (1994a). Chemical and bioactive constituents from *Zanthoxylum simulans*. *Journal of Natural Products, 57*(9), 1206–1211.

Chen, I.-S., Shwu-Jen, W., Yuh-Chwen, L., Ian-Lih, T., Seki, H., Feng-Nien, K., & Che-Ming, T. (1994b). Dimeric 2-quinolone alkaloid and antiplatelet aggregation constituents of *Zanthoxylum simulans*. *Phytochemistry, 36*(1), 237–239.

Chen, J., Chang, Y., Teng, C., Su, C., & Chen, I. (2002a). Quinoline alkaloids and anti-platelet aggregation constituents from the leaves of *Melicope semecarpifolia*. *Planta Medica, 68*(09), 790–793.

Chen, J., Duh, C., Huang, H., & Chen, I. (2003). Furoquinoline alkaloids and cytotoxic constituents from the leaves of *Melicope semecarpifolia*. *Planta Medica, 69*(6), 542–546.

Chen, J., Tang, J. S., Tian, J., Wang, Y. P., & Wu, F. E. (2000). Dasycarine, a new quinoline alkaloid from *Dictamnus dasycarpus*. *Chinese Chemical Letters, 11*(8), 707–708.

Chen, M. J., Wang, X., Zhang, K. Q., & Li, G. H. (2015). Chemical constituents from the bacterium *Wautersiella falsenii* YMF 3.00141. *Applied Biochemistry and Microbiology, 51*, 402–405.

Chen, M., Gan, L., Lin, S., Wang, X., Li, L., Li, Y., Zhu, C., Wang, Y., Jiang, B., & Jiang, J. (2012). Alkaloids from the root of *Isatis indigotica*. *Journal of Natural Products, 75*(6), 1167–1176.

Chen, P., Wu, Y., & Yin, F. (2011). Study on the chemical constituents of vinegar *Schisandra chinensis*. *Journal of Chinese Medicinal Materials, 34*(11), 1728–1729.

Chen, S., Gao, G., Li, Y., Yu, S., & Xiao, P. (2002b). Cytotoxic constituents from *Aquilegia ecalcarata*. *Planta Medica, 68*(06), 554–556.

Cheng, M., Yang, C., Lin, W., Lin, W., Tsai, I., & Chen, I. (2002). Chemical constituents from the leaves of *Zanthoxylum schinifolium*. *Journal of the Chinese Chemical Society, 49*(1), 125–128.

Cheng, M., Lee, K., Tsai, I., & Chen, I. (2005). Two new sesquiterpenoids and anti-HIV principles from the root bark of *Zanthoxylum ailanthoides*. *Bioorganic & Medicinal Chemistry, 13*(21), 5915–5920.

Cho, J., Bae, S., Kim, H., Lee, M., Choi, Y., Jin, B., Lee, H. J., Jeong, H. Y., Lee, Y. G., & Moon, J. (2015). New quinolinone alkaloids from chestnut (*Castanea crenata* Sieb) honey. *Journal of Agricultural and Food Chemistry, 63*(13), 3587–3592.

Choi, Y., Seo, C., Jeong, W., Lee, J. E., Lee, J. Y., Ahn, E., Kang, J., Lee, J., Choi, C. W., & Oh, J. S. (2019). Glycopentanolones AD, four new geranylated quinolone alkaloids from *Glycosmis pentaphylla*. *Bioorganic Chemistry, 87*, 714–719.

Chung, H. S., & Woo, W. S. (2001). A quinolone alkaloid with antioxidant activity from the aleurone layer of anthocyanin-pigmented rice. *Journal of Natural Products, 64*(12), 1579–1580.

Crow, W. D., & Hodgkin, J. H. (1968). Alkaloids of the Australian rutaceae: *Halfordia scleroxyla* and *Halfordia kendack*. IV. Co-occurrence of oxazole and quinoline alkaloids. *Australian Journal of Chemistry, 21*(12), 3075–3077.

Cuca Suarez, L. E., Pattarroyo, M. E., Lozano, J. M., & Delle Monache, F. (2009). Biological activity of secondary metabolites from *Peltostigma guatemalense*. *Natural Product Research, 23*(4), 370–374.

Curini, M., Genovese, S., Menghini, L., Marcotullio, M. C., & Epifano, F. (2008). Phytochemistry and pharmacology of *Boronia pinnata* sm. *Natural Product Communications, 3*(12), 1934578X0800301235.

Cunningham, K. G., & Freeman, G. G. (1953). The isolation and some chemical properties of viridicatin, a metabolic product of *Penicillium viridicatum* westling. *Biochemical Journal, 53*(2), 328.

Dai, C., Li, X., Zhang, K., Li, X., Wang, W., Zang, Y., Chen, X., Li, Q., Wei, M., & Chen, C. (2021). Pesimquinolones I–S, eleven new quinolone alkaloids produced by *Penicillium simplicissimum* and their inhibitory activity on NO production. Bioorganic Chemistry, *108*, 104635.

Darapureddy, C., Prasad, K., & Ch, P. R. (2021). Phytochemical analysis, pharmacological activities, isolation and characterization of bioactive compounds from the roots of *Sterculia urens* roxb. *Turkish Online Journal of Qualitative Inquiry, 12*(8), 7142–7155.

Das, B. P., & Chowdhury, D. N. (1978). Glycosolone: A new quinolone alkaloid from *Glycosmis pentaphylla* (Retz) DC. *Chemistry and Industry*, *8*, 272–273.

Das, T. K., Ghosh, A., Aguinaga, U., Yousufuddin, M., & Kurti, L. (2024). Organocatalytic electrophilic arene amination: Rapid synthesis of 2-quinolones. *ChemRxiv*. doi: 10.26434/chemrxiv-2024-5902g

de Macedo, M. B., Kimmel, R., Urankar, D., Gazvoda, M., Peixoto, A., Cools, F., Torfs, E., Verschaeve, L., Lima, E. S., & Lyčka, A. (2017). Design, synthesis and antitubercular potency of 4-hydroxyquinolin-2(1H)-ones. *European Journal of Medicinal Chemistry, 138*, 491–500.

De Souza, R. C., Fernandes, J. B., Vieira, P. C., da Silva, M. F. d. G., Godoy, M. F., Pagnocca, F. C., Bueno, O. C., Hebling, M. J. A., & Pirani, J. R. (2005). A new imidazole alkaloid and other constituents from *Pilocarpus grandiflorus* and their antifungal activity. *Zeitschrift Für Naturforschung B, 60*(7), 787–791.

Deng, L., Yang, W., Jiang, J., Xu, L., Zhang, J., Liu, C., Ling, J., Kong, W., Li, X., & Li, Y. (2022). Two new anti-tobacco mosaic virus quinoline alkaloids from the stems of *Nicotiana tabacum*. *Chemistry of Natural Compounds, 58*(1), 78–81.

Deng, Y., Ding, T., Deng, L., Hao, X., & Mu, S. (2021). Active constituents of *Zanthoxylum nitidium* from Yunnan Province against leukaemia cells in vitro. *BMC Chemistry, 15*, 1–11.

Detsi, A., Bouloumbasi, D., Prousis, K. C., Koufaki, M., Athanasellis, G., Melagraki, G., Afantitis, A., Igglessi-Markopoulou, O., Kontogiorgis, C., & Hadjipavlou-Litina, D. J. (2007). Design and synthesis of novel quinolinone-3-aminoamides and their α-lipoic acid adducts as antioxidant and anti-inflammatory agents. *Journal of Medicinal Chemistry, 50*(10), 2450–2458.

Di, X., Wang, S., Oskarsson, J. T., Rouger, C., Tasdemir, D., Hardardottir, I., Freysdottir, J., Wang, X., Molinski, T. F., & Omarsdottir, S. (2020). Bromotryptamine and imidazole alkaloids with anti-inflammatory activity from the bryozoan *Flustra foliacea*. *Journal of Natural Products, 83*(10), 2854–2866.

El Euch, I. Z., Frese, M., Sewald, N., Smaoui, S., Shaaban, M., & Mellouli, L. (2018). Bioactive secondary metabolites from new terrestrial *Streptomyces* sp. TN82 strain: Isolation, structure elucidation and biological activity. *Medicinal Chemistry Research, 27*, 1085–1092.

El Sayed, K., Al-Said, M. S., El-Feraly, F. S., & Ross, S. A. (2000). New quinoline alkaloids from *Ruta chalepensis*. *Journal of Natural Products, 63*(7), 995–997.

El-Neketi, M., Ebrahim, W., Lin, W., Gedara, S., Badria, F., Saad, H. A., Lai, D., & Proksch, P. (2013). Alkaloids and polyketides from *Penicillium citrinum*, an endophyte isolated from the Moroccan plant *Ceratonia siliqua*. *Journal of Natural Products, 76*(6), 1099–1104.

Eshiett, I. T., & Taylor, D. (1968). The isolation and structure elucidation of some derivatives of dimethylallyl-coumarin, chromone, quinoline, and phenol from *Fagara* species, and from *Cedrelopsis grevei*. Journal of the Chemical Society C: Organic, *1968*, 481–484.

Eshonov, M. A., & Rasulova, K. A. (2020). Acutinine–A new quinolin-2-one alkaloid from *Haplophyllum acutifolium*. *Chemistry of Natural Compounds, 56*, 509–510.

Eshonov, M. A., Rasulova, K. A., Turgunov, K. K., & Tashkhodzhaev, B. (2020). New quinoline alkaloid acusine and crystal structures of N-methyl-2-phenylquinolin-4-one and pedicine from *Haplophyllum acutifolium*. *Chemistry of Natural Compounds, 56*, 1102–1105.

Faizutdinova, Z. S., Bessonova, I. A., & Yunusov, S. Y. (1967). Structure of folifine and folifidine. *Chemistry of Natural Compounds, 3*(4), 215–217.

Fakhouri, W., Walker, F., Vogler, B., Armbruster, W., & Buchenauer, H. (2001). Isolation and identification of N-mercapto-4-formylcarbostyril, an antibiotic produced by *Pseudomonas fluorescens. Phytochemistry, 58*(8), 1297–1303.

Fan, H., Pan, P., Zhang, Y., & Wang, W. (2018). Synthesis of 2-quinolinones via a hypervalent iodine (III)-mediated intramolecular decarboxylative heck-type reaction at room temperature. *Organic Letters, 20*(24), 7929–7932.

Fang, X., Fu, Z., Zhang, H., Xu, H., & Liang, S. (2018). Chemical constituents of *Garcinia yunnanensis* and their scavenging activity against DPPH radicals. *Chemistry of Natural Compounds, 54*, 232–234.

Feng, X., Dong, Y. F., & Wang, M. (2004). Quinoline alkaloid constituents of *Orixa japonica. Chinese Traditional and Herbal Drugs, 35*(12), 1336–1338.

Fiot, J., Jansen, O., Akhmedjanova, V., Angenot, L., Balansard, G., & Ollivier, E. (2006). HPLC quantification of alkaloids from *Haplophyllum* extracts and comparison with their cytotoxic properties. *Phytochemical Analysis: An International Journal of Plant Chemical and Biochemical Techniques, 17*(5), 365–369.

Friedländer, P. (1882). Ueber o-amidobenzaldehyd. *Berichte Der Deutschen Chemischen Gesellschaft, 15*(2), 2572–2575.

Fu, B., Wang, M., Liu, J., Lin, W., Zhang, C., & Zhao, D. (2020). Secondary metabolites from a marine-derived fungus aspergillus versicolor and their anti-phytopathogenic. *Scientia Agricultura Sinica, 53*(19), 3964–3974.

Fu, Y., Li, Z., Pu, S., & Qian, S. (2013). Two new compounds from *Scolopendra multidens* newport. *Journal of Asian Natural Products Research, 15*(4), 363–367.

Gan, M., Liu, Y., Bai, Y., Guan, Y., Li, L., Gao, R., He, W., You, X., Li, Y., & Yu, L. (2013). Polyketides with New Delhi metallo-β-lactamase 1 inhibitory activity from *Penicillium* sp. *Journal of Natural Products, 76*(9), 1535–1540.

Gao, P., Wang, L., Zhao, L., Zhang, Q., Zeng, K., Zhao, M., Jiang, Y., Tu, P., & Guo, X. (2020). Anti-inflammatory quinoline alkaloids from the root bark of *Dictamnus dasycarpus. Phytochemistry, 172*, 112260.

Garcia, I. A., Pansa, M. F., Pacciaroni, A. D. V., García, M. E., Gonzalez, M. L., Oberti, J. C., Bocco, J. L., Carpinella, M. C., Barboza, G. E., & Nicotra, V. E. (2020). Synthetic lethal activity of benzophenanthridine alkaloids from *Zanthoxylum coco* against BRCA1-deficient cancer cells. *Frontiers in Pharmacology, 11*, 593845.

Ghate, N. B., Chaudhuri, D., Panja, S., Singh, S. S., Gupta, G., Lee, C. Y., & Mandal, N. (2018). In vitro mechanistic study of the anti-inflammatory activity of a quinoline isolated from *Spondias pinnata* bark. *Journal of Natural Products, 81*(9), 1956–1961.

Goodwin, S., & Horning, E. C. (1959). Alkaloids of *Lunasia amara* blanco. Structure of lunacrine. *Journal of the American Chemical Society, 81*(8), 1908–1912.

Goodwin, S., Shoolery, J. N., & Horning, E. C. (1959). Alkaloids of *Lunasia amara* blanco. Hydroxylunacridine. *Journal of the American Chemical Society, 81*(14), 3736–3738.

Guo, C., Lin, X., Liao, S., Yang, B., Zhou, X., Yang, X., Tian, X., Wang, J., & Liu, Y. (2020). Two new aromatic polyketides from a deep-sea fungus *Penicillium* sp. SCSIO 06720. *Natural Product Research, 34*(9), 1197–1205.

Guo, X., Zhao, L., Wang, J., Liu, S., Bi, Q., Wang, Z., & Tan, N. (2018). Chemical constituents from root barks of *Dictamnus dasycarpus* and their cytotoxic activities. *China Journal of Chinese Materia Medica*, 43(24), 4869–4877.

Hanuman, J. B., & Katz, A. (1993). Isolation of quinolinones from ayurvedic processed root tubers of *Aconitum ferox. Natural Product Letters, 3*(3), 227–231.

Haque, M. M., Begum, S., Sohrab, M. H., Ahsan, M., Hasan, C. M., Ahmed, N., & Haque, R. (2013). Secondary metabolites from the stem of *Ravenia spectabilis* lindl. *Pharmacognosy Magazine, 9*(33), 76.

Harrill, J. A., Everett, L. J., Haggard, D. E., Sheffield, T., Bundy, J. L., Willis, C. M., Thomas, R. S., Shah, I., & Judson, R. S. (2021). High-throughput transcriptomics platform for screening environmental chemicals. *Toxicological Sciences, 181*(1), 68–89.

He, X., Li, M., Shang, X., Jia, Z., & Zhang, R. (2011). Composition and anti-inflammatory activities of essential oils from *Gentiana crassicaulis* and *Gentiana scabra. Yaoxue Shijian Zazhi, 29*(4), 274–277.

Heguy, A., Cai, P., Meyn, P., Houck, D., Russo, S., Michitsch, R., Pearce, C., Katz, B., Bringmann, G., & Feineis, D. (1998). Isolation and characterization of the fungal metabolite 3-O-methylviridicatin as an inhibitor of tumour necrosis factor α-induced human immunodeficiency virus replication. *Antiviral Chemistry and Chemotherapy, 9*(2), 149–155.

Hodge, R. P., Harris, C. M., & Harris, T. M. (1988). Verrucofortine, a major metabolite of *Penicillium verrucosum* var. *cyclopium*, the fungus that produces the mycotoxin verrucosidin. *Journal of Natural Products, 51*(1), 66–73.

Hong, W. P., Shin, I., & Lim, H. N. (2020). Recent advances in one-pot modular synthesis of 2-quinolones. *Molecules, 25*(22), 5450.

Houghton, P. J., Woldemariam, T. Z., Watanabe, Y., & Yates, M. (1999). Activity against mycobacterium tuberculosis of alkaloid constituents of angostura bark, *Galipea officinalis. Planta Medica, 65*(03), 250–254.

Huang, C., & Chang, N. (2008). New approach to 2-quinolinones. *Organic Letters, 10*(4), 673–676.

Huang, H., Ishikawa, T., Peng, C., Chen, S., & Chen, I. (2011). Secondary metabolites from the root wood of *Zanthoxylum wutaiense* and their antitubercular activity. *Chemistry & Biodiversity, 8*(5), 880–886.

Huang, H., Ishikawa, T., Peng, C., Tsai, I., & Chen, I. (2008). Constituents of the root wood of *Zanthoxylum wutaiense* with antitubercular activity. *Journal of Natural Products, 71*(7), 1146–1151.

Hubbard, T. D., Hsieh, J., Rider, C. V., Sipes, N. S., Sedykh, A., Collins, B. J., Auerbach, S. S., Xia, M., Huang, R., & Walker, N. J. (2019). Using Tox21 high-throughput screening assays for the evaluation of botanical and dietary supplements. *Applied in Vitro Toxicology, 5*(1), 10–25.

Imai, F., Kinoshita, T., & Sankawa, U. (1989). Constituents of the leaves of *Murraya paniculata* collected in Taiwan. *Chemical and Pharmaceutical Bulletin, 37*(2), 358–362.

Intekhab, J., Aslam, M., & Khalid, H. (2011). Phytochemical study of *Glycosmis mauritiana. American Journal of Plant Sciences, 2*(05), 657.

Iriarte, J., Kincl, F. A., Rosenkranz, G., & Sondheimer, F. (1956). The constituents of *Casimiroa edulis* llave et lex. Part II. The bark. *Journal of the Chemical Society (Resumed), 805*, 4170–4173.

Ishii, H., Koyama, K., Chen, I., Lu, S., & Ishikawa, T. (1982). Studies on the chemical constituents of rutaceous plants. XLVI. The chemical constituents of *Xanthoxylum integrifoliolum* (MERR.) MERR. (*Fagara integrifoliola* MERR.) II. Structural establishment of integriquinolone, a new phenolic quinolone. *Chemical and Pharmaceutical Bulletin, 30*(6), 1992–1997.

Ito, C., Itoigawa, M., Furukawa, A., Hirano, T., Murata, T., Kaneda, N., Hisada, Y., Okuda, K., & Furukawa, H. (2004). Quinolone alkaloids with nitric oxide production inhibitory activity from *Orixa japonica. Journal of Natural Products, 67*(11), 1800–1803.

Ito, C., Kondo, Y., Wu, T., & Furukawa, H. (2000). Chemical constituents of *Glycosmis citrifolia* (willd.) Lindl. Structures of four new acridones and three new quinolone alkaloids. *Chemical and Pharmaceutical Bulletin, 48*(1), 65–70.

Jacquemond-Collet, I., Bessière, J., Hannedouche, S., Bertrand, C., Fourasté, I., & Moulis, C. (2001). Identification of the alkaloids of *Galipea officinalis* by gas chromatography–mass spectrometry. *Phytochemical Analysis: An International Journal of Plant Chemical and Biochemical Techniques, 12*(5), 312–319.

Jacquemond-Collet, I., Hannedouche, S., Fourasté, I., & Moulis, C. (2000). Novel quinoline alkaloid from trunk bark of *Galipea officinalis*. *Fitoterapia, 71*(5), 605–606.

Jain, S. C., Pandey, M. K., Upadhyay, R. K., Kumar, R., Hundal, G., & Hundal, M. S. (2006). Alkaloids from *Toddalia aculeata*. *Phytochemistry, 67*(10), 1005–1010.

Ji, K., Liu, W., Yin, W., Li, J., & Yue, J. (2022). Quinoline alkaloids with anti-inflammatory activity from *Zanthoxylum avicennae*. *Organic & Biomolecular Chemistry, 20*(20), 4176–4182.

Jia, C., Dong, Y., Tu, S., & Wang, G. (2007). Microwave-assisted solvent-free synthesis of substituted 2-quinolones. *Tetrahedron, 63*(4), 892–897.

Jian, J., Fan, Y., Liu, Q., Jin, J., Yuan, C., Gu, W., Hu, Z., Huang, L., & Hao, X. (2023). Quinoline alkaloids from the roots of *Orixa japonica* with the Anti-pathogenic fungi activities. *Chemistry & Biodiversity, 20*(2), e202201097.

Jin, Y., Wei, X., & Shi, Y. (2007). Chemical constituents from *Cynoglossum gansuense*. *Helvetica Chimica Acta, 90*(4), 776–782.

Jurd, L., Benson, M., & Wong, R. Y. (1983). New quinolinone and bis-quinolinone alkaloids from *Euxylophora paraensis*. *Australian Journal of Chemistry, 36*(4), 759–768.

Jurd, L., & Wong, R. Y. (1981). New quinolinone alkaloids from the heartwood of *Euxylophora paraensis*. *Australian Journal of Chemistry, 34*(8), 1625–1632.

Khadem, S., Joseph, R., Rastegar, M., Leek, D. M., Oudatchin, K. A., & Arya, P. (2004). A solution-and solid-phase approach to tetrahydroquinoline-derived polycyclics having a 10-membered ring. *Journal of Combinatorial Chemistry, 6*(5), 724–734.

Khadem, S., & Marles, R. J. (2025). Biological activity of natural 2-quinolinones. *Natural Product Research, 39*(5), 1359–1373.

Khamkhenshorngphanuch, T., Kulkraisri, K., Janjamratsaeng, A., Plabutong, N., Thammahong, A., Manadee, K., Na Pombejra, S., & Khotavivattana, T. (2020). Synthesis and antimicrobial activity of novel 4-hydroxy-2-quinolone analogs. *Molecules, 25*(13), 3059.

Khan, M. A., & Waterman, P. G. (1990). Constituents of *Ravenia spectabilis*. *Fitoterapia, 61*(3), 282.

Khandokar, L., Bari, M. S., Seidel, V., & Haque, M. A. (2021). Ethnomedicinal uses, phytochemistry, pharmacological activities and toxicological profile of *Glycosmis pentaphylla* (Retz.) DC.: A review. *Journal of Ethnopharmacology, 278*, 114313.

Kim, K. H., Lee, I. K., Kim, K. R., Ha, S. K., Kim, S. Y., & Lee, K. R. (2009). New benzamide derivatives and NO production inhibitory compounds from *Limonia acidissima*. *Planta Medica, 75*(10), 1146–1151.

Knorr, L. (1887). Synthetische versuche mit dem acetessigester. *Justus Liebigs Annalen Der Chemie, 238*(1-2), 137–219.

Ko, W., Sohn, J. H., Kim, Y., & Oh, H. (2015). Viridicatol from marine-derived fungal strain *Penicillium* sp. SF-5295 exerts anti-inflammatory effects through inhibiting NF-κB signaling pathway on lipopolysaccharide-induced RAW264. 7 and BV2 cells. *Natural Product Sciences, 21*(4), 240–247.

Kodirova, D. R., Rasulova, K. A., Sagdullaev, S. S., & Aisa, H. A. (2018). *Haplophyllum griffithianum* as a source of quinoline alkaloids. *Chemistry of Natural Compounds, 54*, 213–214.

Korosi, J., Szendrei, K., Novak, I., Reish, J., Blazso, G., Minker, E., & Koltai, M. (1976). Quaternary alkaloids of *Ptelea trifoliata*. II. Examination of the components by an indirect method. *Herba Hungo, 15*, 9–17.

Kostopoulou, I., Tzani, A., Chronaki, K., Prousis, K. C., Pontiki, E., Hadjiplavlou-Litina, D., & Detsi, A. (2023). Novel multi-target agents based on the privileged structure of 4-hydroxy-2-quinolinone. *Molecules, 29*(1), 190.

Kowal, N. M., Di, X., Omarsdottir, S., & Olafsdottir, E. S. (2023). Flustramine Q, a novel marine origin acetylcholinesterase inhibitor from *Flustra foliacea*. *Future Pharmacology, 3*(1), 38–47.

Kumar, A., AlGhamdi, K. M., Khan, A. A., Ahamad, R., Ghadeer, A., & Bari, A. (2023). *Psoralea corylifolia* (babchi) seeds enhance proliferation of normal human cultured melanocytes: GC–MS profiling and biological investigation. *Open Chemistry, 21*(1), 20220292.

Kumar, P., Das, B. P., & Sinha, S. (1986). Homo-glycosolone-a new quinolone alkaloid from *Glycosmis-pentaphylla* (Retz) DC. *Chemistry & Industry*, (19), 669–670.

Li, C., Zhang, F., Gan, D., Wang, C., Zhou, H., Yin, T., & Cai, L. (2023). Secondary metabolites isolated from *penicillium expansum* and their chemotaxonomic value. *Biochemical Systematics and Ecology, 107*, 104584.

Li, D., Oku, N., Shinozaki, Y., Kurokawa, Y., & Igarashi, Y. (2020). 4-hydroxy-3-methyl-2(1H)-quinolone, originally discovered from a brassicaceae plant, produced by a soil bacterium of the genus *Burkholderia* sp.: Determination of a preferred tautomer and antioxidant activity. *Beilstein Journal of Organic Chemistry, 16*(1), 1489–1494.

Li, P., Fan, Y., Chen, H., Chao, Y., Du, N., & Chen, J. (2016). Phenylquinolinones with antitumor activity from the Indian Ocean-derived fungus *Aspergillus versicolor* Y31-2. *Chinese Journal of Oceanology and Limnology, 34*(5), 1072–1075.

Li, S. (2017). Chemical constituents from *Melicope ptelefolia*. *Chinese Traditional and Herbal Drugs, 48*(6), 1076–1079.

Liang, P., Zhang, Y. Y., Yang, P., Grond, S., Zhang, Y., & Qian, Z. (2019). Viridicatol and viridicatin isolated from a shark-gill-derived fungus *Penicillium polonicum* AP2T1 as MMP-2 and MMP-9 inhibitors in HT1080 cells by MAPKs signaling pathway and docking studies. *Medicinal Chemistry Research, 28*, 1039–1048.

Liang, S., Chen, C., Chen, R., Li, R., Chen, W., Jiang, G., & Du, L. (2022). Michael acceptor molecules in natural products and their mechanism of action. *Frontiers in Pharmacology, 13*, 1033003.

Liao, N., Li, M., Qin, F., Jiang, J., Luo, R., Chen, X., Zhou, M., Zhu, Y., & Wang, H. (2021). A new phenolic acid from *Zanthoxylum nitidum* var. *tomentosum* (Rutaceae) and its chemotaxonomic significance. *Biochemical Systematics and Ecology, 99*, 104351.

Liu, B., Qui, Y., Yan, Z., Wu, X., Qiu, Y., & Yang, J. (2021b). Research on secondary metabolites of marine-derived *penicillium polonicum* H92. *Chinese Marine Drugs, 40*(4), 27–32. doi: 10.13400/j.cnki.cjmd.2021.04.004

Liu, S., Tang, X., He, F., Jia, J., Hu, H., Xie, B., Li, M., & Qiu, Y. (2022). Two new compounds from a mangrove sediment-derived fungus *penicillium polonicum* H175. *Natural Product Research, 36*(9), 2370–2378.

Liu, X., Li, J., Kong, W., Mi, Q., Xu, Y., Xu, X., Huang, H., Yang, G., Yang, Y., & Wang, J. (2021a). Patent# CN11308766. China Tobacco Yunnan Industrial Co., Ltd., China (Ed.), *Isolation of quinoline alkaloid from tobacco stem for resisting tobacco mosaic virus* (pp. 1–11). China.

Lookpan, T., Voravuthikunchai, S. P., Sitthisuk, P., Poorahong, W., Watanapokasin, R., & Chakthong, S. (2024). A new alkaloid and a new benzaldehyde derivative from the twig of *Zanthoxylum rhetsa* (Roxb.) DC. *Natural Product Research, 38*(3), 463–469.

Lu, Q., Ma, R., Yang, Y., Mo, Z., Pu, X., & Li, C. (2020). *Zanthoxylum nitidum* (Roxb.) DC: Traditional uses, phytochemistry, pharmacological activities and toxicology. *Journal of Ethnopharmacology, 260*, 112946.

Luo, X. M., Qi, S. H., Yin, H., Gao, C. H., & Zhang, S. (2009). Alkaloids from the stem bark of *Micromelum falcatum*. *Chemical and Pharmaceutical Bulletin, 57*(6), 600–602.

Machado, D. d. N. M., Palmeira Júnior, S. F., Conserva, L. M., & de Lyra Lemos, R. P. (2005). Quinoline alkaloids from *Sebastiania corniculata* (euphorbiaceae). *Biochemical Systematics and Ecology, 33*(5), 555–558.

Mahmoud, M. A. S. (2005). Bioactive secondary metabolites from marine and terrestrial bacteria: Isoquinolinequinones, bacterial compounds with a novel pharmacophor (Dissertation), Fakultät für Chemie (inkl. GAUSS), Sweden.

Marcaccini, S., Pepino, R., Pozo, M. C., Basurto, S., García-Valverde, M., & Torroba, T. (2004). One-pot synthesis of quinolin-2-(1H)-ones *via* tandem Ugi–Knoevenagel condensations. *Tetrahedron Letters, 45*(21), 3999–4001.

Matsushima, H., Fukumi, H., & Arima, K. (1973). Isolation of zeanic acid, a natural plant growth-regulator from corn steep liquor and its chemical structure. *Agricultural and Biological Chemistry, 37*(8), 1865–1871.

Mbaveng, A. T., Noulala, C. G., Samba, A. R., Tankeo, S. B., Fotso, G. W., Happi, E. N., Ngadjui, B. T., Beng, V. P., Kuete, V., & Efferth, T. (2021). Cytotoxicity of botanicals and isolated phytochemicals from *Araliopsis soyauxii* Engl.(Rutaceae) towards a panel of human cancer cells. *Journal of Ethnopharmacology, 267*, 113535.

McCormick, J. L., McKee, T. C., Cardellina, J. H., & Boyd, M. R. (1996). HIV inhibitory natural products. 26. Quinoline alkaloids from *Euodia roxburghiana*. *Journal of Natural Products, 59*(5), 469–471.

Min, K. (1998). Antifungal activity of the extracts of *Zanthoxylum schinifolium* Sieb. et Zucc. against dermatophytes. *Journal of the Korean Wood Science and Technology, 26*(4), 78–85.

Min, Y. D., Kwon, H. C., Yang, M. C., Lee, K. H., Choi, S. U., & Lee, K. R. (2007). Isolation of limonoids and alkaloids from *Phellodendron amurense* and their multidrug resistance (MDR) reversal activity. *Archives of Pharmacal Research, 30*, 58–63.

Mirek, J., & Syguła, A. (1982). Semiempirical MNDO and UV absorption studies on tautomerism of 2-quinolones. *Zeitschrift Für Naturforschung A, 37*(11), 1276–1283.

Mohammadhosseini, M., Venditti, A., Frezza, C., Serafini, M., Bianco, A., & Mahdavi, B. (2021). The genus *Haplophyllum juss.*: Phytochemistry and bioactivities—A review. *Molecules, 26*(15), 4664.

Moreira, W. A., Lima, M. d. P., Ferreira, A. G., Ferreira, I. C., & Nakamura, C. V. (2009). Chemical constituents from the roots of *Spathelia excelsa* and their antiprotozoal activity. *Journal of the Brazilian Chemical Society, 20*, 1089–1094.

Moulis, C., Wirasutisna, K. R., Gleye, J., Loiseau, P., Stanislas, E., & Moretti, C. (1983). A 2-quinolone alkaloid from *Almeidea guyanensis. Phytochemistry, 22*(9), 2095–2096.

Moussaoui, O., Chakroune, S., Rodi, Y. K., & Hadrami, E. M. E. (2022). 2-Quinolone-based derivatives as antibacterial agents: A review. *Mini-Reviews in Organic Chemistry, 19*(3), 331–351.

Na, M. W., Jeong, S. Y., Ko, Y., Kang, D., Pang, C., Ahn, M., & Kim, K. H. (2022). Chemical investigation of *Tetradium ruticarpum* fruits and their antibacterial activity against helicobacter pylori. *ACS Omega, 7*(27), 23736–23743.

Nayar, M., Sutar, C. V., & Bhan, M. K. (1971). Alkaloids of the stem bark of *Hesperethusa crenulata. Phytochemistry, 10*(11), 2843–2844.

Nazrullaev, S. S., Bessonova, I. A., & Akhmedkhodzhaeva, K. S. (2001). Estrogenic activity as a function of chemical structure in *Haplophyllum* quinoline alkaloids. *Chemistry of Natural Compounds, 37*, 551–555.

Nesmelova, E. F., Bessonova, I. A., & Yunusov, S. Y. (1978). Alkaloids of *Haplophyllum latifolium* the structure of haplatine. *Chemistry of Natural Compounds, 14*, 645–650.

Ngadjui, B. T., Ayafor, J. F., Sondengam, B. L., & Connolly, J. D. (1989). Quinolone and carbazole alkaloids from *Clausena anisata. Phytochemistry, 28*(5), 1517–1519.

Nguyen, T. H. V., Tran, T. T., Cam, T. I., Pham, M. Q., Pham, Q. L., Vu, D. H., Nguyen, X. N., Chau, V. M., & Van, K. P. (2019). Alkaloids from *Zanthoxylum nitidum* and their cytotoxic activity. *Natural Product Communications, 14*(5), 1934578X19844133.

Nhiem, N. X., Quan, P. M., & Van, N. T. H. (2020). Alkaloids and their pharmacology effects from *Zanthoxylum* genus. *Bioactive compounds in nutraceutical and functional food for good human health*. IntechOpen. www.intechopen.com/chapters/71424

Niu, X., Li, S., Peng, L., Lin, Z., Rao, G., & Sun, H. (2001). Constituents from *Limonia crenulata. Journal of Asian Natural Products Research, 3*(4), 299–311.

Noshita, T., Tando, M., Suzuki, K., Murata, K., & Funayama, S. (2001). New quinoline alkaloids from the leaves and stems of *Orixa japonica*, orijanone, isopteleflorine and 3′-O-methylorixine. *Bioscience, Biotechnology, and Biochemistry, 65*(3), 710–713.

Noulala, C. G. T., Fotso, G. W., Rennert, R., Lenta, B. N., Sewald, N., Arnold, N., Happi, E. N., & Ngadjui, B. T. (2020). Mesomeric form of quaternary indoloquinazoline alkaloid and other constituents from the Cameroonian Rutaceae *Araliopsis soyauxii* engl. *Biochemical Systematics and Ecology, 91*, 104050.

Novâk, I., Szendrei, K., Pàpay, V., Minker, E., & Koltai, M. (1970). Hazai Rutaceae-Fajok vizsgâlata *Ptelea trifoliata* L. *Herba Hungarica, 9*, 23–31.

Pan, Y., Lau, K., Al-Mogren, M. M., Mahjoub, A., & Hochlaf, M. (2014). Theoretical studies of 2-quinolinol: Geometries, vibrational frequencies, isomerization, tautomerism, and excited states. *Chemical Physics Letters, 613*, 29–33.

Pang, X., Cai, G., Lin, X., Salendra, L., Zhou, X., Yang, B., Wang, J., Wang, J., Xu, S., & Liu, Y. (2019). New alkaloids and polyketides from the marine sponge-derived fungus *Penicillium* sp. SCSIO41015. *Marine Drugs, 17*(7), 398.

Pastukhova, V. I., Sidyakin, G. P., & Yunusov, S. Y. (1965). *Haplophyllum* alkaloids. IV. Structure of foliosidine. *Chemistry of Natural Compounds, 1*, 20–24.

Paul, B. D., & Bose, P. K. (1968). New quinolone alkaloids from *Ravenia spectabilis*. *English Journal of the Indian Chemical Society, 45*, 552–553.

Peng, K., Zhou Xiao-Li, A, P., & Zhou, X. (2014). Chemical constituents of the essential oil from *Nuphar pumilum* (hoffm.) DC. *Lishizhen Medicine and Materia Medica Research, 25*(6), 1312–1313.

Perrett, S., & Whitfield, P. J. (1995). Atanine (3-dimethylallyl-4-methoxy-2-quinolone), an alkaloid with anthelmintic activity from the Chinese medicinal plant, *Evodia rutaecarpa*. *Planta Medica, 61*(03), 276–278.

Pinto, B. N., Alvarenga, E. S., Santos, A. R., Oliveira, W. F., de Paula, V. F., Oliveira, M. N., Junior, J. M., & de L. Batista, A. N. (2022a). Structural elucidation by NMR analysis assisted by DFT calculations of a novel natural product from *Conchocarpus mastigophorus* (rutaceae). *Asian Journal of Organic Chemistry, 11*(6), e202200182.

Pinto, B. N., Alvarenga, E. S., Santos, A. R., Oliveira, W. F., de Paula, V. F., Oliveira, M. N., Junior, J. M., & de L. Batista, A. N. (2022b). Structural elucidation by NMR analysis assisted by DFT calculations of a novel natural product from *Conchocarpus mastigophorus* (rutaceae). *Asian Journal of Organic Chemistry, 11*(6), e202200182.

Prescott, T. A., Sadler, I. H., Kiapranis, R., & Maciver, S. K. (2007). Lunacridine from *Lunasia amara* is a DNA intercalating topoisomerase II inhibitor. *Journal of Ethnopharmacology, 109*(2), 289–294.

Price, J. R. (1959). Alkaloids of the Australian Rutaceae: *Lunasia quercifolia*. II. The Nature of Lunasine. *Australian Journal of Chemistry, 12*(3), 458–467.

Puricelli, L., Innocenti, G., Monache, G. D., Caniato, R., Filippini, R., & Cappelletti, E. M. (2002). In vivo and in vitro production of alkaloids by *Haplophyllum patavinum*. *Natural Product Letters, 16*(2), 95–100.

Purwantiningsih, H. A., & Chan, K. L. (2011). Free radical scavenging activity of the standardized ethanolic extract of *Eurycoma longifolia* (TAF-273). *International Journal of Pharmaceutical Sciences, 3*, 343–347.

Qin, F., Wang, C. Y., Wang, C., Chen, Y., Li, J., Li, M., Zhu, Y., Lee, S. K., & Wang, H. (2023). Undescribed isoquinolines from *Zanthoxylum nitidum* and their antiproliferative effects against human cancer cell lines. *Phytochemistry, 205*, 113476.

Qin, Y., Quan, H., Zhou, X., Chen, S., Xia, W., Li, H., Huang, H., Fu, X., & Dong, L. (2021). The traditional uses, phytochemistry, pharmacology and toxicology of *Dictamnus dasycarpus*: A review. *Journal of Pharmacy and Pharmacology, 73*(12), 1571–1591.

Quaglia, M., Santinelli, M., Sulyok, M., Onofri, A., Covarelli, L., & Beccari, G. (2020). *Aspergillus, Penicillium* and *Cladosporium* species associated with dried date fruits collected in the Perugia (Umbria, Central Italy) market. *International Journal of Food Microbiology, 322,* 108585.

Randrianarivelojosia, M., Rasidimanana, V. T., Rabarison, H., Cheplogoi, P. K., Ratsimbason, M., Mulholland, D. A., & Mauclère, P. (2003). Plants traditionally prescribed to treat *tazo* (malaria) in the eastern region of Madagascar. *Malaria Journal, 2,* 1–9.

Ranieri Cortez, L. E., Garcia Cortez, D. A., Fernandes, J. B., Vieira, P. C., Ferreira, A. G., & da Silva, M. (2009). New alkaloids from conchocarpus gaudichaudianus. *Heterocycles, 78*(8), 2053.

Rapoport, H., & Holden, K. G. (1959). Isolation of alkaloids from *Balfourodendron riedelianum*. The structure of balfourodine. *Journal of the American Chemical Society, 81*(14), 3738–3743.

Rastogi, K., Kapil, R. S., & Popli, S. P. (1980). New alkaloids from *Glycosmis mauritiana. Phytochemistry, 19*(5), 945–948.

Rasulova, K. A., & Bessonova, I. A. (1992). Alkaloids of *Haplophyllum perforatum. Chemistry of Natural Compounds, 28,* 214–216.

Rasulova, K. A., Bobakulov, K. M., Eshonov, M. A., & Abdullaev, N. D. (2019). Pedicine, a new alkaloid from *Haplophyllum pedicellatum. Chemistry of Natural Compounds, 55,* 709–711.

Razakova, D. M., Bessonova, I. A., & Yunusov, S. Y. (1976). Alkaloids of *Haplophyllum perforatum. Chemistry of Natural Compounds, 12*(5), 618.

Razakova, D. M., Bessonova, I. A., & Yunusov, S. Y. (1984). Components of *Haplophyllum obtusifolium. Chemistry of Natural Compounds, 20,* 599–600.

Razzakova, D. M., Bessonova, I. A., & Yunusov, S. Y. (1972). Folimine—A new alkaloid from *Haplophyllum foliosum. Chemistry of Natural Compounds, 8*(1), 139.

Reisch, J., Szendrei, K., Novák, I., Minker, E., & Pàpay, V. (1969). Inhaltsstoffe der blüten von *ptelea trifoliata*: Arctigenin-methyläther,()-hydroxylunin und ptelefolin. *Tetrahedron Letters, 10*(43), 3803–3806

Reisch, J., Körösi, J., Szendrei, K., Novak, I., & Minker, E. (1975). Cumarine und chinolin-alkaloide aus der wurzelrinde von *Ptelea trifoliata. Phytochemistry, 14*(7), 1678–1679.

Reisch, J., Szendrei, K., Novak, I., Minker, E., Körösi, J., & Csedö, K. (1972). Drei neue chinolon-(2)-alkaloide aus *Ptelea trifoliata. Tetrahedron Letters, 13*(5), 449–452.

Rodríguez-Guzmán, R., Radwan, M. M., Burandt, C. L., Williamson, J. S., & Ross, S. A. (2010). Xenobiotic biotransformation of 4-methoxy-N-methyl-2-quinolone, isolated from *Zanthoxylum monophyllum. Natural Product Communications, 5*(9), 1934578X1000500923.

Rüegger, A., & Stauffacher, D. (1963). Lunidin und lunidonin, zwei neue alkaloide aus *lunasia amara* BLANCOvar. repanda (LAUTERB. et K. SCHUM.) LAUTERB. *Helvetica Chimica Acta, 46*(6), 2329–2336.

Ryu, M., & Chung, H. (2010). Isolation of alkaloids with immune stimulating activity from *Oryza sativa* cv. heugnambyeo. *Journal of the Korean Chemical Society, 54*(1), 65–70.

Santiago-Brugnoli, L., Rosquete-Porcar, C., Pouységu, L., & Quideau, S. (2013). *Zanthoxyfolina*, un nuevo meroterpenoide y derivados del labdano aislados de las hojas de *Zanthoxylum rhoifolium* LAM (Rutaceae). *Avances En Química, 8*(2), 85–88.

Sarker, S. D., Waterman, P. G., & Armstrong, J. A. (1995). 3, 4, 8-trimethoxy-2-quinolone: A new alkaloid from *Eriostemon gardneri. Journal of Natural Products, 58*(4), 574–576.

Sartor, C. F., Lima, M. d. P., Silva, M. F. G. d., Fernandes, J. B., Forim, M. R., & Pirani, J. R. (2019). New limonoids from *Dictyoloma vandellianum* and *Sohnreyia excelsa*: Chemosystematic considerations. *Journal of the Brazilian Chemical Society, 30*(11), 2464–2476.

Schmid, H., & Karrer, P. (1945). Über wasserlösliche inhaltsstoffe von *papaver somniferum* L. *Helvetica Chimica Acta, 28*(1), 722–740.

Selvaraj, C., Chandra, I., & Singh, S. K. (2021). Artificial intelligence and machine learning approaches for drug design: Challenges and opportunities for the pharmaceutical industries. *Molecular Diversity, 26*, 1–21.

Sener, B., Mutlugil, A., Noyanalpan, N., & Lewis, J. R. (1990). Alkaloids from *Haplophyllum myrtifolium* Boiss. Journal of Faculty of Pharmacy of Gazi University 7, 17–24.

Severino, V. G., De Freitas, S. D., Braga, P. A., Forim, M. R., Da Silva, M. F. d. G., Fernandes, J. B., Vieira, P. C., & Venâncio, T. (2014). New limonoids from *Hortia oreadica* and unexpected coumarin from *H. superba* using chromatography over cleaning sephadex with sodium hypochlorite. *Molecules, 19*(8), 12031–12047.

Shi, X., Ye, G., Tang, W., & Zhao, W. (2010). A new coumarin and carbazole alkaloid from *Clausena vestita* D. D. T_{AO}. *Helvetica Chimica Acta, 93*(5), 985–990.

Silva, V. L., Pinto, D., Santos, C. M., & Rocha, D. (2022). Quinolinones and related systems (update 2022). *Science of Synthesis Knowledge Updates, 3*, 183–442.

Silva, V. L., & Silva, A. M. (2019). Palladium-catalysed synthesis and transformation of quinolones. *Molecules, 24*(2), 228.

Singh, Y., Bhatia, N., Biharee, A., Kulkarni, S., Thareja, S., & Monga, V. (2023). Developing our knowledge of the quinolone scaffold and its value to anticancer drug design. *Expert Opinion on Drug Discovery, 18*(10), 1151–1167.

Soares, T. A., Nunes-Alves, A., Mazzolari, A., Ruggiu, F., Wei, G., & Merz, K. (2022). The (Re)-Evolution of Quantitative Structure–Activity Relationship (QSAR) studies propelled by the surge of machine learning methods. *Journal of Chemical Information and Modeling, 62*(22), 5317–5320.

Song, W., Zeng, G., Peng, W., Chen, K., & Tan, N. (2014). Cytotoxic amides and quinolones from *Clausena lansium. Helvetica Chimica Acta, 97*(2), 298–305.

Sripisut, T., Phakhodee, W., Ritthiwigrom, T., Cheenpracha, S., Prawat, U., Deachathai, S., Machan, T., & Laphookhieo, S. (2013). Alkaloids from *Glycosmis cochinchinensis* twigs. *Phytochemistry Letters, 6*(3), 337–339.

Stucki, A. O., Barton-Maclaren, T. S., Bhuller, Y., Henriquez, J. E., Henry, T. R., Hirn, C., Miller-Holt, J., Nagy, E. G., Perron, M. M., & Ratzlaff, D. E. (2022). Use of new approach methodologies (NAMs) to meet regulatory requirements for the assessment of industrial chemicals and pesticides for effects on human health. *Frontiers in Toxicology, 4*, 964553.

Su, Y., Luo, Y., Guo, C., & Guo, D. (2004). Two new quinoline glycoalkaloids from *Echinops gmelinii. Journal of Asian Natural Products Research, 6*(3), 223–227.

Sultana, N., Choudhary, M. I., & Akhter, F. (2007). X-ray diffraction studies on inhibitor of platelet aggregation dictamnine. *Journal of the Chemical Society of Pakistan, 29*(2), 194–197.

Sun, J., Wang, P., Liang, J., & Chen, L. (2016). Phytochemical and chemotaxonomic study on *Dictamnus angustifolius* G. Don ex Sweet (Rutaceae). *Biochemical Systematics and Ecology, 68*, 74–76.

Sun, M., & Wang, T. (2011). Insecticidal activities and chemical components of alcohol extract from leaves of *Rhodendron dauricum* L. *Journal of Forestry Research, 22*, 133–135.

Tangjitjaroenkun, J., Chantarasriwong, O., & Chavasiri, W. (2012). Chemical constituents of the stems of *Zanthoxylum limonella* alston. *Phytochemistry Letters, 5*(3), 443–445.

Tani, M., Gyobu, Y., Sasaki, T., Takenouchi, O., Kawamura, T., Kamimura, T., & Harada, T. (2004a). SF2809 compounds, novel chymase inhibitors from *Dactylosporangium* sp. 1. Taxonomy, fermentation, isolation and biological properties. *The Journal of Antibiotics, 57*(2), 83–88.

Tani, M., Harimaya, K., Gyobu, Y., Sasaki, T., Takenouchi, O., Kawamura, T., Kamimura, T., & Harada, T. (2004b). SF2809 compounds, novel chymase inhibitors from *Dactylosporangium* sp. 2. Structural elucidation. *The Journal of Antibiotics, 57*(2), 89–96.

Tanjung, M., Saputri, R. D., Wahjoedi, R. A., & Tjahjandarie, T. S. (2017). 4-methoxy-3-(3-methylbut-2-en-1-yl)-7-[(3-methylbut-2-en-1-yl)oxy]quinolin-2(1*H*)-one from *Melicope moluccana* TG hartley. *Molbank, 2017*(2), M939.

Tarus, P. K., Coombes, P. H., Crouch, N. R., & Mulholland, D. A. (2006). Benzo [c] phenanthridine alkaloids from stem bark of the forest knobwood, *Zanthoxylum davyi* (rutaceae). *South African Journal of Botany, 72*(4), 555–558.

Tashima, T. (2015). The structural use of carbostyril in physiologically active substances. *Bioorganic & Medicinal Chemistry Letters*, 25(17), 3415–3419.

Tateishi, K., Matsushima, Y., & Shibata, H. (1989). Changes in the IAA metabolic content of *Zea mays* L kernels during maturation. *Agricultural and Biological Chemistry, 53*(10), 2545–2551.

Tateishi, K., Shibata, H., Matsushima, Y., & Iijima, T. (1987). Isolation of two new IAA conjugates, 7'O-β-d-glucopyranoside of 3, 7-dihydroxy-2-indolinone-3-acetic acid and 8'-O-d-glucopyranoside of 8-hydroxy-2-quinolone-4-carboxylic acid, from immature sweet corn kernels (*Zea mays* L). *Agricultural and Biological Chemistry, 51*(12), 3445–3447.

Teixeira, T. R., Rangel, K. C., Tavares, R. S. N., Kawakami, C. M., Dos Santos, G. S., Maria-Engler, S. S., Colepicolo, P., Gaspar, L. R., & Debonsi, H. M. (2021). In vitro evaluation of the photoprotective potential of quinolinic alkaloids isolated from the antarctic marine fungus *Penicillium echinulatum* for topical use. *Marine Biotechnology, 23*, 357–372.

Tian, E. (2014). Chemical constituents from stems of *Glycosmis pentaphylla. Chinese Traditional and Herbal Drugs, 45*(10), 1359–1362.

Tsai, I. L., Chang, R. G., Fang, S. C., Ishikawa, T., & Chen, I. S. (1996). Chemical constituents from the root bark of *Formosan toddalia asiatica. Chinese Pharmaceutical Journal, 48*, 63–75.

Ubaidullaev, K., Bessonova, I. A., & Yunusov, S. Y. (1972). Alkaloids of *Haplophyllum pedicellatum, H. obtusifolium*, and *H. bucharicum*. structure of bucharamine. *Chemistry of Natural Compounds, 8*(3), 337–339.

Volle, J., Mävers, U., & Schlosser, M. (2008). The tautomeric persistence of electronically and sterically biased 2-quinolinones*, European Journal of Organic Chemistry*, 2430–2438.

Wang, C., Huang, X., Xiao, H., Hao, Y., Xu, L., Yan, Q., Zou, Z., Xie, C., Xu, Y., & Yang, X. (2021). Chemical constituents of the marine fungus *Penicillium* sp. MCCC 3A00228. *Chemistry & Biodiversity, 18*(10), e2100697.

Wang, J., He, W., Huang, X., Tian, X., Liao, S., Yang, B., Wang, F., Zhou, X., & Liu, Y. (2016). Antifungal new oxepine-containing alkaloids and xanthones from the deep-sea-derived fungus *Aspergillus versicolor* SCSIO 05879. *Journal of Agricultural and Food Chemistry, 64*(14), 2910–2916.

Wang, R., Yang, X., Ma, C., Shang, M., Liang, J., Wang, X., Cai, S., & Shoyama, Y. (2003). Alkaloids from the seeds of *Sterculia lychnophora* (Pangdahai). *Phytochemistry, 63*(4), 475–478.

Wang, Y. F., Zhang, C. G., Yao, R. R., & Zhou, W. S. (1982). Studies on chemical constituents of *Parthenocissus tricuspidata* (Sieb et Zucc) planch. *Yao Xue Xue Bao= Acta Pharmaceutica Sinica, 17*(6), 466–468.

Wei, M., Yang, R., Shao, C., Wang, C., Deng, D., She, Z., & Lin, Y. (2011). Isolation, structure elucidation, crystal structure, and biological activity of a marine natural alkaloid, viridicatol. *Chemistry of Natural Compounds, 47*, 322–325.

Wijeratne, E. K., Bandara, B. R., Gunatilaka, A. L., Tezuka, Y., & Kikuchi, T. (1992). Chemical constituents of three rutaceae species from Sri Lanka. *Journal of Natural Products, 55*(9), 1261–1269.

Wondimu, A., Dagne, E., & Waterman, P. G. (1988). Quinoline alkaloids from the leaves of *Teclea simplicifolia. Phytochemistry, 27*(3), 959–960.

Wu, Q., Zhang, N., Gong, X., Ren, M., Xu, Y., & Chen, Y. (2023). Transition metal-free photocatalytic radical annulation of 2-cyanoaryl acrylamides with difluoromethyl radicals to assemble 4-amino-quinolinone derivatives. *Green Chemistry, 25*(16), 6188–6193.

Wu, T., Yeh, J., & Wu, P. (1995). The heartwood constituents of *Tetradium glabrifolium. Phytochemistry, 40*(1), 121–124.

Wurz, G., Hofer, O., & Greger, H. (1993). Structure and synthesis of phenaglydon, a new quinolone derived phenanthridine alkaloid from *Glycosmis cyanocarpa. Natural Product Letters, 3*(3), 177–182.

Xu, L., Han, T., Wu, J., Zhang, Q., Zhang, H., Huang, B., Rahman, K., & Qin, L. (2009). Comparative research of chemical constituents, antifungal and antitumor properties of ether extracts of *Panax ginseng* and its endophytic fungus. *Phytomedicine, 16*(6–7), 609–616.

Yan, Y., Liu, X., Zhang, L., Wang, Y., Chen, Q., Chen, Z., Xu, L., & Liu, T. (2020). Chemical constituents from *Dictamnus dasycarpus* turcz. *Biochemical Systematics and Ecology, 93*, 104134.

Yang, C., Cheng, M., Lee, S., Yang, C., Chang, H., & Chen, I. (2009). Secondary metabolites and cytotoxic activities from the stem bark of *Zanthoxylum nitidum. Chemistry & Biodiversity, 6*(6), 846–857.

Yang, G., Wu, Y., & Chen, Y. (2012a). Alkaloids from the stems of *Glycosmis pentaphylla. Helvetica Chimica Acta, 95*(8), 1449–1454.

Yang, S., Sun, F., Ruan, J., Yan, J., Huang, P., Wang, J., Han, L., Zhang, Y., & Wang, T. (2019). Anti-inflammatory constituents from *Cortex dictamni. Fitoterapia, 134*, 465–473.

Yang, X. W., Hattori, M., & Namba, T. (1996a). Studies on the anti-lipid peroxidative actions of the methanolic extract of the root of *Aegle marmelos* and its constituents in vivo and in vitro. *Journal of Chinese Pharmaceutical Sciences, 5*(3), 132–140.

Yang, X., & Yang, J. (2008). Chemical constituents from the mycelia of *Alternaria* sp. YD-01. *Yao Xue Xue Bao= Acta Pharmaceutica Sinica, 43*(11), 1116–1118.

Yang, X., Feng, Y., Duffy, S., Avery, V. M., Camp, D., Quinn, R. J., & Davis, R. A. (2011). A new quinoline epoxide from the Australian plant *Drummondita calida. Planta Medica, 77*(14), 1644–1647.

Yang, X., Hattori, M., & Namba, T. (1996b). Two new coumarins from the roots of *Aegle marmelos. Journal of Chinese Pharmaceutical Sciences, 5*, 68–73.

Yang, Y. H., Ye, Y., Li, C. Y., Zeng, Y., & Zhao, P. J. (2013). Chemical constituents of endophyte *Penicillium* sp. DCS82 from *Daphniphyllum Longeracemosum. Guihaia, 4*, 30.

Yang, Z., Zhang, D., Ren, J., & Yang, M. (2012b). Skimmianine, a furoquinoline alkaloid from *Zanthoxylum nitidum* as a potential acetylcholinesterase inhibitor. *Medicinal Chemistry Research, 21*, 722–725.

Yao, G., Chen, X., Zheng, H., Liao, D., Yu, Z., Wang, Z., & Chen, J. (2021). Genomic and chemical investigation of bioactive secondary metabolites from a marine-derived fungus *Penicillium steckii* P2648. *Frontiers in Microbiology, 12*, 600991.

Ye, S., Liu, Y., Wang, S., Guo, M., Wu, J., Hao, Z., Pan, J., Guan, W., Kuang, H., & Yang, B. (2024). Two new alkaloids isolated from *Dictamnus dasycarpus* turcz. Natural Product Research, *39*(13), 1–8.

Yin, Y. F., & Hu Li Hong, H. L. (2005). Two DNA topoisomerase I inhibitors from *Camptotheca acuminata* Decne. (Nyssaceae). *Chinese Journal of Natural Medicines, 3*(1), 21–24.

Yoon, J. S., Jeong, E. J., Yang, H., Kim, S. H., Sung, S. H., & Kim, Y. C. (2012). Inhibitory alkaloids from *dictamnus dasycarpus* root barks on lipopolysaccharide-induced nitric oxide production in BV2 cells. *Journal of Enzyme Inhibition and Medicinal Chemistry, 27*(4), 490–494.

Yu, H., Li, Y., Zhang, M., Zou, Z., Hao, Y., Xie, M., Li, L., Meng, D., & Yang, X. (2023). Chemical constituents of the deep-sea gammarid shrimp-derived fungus *Penicillium citrinum* XIA-16. *Chemistry & Biodiversity, 20*(11), e202301507.

Yu, Y. N., & Li, X. (1990). Studies on the chemical constituents of *Brucea javanica* (L.) Merr. *Yao Xue Xue Bao= Acta Pharmaceutica Sinica, 25*(5), 382–386.

Yue, Y., Yu, H., Suo, Q., Li, R., Liu, S., Xing, R., Zhang, Q., & Li, P. (2022). Discovery of a novel jellyfish venom metalloproteinase inhibitor from secondary metabolites isolated from jellyfish-derived fungus *Aspergillus versicolor* SmT07. *Chemico-Biological Interactions, 365*, 110113.

Zhang, C., Ding, S., Shi, W., Cao, F., Zhu, H., & Wen, M. (2017b). A new quinolinone from freshwater lake-derived fungus *Myrothecium verrucaria*. *Natural Product Research, 31*(1), 99–103.

Zhang, D. Q., Peng, K., Guo, L. L., & Zhang, H. Y. (2011). GC/MS determination of bioactive components of waste leaves from *Platanus×acerifolia* (Ait.) Willd. *Key Engineering Materials, 480*, 502–506.

Zhang, D., Shi, Y., Li, J., Ruan, D., Jia, Q., Zhu, W., Chen, K., Li, Y., & Wang, R. (2019). Alkaloids with nitric oxide inhibitory activities from the roots of *Isatis tinctoria*. *Molecules, 24*(22), 4033.

Zhang, P., Wei, Q., Yuan, X., & Xu, K. (2020). Newly reported alkaloids produced by marine-derived *Penicillium* species (covering 2014–2018). *Bioorganic Chemistry, 99*, 103840.

Zhang, Y., Mu, J., Essmann, F., Feng, Y., Kramer, M., Bao, H., & Grond, S. (2017a). A new quinolinone and its natural/artificial derivatives from a shark gill-derived fungus *Penicillium polonicum* AP2T1. *Natural Product Research, 31*(9), 985–989.

Zhang, Z., & Li, R. (2020). Recent advances in the synthesis of quinolin-2-ones and phenanthridin-6-ones by direct carbonylation (microreview). *Chemistry of Heterocyclic Compounds, 56*, 509–511.

Zheng, C., Xiong, W., Zhang, L., Jin, W., Zhang, J., Li, Y., Hu, Q., Min, Z., Kong, G., & Ye, Y. (2023). Two new anti-tobacco mosaic virus quinolin-2(1*H*)-ones from the twigs of *Cassia auriculata*. *Chemistry of Natural Compounds, 59*(1), 107–110.

Zhong, M., Kang, H., Liu, W., Ma, L., & Liu, D. (2023). Alkaloid diversity expansion of a talent fungus *Penicillium raistrichii* through OSMAC-based cultivation. *Frontiers in Microbiology, 14*, 1279140.

Zhu, J., Song, L., Shen, S., Fu, W., Zhu, Y., & Liu, L. (2023a). Bioactive alkaloids as secondary metabolites from plant endophytic *Aspergillus* genus. *Molecules, 28*(23), 7789.

Zhu, M., Wei, P., Peng, Q., Qin, S., Zhou, Y., Zhang, R., Zhu, C., & Zhang, L. (2019). Simultaneous qualitative and quantitative evaluation of *Toddalia asiatica* root by using HPLC-DAD and UPLC-QTOF-MS/MS. *Phytochemical Analysis, 30*(2), 164–181.

Zhu, Q., Liu, S., Zhai, M., Qiao, X., Chen, Q., & Zhu, N. (2023b). Two new quinolinone glycoalkaloids from *Chrysanthemum indicum* L. and their antimicrobial activity. *Natural Product Research, 39*(2), 1–8.

Zohora, F. T., Azam, A. Z., Ahmed, S., Rahman, K. M., Halim, M. A., Anwar, M. R., Sohrab, M. H., Tabassum, F., Hasan, C. M., & Ahsan, M. (2022). Isolation and in silico prediction of potential drug-like compounds with a new dimeric prenylated quinolone alkaloid from *Zanthoxylum rhetsa* (Roxb.) root extracts targeted against SARS-CoV-2 (Mpro). *Molecules, 27*(23), 8191.

6 4-Quinolinones

6.1 INTRODUCTION

The 4-quinolinone (also called 4-quinolone) skeleton is significant in biology, medicine, and ecology because of its unique features and the various activities it provides to different compounds. This structure is found in many active compounds, such as natural products, synthetic antibiotics, and specialized metabolites. As a result, it is widely used in drug discovery, agriculture, and studying microbial ecology. Numerous review articles have been published highlighting the significance of 4-quinolinone compounds (e.g. Larsen 2005; Heeb et al. 2011; Daneshtalab and Ahmed 2012; Dhiman et al. 2019; Aly et al. 2021; Dube et al. 2023; Gach-Janczak et al. 2025). The following are the key aspects highlighting the importance of 4-quinolinones:

6.1.1 PHARMACOLOGICAL AND ANTIMICROBIAL PROPERTIES OF 4-QUINOLINONES

4-Quinolinones are important in treating several bacterial infections (Naeem et al. 2016). Many of them are successful in treating Gram-positive and Gram-negative bacterial infections. The essential framework of 4-quinolinones is a bicyclic structure with nitrogen at the first position and a carbonyl group at the fourth position (Figure 6.1).

The structure of 4-quinolinone enhances its stability and gives it water-repellent properties. Quinolinones can penetrate the cell wall of bacteria. For instance, bonding fluoroquinolones to bacterial enzymes helps kill the bacteria, resulting in antibiotic properties. There are many fluoroquinolones, each with a different character; therefore, many physicians use them to treat infections of various body parts. The first quinolinone antibiotic, nalidixic acid (Figure 6.2), which is made from chloroquine (an antimalarial drug), is a 4-quinolinone whose discovery started in the 1960s. Urinary tract infection was treated effectively with nalidixic acid. However, it had some drawbacks, such as poor absorption in the body and a narrow range.

Fluoroquinolones were developed, which could treat more types of infections. Ciprofloxacin, levofloxacin and moxifloxacin are well-known fluoroquinolones. These drugs work well against many germs, inhibiting two critical bacterial enzymes, DNA gyrase and topoisomerase IV. They are highly regarded for their oral administration, long-lasting action, and distribution throughout the body's tissues.

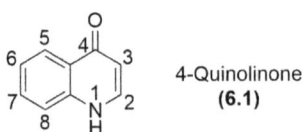

FIGURE 6.1 Chemical structure and numbering system for 4-quinolinone.

DOI: 10.1201/9781003534600-6

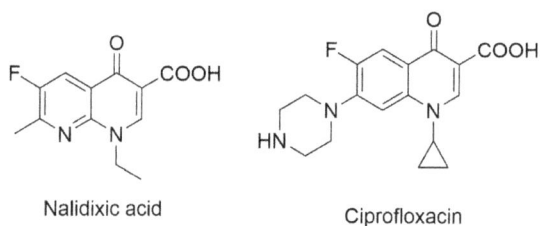

FIGURE 6.2 Chemical structures of Nalidixic acid and Ciprofloxacin.

These qualities make them an ideal option for administering to patients at home and reducing intravenous antibiotics, usually required for severe infections or in hospitals. Fluoroquinolones can also prevent the intracellular organisms *Mycoplasma pneumoniae* and *Legionella pneumophila*. Although fluoroquinolones' uses are not limited to usual bacteria, their broad spectrum of activity has drawbacks. It helps fight infections associated with a variety of bacteria, but in doing so, it also affects the healthy bacteria in our body. Investigations revealed a few quinolinone derivatives that could be effective against tuberculosis (e.g. levofloxacin and moxifloxacin) and malaria (e.g. ELQ-300 in Figure 6.2). In addition, several quinolinone derivatives, mostly 4-quinolinones, were found to be antifungal (Zhang 2019). As a result, it is all part of a new trend. It involves repurposing existing drugs for other purposes. However, this trend is catching on as developing new drugs can be highly expensive and risky. 4-Quinolinones revolutionized the management of infections through their broad-spectrum activity, convenient dosing, and patient compliance.

6.1.2 Ecological Role in Microbial Signalling and Competition

Microbes in nature interact in complex ways to survive. They often use chemical signals and compete with each other. The 4-quinolinone structure is key in these interactions. Its derivatives serve as signalling molecules and competitors in microbial communities. These molecules are especially significant in quorum sensing, a system that allows bacteria to communicate based on their population density (Diggle et al. 2006a; Heeb et al. 2010; Huse and Whiteley 2011). This communication helps the group work together. They form structures called biofilms, improve their ability to cause disease, and protect themselves from other species. A key example of a 4-quinolinone derivative is the Pseudomonas quinolinone signal (PQS) [(6.29) in Table 6.1]. *Pseudomonas aeruginosa* produces PQS and influences gene expression in bacteria. The ability of *P. aeruginosa* to form biofilms through PQS-based quorum sensing is closely tied to its ability to cause ongoing infections, especially in patients with cystic fibrosis and those with weakened immune systems. PQS and similar compounds can stop some Gram-positive bacteria from growing by interfering with their respiratory enzymes (Saalim et al. 2020). This ability to outcompete others is crucial for survival in diverse settings.

Another compound, HQNO [(6.117) in Table 1], also plays a role in these interactions. HQNO explicitly targets a part of the bacteria's energy-making process called the electron transport chain. It disrupts energy production in specific competitors,

TABLE 6.1
The Chemical Identity of 4-Quinolinone Alkaloids

Structure (Number)	Systematic Name	Common Name	CAS #
(6.1)	4(1H)-Quinolinone	4-Quinolinone 4-Quinolinone	529-37-3
(6.2)	2-Methyl-4(1H)-quinolinone	2-Methyl-4-quinolinone	5660-24-2
(6.3)	2,3-Dimethyl-4(1H)-quinolinone	—	58596-45-5
(6.4)	1,4-Dihydro-4-oxo-3-quinolinecarboxamide	—	103914-79-0
(6.5)	1,4-Dihydro-4-oxo-2-quinolinecarboxylic acid	Transtorine	13593-94-7

(Continued)

TABLE 6.1 (Continued)
The Chemical Identity of 4-Quinolinone Alkaloids

Structure (Number)	Systematic Name	Common Name	CAS #
(6.6)	7-Hydroxy-8-methoxy-4(1H)-quinolinone	Hymoquinolinone	98267-21-1
(6.7)	3,5,8-Trihydroxy-4(1H)-quinolinone	Uranidine	92264-09-0
(6.8)	1-Methyl-2-propyl-4(1H)-quinolinone	Leptomerine	22048-97-1
(6.9)	2-(1H-Pyrrol-2-yl)-4(1H)-quinolinone	Penicinoline E	1462328-43-3

(Continued)

TABLE 6.1 (Continued)
The Chemical Identity of 4-Quinolinone Alkaloids

Structure (Number)	Systematic Name	Common Name	CAS #
(6.10)	1,4-Dihydro-8-methoxy-4-oxo-2-quinolinecarboxylic acid	—	93445-77-3
(6.11)	1,4-Dihydro-5,8-dihydroxy-4-oxo-2-quinolinecarboxylic acid	—	115525-96-7
(6.12)	7-Chloro-1,4-dihydro-4-oxo-2-quinolinecarboxylic acid	Ageloline A	77474-31-8
(6.13)	2-(1-Ethylpropyl)-1-methyl-4(1H)-quinolinone	Leiokinine B	126365-17-1
(6.14)	2-(2,3-Dimethylbutyl)-4(1H)-quinolinone	—	2030403-74-6

(Continued)

TABLE 6.1 (*Continued*)
The Chemical Identity of 4-Quinolinone Alkaloids

Structure (Number)	Systematic Name	Common Name	CAS #
(6.15)	2-Hexyl-4(1H)-quinolinone	Pseudane VI HXQ	18813-68-8
(6.16)	1-Methyl-2-pentyl-4(1H)-quinolinone	–	22048-98-2
(6.17)	3-Methoxy-1-methyl-2-propyl-4(1H)-quinolinone	Leiokinine A	132587-63-4
(6.18)	1-Methyl-2-phenyl-4(1H)-quinolinone	–	17182-60-4
(6.19)	2-(4Z)-4-Hepten-1-yl-4(1H)-quinolinone	Acutine	36150-05-7

(Continued)

TABLE 6.1 (Continued)
The Chemical Identity of 4-Quinolinone Alkaloids

Structure (Number)	Systematic Name	Common Name	CAS #
(6.20)	2-(2E)-2-Hepten-1-yl-4(1H)-quinolinone	—	2509157-19-9
(6.21)	2-Heptyl-4(1H)-quinolinone	Pseudane VII HHQ	40522-46-1
(6.22)	1,4-Dihydro-4-oxo-2-(1H-pyrrol-2-yl)-3-quinolinecarboxylic acid	Penicinoline Marinamide	1214268-60-6
(6.23)	2-(2-Cyclopentylethyl)-1-methyl-4(1H)-quinolinone	—	2102522-27-8
(6.24)	2-(4E)-4-Hepten-1-yl-1-methyl-4(1H)-quinolinone	Avicenine F	28839758-29-9

(Continued)

TABLE 6.1 (*Continued*)
The Chemical Identity of 4-Quinolinone Alkaloids

Structure (Number)	Systematic Name	Common Name	CAS #
(6.25)	2-(2E)-2-Hepten-1-yl-3-methyl-4(1H)-quinolinone	PSC-B	178956-00-8
(6.26)	2-(4-Methylheptyl)-4(1H)-quinolinone	–	2115700-00-8
(6.27)	2-Heptyl-3-methyl-4(1H)-quinolinone	PSC-C	172484-87-6
(6.28)	2-Heptyl-1-methyl-4(1H)-quinolinone	Schinifoline	80554-58-1
(6.29)	2-Heptyl-3-hydroxy-4(1H)-quinolinone	PQS Pseudomonas quinolinone signal	108985-27-9

(Continued)

TABLE 6.1 (Continued)

The Chemical Identity of 4-Quinolinone Alkaloids

Structure (Number)	Systematic Name	Common Name	CAS #
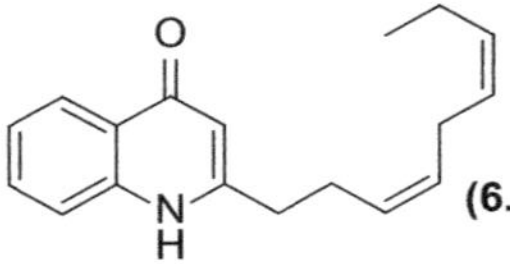 (6.30)	2-(3Z,6Z)-3,6-Nonadien-1-yl-4(1H)-quinolinone	Haplacutine E	1185855-66-6
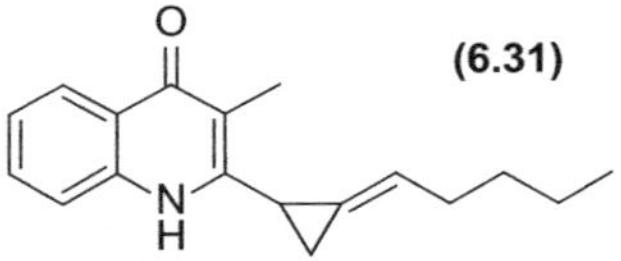 (6.31)	(+)-3-Methyl-2-(pentylidenecyclopropyl)-4(1H)-quinolinone	Antibiotic G 1499-2	68978-12-1
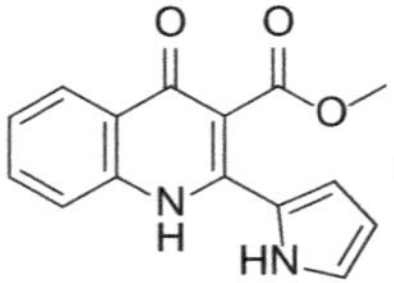 (6.32)	Methyl 1,4-dihydro-4-oxo-2-(1H-pyrrol-2-yl)-3-quinolinecarboxylate	Methyl penicinoline Methyl marinamide	1336924-88-9
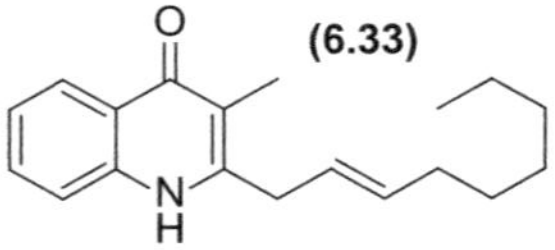 (6.33)	3-Methyl-2-(2E)-2-nonen-1-yl-4(1H)-quinolinone	PSC-D	130772-03-1
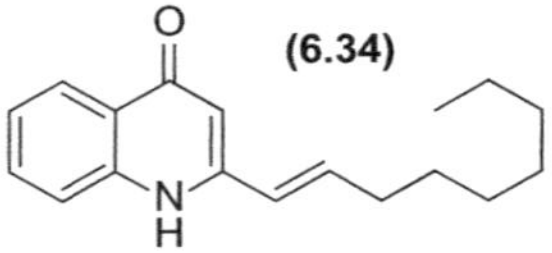 (6.34)	2-(1E)-1-Nonen-1-yl-4(1H)-quinolinone	Δ¹-Pseudene IX	60783-01-9

(Continued)

TABLE 6.1 (Continued)
The Chemical Identity of 4-Quinolinone Alkaloids

Structure (Number)	Systematic Name	Common Name	CAS #
(6.35)	3-Methyl-2-(2E)-2-octen-1-yl-4(1H)-quinolinone	Burkholone	1003292-33-8
(6.36)	2-Nonyl-4(1H)-quinolinone	Pseudane IX	55396-45-7
(6.37)	1-Methyl-2-octyl-4(1H)-quinolinone	—	80554-61-6
(6.38)	2-(1,3-Benzodioxol-5-yl)-1-methyl-4(1H)-quinolinone	Graveoline Rutamine	485-61-0
(6.39)	2-[(2E)-3,7-Dimethyl-2,6-octadien-1-yl]-4(1H)-quinolinone	CJ 13565	189372-51-8

(Continued)

TABLE 6.1 (*Continued*)

The Chemical Identity of 4-Quinolinone Alkaloids

Structure (Number)	Systematic Name	Common Name	CAS #
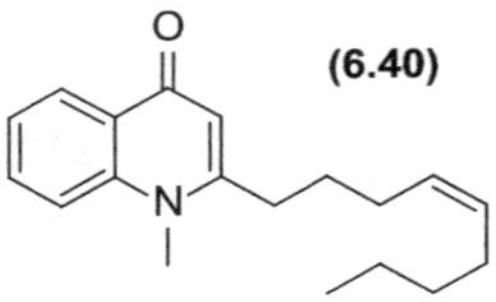 (6.40)	1-Methyl-2-(4Z)-4-nonen-1-yl-4(1H)-quinolinone	—	1422034-91-0
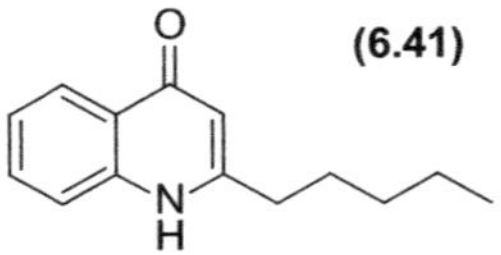 (6.41)	2-Pentyl-4(1H)-quinolinone	Pseudane V PHQ	109072-26-6
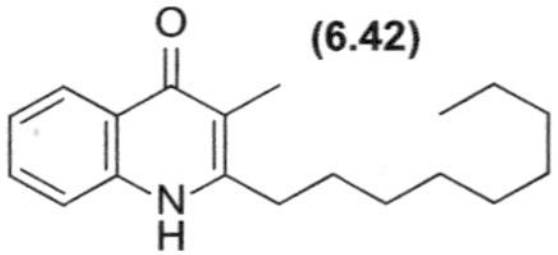 (6.42)	3-Methyl-2-nonyl-4(1H)-quinolinone	PSC-E	172484-88-7
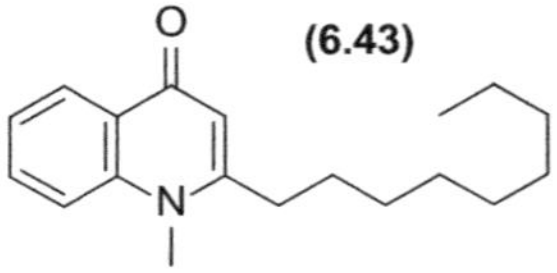 (6.43)	1-Methyl-2-nonyl-4(1H)-quinolinone	—	68353-24-2

(Continued)

TABLE 6.1 (Continued)
The Chemical Identity of 4-Quinolinone Alkaloids

Structure (Number)	Systematic Name	Common Name	CAS #
(6.44)	3-Hydroxy-2-nonyl-4(1H)-quinolinone	–	1259944-03-0
(6.45)	1-Hydroxy-2-nonyl-4(1H)-quinolinone	–	185855-02-1
(6.46)	3,6-Dimethoxy-1-methyl-2-phenyl-4(1H)-quinolinone	Japonine	30426-61-0
(6.47)	2-(1,3-Benzodioxol-5-yl)-5-hydroxy-1-methyl-4(1H)-quinolinone	–	1345406-84-9

(Continued)

TABLE 6.1 (Continued)
The Chemical Identity of 4-Quinolinone Alkaloids

Structure (Number)	Systematic Name	Common Name	CAS #
(6.48)	2-[(2E)-3,7-Dimethyl-2,6-octadien-1-yl]-1-methyl-4(1H)-quinolinone	CJ 13566	189372-53-0
(6.49)	2-[(2E)-3,7-Dimethyl-2,6-octadien-1-yl]-3-methyl-4(1H)-quinolinone	CJ 13136	189372-40-5
(6.50)	2-(1-Undecen-1-yl)-4(1H)-quinolinone	—	250717-34-1
(6.51)	2-Undecyl-4(1H)-quinolinone	—	56183-46-1
(6.52)	2-Decyl-1-methyl-4(1H)-quinolinone	—	213138-50-2

(Continued)

TABLE 6.1 (*Continued*)
The Chemical Identity of 4-Quinolinone Alkaloids

Structure (Number)	Systematic Name	Common Name	CAS #
(6.53)	1-Hydroxy-3-methyl-2-(2E)-2-nonen-1-yl-4(1H)-quinolinone	YL 02729S YM 30059	162382-62-9
(6.54)	3-Methoxy-2-methyl-5-octyl-4(1H)-quinolinone	8-demethoxy-waltherione F	2986324-73-4
(6.55)	2-(1,3-Benzodioxol-5-yl)-7-methoxy-1-methyl-4(1H)-quinolinone	Lunamarine Punarnavine	483-52-3
(6.56)	2-[(2E)-3,7-Dimethyl-2,6-octadien-1-yl]-1,3-dimethyl-4(1H)-quinolinone	CJ 13217	189372-43-8

(*Continued*)

TABLE 6.1 (*Continued*)

The Chemical Identity of 4-Quinolinone Alkaloids

Structure (Number)	Systematic Name	Common Name	CAS #
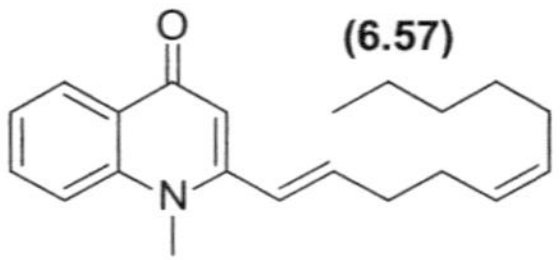 (6.57)	1-Methyl-2-(1E,5Z)-1,5-undecadien-1-yl-4(1H)-quinolinone	–	1422035-35-5
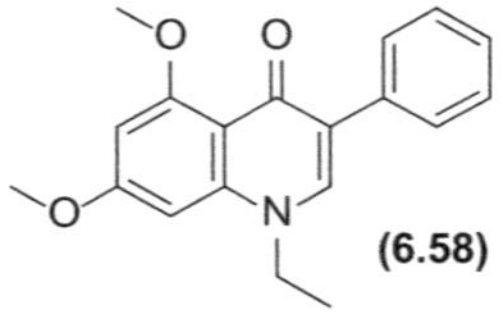 (6.58)	1-Ethyl-5,7-dimethoxy-3-phenyl-4(1H)-quinolinone	Crataemine	1426047-79-1
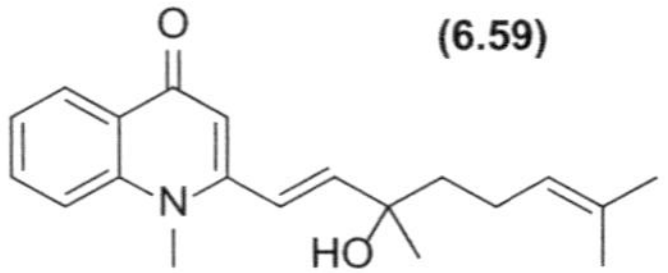 (6.59)	2-[(1E)-3-Hydroxy-3,7-dimethyl-1,6-octadien-1-yl]-1-methyl-4(1H)-quinolinone	–	2047198-45-6
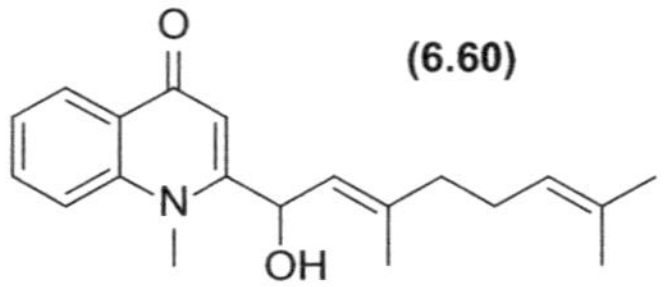 (6.60)	2-[(2E)-1-Hydroxy-3,7-dimethyl-2,6-octadien-1-yl]-1-methyl-4(1H)-quinolinone	–	2047198-42-3

(*Continued*)

TABLE 6.1 (*Continued*)
The Chemical Identity of 4-Quinolinone Alkaloids

Structure (Number)	Systematic Name	Common Name	CAS #
(6.61)	1-Methyl-2-(1E)-1-undecen-1-yl-4(1H)-quinolinone	—	1265227-00-6
(6.62)	1-Methyl-2-(1Z)-1-undecen-1-yl-4(1H)-quinolinone	—	—
(6.63)	1-Methyl-2-(6Z)-6-undecen-1-yl-4(1H)-quinolinone	—	120693-49-4
(6.64)	1-Methyl-2-(5Z)-5-undecen-1-yl-4(1H)-quinolinone	—	182056-11-7

(*Continued*)

TABLE 6.1 (Continued)
The Chemical Identity of 4-Quinolinone Alkaloids

Structure (Number)	Systematic Name	Common Name	CAS #
(6.65)	2-(10-Oxoundecyl)-4(1H)-quinolinone	—	29479-53-6
(6.66)	1-Methyl-2-undecyl-4(1H)-quinolinone	—	59443-02-6
(6.67)	2-[4-(1,3-Benzodioxol-5-yl)butyl]-4(1H)-quinolinone	—	17889-77-9
(6.68)	2-[(2E)-5-(3,3-Dimethyl-2-oxiranyl)-3-methyl-2-penten-1-yl]-1,3-dimethyl-4(1H)-quinolinone	CJ 13564	189372-48-3

(Continued)

TABLE 6.1 (Continued)
The Chemical Identity of 4-Quinolinone Alkaloids

Structure (Number)	Systematic Name	Common Name	CAS #
(6.69)	1-Methyl-2-[(7E)-6-oxo-7-undecen-1-yl]-4(1H)-quinolinone	Euocarpine B	1605307-20-7
(6.70)	1-Methyl-2-[(5E)-7-oxo-5-undecen-1-yl]-4(1H)-quinolinone	Euocarpine A	1608506-96-2
(6.71)	2-Dodecyl-1-methyl-4(1H)-quinolinone	—	182055-94-3
(6.72)	3,8-Dimethoxy-2-methyl-5-octyl-4(1H)-quinolinone	Waltherione F	1632043-41-4

(Continued)

TABLE 6.1 (Continued)

The Chemical Identity of 4-Quinolinone Alkaloids

Structure (Number)	Systematic Name	Common Name	CAS #
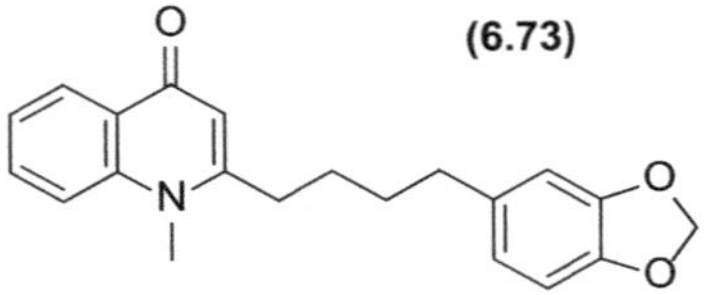 (6.73)	2-[4-(1,3-Benzodioxol-5-yl) butyl]-1-methyl-4(1H)-quinolinone	—	1803329-88-5
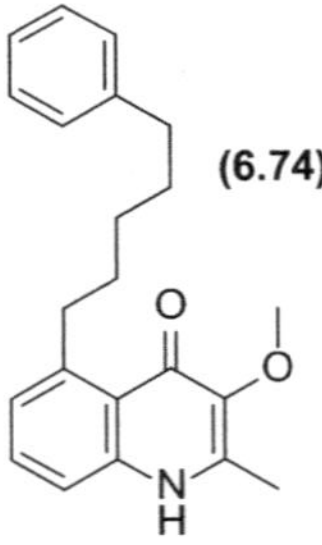 (6.74)	3-Methoxy-2-methyl-5-(5-phenylpentyl)-4(1H)-quinolinone	—	—
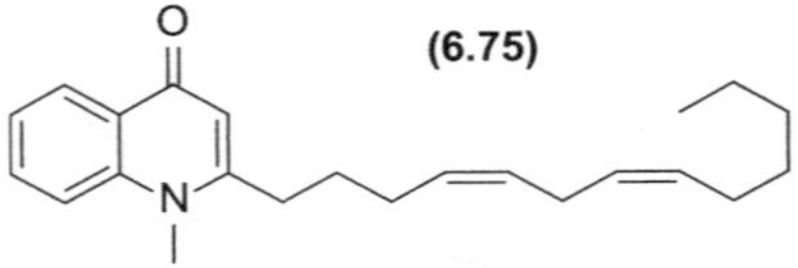 (6.75)	1-Methyl-2-(4Z,7Z)-4,7-tridecadien-1-yl-4(1H)-quinolinone	—	120693-53-0
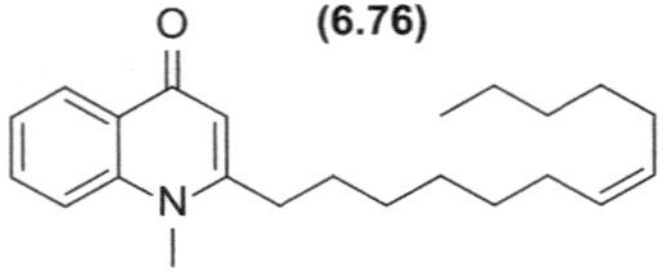 (6.76)	1-Methyl-2-(7Z)-7-tridecen-1-yl-4(1H)-quinolinone	—	182056-19-5

(Continued)

TABLE 6.1 (*Continued*)
The Chemical Identity of 4-Quinolinone Alkaloids

Structure (Number)	Systematic Name	Common Name	CAS #
(6.77)	1-Methyl-2-(8Z)-8-tridecen-1-yl-4(1H)-quinolinone	Evocarpine Euocarpine	15266-38-3
(6.78)	1-Methyl-2-(8E)-8-tridecen-1-yl-4(1H)-quinolinone	—	98393-27-2
(6.79)	2-[(2Z,4E)-6-[2-(1E)-1-Propen-1-ylphenyl]-2,4-hexadien-1-yl]-4(1H)-quinolinone	Marinoquinolinone B	3033870-38-8
(6.80)	1-Methyl-2-tridecyl-4(1H)-quinolinone	Dihydroevocarpine	15266-35-0

(*Continued*)

TABLE 6.1 (Continued)
The Chemical Identity of 4-Quinolinone Alkaloids

Structure (Number)	Systematic Name	Common Name	CAS #
(6.81)	2-(6Z,9Z)-6,9-Pentadecadien-1-yl-4(1H)-quinolinone	—	1245657-04-8
(6.82)	5,8-Dimethoxy-2-(3-methoxyphenyl)-3-propyl-4(1H)-quinolinone	—	1010070-87-7
(6.83)	1-Methyl-2-[(8E)-7-oxo-8-tridecen-1-yl]-4(1H)-quinolinone	—	1422377-36-3
(6.84)	1-Methyl-2-[(7E)-9-oxo-7-tridecen-1-yl]-4(1H)-quinolinone	Euocarpine C	1605306-99-7

(Continued)

TABLE 6.1 (*Continued*)
The Chemical Identity of 4-Quinolinone Alkaloids

Structure (Number)	Systematic Name	Common Name	CAS #
(6.85)	1-Methyl-2-[(9E)-8-oxo-9-tridecen-1-yl]-4(1H)-quinolinone	Euocarpine D	1605300-29-5
(6.86)	1-Methyl-2-[(7E)-6-oxo-7-tridecen-1-yl]-4(1H)-quinolinone	—	1642155-01-8
(6.87)	2-[(2E)-3,7-Dimethyl-2,6-octadien-1-yl]-3-methyl-1-[(methylthio)methyl]-4(1H)-quinolinone	CJ 13536	189372-45-0
(6.88)	1-Methyl-2-tetradecyl-4(1H)-quinolinone	—	536999-65-2

(*Continued*)

TABLE 6.1 (*Continued*)

The Chemical Identity of 4-Quinolinone Alkaloids

Structure (Number)	Systematic Name	Common Name	CAS #
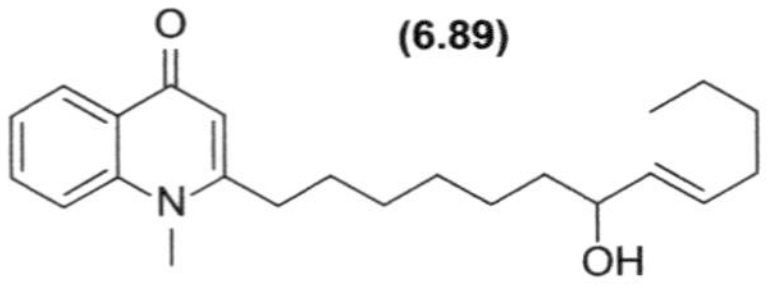 (6.89)	2-[(8E)-7-Hydroxy-8-tridecen-1-yl]-1-methyl-4(1H)-quinolinone	—	1422377-39-6
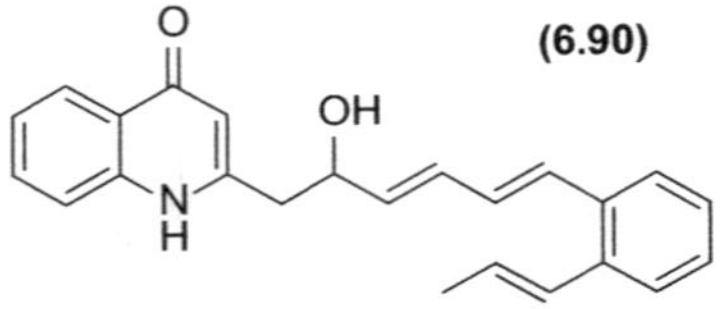 (6.90)	2-((3E,5E)-2-Hydroxy-6-(2-((E)-prop-1-en-1-yl)phenyl)hexa-3,5-dien-1-yl)-4(1H)-quinolinone	Marinoquinolinone A	3033870-37-7
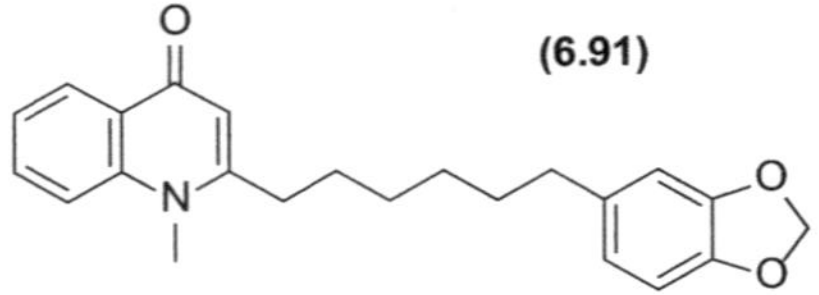 (6.91)	2-[6-(1,3-Benzodioxol-5-yl)hexyl]-1-methyl-4(1H)-quinolinone	—	485803-65-4
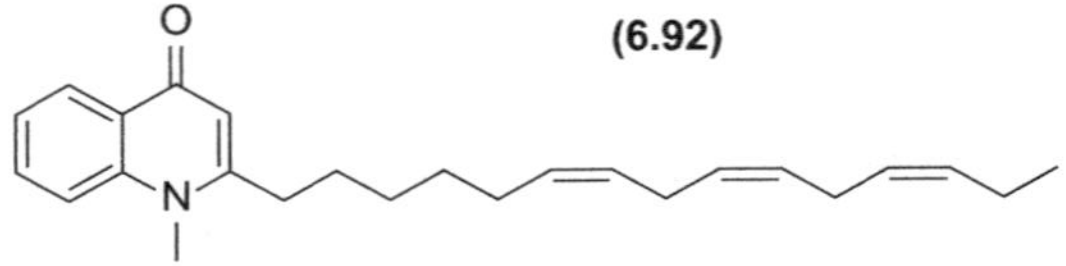 (6.92)	1-Methyl-2-(6Z,9Z,12Z)-6,9,12-pentadecatrien-1-yl-4(1H)-quinolinone	—	2505299-46-5

(Continued)

TABLE 6.1 (Continued)
The Chemical Identity of 4-Quinolinone Alkaloids

Structure (Number)	Systematic Name	Common Name	CAS #
(6.93)	1-Methyl-2-((7Z,9Z,12E)-pentadeca-7,9,12-trien-1-yl)-4(1H)-quinolinone	—	—
(6.94)	1-Methyl-2-((7Z,9Z,11Z)-pentadeca-7,9,11-trien-1-yl)-4(1H)-quinolinone	—	—
(6.95)	2-Methyl-3-[(2E,6E)-3,7,11-trimethyl-2,6,10-dodecatrien-1-yl]-4(1H)-quinolinone	Aurachin D	108354-13-8
(6.96)	1-Methyl-2-(6Z,9Z)-6,9-pentadecadien-1-yl-4(1H)-quinolinone	—	120693-52-9
(6.97)	1-Methyl-2-(6,9-pentadecadien-1-yl)-4(1H)-quinolinone	—	188113-99-7

(Continued)

TABLE 6.1 (Continued)
The Chemical Identity of 4-Quinolinone Alkaloids

Structure (Number)	Systematic Name	Common Name	CAS #
(6.98)	3,8-Dimethoxy-2-methyl-5-(5-phenylpentyl)-4(1H)-quinolinone	Waltherione R	2987384-51-8
(6.99)	1-Methyl-2-(6Z)-6-pentadecen-1-yl-4(1H)-quinolinone	—	120693-51-8
(6.100)	1-Methyl-2-(10Z)-10-pentadecen-1-yl-4-(1H)-quinolinone	—	120693-50-7
(6.101)	1-Methyl-2-(5Z)-5-pentadecen-1-yl-4(1H)-quinolinone	—	1265226-87-6

(Continued)

TABLE 6.1 (*Continued*)
The Chemical Identity of 4-Quinolinone Alkaloids

Structure (Number)	Systematic Name	Common Name	CAS #
(6.102)	1-Methyl-2-pentadecyl-4(1H)-quinolinone	—	59443-03-7
(6.103)	2-(2-Hydroxy-4-methoxyphenyl)-5,8-dimethoxy-3-propyl-4(1H)-quinolinone	—	1010070-86-6
(6.104)	5,6-Dimethoxy-2-(2,3,5-trimethoxyphenyl)-4(1H)-quinolinone	—	1010070-89-9
(6.105)	—	Aurachin R	—

(Continued)

TABLE 6.1 (Continued)
The Chemical Identity of 4-Quinolinone Alkaloids

Structure (Number)	Systematic Name	Common Name	CAS #
(6.106)	1-Hydroxy-2-methyl-3-[(2E,6E)-3,7,11-trimethyl-2,6,10-dodecatrien-1-yl]-4(1H)-quinolinone	Aurachin C	108354-14-9
(6.107)	1-Hydroxy-3-[(2E,6E)-9-hydroxy-3,7,11-trimethyl-2,6,10-dodecatrien-1-yl]-2-methyl-4(1H)-quinolinone	Aurachin RE	1208334-78-4
(6.108)	2-(3,4-Dimethoxyphenyl)-5,8-dimethoxy-3-propyl-4(1H)-quinolinone	–	1010070-88-8
(6.109)	3,7,8-Trimethoxy-2-methyl-5-(5-phenylpentyl)-4(1H)-quinolinone	Melovinone	69905-17-5

(Continued)

TABLE 6.1 (Continued)
The Chemical Identity of 4-Quinolinone Alkaloids

Structure (Number)	Systematic Name	Common Name	CAS #
(6.110)	—	Intervenolin	1430330-16-7
(6.111)	(5S,10R)-7,9-Dibromo-N-[2-[1-(1,4-dihydro-3,5,8-trihydroxy-4-oxo-6-quinolinyl)-1H-imidazol-4-yl]ethyl]-10-hydroxy-8-methoxy-1-oxa-2-azaspiro[4.5]deca-2,6,8-triene-3-carboxamide	Ceratinadin A	1245784-80-8
(6.112)	(5R,10S)-N-[2-[2-Amino-5-(1,4-dihydro-3,5,8-trihydroxy-4-oxo-6-quinolinyl)-1H-imidazol-4-yl]ethyl]-7,9-dibromo-10-hydroxy-8-methoxy-1-oxa-2-azaspiro[4.5]deca-2,6,8-triene-3-carboxamide	—	342632-24-0

(Continued)

TABLE 6.1 (Continued)
The Chemical Identity of 4-Quinolinone Alkaloids

Structure (Number)	Systematic Name	Common Name	CAS #
(6.113)	(5S,10R)-N-[2-[2-Amino-5-(1,4-dihydro-3,5,8-trihydroxy-4-oxo-6-quinolinyl)-1H-imidazol-4-yl]ethyl]-7,9-dibromo-10-hydroxy-8-methoxy-1-oxa-2-azaspiro[4.5]deca-2,6,8-triene-3-carboxamide	Ceratinadin B	1245784-86-4
(6.114)	2-(15-Hydroxypentadecyl)-1-methyl-4(1H)-quinolinone	—	1642155-02-9
(6.115)	2-(13-Hydroxytridecyl)-1-methyl-4(1H)-quinolinone	—	1642155-03-0

(Continued)

TABLE 6.1 (Continued)
The Chemical Identity of 4-Quinolinone Alkaloids

Structure (Number)	Systematic Name	Common Name	CAS #
(6.116)	2-Octyl-4(1H)-quinolinone	Pseudane VIII OQ	80554-60-5
(6.117)	2-Heptyl-1-hydroxy-4(1H)-quinolinone	HQNO Pyo II	341-88-8
(6.118) = (6.45)!	2-Nonyl-1-hydroxy-4(1H)-quinolinone	NQNO	316-66-5
(6.119)	Butyl 1,4-dihydro-4-oxo-2-quinolinecarboxylate	—	1357743-49-7

(Continued)

TABLE 6.1 (Continued)
The Chemical Identity of 4-Quinolinone Alkaloids

Structure (Number)	Systematic Name	Common Name	CAS #
(6.120)	5-hydroxy-6-methoxy-1-methyl-2-phenylquinolin-4(1H)-one	—	—
(6.121)	2-[(1E)-2-(Methylthio)ethenyl]-4(1H)-quinolinone	—	2509156-71-0
(6.122)	2-(4-Hydroxyphenyl)-5-methoxy-1-methylquinolin-4(1H)-one	—	—
(6.123)	5-Hydroxy-6-methoxy-2-(4-methoxyphenyl)-1-methylquinolin-4(1H)-one	—	—

(Continued)

TABLE 6.1 (*Continued*)
The Chemical Identity of 4-Quinolinone Alkaloids

Structure (Number)	Systematic Name	Common Name	CAS #
(6.124)	8-Hydroxy-3-methoxy-2-methyl-5-(4-oxo-octyl)quinolin-4(1H)-one	8-Demethyl-14-oxo-waltherione F	–
(6.125)	(R)-8-hydroxy-5-(4-hydroxyoctyl)-3-methoxy-2-methylquinolin-4(1H)-one	(R)-8-demethyl-14-hydroxywaltherione F	–
(6.126)	1-Methyl-2-(tetradec-13-en-1-yl)quinolin-4(1H)-one	–	–

(*Continued*)

TABLE 6.1 (*Continued*)
The Chemical Identity of 4-Quinolinone Alkaloids

Structure (Number)	Systematic Name	Common Name	CAS #
(6.127)	1-Methyl-2-(tridec-12-en-1-yl)quinolin-4(1H)-one	—	—
(6.128)	1-Methyl-2-[(9E,13E)-heptadecadienyl]-4(1H)-quinolinone	—	—
(6.129)	2-(12-Hydroxytridecyl)-1-methylquinolin-4(1H)-one	—	—
(6.130)	2-Hexadecyl-1-methylquinolin-4(1H)-one	—	—

(*Continued*)

TABLE 6.1 (Continued)
The Chemical Identity of 4-Quinolinone Alkaloids

Structure (Number)	Systematic Name	Common Name	CAS #
(6.131)	(S)-5-(1-hydroxyoctyl)-3,8-dimethoxy-2-methylquinolin-4(1H)-one	11(S)-hydroxy waltherione F	–
(6.132)	(R)-5-(1-hydroxyoctyl)-3,8-dimethoxy-2-methylquinolin-4(1H)-one	11(R)-hydroxy waltherione F	–
(6.133)	3,8-Dimethoxy-2-methyl-5-octanoylquinolin-4(1H)-one	–	

(Continued)

TABLE 6.1 (*Continued*)
The Chemical Identity of 4-Quinolinone Alkaloids

Structure (Number)	Systematic Name	Common Name	CAS #
(6.134)	3,8-Dimethoxy-2-methyl-5-(7-oxooctyl)quinolin-4(1H)-one	—	—
(6.135)	8-Hydroxy-3-methoxy-2-methyl-5-(5-phenylpentyl)quinolin-4(1H)-one	Walindicaone D	—

(Continued)

TABLE 6.1 (Continued)
The Chemical Identity of 4-Quinolinone Alkaloids

Structure (Number)	Systematic Name	Common Name	CAS #
(6.136)	3-Methoxy-2-methyl-5-(5-phenylpentyl)-quinolin-4(1H)-one	8-Demethoxy-waltherione R	—
(6.137)	3,8-Dimethoxy-2-methyl-5-(5-phenylpentanoyl)quinolin-4(1H)-one	Walindicaone E	—
(6.138)	(Z)-1-Methyl-2-(non-4-en-1-yl)quinolin-4(1H)-one	—	—

(Continued)

TABLE 6.1 (Continued)
The Chemical Identity of 4-Quinolinone Alkaloids

Structure (Number)	Systematic Name	Common Name	CAS #
(6.139)	1-Methyl-2-(1Z)-1-undecen-1-yl-4(1H)-quinolinone	—	—
(6.140)	2-methyl-3-((2E,4E,6E)-3,7,11-trimethyldodeca-2,4,6,10-tetraen-1-yl)quinolin-4(1H)-one	Aurachin Q	—
(6.141)	1-Methyl-4(1H)-quinolinone	Echinopsine	83-54-5
(6.142)	7-Methoxy-1-methyl-2-phenyl-4(1H)-quinolinone	Eduleine	483-51-2

(Continued)

TABLE 6.1 (Continued)
The Chemical Identity of 4-Quinolinone Alkaloids

Structure (Number)	Systematic Name	Common Name	CAS #
(6.143)	4-oxo-1,4-dihydroquinoline-2-carboxylic acid (4-Hydroxy-2-quinolinecarboxylic acid)	Kynurenic acid	492-27-3
(6.144)	3-(2-hydroxy-3-methylbutyl)-8-methoxy-1-methylquinolin-4(1H)-one	Pilokeanine	1360-43-6
(6.145)	2-n-Octyl-4-hydroxyquinoline N-oxide	–	2503-84-6
(6.146)	1,2-Dimethyl-4(1H)-quinolinone	–	6760-40-3

(Continued)

TABLE 6.1 (Continued)
The Chemical Identity of 4-Quinolinone Alkaloids

Structure (Number)	Systematic Name	Common Name	CAS #
(6.147)	6-Methoxy-1-methyl-2-phenyl-4-(1H)-quinolinone	Eduline	6878-08-6
(6.148)	1-Ethyl-4(1H)-quinolinone	—	13720-89-3
(6.149)	2-Tridecyl-4(1H)-quinolinone	—	14427-54-4
(6.150)	2-Propyl-4(1H)-quinolinone	—	18813-67-7

(Continued)

TABLE 6.1 (*Continued*)
The Chemical Identity of 4-Quinolinone Alkaloids

Structure (Number)	Systematic Name	Common Name	CAS #
(6.151)	5-Hydroxy-1-methyl-2-phenyl-4(1H)-quinolinone	—	19843-07-3
(6.152)	7-Chloro-4(1H)-quinolinone	Chloroxoquinoline	23833-97-8
(6.153)	2-(9-Hydroxynonyl)-4(1H)-quinolinone	—	29479-52-5
(6.154)	2-(3,6-Nonadien-1-yl)-4(1H)-quinolinone	—	29479-51-4
(6.155)	2-Methoxy-1-methyl-4(1H)-quinolinone	—	40335-01-1

(Continued)

TABLE 6.1 (Continued)
The Chemical Identity of 4-Quinolinone Alkaloids

Structure (Number)	Systematic Name	Common Name	CAS #
(6.156)	2-(4-Hydroxy-3-methoxyphenyl)-1-methyl-4(1H)-quinolinone	Folimidine	40444-99-3
(6.157)	3-Methoxy-1-methyl-4(1H)-quinolinone	—	52381-20-1
(6.158)	8-Hydroxy-4(1H)-quinolinone	—	53846-46-1
(6.159)	2-(1E)-1-Hepten-1-yl-4(1H)-quinolinone	Δ¹-Pseudene VII	60783-02-0

(Continued)

TABLE 6.1 (Continued)
The Chemical Identity of 4-Quinolinone Alkaloids

Structure (Number)	Systematic Name	Common Name	CAS #
(6.160)	2-Hydroxy-4(1H)-quinolinone	—	70254-44-3
(6.161)	2-(3-Undecen-1-yl)-4(1H)-quinolinone	—	71932-11-1
(6.162)	2-(1,3-Benzodioxol-5-yl)-4(1H)-quinolinone	Norgraveoline	74054-38-9
(6.163)	1,4-Dihydro-7-methoxy-4-oxo-2-quinolinecarboxylic acid	Ephedralone	77474-33-0

(Continued)

TABLE 6.1 (Continued)
The Chemical Identity of 4-Quinolinone Alkaloids

Structure (Number)	Systematic Name	Common Name	CAS #
(6.164)	2-Decyl-4(1H)-quinolinone	Pseudane X	80554-64-9
(6.165)	2-Dodecyl-4(1H)-quinolinone	Pseudane XII	80554-65-0
(6.166)	2-(6E,9E,12E)-6,9,12-pentadecatrienyl-4(1H)-quinolinone	Hapovine	80981-97-1
(6.167)	1,4-Dihydro-4-oxo-2-quinolinehexanoic acid	Malatyamine	98752-01-3

(Continued)

TABLE 6.1 (Continued)
The Chemical Identity of 4-Quinolinone Alkaloids

Structure (Number)	Systematic Name	Common Name	CAS #
(6.168)	2-(Nonadienyl)-4(1H)-quinolinone	–	104418-56-6
(6.169)	2-(4-Hydroxyphenyl)-1-methyl-4(1H)-quinolinone	Reevesianine A	109030-96-8
(6.170)	2-(4-Hydroxyphenyl)-6-methoxy-1-methyl-4(1H)-quinolinone	Reevesianine B	109030-98-0
(6.171)	2-Ethyl-4(1H)-quinolinone	–	109072-25-5

(Continued)

TABLE 6.1 (Continued)
The Chemical Identity of 4-Quinolinone Alkaloids

Structure (Number)	Systematic Name	Common Name	CAS #
(6.172)	2-(2-Hepten-1-yl)-3-methyl-4(1H)-quinolinone	—	124201-44-1
(6.173)	7-Bromo-1,4-dihydro-4-oxo-2-quinolinecarboxylic acid	Caelestine A	130064-00-5
(6.174)	2-Butyl-4(1H)-quinolinone	Pseudane IV	135015-64-4
(6.175)	2-Ethyl-1-methyl-4(1H)-quinolinone	—	139710-76-2

(Continued)

TABLE 6.1 *(Continued)*
The Chemical Identity of 4-Quinolinone Alkaloids

Structure (Number)	Systematic Name	Common Name	CAS #
(6.176)	1-[(Acetyloxy)methyl]-2-methyl-4(1H)-quinolinone	—	151077-58-6
(6.177)	1-[(Acetyloxy)methyl]-2,3-dimethyl-4(1H)-quinolinone	—	151077-57-5
(6.178)	1-[(Acetyloxy)methyl]-2-nonyl-4(1H)-quinolinone	—	156547-73-8

(Continued)

TABLE 6.1 (Continued)
The Chemical Identity of 4-Quinolinone Alkaloids

Structure (Number)	Systematic Name	Common Name	CAS #
(6.179)	2-Heptyl-8-methoxy-1-methyl-4(1H)-quinolinone	—	159979-57-4
(6.180)	2-Hexyl-8-methoxy-1-methyl-4(1H)-quinolinone	—	159979-56-3
(6.181)	8-Methoxy-1-methyl-2-pentyl-4(1H)-quinolinone	—	159979-55-2
(6.182)	1-Methyl-2-[(Z)-9-pentadecenyl]-4(1H)-quinolinone	—	170661-70-8

(Continued)

TABLE 6.1 (*Continued*)
The Chemical Identity of 4-Quinolinone Alkaloids

Structure (Number)	Systematic Name	Common Name	CAS #
(6.183)	2-(7-Methoxy-1,3-benzodioxol-5-yl)-1-methyl-4(1H)-quinolinone	—	173180-39-7
(6.184)	8-Methoxy-2-(7-methoxy-1,3-benzodioxol-5-yl)-1-methyl-4(1H)-quinolinone	—	173180-40-0
(6.185)	3-Methyl-2-pentyl-4(1H)-quinolinone	PSC-A	178955-99-2
(6.186)	2,3-Dimethyl 1,4-dihydro-1-methyl-4-oxo-2,3-quinolinedicarboxylate	—	181289-63-4

(*Continued*)

TABLE 6.1 (Continued)
The Chemical Identity of 4-Quinolinone Alkaloids

Structure (Number)	Systematic Name	Common Name	CAS #
(6.187)	3-Methyl-2-octyl-4(1H)-quinolinone	—	185854-82-4
(6.188)	(−)-1-[(Acetyloxy)methyl]-2-[10-(acetyloxy)undecyl]-4(1H)-quinolinone	—	189066-88-4
(6.189)	(−)-2-[(1E)-3-Hydroxy-3,7-dimethyl-1,6-octadien-1-yl]-1-methyl-4(1H)-quinolinone	CJ 13568	189372-57-4
(6.190)	(+)-2-[(2E)-1-Hydroxy-3,7-dimethyl-2,6-octadien-1-yl]-1-methyl-4(1H)-quinolinone	CJ 13567	189372-55-2

(Continued)

TABLE 6.1 (*Continued*)
The Chemical Identity of 4-Quinolinone Alkaloids

Structure (Number)	Systematic Name	Common Name	CAS #
(6.191)	1-Methyl-2-(4Z)-4-tridecen-1-yl-4(1H)-quinolinone	—	250293-82-4
(6.192)	1-Methyl-2-(8-oxononyl)-4(1H)-quinolinone	—	263026-23-9
(6.193)	2-(8-Oxononyl)-4(1H)-quinolinone	—	263026-21-7
(6.194)	1-Methyl-2-(9-oxodecyl)-4(1H)-quinolinone	—	263026-24-0

(Continued)

TABLE 6.1 (*Continued*)
The Chemical Identity of 4-Quinolinone Alkaloids

Structure (Number)	Systematic Name	Common Name	CAS #
(6.195)	(+)-Methyl 1,4-dihydro-α-methoxy-3-(methoxycarbonyl)-1-methyl-4-oxo-2-quinolineacetate	Sarcomejine	284493-49-8
(6.196)	2-[6-(1,3-Benzodioxol-5-yl)hexyl]-4(1H)-quinolinone	—	284674-81-3
(6.197)	Methyl 1,4-dihydro-7-hydroxy-3,8-dimethoxy-1-methyl-4-oxo-2-quinolinecarboxylate	Megistonine II	321553-09-7
(6.198)	Methyl 1,4-dihydro-7-hydroxy-3-methoxy-1-methyl-8-(3-methyl-2-buten-1-yl)-4-oxo-2-quinolinecarboxylate	Megistonine I	321553-08-6

(*Continued*)

TABLE 6.1 (Continued)
The Chemical Identity of 4-Quinolinone Alkaloids

Structure (Number)	Systematic Name	Common Name	CAS #
(6.199)	5,6-Dimethoxy-2-(3-methoxyphenyl)-4(1H)-quinolinone	—	437659-23-9
(6.200)	5,6-Dimethoxy-2-(2,3,6-trimethoxyphenyl)-4(1H)-quinolinone	—	437659-25-1
(6.201)	2-(3,4-Dimethoxyphenyl)-5,6-dimethoxy-4(1H)-quinolinone	—	437659-24-0
(6.202)	2-(12-Hydroxy-12-methyltridecyl)-4(1H)-quinolinone	—	571203-45-7

(Continued)

TABLE 6.1 (*Continued*)
The Chemical Identity of 4-Quinolinone Alkaloids

Structure (Number)	Systematic Name	Common Name	CAS #
(6.203)	2-(12-Hydroxy-12-methyltridecyl)-3-methoxy-4(1H)-quinolinone	—	571203-44-6
(6.204)	(+)-2-(14-Hydroxy-14,15-dimethylhexadecyl)-3-methoxy-4(1H)-quinolinone	—	572918-35-5
(6.205)	(+)-2-(14-Hydroxy-14,15-dimethylhexadecyl)-4(1H)-quinolinone	—	572918-34-4
(6.206)	(-)-2-(10-Hydroxy-10-methyldodecyl)-3-methoxy-4(1H)-quinolinone	—	863332-60-9
(6.207)	3-Methoxy-2-(12-oxotridecyl)-4(1H)-quinolinone	—	863332-59-6

TABLE 6.1 (Continued)
The Chemical Identity of 4-Quinolinone Alkaloids

Structure (Number)	Systematic Name	Common Name	CAS #
(6.208)	6-Hydroxy-2-(3-hydroxy-3-methylbutyl)-4(1H)-quinolinone	—	863332-63-2
(6.209)	7-Hydroxy-2-(3-hydroxy-3-methylbutyl)-4(1H)-quinolinone	—	863332-62-1
(6.210)	2-(11-Hydroxy-11-methyldodecyl)-3-methoxy-4(1H)-quinolinone	—	863332-61-0
(6.211)	(+)-2-(12-Hydroxytridecyl)-3-methoxy-4(1H)-quinolinone	—	863392-78-3

(Continued)

TABLE 6.1 (Continued)
The Chemical Identity of 4-Quinolinone Alkaloids

Structure (Number)	Systematic Name	Common Name	CAS #
(6.212)	3-(Aminocarbonyl)-1,4-dihydro-1-methyl-4-oxo-2-quinolinecarboxylic acid	Quinolonic acid	888728-71-0
(6.213)	Methyl 1,4-dihydro-1-methyl-4-oxo-2-(1,2,3-trimethoxy-3-oxo-1-propen-1-yl)-3-quinolinecarboxylate	Sarcomejine B	939403-86-8
(6.214)	1-Methyl-2-(10-oxoundecyl)-4(1H)-quinolinone	—	946131-56-2
(6.215)	7-Hydroxy-4(1H)-quinolinone	—	955938-89-3

(Continued)

TABLE 6.1 (*Continued*)
The Chemical Identity of 4-Quinolinone Alkaloids

Structure (Number)	Systematic Name	Common Name	CAS #
(6.216)	7-Bromo-4(1H)-quinolinone	—	956268-33-0
(6.217)	1-Methyl-2-(1-methylpropyl)-4(1H)-quinolinone	—	1008099-97-5
(6.218)	(αE)-N-[2-[2-Amino-5-(1,4-dihydro-3,5,8-trihydroxy-4-oxo-6-quinolinyl)-1H-imidazol-4-yl]ethyl]-4-(3-aminopropoxy)-3,5-dibromo-α-(hydroxyimino)benzenepropanamide	Tyrokeradine B	1139903-36-8

(*Continued*)

TABLE 6.1 (*Continued*)
The Chemical Identity of 4-Quinolinone Alkaloids

Structure (Number)	Systematic Name	Common Name	CAS #
(6.219)	3-[4-[(2E)-3-[[2-[2-Amino-5-(1,4-dihydro-3,5,8-trihydroxy-4-oxo-6-quinolinyl)-1H-imidazol-4-yl]ethyl]amino]-2-(hydroxyimino)-3-oxopropyl]-2,6-dibromophenoxy]-N,N,N-trimethyl-1-propanaminium	Tyrokeradine A	1139903-30-2
(6.220)	2-[(4E,6E)-3-Hydroxy-4,6-nonadien-1-yl]-4(1H)-quinolinone	Haplacutine C	1185855-68-8
(6.221)	2-(3Z)-3-Nonen-1-yl-4(1H)-quinolinone	Haplacutine F	1185855-67-7

(*Continued*)

TABLE 6.1 (*Continued*)
The Chemical Identity of 4-Quinolinone Alkaloids

Structure (Number)	Systematic Name	Common Name	CAS #
(6.222)	2-[(4E,6Z)-3-Oxo-4,6-nonadien-1-yl]-4(1H)-quinolinone	Haplacutine D	1185855-65-5
(6.223)	2-[(4E,6Z)-3-Hydroxy-4,6-nonadien-1-yl]-4(1H)-quinolinone	Haplacutine B	1185855-64-4
(6.224)	2-[(3Z,5E)-7-Hydroxy-3,5-nonadien-1-yl]-4(1H)-quinolinone	Haplacutine A	1185855-63-3

(*Continued*)

TABLE 6.1 (*Continued*)

The Chemical Identity of 4-Quinolinone Alkaloids

Structure (Number)	Systematic Name	Common Name	CAS #
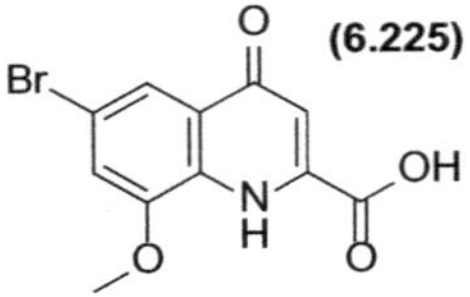 (6.225)	6-Bromo-1,4-dihydro-8-methoxy-4-oxo-2-quinolinecarboxylic acid	Caelestine C	1242974-24-8
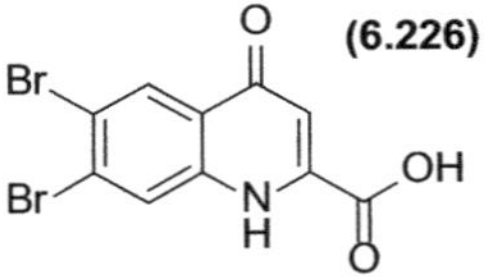 (6.226)	6,7-Dibromo-1,4-dihydro-4-oxo-2-quinolinecarboxylic acid	Caelestine B	1242974-22-6
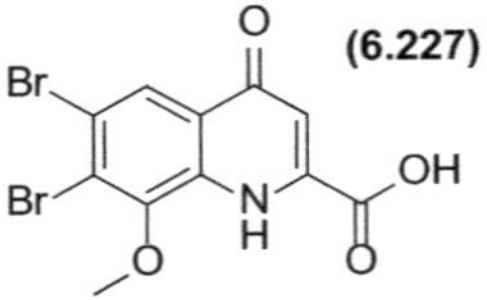 (6.227)	6,7-Dibromo-1,4-dihydro-8-methoxy-4-oxo-2-quinolinecarboxylic acid	Caelestine D	1242974-26-0
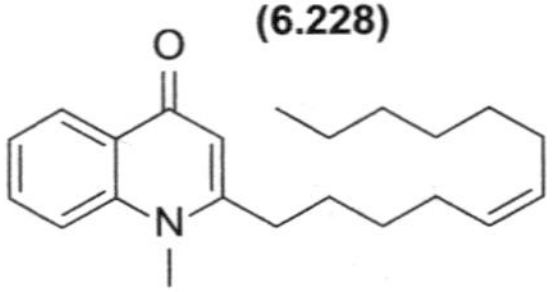 (6.228)	2-(5Z)-5-Dodecen-1-yl-1-methyl-4(1H)-quinolinone	–	1265226-92-3

(*Continued*)

TABLE 6.1 (*Continued*)
The Chemical Identity of 4-Quinolinone Alkaloids

Structure (Number)	Systematic Name	Common Name	CAS #
(6.229)	1-Methyl-2-phenoxy-4(1H)-quinolinone	–	1360829-17-9
(6.230)	(-)-6-Hydroxy-2-(2-hydroxypropyl)-1-methyl-4(1H)-quinolinone	Hystrolinone	1423633-79-7
(6.231)	3-((1,2-Dihydroxy-1-propen-1-yl)-5-hydroxy-1-methyl-4(1H)-quinolinone	–	1460342-45-3
(6.232)	Methyl 1,4-dihydro-1-methyl-4-oxo-2-quinolinepentanoate	Evollionine C	1639421-38-7

(*Continued*)

TABLE 6.1 (Continued)
The Chemical Identity of 4-Quinolinone Alkaloids

Structure (Number)	Systematic Name	Common Name	CAS #
(6.233)	1-Methyl-2-[(4E)-6-oxo-4-undecen-1-yl]-4(1H)-quinolinone	—	1642154-99-1
(6.234)	1-Methyl-2-(6-phenylhexyl)-4(1H)-quinolinone	—	1803281-78-8
(6.235)	1-Methyl-2-(7-methylnonyl)-4(1H)-quinolinone	—	1803281-75-5
(6.236)	2-(9-Methylundecyl)-4(1H)-quinolinone	—	1803281-60-8
(6.237)	2-(7-Methylnonyl)-4(1H)-quinolinone	—	1803281-59-5

(Continued)

TABLE 6.1 (*Continued*)
The Chemical Identity of 4-Quinolinone Alkaloids

Structure (Number)	Systematic Name	Common Name	CAS #
(6.238)	2-(6-Phenylhexyl)-4(1H)-quinolinone	–	1803329-87-4
(6.239)	2-(11-Methyldodecyl)-4(1H)-quinolinone	–	1803329-86-3
(6.240)	2-(9-Methyldecyl)-4(1H)-quinolinone	–	1803329-85-2
(6.241)	1-Methyl-2-(9-methyldecyl)-4(1H)-quinolinone	–	1803281-21-1
(6.242)	1-Methyl-2-(9-methylundecyl)-4(1H)-quinolinone	–	1803281-06-2

(*Continued*)

TABLE 6.1 (Continued)
The Chemical Identity of 4-Quinolinone Alkaloids

Structure (Number)	Systematic Name	Common Name	CAS #
(6.243)	2-(3-Methylbutyl)-4(1H)-quinolinone	—	2030403-73-5
(6.244)	2-(5-Methylheptyl)-4(1H)-quinolinone	—	2115706-60-8
(6.245)	2-(6-Methylheptyl)-4(1H)-quinolinone	—	2117680-25-6
(6.246)	2-(2E)-2-Nonen-1-yl-4(1H)-quinolinone	—	2118289-15-7
(6.247)	1-Methyl-2-(1E)-1-propen-1-yl-4(1H)-quinolinone	—	2123489-24-5

(Continued)

TABLE 6.1 (*Continued*)
The Chemical Identity of 4-Quinolinone Alkaloids

Structure (Number)	Systematic Name	Common Name	CAS #
(6.248)	1-Methyl-2-(1-pentadecen-1-yl)-4(1H)-quinolinone	—	2180947-11-7
(6.249)	1-Methyl-2-(6Z,9Z,12E)-6,9,12-pentadecatrien-1-yl-4(1H)-quinolinone	—	2271054-66-9
(6.250)	Methyl 1,4-dihydro-2-[(3S)-3-hydroxy-2-oxo-1-piperidinyl]-4-oxo-3-quinolinecarboxylate	Aspergillspin C	2351061-29-3
(6.251)	Methyl 1,4-dihydro-1-methyl-4-oxo-2-quinolinepropanoate	Quinolinone A	2408205-95-6

(*Continued*)

TABLE 6.1 (Continued)
The Chemical Identity of 4-Quinolinone Alkaloids

Structure (Number)	Systematic Name	Common Name	CAS #
(6.252)	2-(Phenylmethyl)-4(1H)-quinolinone	–	2509157-11-1
(6.253)	8-Methoxy-1-methyl-2-phenyl-4(1H)-quinolinone	–	2579174-04-0
(6.254)	1,4-Dihydro-2-(hydroxymethyl)-4-oxo-3-quinolinecarboxylic acid	Penicinoline G	2765218-70-8
(6.255)	N-[[2-(1,1-Dimethyl-2-propen-1-yl)-1,4-dihydro-4-oxo-3-quinolinyl]methyl]-L-glutamine	Solitumine B	2773388-21-7
(6.256)	1-Methyl-2-(4-oxoheptyl)-4(1H)-quinolinone	Avicenine E	2839758-28-8

(Continued)

TABLE 6.1 (Continued)
The Chemical Identity of 4-Quinolinone Alkaloids

Structure (Number)	Systematic Name	Common Name	CAS #
(6.257)	1-Methyl-2-(4-oxopentyl)-4(1H)-quinolinone	Avicenine D	2839758-27-7
(6.258)	2-(3-Hydroxypropyl)-1-methyl-4(1H)-quinolinone	Avicenine C	2839758-26-6
	2-[2-(4-Hydroxy-3-methoxyphenyl)ethyl]-1-methyl-4(1H)-quinolinone	Avicenine B	2839758-25-5
(6.260)	2-(5-Methylhexyl)-4(1H)-quinolinone	—	2986911-05-9
(6.261)	2-(2-Hydroxynonyl)-4(1H)-quinolinone	—	3033542-48-9

(Continued)

TABLE 6.1 (Continued)
The Chemical Identity of 4-Quinolinone Alkaloids

Structure (Number)	Systematic Name	Common Name	CAS #
(6.262)	2-[(2E)-3-methyl-2-penten-1-yl]-4(1H)-quinolinone	—	303542-47-8
(6.263)	1-Methyl-2-[7-hydroxy-(Z)-8-tridecenyl]-4-(1H)-quinolinone		
(6.264)	1-Methyl-2-[10-hydroxy-(Z)-8-tridecenyl]-4(1H)-quinolinone		
(6.265)	1-Methyl-2-[8-hydroxy-(E)-9-tridecenyl]-4(1H)-quinolinone		

making it harder for them to survive. This selective killing helps *Pseudomonas* species thrive in various environments, including soil and natural water, by creating unfavourable conditions for other bacteria. PQS helps control genes that are important for forming biofilms and causing diseases. These genes include those that produce elastase, rhamnolipids, and pyocyanin. Biofilms protect bacteria and create environments that resist antibiotics, making treatment more difficult. Besides their protective functions, quinolinone signalling may also help with nutrient cycling by affecting the populations of microbes. 4-Quinolinone derivatives have also been found in marine cyanobacteria and some sponges. In the ocean, where nutrients change and predators are common, quinolinone compounds help organisms fight harmful pathogens and predators. For example, cyanobacteria produce quinolinone substances that stop the growth of invasive species and harmful microorganisms.

6.1.3 BIOLOGICAL ACTIVITY IN ANTICANCER AND ANTI-INFLAMMATORY AGENTS

Many studies have shown that quinolinone-based compounds show anticancer and anti-inflammatory properties, making them a potential source of drug development (Dube et al. 2022). Due to their structure, they interact with different cellular targets, generally key enzymes and receptors instrumental in the processes that govern cell growth, death, and inflammation. Some 4-quinolinone compounds act on the bacterial DNA topoisomerase I and II to kill tumour cells in cancer cells via targeting similar to the drugs (Nand et al. 2016). Quinolactacin A is a compound that occurs naturally in a marine organism that fights cancer. In laboratory tests, it kills cancer cells via apoptosis (programmed cell death). Quinolinone derivatives can interrupt the cell cycle and cause cellular death through interference with DNA synthesis, reducing tumour growth. That characteristic makes them helpful in treating rheumatoid arthritis, inflammatory bowel disease, or asthma. For example, these quinolinones help to eliminate harmful pathways in our bodies, leading to inflammation, such as the nuclear factor kappa B (NF-κB) and cyclooxygenase (COX) pathways. 4-Quinolinone derivatives help curb inflammation and tissue damage by targeting relevant pathways. These compounds can lower two inflammatory markers, tumour necrosis factor-alpha and interleukin-6. Lowering these markers controls inflammation and eases the symptoms of autoimmune diseases. Also, 4-quinolinone-based anti-inflammatories could be less harmful than traditional nonsteroidal anti-inflammatory drugs, as they do not impact COX-1-mediated pathways and are not as likely to cause stomach problems.

6.1.4 CHEMICAL STABILITY AND MODIFIABILITY FOR DRUG DEVELOPMENT

The 4-quinolinone stability helps it resist breakdown in the body. It means that drugs based on quinolinone can maintain their structure and effectiveness over time, making them reliable treatments with predictable effects. One key feature of the 4-quinolinone structure is that it can be easily modified. Chemists can add different groups to change their pharmacological properties. Changing the type of functional group at the nitrogen atom at position 1 or the carbonyl group at position 4 can influence solubility, permeability, and binding to a biological target. Installing a fluorine atom at position 6 leads to the creation of fluoroquinolones, which increases the drug's efficacy and range of actions. Fluorine makes the molecule more lipophilic, which aids entry through cell membranes and increases accumulation inside the cell. This alteration has been essential in enhancing quinolinones' potency against systemic infections; it permits these drugs to achieve the desired quantities in challenging places like the central nervous system and intracellularly within cells. Substituted 4-quinolinones can be obtained, which will aid in drug discovery, especially high-throughput screening (see Structure–Activity Relationship section). Researchers create libraries of quinolinone derivatives with different structures (Sharma et al. 2022; Nery et al. 2023). They can then quickly test these compounds against various biological targets to find candidates with desired properties for further development. Quinolinone libraries play a key role in hit-to-lead optimization. It is vital to drug discovery, where researchers improve promising compounds for better binding affinity, potency, and safety. For example, changing position 7 on the quinolinone core can affect antibacterial strength. Larger groups at this position often make the compounds more effective against Gram-negative bacteria. Researchers focus on making compounds that can avoid bacterial resistance and stay stable in the human body. The 4-quinolinone structure is helpful in various chemical reactions, which help create hybrid molecules that mix the quinolinone core with other functional groups. These hybrid drugs can target different pathways or organisms. For example, researchers have developed hybrid quinolinone derivatives that include antimalarial components, making them effective against bacterial infections and malaria (Asakawa and Manetsch 2021). These drugs are essential in areas where people often suffer from more than one infection and have limited treatment options. The flexibility of the quinolinone structure also supports the creation of prodrugs. Prodrug forms of quinolinones remain inactive or less potent until the body metabolizes them. It can enhance the drug's availability and focus as the prodrug becomes active only in targeted tissues or cells. It can also lessen side effects on healthy tissues.

6.1.5   POTENTIAL ROLE IN TACKLING ANTIBIOTIC RESISTANCE

Antibiotic resistance is a serious global health problem. Molecular evolution is creating superbugs that are resistant to antibiotics. Researchers have witnessed increasing bacterial resistance to antibiotics. Consequently, they investigate 4-quinolinone as a possible foundation for new antibiotics (e.g. Pham et al. 2019). Fluoroquinolones are widely used for their activity against a variety of pathogens. However, as the bacteria become resistant now, scientists are working to modify the structure of quinolinone to bring out new ones that can escape resistance. Bacteria mainly gain resistance to quinolinones through mutations in the genes for DNA gyrase and topoisomerase IV. The ability of enzymes to bind with the quinolinone gets compromised due to such mutations, and subsequently, the bacteria will start growing despite the drug. Scientists alter the quinolinone structure to enhance its binding to mutated enzymes. Flexibility helps the antibiotic bind to the bacteria, which have modified binding sites due to resistance. Bacteria can use efflux pumps to resist quinolinone antibiotics. Quinolinone drugs are either entirely removed by these pumps or their concentrations are lowered to a level that would not kill the bacteria. This allows the bacteria to survive and multiply. Scientists are working on newer versions of quinolinones that will either evade or block these pumps. Changing the structure of the quinolinone at position 7, for instance, can help the drug escape recognition by the efflux pumps. Using efflux pump inhibitors with quinolinones may enhance the efficacy of those antibiotics as a combination therapy against this type of resistance. One way to fight against quinolinone resistance is a medication that targets more than one site in bacteria: quinolinone derivatives that target two bacterial enzymes at once. This strategy looks good for fighting resistance since it makes things much more challenging for the bacteria. To survive, the bacteria must make many mutations in different pathways. Polypharmacology can extend the effective use of antibiotics and is essential for dealing with the issue of drug-resistant infections. A great way to fight antibiotic resistance is to develop new quinolinone drugs with new resistant mechanisms. For instance, researchers look at different quinolinone derivatives that will not stop DNA replication but will stop bacterial cell wall synthesis or something similar. These alternative mechanisms, in combination with traditional quinolinones, enhance their therapeutic efficacy and prevent resistant strains. Using nanoparticles to deliver quinolinones directly to the site of infection is another promising method to fight antibiotic resistance. This method allows the drug to build a greater concentration in the area where it is needed. Scientists are also investigating liposomes and controlled-release systems of quinolinones to lengthen their efficacy and diminish their resistance.

6.1.6 Broad Utility in Agriculture and Veterinary Medicine

Agriculture and animal health frequently employ the 4-quinolinone structure. Quinolinone products control infectious diseases, promote animal health, and increase agricultural productivity. Quinolinones are proactive against various pathogens that harm livestock, poultry, and crops (Shakirov et al. 1996; Dube et al. 2023). In farming, antibiotics can help prevent and treat illness in farm animals. These infections are possibly respiratory, gastrointestinal, or systemic. The diseases are caused by *Escherichia coli*, *Salmonella*, and *Pasteurella multocida*. Oral or animal feed administration of quinolinone antibiotics is especially effective. They help farm animals live longer and healthier lives while improving growth rates. Farmers treat sick fish and other aquatic animals infected with bacteria with quinolinone drugs. For example, fish farms have observed the spread of *Aeromonas* species. It causes the death of many fish and food insecurity. Quinolinones easily mix with water and get inside fish tissue. Regardless, fish farmers must keep a close eye on them to prevent the risk of water pollution or antibiotic resistance. Numerous countries have regulations that limit the use of quinolinones in aquaculture practices. Certain pesticides based on quinolinone assist in controlling bacterial infections in crops. Diseases from bacterial infections of *Xanthomonas* or *Pseudomonas syringae* reduce crop quality and yield. By using quinolinone-based pesticides, farmers can reduce the risk of infection by these bacteria and boost crop yields. The negative impact of quinolinones on the environment has been a concern for many years because it has residual effects in the soil for a long time and impacts non-target organisms. Consequently, the law does not allow quinolinone-related pesticides. These regulations aim to strike a balance between agriculture and safety. According to veterinarians, farmers give quinolinone antibiotics as feed to prevent infection. In dairy and poultry farms, animals are not far from each other. Young or weak animals are more prone to sickness. Dairy calves are treated with quinolinones to manage respiratory diseases, a significant cause of morbidity and mortality. Poultry farms use quinolinones to manage avian colibacillosis outbreaks. The disease has the potential to affect the chickens as well and can cause severe health complications and death ultimately. Quinolinones are extensively utilized in farm animals for food, creating pressure for resistance development. These resistant strains can be transmitted to humans through contact with an infected human, exposure to the environment, or consumption of the product of the animal. As a result, various regulatory bodies have implemented restrictions on the preventive use of quinolinones in animals. They recommend developing alternative disease prevention strategies, such as improved biosecurity, vaccines, and probiotics, to improve gut bacteria. Based on certain studies, feed efficiency and growth rates can be enhanced by quinolinones. This can occur by altering the gut microbiome, reducing harmful bacteria, and increasing nutrient absorption.

6.1.7 Environmental Persistence and Impact on Ecosystems

Using 4-quinolinones in medicine, agriculture, and aquaculture raises significant concerns for environmental fate and impacts. The molecules may have adverse effects on the environment. Quinolinone compounds remain stable in the environment and thus resist degradation by bacteria. They can build up and cause harm to microorganisms in the soil, water, and sediments. Furthermore, they can harm aquatic creatures and interfere with the food chain. Animals raised for farming (like poultry) and animals kept as pets are given antibiotics based on quinolinone. The bulk of the dose is not metabolized but is excreted unchanged. Often, it ends up in the environment via manure, slurry, and runoff from fields. When fish are treated, quinolinones are directly excreted into water and persist therein. The lengthy existence of quinolinones can severely impact the local environment and, more importantly, threaten human beings. Water bodies may get contaminated with quinolinone due to runoff. Quinolinones in water would kill some bacteria while allowing others to flourish. They affect the function of microbial communities. Microbial diversity can be affected when the species composition of microbes affects ecosystem processes such as nutrient cycling and degradation. Aquatic bacteria can develop antibiotic-resistant traits due to the presence of quinolinone residues. Human pathogens can acquire antibiotic resistance due to the transfer of resistance traits from bacteria. Research indicates quinolinones can adversely affect aquatic life's growth, reproduction, and survival. Exposure to fluoroquinolones, for example, obstructs endocrine signalling, resulting in developmental and reproductive disorders in fish. Algae are essential for photosynthesis and growth. Quinolinones are capable of reducing these vital processes. Residues of quinolinone, which stay in the soil, can threaten important bacteria necessary for bacteria and soil fertility. Soil bacteria help in fixing nitrogen and decaying organic matter. As a result of the presence of quinolinones, soil productivity will decrease, and nutrient transformation within the soil will be altered. To prevent the damage of quinolinones, improved wastewater treatment methods are required so that they do not enter natural water sources. Additional processes used include advanced oxidation treatment, membrane filtration, and adsorption for wastewater treatment. Using precise or smaller doses, improving safety measures, and exploring alternative therapies can help reduce the release of quinolinones into the environment.

6.1.8 Chemical Identity

The chemical identity (structure, name, and CAS number) of 4-quinolinone alkaloids in this chapter is shown in Table 6.1. Some of the reliable literature sources were reviewed to verify the accuracy of this data (e.g. Buckingham et al. 2010; Buckingham 2023).

6.1.9 Structural Features-Tautomerism

The tautomerism of 4-quinolinone and 4-hydroxyquinoline is important for their properties and reactivity. This equilibrium is between a keto form, 4-quinolinone, and the enol form, 4-hydroxyquinoline (Figure 6.3).

The process happens by the transfer of a proton; the hydrogen transfers from nitrogen to oxygen. At the same time, π-electrons are rearranged. The tautomeric relationship is particularly sensitive to external factors such as pH, solvent polarity, temperature, and substituents at the quinoline ring. The equilibrium between these tautomers is almost energetically degenerate, usually less than 2 kcal/mol, allowing them to interchange in solution seamlessly. Due to the small energy gap, the predominant tautomer depends on various factors and conditions. For example, due to a hydrogen bond with the carbonyl oxygen, the equilibrium of 4-quinolinone, and 4-hydroxyquinoline would favour 4-quinolinone in protic solvents such as water. On the other hand, in aprotic solvents, the equilibrium shifts towards forming 4-hydroxyquinoline (Heidarnezhad et al. 2013). The pH of the environment is also an important parameter. Acidic conditions usually favour the hydroxyquinoline tautomer via protonation of the nitrogen ring. Basic conditions, in turn, may lead to deprotonation and the formation of a resonance-stabilized quinolinone anion (Seixas et al. 2011). The substituents on the quinoline ring and their nature have been shown to dramatically affect the tautomeric equilibrium (Mphahlele and El-Nahas 2004; Kurasawa et al. 2012; Horta et al. 2015). Groups that withdraw electrons at the positions 3, 5, or 7 enhance the acidity of the hydroxyl proton. Consequently, the resulting 4-hydroxyquinoline would be stabilized. On the other hand, groups that donate

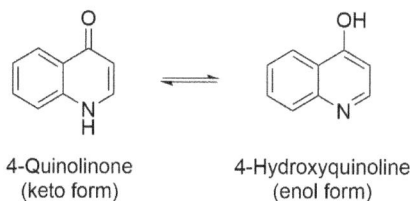

4-Quinolinone
(keto form)

4-Hydroxyquinoline
(enol form)

FIGURE 6.3 Tautomeric forms of 4-quinolinone.

electrons will enhance the basicity of the nitrogen. Thus, the 4-quinolinone form can be favoured by electron-donating groups at positions 5 or 7. Substituents also greatly affect the compounds' overall reactivity and properties (Kang et al. 2019). The tautomerism of 4-quinolinone and 4-hydroxyquinoline is essential in medicinal chemistry and pharmaceuticals. By being tautomeric, many quinolinone derivatives have antibiotic activity. The tautomeric nature affects drug-target interactions, metabolism, and pharmacokinetics. Thus, understanding this property is crucial for designing drugs. Furthermore, tautomerism represents both problems and possibilities in analytical chemistry. It might impact their spectroscopic properties, so special care should be taken when interpreting Nuclear Magnetic Resonance (NMR), Infrared (IR), and X-ray data (Nasiri et al. 2006). Awareness of the tautomeric relationship (structure) of products for industrial and synthetic applications helps set up effective synthetic routes, control product quality, and patenting. Computational studies, including quantum mechanical calculations and molecular dynamics simulations, have provided insight into the relative stabilities of tautomers and their behaviour in various environments.

6.1.10   SYNTHESIS

Over the years, numerous synthetic methods have been developed to construct the 4-quinolinone scaffold. These methods can be broadly classified into classical approaches, transition metal-catalysed reactions, modern techniques, and advanced approaches. We will provide a brief overview of these methods, as several review articles have already been published on this topic (Boteva and Krasnykh 2009; Naeem et al. 2016; Dhiman et al. 2019; Shen et al. 2019; Ghosh and Das 2019; Dine et al. 2023; Gach-Janczak et al. 2025).

6.1.10.1   Classical Approaches

Synthesis of 4-quinolinones by classical approaches mainly involves thermal cyclization reactions under harsh conditions with multi-step sequences. Gould–Jacobs and Conrad–Limpach reactions are two well-known examples. The first report of the Gould–Jacobs reaction was published in 1939, a thermal cyclization method to synthesize 4-quinolinones. The aniline reacts with diethyl ethoxymethylenemalonate and is heated, causing cyclization, acid (hydrolysis), and decarboxylation to get 4-quinolinone. The synthesis of nalidixic acid, oxolinic acid, and norfloxacin, several drugs currently on the market, employs the Gould–Jacobs reaction. One more well-known technique for the synthesis of 4-quinolinone is the Conrad–Limpach reaction. This method involves condensing an aniline with a β-ketoester and cyclizing it using an acid. Although the Gould–Jacobs and Conrad–Limpach reactions have been helpful in the synthesis of 4-quinolinones, they possess certain drawbacks. Both of these reactions require high temperatures; for instance, the Gould–Jacobs reaction requires temperatures greater than 250°C during cyclization. The decomposition of the product and the undesirable side reaction are consequences of high temperature. Also, asymmetric substituents at anilines in the Gould–Jacobs reaction have regioselectivity problems and yield a mixture of products. The use of strong acids, additionally, puts limits on functional group tolerance.

6.1.10.2 Transition Metal-Catalysed Reactions

Different transition metals, such as palladium, copper, cobalt, gold, and nickel (or their derivatives), are used in the synthesis methods to catalyse the formation of 4-quinolinone derivatives. Palladium and copper have been used widely to allow milder conditions and structural diversity (Silva and Silva 2019; Mandal and Khan 2024). Palladium is used as a catalyst to make 4-Quinoline by carbonylative cyclization-coupling with the 2-haloanilines and alkynes with the help of carbon monoxide (CO) (Guo et al. 2024). In this method, 2-haloaniline is coupled with an alkyne, followed by CO insertion and intramolecular cyclization under the influence of a palladium catalyst. However, gaseous CO is dangerous and requires precise pressure regulation for use. As an alternative source of CO, various researchers have attempted to use molybdenum hexacarbonyl, $Mo(CO)_6$, and formic acid (Ferretti et al. 2024). $Mo(CO)_6$ is a solid source of CO, while the formic acid/acetic anhydride combination is the CO surrogate. These alternatives allow for safer and more controlled reactions. Copper catalysts have also been used successfully in the synthesis of 4-quinolinones. It refers to a Cu(I)-catalysed direct cyclization of anilines and alkynes to form diverse 4-quinolinones (Organic Chemistry Portal 2025). This method is effective and atom-economical for constructing this 4-quinolinone scaffold (Tummanapalli et al. 2024).

6.1.10.3 Modern Techniques

Modern techniques for synthesizing 4-quinolinones utilize diverse strategies to improve efficiency, selectivity, and sustainability.

a. **Microwave-Assisted Synthesis**: Microwave irradiation has emerged as a valuable tool in organic synthesis, enabling rapid and efficient reactions. Microwave-assisted synthesis has been applied to 4-quinolinone synthesis, significantly reducing reaction times and improving yields (Organic Chemistry Portal 2025).

b. **Green Chemistry Approaches**: With growing environmental concerns, there is an increasing emphasis on developing sustainable and environmentally friendly synthetic methods. Green chemistry approaches, such as using water or ethanol as solvents, have been explored in 4-quinolinone synthesis (Jaiswal et al. 2023; Ahmed and Akter 2024).

c. **Multicomponent Reactions (MCRs)**: **Multicomponent reactions** involve the reaction of three or more starting materials in a single step to generate complex molecules. They offer several advantages, including high atom economy, synthetic versatility, and the ability to incorporate structural diversity into the final products (Mandal and Khan 2024). These reactions have been successfully employed in 4-quinolinone synthesis, providing efficient and convergent routes to diverse 4-quinolinone derivatives.

d. **Cascade Transformations**: A series of reactions in one pot is known as a cascade transformation (Nammalwar and Bunce 2013). These processes lead to the formation of complex molecules, mostly from more readily available starting materials. One example is synthesizing N-fused quinolinone-4 tetracyclic scaffolds from 2,2-disubstituted indolin-3-ones (Arutiunov et al. 2024). In this method, 2-(2-aryl-3-oxoindolin-2-yl)-2-phenylacetonitriles are converted by base-assisted intramolecular nucleophilic aromatic substitution (SNAr) cyclization to 2-[2-oxo-2-aryl(alkyl)ethyl]-2-phenylindolin-3-ones to give the tetracyclic quinolinones.

e. **Metal-Free Synthesis**: Methods of formation that do not employ metals are appreciated for their simplicity. By avoiding metal catalysts, these methods are eco-friendly and safer for large-scale production. Dine et al. 2023 report on an N-substituted 4-quinolinone synthesis that uses pentane as a solvent and a 2-(tetrahydro-2*H*-pyran-4-yl)-1*H*-indole precursor.

f. **Functionalization of 4-Quinolinone**: Functionalization of 4-quinolinone involves introducing new functional groups or fusing polycyclic rings to the quinolinone core, enhancing its biological and material science applications. Various functionalization techniques are employed to achieve these modifications, including C–C cross-coupling reactions, C-hetero cross-coupling, C–H bond activation, nitration, and regioselective electrophilic insertion (Ghosh and Das 2019).

6.1.10.4   Recent Advances

Recent advances in 4-quinolinone synthesis have focused on developing milder reaction conditions, improving selectivity, and exploring new catalytic systems. Some notable developments include: (a) Non-gaseous CO sources: The use of carbon monoxide-releasing molecules like $Mo(CO)_6$ and formic acid has enabled safer and more controlled carbonylative cyclization reactions (Guo et al. 2024). These alternatives avoid the hazards of handling gaseous CO; (b) Dual-base systems: Employing a combination of bases, such as piperazine and triethylamine, enhances selectivity and yields in 4-quinolinone synthesis. The dual-base system fine-tunes the kinetics of carbonylation and cyclization, which leads to improved reaction outcomes; (c) Electrochemical methods: Electrochemical techniques have been used for C_3–H halogenation of 4-quinolinones, providing access to halogenated derivatives with unique regioselectivity (Organic Chemistry Portal 2025). This approach offers an alternative to traditional halogenation methods that may require harsh reagents or pre-functionalization; and (d) Metal-free arylation: Transition-metal-free C-3-arylation of 4-quinolinones has been achieved using arylhydrazines as aryl radical sources and air as an oxidant (Organic Chemistry Portal 2025). This method avoids the use of expensive or potentially toxic transition metal catalysts.

6.1.11   STRUCTURE–ACTIVITY RELATIONSHIP

The 4-quinolinone scaffold offers a versatile platform for chemical modifications, each impacting its biological activity and safety profile. Numerous review articles have already been published on this subject (Wang et al. 2019; Sharma et al. 2022; Nery et al. 2023; Gach-Janczak et al. 2025). Here, we discuss the most critical positions and substituents in the 4-quinolinone skeleton that contribute to antimicrobial and anticancer activities.

6.1.11.1 Antimicrobial Activity

The N-1 position must be substituted either with an alkyl or with an aryl group for its antibacterial activity. A small cyclopropyl group, such as that in ciprofloxacin (Figure 6.2) and moxifloxacin (Figure 6.4), often works well.

This group aids the drug's strong binding to DNA gyrase, thus increasing its efficacy against many bacteria. Most quinolinone antibiotics with a carboxylic acid at C3 and a carbonyl at C4 are effective (Peterson 2001). A combination of two significant molecular features is required to get through the bacteria's membranes and to the target site in the mineral-associated enzyme-DNA complex. If these groups are removed or altered, the antibacterial activity will be lost significantly. Minor adjustments to C5 can enhance potency against Gram-positive bacteria. For example, adding an amino group (e.g., sparfloxacin; Figure 6.4) or a methyl group (e.g., grepafloxacin; Figure 6.4) at this position enhances potency. However, larger groups at C5 tend to limit effectiveness through steric hindrance (Domagala 1994). The C6 position is also essential. Fluoroquinolones typically have fluorine here, which increases lipophilicity, improves penetration of bacterial cells, and enhances potency by more effectively inhibiting DNA gyrase. This characteristic is present in most second- to fourth-generation quinolinones. Scientists are trying out non-fluorinated quinolinones to prevent toxic effects, although they require other modifications to maintain activity. Regarding the C7 position, installing a ring system may alter the biological activity and pharmacokinetic properties. For instance, a piperazine ring at C7, as in norfloxacin (Figure 6.4) and ciprofloxacin, is effective against Gram-negative bacteria. Moxifloxacin has better efficacy against Gram-positive bacteria due to a pyrrolidine ring. The ring type impacts the uptake of the drug in bacteria and its affinity to achieve the action. Minor modifications to the C7 ring enhance water

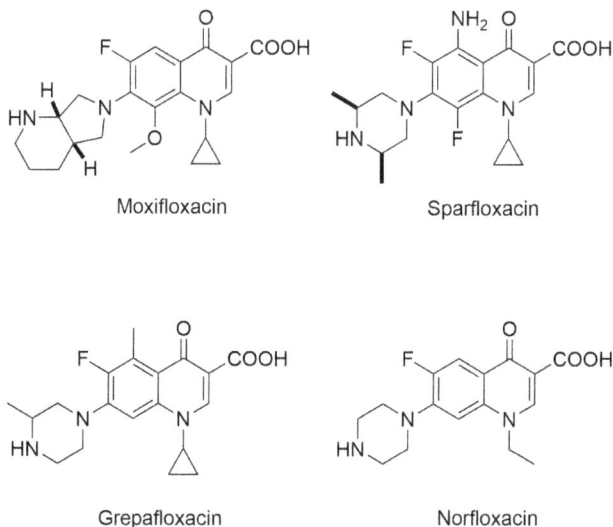

Moxifloxacin   Sparfloxacin

Grepafloxacin   Norfloxacin

FIGURE 6.4 Chemical structures of Moxifloxacin, Sparfloxacin, Grepafloxacin, and Norfloxacin.

solubility and reduce undesired actions in the central nervous system. Undesirably, a simple piperazine ring can increase the risk of neurotoxic side effects, especially when taken with nonsteroidal anti-inflammatory drugs. The C8 chemical group additives to the chemical structure affect the antibacterial spectrum and the pharmacokinetics (Peterson 2001; Domagala 1994). For instance, a methoxy group at the C8 position found in moxifloxacin makes it more effective against anaerobic bacteria and reduces phototoxicity associated with C8 halogens. Historically, C8 halogens (as in sparfloxacin), fluorine, or chlorine were included to increase antibacterial activity, but this led to serious photosensitivity reactions. Because of this, a methoxy substitution at C8 is an important feature of newer quinolinones. It allows the drug to retain its effectiveness against pathogenic microorganisms whilst reducing the side effects and potentially delaying resistance.

6.1.11.2 Anticancer Activity

Adjustments to the nitrogen atom, in combination with a carbonyl functional group at the fourth position, are fundamental to biological effectiveness. Replacing N1–H with a cyclopropyl unit has shown greater activity than an ethyl group. Substituted phenyl or thiazole rings at this site also improve performance. At position 2, alkyl groups have demonstrated superior potential for antineoplastic properties compared to their aryl counterparts. The substituent located at position 3 is most effective when aligned in a coplanar fashion with the quinoline core, ensuring maximum biological impact. The specific chemical groups attached at C5 significantly influence cellular uptake and interaction with the intended biological target. A fluorine atom at position 6 is the most effective for enhancing activity. The substituent introduced at C7 is particularly influential in establishing direct engagement with topoisomerase II in conjunction with DNA. The presence of aromatic systems at this location (i.e. C7) has been linked to a notable increase in antitumour potential. Functionalization at the C8, particularly with a methoxy group, has been associated with modulation of enzymatic activity and improved anticancer effects. Ultimately, the deliberate positioning of lipophilic, planar, and electron-affecting groups is fundamental in optimizing 4-quinolinone derivatives for enhanced anticancer properties (Gach-Janczak et al. 2025).

6.2 NATURAL OCCURRENCE

6.2.1 Animal Sources

Table 6.2 lists fifteen 4-quinolinones from various animals, especially soft corals and sponges *Aplysinidae, Alcyoniidae,* and ascidians from the Polyclinidae family. The different species here evolved massive chemical diversity to adapt to ecological pressures like predation and competition. Genera such as *Sinularia* (a soft coral) and *Verongia* (a sponge) produce similar compounds. It indicates there are possible specialized biochemical pathways in these genera to counteract predators. Interestingly, compound 6.6, which is produced by both *Sinularia polydactyla* and *Nephthea chabroli,* follows similar defence mechanisms. The continuous presence of 4-quinolinones at the genus level shows that species in close relation have evolved together. In addition, 4-quinolinones can be found in some terrestrial invertebrates, such as *Scolopendra*

TABLE 6.2

Natural Occurrence of 4-Quinolinone Alkaloids

Kingdom	Family	Genus	Species	Compound(s)	References
Animalia	Alcyoniidae	*Sinularia*	*microclavate*	6.6	Liu et al. (1990)
Animalia	Alcyoniidae	*Sinularia*	*polydactyla*	6.6	Long et al. (1984); Selim (2012)
Animalia	Aplysinidae	*Verongia (Aplysina)*	*aerophoba*	6.7	Cimino et al. (1984); Carnovali et al. (2022); Teeyapant et al. (1993); Thoms et al. (2004)
Animalia	Aplysinidae	*Verongia (Aplysina)*	Sp.	6.7	Loya et al. (1994)
Animalia	Cryptosulidae	*Cryptosula*	*pallasiana*	6.216	Tian et al. (2015)
Animalia	Darwinellidae	*Dendrilla*	*membranosa*	6.11	Molinski and Faulkner (1988)
Animalia	Enteroctopodidae	*Enteroctopus*	*dofleini*	6.158	Siuda (1974)
Animalia	Nephtheidae	*Nephthea*	*chabroli*	6.6	Zhang et al. (2003)
Animalia	Pentatomidae	*Aspongopus*	*chinensis*	6.5	Di et al. (2015)
Animalia	Phloeodictyidae	*Oceanapia*	Sp. 7	6.7, 6.112	Zhang et al. (2022a)
Animalia	Phloeodictyidae	*Oceanapia*	*sp.*	6.7, 6.112	Fahey et al. (2010)
Animalia	Polyclinidae	*Aplidium*	*caelestis*	6.173, 6.225, 6.226, 6.227	Nicholas et al. (2001)
Animalia	Pseudoceratinidae	*Pseudoceratina*	*sp. SS-214*	6.111, 6.113	Yin et al. (2010)
Animalia	Scolopendridae	*Scolopendra*	*mutilans*	6.158	Kon et al. (2010)
Animalia	Tenebrionidae	*Blaps*	*japanensis*	6.5	Lee et al. (2016)
Animalia	Tylodinidae	*Tylodina*	*perversa*	6.7	Yan et al. (2019)
Animalia	Verongida (order)	—	*SS-301*	6.218, 6.219	Teeyapant et al. (1993)
Bacteria	Alteromonadaceae	*Marinobacterium*	*sp. C17-8*	6.79, 6.90	Mukai et al. (2009)
Bacteria	Alteromonadaceae	*Alteromonas*	*sp. KNS-16*	6.50, 6.51	Karim et al. (2022)
					Cho (2012b)

(Continued)

TABLE 6.2 (Continued)
Natural Occurrence of 4-Quinolinone Alkaloids

Kingdom	Family	Genus	Species	Compound(s)	References
Bacteria	Bacillaceae	*Bacillus*	*sp. RAR-M1-45*	6.25, 6.33	Dat et al. (2022)
Bacteria	Burkholderiaceae	*Burkholderia*	*pseudomalleii (strains K96243, 576, 10276, 844)*	6.21	Diggle et al. (2006a, 2006b)
Bacteria	Burkholderiaceae	*Burkholderia*	*Thailandensis E30*	6.21	Diggle et al. (2006a)
					Diggle et al. (2006b)
Bacteria	Burkholderiaceae	*Burkholderia*	*Cenocepacia J415*	6.21	Diggle et al. (2006a, 2006b)
Bacteria	Burkholderiaceae	*Burkholderia*	*cepacia strain Cs5*	6.25, 6.27	Kilani-Feki et al. (2011a)
Bacteria	Burkholderiaceae	*Burkholderia*	*Cepacian (strain collection)*	6.25, 6.27	Kilani-Feki et al. (2011b)
Bacteria	Burkholderiaceae	*Burkholderia*	*strain MBAF1239*	6.21, 6.33, 6.27, 6.25, 6.246, 6.20	Li et al. (2018)
Bacteria	Burkholderiaceae	*Burkholderia*	*sp. QN15488*	6.35	Mori et al. (2007)
Bacteria	Burkholderiaceae	*Burkholderia*	*cepacia VY81*	6.25	Toi et al. (2022)
Bacteria	Cytophagaceae	*Cytophaga*	*Johnsonii ATCC 21123*	6.31	Evans et al. (1978)
Bacteria	Erwiniaceae	*Pantoea*	*ananatis*	6.29	Vivero-Gomez et al. (2021)
Bacteria	Micrococcaceae	*Arthrobacter*	*sp. YL-02729S*	6.53	Kamigiri et al. (1996)
Bacteria	Myxococcaceae	*Stigmatella*	*aurantiaca Sg a15*	6.95, 6.106	Kunze et al. (1987)
Bacteria	Myxococcaceae	*Stigmatella*	*aurantiaca Sg a15*	6.106, 6.95	Sandmann et al. (2007)
Bacteria	Nocardiaceae	*Nocardia*	*sp. ML96-86F2*	6.56, 6.49, 6.110	Kawada et al. (2013)
Bacteria	Nocardiaceae	*Rhodococcus*	*erythropolis JCM 6824*	6.107	Kitagawa and Tamura (2008)
Bacteria	Nocardiaceae	*Rhodococcus*	*sp. Acta 2259*	6.140, 6.105, 6.106, 6.95	Nachtigall et al. (2010)
Bacteria	Nocardiopsaceae	*Nocardiopsis*	*terrae YIM 90022*	6.4	Tian et al. (2014)
Bacteria	Paenibacillaceae	*Paenibacillus*	*polymyxa SK1*	6.29	Khan et al. (2020)

(Continued)

TABLE 6.2 (Continued)
Natural Occurrence of 4-Quinolinone Alkaloids

Kingdom	Family	Genus	Species	Compound(s)	References
Bacteria	Pseudoalteromonadaceae	Pseudoalteromonas	sp. M2	6.15, 6.150, 6.21, 6.36, 6.51, 6.164, 6.116, 6.41, 6.174, 6.14, 6.243, 6.21, 6.174	Kim et al. (2016)
Bacteria	Pseudoalteromonadaceae	Pseudoalteromonas	galatheae S-4498	6.15, 6.21, 6.36, 6.116, 6.41, 6.171, 6.174	Paulsen et al. (2020)
Bacteria	Pseudomonadaceae	Pseudomonas	aeruginosa	6.118, 6.117, 6.21, 6.36, 6.34, 6.161	Budzikiewicz et al. (1979)
Bacteria	Pseudomonadaceae	Pseudomonas	aeruginosa	6.29	Pesci et al. (1999)
Bacteria	Pseudomonadaceae	Pseudomonas	aeruginosa	6.50, 6.51	Youn et al. (2018)
Bacteria	Pseudomonadaceae	Pseudomonas	1531-E7	6.118, 6.21, 6.36, 6.50	Bultel-Ponce et al. (1999)
Bacteria	Pseudomonadaceae	Pseudomonas	Pyocyanea (aeruginosa)	6.118, 6.117, 6.145, 6.51	Cornforth and James (1956)
Bacteria	Pseudomonadaceae	Pseudomonad	E14, E15	6.1, 6.21, 6.36, 6.34	Debitus et al. (1998)
Bacteria	Pseudomonadaceae	Pseudomonas	Putida KT2440	6.21	Diggle et al. (2006a)
Bacteria	Pseudomonadaceae	Pseudomonas	Aeruginosa A197	6.21, 6.36, 6.34, 6.159	Kozlovskii et al. (1976)
Bacteria	Pseudomonadaceae	Pseudomonas	Cepacian PC-II	6.33, 6.42, 6.27, 6.185, 6.25, 6.42, 6.27, 6.25	Moon et al. (1996)
Bacteria	Pseudomonadaceae	Pseudomonas	cepacia	6.172	Roitman et al. (1990)
Bacteria	Pseudomonadaceae	Pseudomonas	Aeruginosa BD06-03	6.121, 6.34, 6.252, 6.260, 6.261, 6.262	Li et al. (2020)
Bacteria	Pseudonocardiaceae	–	–	6.87	Ahmad et al. (2024)
Bacteria	Pseudonocardiaceae	Pseudonocardia	sp. CL38489	6.189, 6.39, 6.190, 6.48, 6.56, 6.49, 6.87	Dekker et al. (1998)
Bacteria	Pseudonocardiaceae	Saccharomonospora	sp. CNQ-490	6.152	Kim et al. (2021)
Bacteria	Streptomycetaceae	Streptomyces	sp. SBT345	6.12	Cheng et al. (2016)
Bacteria	Streptomycetaceae	Streptomyces	sindenensis OUCMDZ-1368	6.2, 6.21, 6.36, 6.51, 6.34, 6.29, 6.44	Qing-yun et al. (2015)
Bacteria	Streptomycetaceae	Streptomyces	sp. NA04227	6.95, 6.106	Zhang et al. (2017)

(Continued)

TABLE 6.2 (Continued)

Natural Occurrence of 4-Quinolinone Alkaloids

Kingdom	Family	Genus	Species	Compound(s)	References
Fungi	Aspergillaceae	—	—	6.22	Nand et al. (2016)
Fungi	Aspergillaceae	Aspergillus	sp. SCSIO 41501	6.5, 6.22, 6.32, 6.250	Ma et al. (2019b)
Fungi	Aspergillaceae	Aspergillus	Sp. FSY-01 and FSW-02	6.22, 6.32	Zhu et al. (2013)
Fungi	Aspergillaceae	Penicillium	Citrinum IFM 53298	6.157	Wakana et al. (2006)
Fungi	Aspergillaceae	Penicillium	citrinum strain MST-F10130	6.212	Clark et al. (2006)
Fungi	Aspergillaceae	Penicillium	sp. SCSIO 41015	6.212	Pang et al. (2019)
Fungi	Aspergillaceae	Penicillium	steckii SCSIO 41025	6.22, 6.32, 6.4, 6.9, 6.254	Chen et al. (2021)
Fungi	Aspergillaceae	Penicillium	sp. ghq208	6.22, 6.32, 6.9	Gao et al. (2012)
Fungi	Aspergillaceae	Penicillium	sp.	6.22	Shao et al. (2010)
Fungi	Aspergillaceae	Penicillium	sp. 1P	6.32	Hamed et al. (2017)
Fungi	Aspergillaceae	Penicillium	sp. (SF-5995)	6.32	Kim et al. (2014)
Fungi	Aspergillaceae	Penicillium	solitum IS1-A	6.255	Rodríguez et al. (2020)
Fungi	Cordycipitaceae	Cordyceps	militaris	6.158	Sun et al. (2017)
Fungi	Didymellaceae	Epicoccum	sorghinum	6.5	Wu and Ge (2023)
Fungi	Endophytic	Endophytic	—	6.55	Okonkwo-Uzor et al. (2022)
Fungi	Onygenaceae	Auxarthron	reticulatum	6.32	Elsebai et al. (2011)
Fungi	Polyporaceae	Trametes	trogii TGC-P-3	6.1	Bian et al. (2016)
Fungi	Trichocomaceae	Neosartorya	hiratsukae	6.141	Paluka et al. (2020)
Plantae	Amaryllidaceae	Crinum	firmifolium	6.19, 6.21, 6.116, 6.244, 6.245	Presley et al. (2017)
Plantae	Amaryllidaceae	Allium	tenuissimum	6.35	Zhang et al. (2022b)
Plantae	Apiaceae	Notopterygium	incisum	6.5	Xu et al. (2012)
Plantae	Apocynaceae	Ervatamia (Tabernaemontana)	chinensis	6.77, 6.80, 6.43, 6.63	Guo et al. (2012)
Plantae	Apocynaceae	Gongronema	latifolium	6.55	Osuagwu et al. (2023)

(Continued)

TABLE 6.2 (Continued)
Natural Occurrence of 4-Quinolinone Alkaloids

Kingdom	Family	Genus	Species	Compound(s)	References
Plantae	Apocynaceae	*Winchia*	*calophylla*	6.100	Zhu et al. (2005)
Plantae	Araceae	*Scindapsus*	*officinalis*	6.1	Yu et al. (2018)
Plantae	Araliaceae	*Hedera*	*Helix (Ivy honey)*	6.1	Makowicz et al. (2018)
Plantae	Asparagaceae	*Polygonatum*	*sibiricum*	6.63	Cheng et al. (2023)
Plantae	Asteraceae	*Echinops*	*sphaerocephalus*	6.141	Adams et al. (2004)
Plantae	Asteraceae	*Echinops*	*echinatus*	6.141	Chaudhuri (1987)
Plantae	Asteraceae	*Echinops*	*ritro*	6.141	Greshoff (1900)
Plantae	Asteraceae	*Echinops*	*albicaulis*	6.141	Datkhayev (2017)
					Kiyekbayeva et al. (2018)
Plantae	Astraceae	*Echinops*	*nanus*	6.141	Nakano et al. (2012)
Plantae	Astraceae	*Echinops*	*spinosus*	6.141	Halim et al. (2011)
Plantae	Bignoniaceae	*Newbouldia*	*laevis*	6.55	Forghe and Nna (2020)
Plantae	Boraginaceae	*Lappula*	*echinata*	6.10	Zhang et al. (2005)
Plantae	Capparaceae	*Crataeva*	*nurvala*	6.58	Sinha et al. (2012)
Plantae	Cucurbitaceae	*Luffa*	*acutangula var. amara*	6.187	Santosh and Sunil (2019)
Plantae	Cucurbitaceae	*Citrullus*	*colocynthis*	6.193	Salama (2012)
Plantae	Cyperaceae	*Cyperus*	*esculentus*	6.55	Ihenetu et al. (2021)
Plantae	Ephedraceae	*Ephedra*	*transitoria*	6.5	Al-Khalil et al. (1998)
Plantae	Ephedraceae	*Ephedra*	*aphylla*	6.163	Hussein et al. (1997)
Plantae	Ephedraceae	*Ephedra*	*alata*	6.163	Nawwar et al. (1985)
Plantae	Euphorbiaceae	*Jatropha*	*curcas*	6.55	Felix et al. (2022)
Plantae	Euphorbiaceae	*Croton*	*tiglium*	6.1	Su et al. (2016)
Plantae	Fabaceae	*Arachis*	*hypogaea*	6.76, 6.77	Deng et al. (2016)
Plantae	Fagaceae	*Castanea*	*spp. (chestnut flower)*	6.119	Gao et al. (2010)
Plantae	Fagaceae	*Castanea*	*crenata*	6.1, 6.143	Cho et al. (2015)

(Continued)

TABLE 6.2 (Continued)
Natural Occurrence of 4-Quinolinone Alkaloids

Kingdom	Family	Genus	Species	Compound(s)	References
Plantae	Fagaceae	Castanea	mollissima	6.119	Gao et al. (2010)
Plantae	Fagaceae	Castanea	mollissima	6.5	Tang et al. (2004)
Plantae	Fagaceae	Castanea	Sativa (honey)	6.5	Truchado et al. (2009)
Plantae	Lamiaceae	Rabdosia	rubescens	6.215	Feng et al. (2007)
Plantae	Lamiaceae	Vitex	rotundifolia	6.1	Luo et al. (2022)
Plantae	Malvaceae	Waltheria	indica	6.72	Cretton et al. (2014, 2016); de Medeiros Silva et al. (2024)
Plantae	Malvaceae	Waltheria	indica	6.109, 6.54, 6.98	Cretton et al. (2020)
Plantae	Malvaceae	Sida	corymbosa	6.55	Dike et al. (2018)
Plantae	Malvaceae	Waltheria	indica	6.124, 6.125, 6.72	Hua et al. (2023)
Plantae	Malvaceae	Melochia	tomentosa	6.109	Kapadia et al. (1978)
Plantae	Malvaceae	Waltheria	indica	6.132, 6.131, 6.133, 6.134, 6.72, 6.135, 6.136, 6.137	Liu et al. (2023)
Plantae	Malvaceae	Waltheria	indica	6.72	Monteillier et al. (2017)
Plantae	Malvaceae	Sida	linifolia	6.55	Nwankwo et al. (2022)
Plantae	Malvaceae	Melochia	umbellata var. deglabrata	6.72	Rahim et al. (2020)
Plantae	Malvaceae	Urena	lobata	6.160	Shrestha et al. (2016)
Plantae	Nyctaginaceae	Boerhavia	diffusa	6.55	Saini et al. (2022)
Plantae	Poaceae	Oryza	Sativa (bran)	6.5	Sahashi (1925)
Plantae	Rhodomelaceae	Acanthophora	muscoides	6.148	Sunarpi et al. (2018)
Plantae	Rosaceae	Pyrus	pyrifolia	6.5	Lee et al. (2013)

(Continued)

TABLE 6.2 (Continued)
Natural Occurrence of 4-Quinolinone Alkaloids

Kingdom	Family	Genus	Species	Compound(s)	References
Plantae	Rubiaceae	Dictyandra (Leptactina)	arborescens	6.55	Enenebeaku et al. (2022)
Plantae	Rutaceae	—	—	6.51	Ahmad et al. (2024)
Plantae	Rutaceae	—	—	6.76, 6.77	Raj et al. (2021)
Plantae	Rutaceae	Acronychia	baueri	6.146	Lamberton (1966)
Plantae	Rutaceae	Boronia	adulata	6.41, 6.150	Agier et al. (2007)
Plantae	Rutaceae	Boronia	ternata var elongata	6.150, 6.16, 6.51, 6.165, 6.41, 6.217	Agier et al. (2007)
Plantae	Rutaceae	Boronia	Lanceolata	6.3, 6.176, 6.177	Ahsan et al. (1993)
Plantae	Rutaceae	Boronia	bowmanii	6.36, 6.3, 6.43, 6.176, 6.177, 6.178	Ahsan et al. (1994)
Plantae	Rutaceae	Boronia	bowmanii	6.3, 6.43, 6.36	Nazrul Islam et al. (2002)
Plantae	Rutaceae	Boronia	ternata	6.150	Duffield and Jefferies (1963)
Plantae	Rutaceae	Boronia	algida	6.102	Sarker et al. (1995)
Plantae	Rutaceae	Casimiroa	edulis	6.104, 6.108, 6.82, 6.103	Awaad et al. (2007)
Plantae	Rutaceae	Casimiroa	edulis	6.142	Iriarte et al. (1956)
Plantae	Rutaceae	Casimiroa	edulis	6.151	Ito et al. (1998)
Plantae	Rutaceae	Casimiroa	edulis	6.199, 6.200, 6.201	Khaleel (2002)
Plantae	Rutaceae	Casimiroa	edulis	6.147	Kincl et al. (1956)
Plantae	Rutaceae	Casimiroa	edulis	6.18, 6.142	Rizvi et al. (1985)
Plantae	Rutaceae	Citrus	hystrix	6.230	Panthong et al. (2013)
Plantae	Rutaceae	Conchocarpus	heterophyllus	6.18	Ambrozin et al. (2008)
Plantae	Rutaceae	Conchocarpus	inopinatus	6.18, 6.142	Bellete et al. (2012)
Plantae	Rutaceae	Conchocarpus	marginatus	6.18, 6.142	Bellete et al. (2012)
Plantae	Rutaceae	Conchocarpus	fontanesianus	6.18	Cabral et al. (2016)
Plantae	Rutaceae	Dictamnus	dasycarpus	6.141	Ye et al. (2024)

(Continued)

TABLE 6.2 (Continued)
Natural Occurrence of 4-Quinolinone Alkaloids

Kingdom	Family	Genus	Species	Compound(s)	References
Plantae	Rutaceae	*Dictyoloma*	*vandellianum*	6.149, 6.202, 6.203, 6.204, 6.205	Sartor et al. (2003)
Plantae	Rutaceae	*Esenbeckia*	*Leiocarpa*	6.8, 6.17	Cardoso-Lopes et al. (2010)
Plantae	Rutaceae	*Esenbeckia*	*Leiocarpa*	6.13, 6.17	Nakatsu et al. (1990)
Plantae	Rutaceae	*Esenbeckia*	*almawillia*	6.183, 6.184	Oliveira et al. (1996)
Plantae	Rutaceae	*Esenbeckia*	*almawillia*	6.179, 6.180, 6.181	Guilhon et al. (1994)
Plantae	Rutaceae	*Esenbeckia*	*pentaphylla*	6.147	Simpson and Jacobs (2005)
Plantae	Rutaceae	*Euodia*	*rutaecapra var. officinalis*	6.75, 6.52, 6.191	Chuang et al. (1999)
Plantae	Rutaceae	*Tetradium*	*ruticarpum*	6.139, 6.75, 6.76	Ma et al. (2018)
Plantae	Rutaceae	*Tetradium*	*ruticarpum*	6.263, 6.264, 6.265	Matsuo et al. (2025)
Plantae	Rutaceae	*Euodia*	*ailanthifolia*	6.43, 6.37	Yong et al. (2024)
Plantae	Rutaceae	*Euodia*	*austrosinensis*	6.77, 6.80, 6.66, 6.43, 6.71 6.76, 6.64, 6.86	Yong et al. (2024)
Plantae	Rutaceae	*Euodia*	*compacta*	6.149, 6.102, 6.37, 6.96, 6.75, 6.64, 6.88, 6.83, 6.114	Yong et al. (2024)
Plantae	Rutaceae	*Euodia*	*daniellii*	6.66, 6.99, 6.71, 6.64, 6.88, 6.115	Yong et al. (2024)
Plantae	Rutaceae	*Euodia*	*delavayi*	6.149	Yong et al. (2024)
Plantae	Rutaceae	*Euodia*	*fargesii*	6.149, 6.77, 6.80, 6.102, 6.66, 6.96, 6.100, 6.75, 6.71, 6.76, 6.64, 6.52, 6.88, 6.83, 6.85, 6.233, 6.114, 6.115	Yong et al. (2024)
Plantae	Rutaceae	*Euodia*	*fraxinifolia*	6.149, 6.36, 6.102, 6.43, 6.96, 6.75, 6.52, 6.88	Yong et al. (2024)

(Continued)

TABLE 6.2 (Continued)
Natural Occurrence of 4-Quinolinone Alkaloids

Kingdom	Family	Genus	Species	Compound(s)	References
Plantae	Rutaceae	Euodia	glabrifolia	6.77, 6.80, 6.36, 6.102, 6.66, 6.43, 6.37, 6.96, 6.99 6.75, 6.71, 6.76, 6.64, 6.52 6.88, 6.83, 6.86, 6.115	Yong et al. (2024)
Plantae	Rutaceae	Euodia	lepta	6.36, 6.43	Yong et al. (2024)
Plantae	Rutaceae	Tetradium	ruticarpum	6.77, 6.66, 6.233, 6.114, 6.86, 6.115	Zhao et al. (2015)
Plantae	Rutaceae	Tetradium	ruticarpum	6.77, 6.43, 6.96, 6.63, 6.75	Adams et al. (2004)
Plantae	Rutaceae	Tetradium	ruticarpum	6.77, 6.43, 6.96, 6.63, 6.75	Adams et al. (2005)
Plantae	Rutaceae	Tetradium	ruticarpum	6.77, 6.80, 6.66	Adams et al. (2007)
Plantae	Rutaceae	Tetradium	ruticarpum	6.76, 6.77	Hamasaki et al. (2000)
Plantae	Rutaceae	Tetradium	ruticarpum	6.77, 6.80, 6.66, 6.43, 6.96, 6.63	Han et al. (2007)
Plantae	Rutaceae	Tetradium	ruticarpum	6.77, 6.80, 6.102, 6.66, 6.43, 6.96, 6.99, 6.100, 6.63, 6.75, 6.71, 6.52, 6.88, 6.61, 6.40, 6.57, 6.89, 6.83	Huang et al. (2012)
Plantae	Rutaceae	Tetradium	ruticarpum	6.77, 6.80, 6.36, 6.51, 6.102, 6.66, 6.43, 6.96, 6.75	Jin et al. (2004)
Plantae	Rutaceae	Tetradium	ruticarpum	6.80, 6.102, 6.66	Kamikado et al. (1976, 1978)
Plantae	Rutaceae	Tetradium	ruticarpum	6.43	Kamikado et al. (1978)
Plantae	Rutaceae	Tetradium	ruticarpum	6.77, 6.96, 6.75, 6.88	Ko et al. (2002)
Plantae	Rutaceae	Tetradium	ruticarpum	6.77, 6.96, 6.75	Lee et al. (1998)
Plantae	Rutaceae	Tetradium	ruticarpum	6.66	Lee et al. (2013)
Plantae	Rutaceae	Tetradium	ruticarpum	6.232	Li et al. (2014)
Plantae	Rutaceae	Tetradium	ruticarpum	6.77, 6.80, 6.66, 6.43, 6.96, 6.75, 6.182, 6.64, 6.248	Li et al. (2016)

(Continued)

TABLE 6.2 (Continued)

Natural Occurrence of 4-Quinolinone Alkaloids

Kingdom	Family	Genus	Species	Compound(s)	References
Plantae	Rutaceae	Tetradium	ruticarpum	6.102, 6.149	Li et al. (2022)
Plantae	Rutaceae	Tetradium	ruticarpum	6.126, 6.127, 6.128, 6.129, 6.130, 6.249	Ling et al. (2016)
Plantae	Rutaceae	Tetradium	ruticarpum	6.77, 6.96, 6.71	Liu et al. (2005)
Plantae	Rutaceae	Tetradium	ruticarpum	6.138	Liu et al. (2023)
Plantae	Rutaceae	Tetradium	ruticarpum	6.93	Qin et al. (2021)
Plantae	Rutaceae	Tetradium	ruticarpum	6.94	Qin et al. (2021)
Plantae	Rutaceae	Tetradium	ruticarpum	6.77, 6.80, 6.102, 6.66, 6.96, 6.99, 6.100, 6.63, 6.75	Sugimoto et al. (1988)
Plantae	Rutaceae	Tetradium	ruticarpum	6.149, 6.77, 6.80, 6.102, 6.66, 6.43, 6.100, 6.63, 6.182, 6.71, 6.76, 6.64	Tang et al. (1996)
Plantae	Rutaceae	Tetradium	ruticarpum	6.77, 6.76	Tominaga et al. (2002)
Plantae	Rutaceae	Euodia	sutchuenensis	6.36, 6.43	Lien et al. (2002)
Plantae	Rutaceae	Tetradium	ruticarpum	6.149, 6.77, 6.80	Tschesche and Werner (1967)
Plantae	Rutaceae	Tetradium	ruticarpum	6.28, 6.77, 6.80, 6.16, 6.102, 6.66, 6.43, 6.37, 6.28, 6.96, 6.175, 6.71, 6.52, 6.88, 6.228, 6.101, 6.84, 6.85, 6.69, 6.70	Wang et al. (2013)
Plantae	Rutaceae	Tetradium	ruticarpum	6.66, 6.43	Wu et al. (2022)
Plantae	Rutaceae	Euodia	officinalis	6.77, 6.80, 6.100	Xu et al. (2006)
Plantae	Rutaceae	Tetradium	ruticarpum	6.77	Yamahara et al. (1988)
Plantae	Rutaceae	Tetradium	ruticarpum var. officinalis	6.102, 6.43, 6.71	Yan et al. (2019)
Plantae	Rutaceae	Tetradium	ruticarpum var. bodinaieri	6.65, 6.51, 6.66, 6.214	Yang et al. (2006)

(Continued)

TABLE 6.2 (Continued)
Natural Occurrence of 4-Quinolinone Alkaloids

Kingdom	Family	Genus	Species	Compound(s)	References
Plantae	Rutaceae	*Tetradium*	*ruticarpum*	6.23	Zhuang et al. (2017)
Plantae	Rutaceae	*Tetradium*	*ruticarpum*	6.96	Chuang et al. (1996)
Plantae	Rutaceae	*Tetradium*	*ruticarpum*	6.77, 6.80, 6.66, 6.43, 6.96, 6.63	Lee et al. (1996)
Plantae	Rutaceae	*Tetradium*	*ruticarpum*	6.77, 6.80, 6.66, 6.99, 6.100	Lee et al. (2010)
Plantae	Rutaceae	*Tetradium*	*ruticarpum*	6.77, 6.80, 6.102, 6.66, 6.96, 6.75	Rho et al. (1999)
Plantae	Rutaceae	*Tetradium*	*ruticarpum*	6.77, 6.80, 6.66, 6.43, 6.99, 6.100, 6.63	Chuang et al. (1996)
Plantae	Rutaceae	*Feronia*	*Limonia*	6.231	Pitchai et al. (2012)
Plantae	Rutaceae	*Flindersia*	*dissosperma*	6.36, 6.51, 6.66, 6.43, 6.164	Robertson et al. (2018)
Plantae	Rutaceae	*Flindersia*	*maculosa*	6.36, 6.51, 6.66, 6.43, 6.164, 6.52	Robertson et al. (2018)
Plantae	Rutaceae	*Flindersia*	*fournieri*	6.18	Tillequin et al. (1980)
Plantae	Rutaceae	*Haplophyllum*	*Leptomerum*	6.8, 6.18	Akhmedzhanova et al. (1986)
Plantae	Rutaceae	*Haplophyllum*	*Leptomerum*	6.19, 6.21	Akhmedzhanova et al. (2010)
Plantae	Rutaceae	*Haplophyllum*	*cappadocicum*	6.167	Arar et al. (1985)
Plantae	Rutaceae	*Haplophyllum*	*acutifolium*	6.18	Eshonov et al. (2020)
Plantae	Rutaceae	*Haplophyllum*	*acutifolium*	6.19	Gulyamova et al. (1971)
Plantae	Rutaceae	*Haplophyllum*	*sp.*	6.18, 6.19	Nazrullaev et al. (2001)
Plantae	Rutaceae	*Haplophyllum*	*laeviusculum*	6.55	Parimah Parhoodeh et al. (2012)
Plantae	Rutaceae	*Haplophyllum*	*dubium*	6.38, 6.162	Razakova et al. (1979)
Plantae	Rutaceae	*Haplophyllum*	*popovii*	6.166	Razakova and Bessonova (1981)
Plantae	Rutaceae	*Haplophyllum*	*acutifolium*	6.19, 6.21, 6.36, 6.168	Razakova et al. (1986)
Plantae	Rutaceae	*Haplophyllum*	*foliosum*	6.156	Razzakova et al. (1972)
Plantae	Rutaceae	*Haplophyllum*	*acutifolium*	6.19	Razzakova et al. (1973)

(Continued)

TABLE 6.2 (Continued)
Natural Occurrence of 4-Quinolinone Alkaloids

Kingdom	Family	Genus	Species	Compound(s)	References
Plantae	Rutaceae	Haplophyllum	acutifolium	6.19, 6.36, 6.220, 6.221, 6.30, 6.222, 6.223, 6.224	Staerk et al. (2009)
Plantae	Rutaceae	Lunasia	amara	6.55, 6.142	Goodwin et al. (1959)
Plantae	Rutaceae	Lunasia	quercifolia	6.151	Hart et al. (1968)
Plantae	Rutaceae	Lunasia	quercifolia	6.142	Johnstone et al. (1958)
Plantae	Rutaceae	Lunasia	amara	6.38, 6.55	Saputra et al. (2024)
Plantae	Rutaceae	Lunasia	amara	6.55	Steldt and Chen (1943)
Plantae	Rutaceae	Lunasia	amara	6.55, 6.47	Subehan (2010)
Plantae	Rutaceae	Orixa	japonica	6.46, 6.120	Jian et al. (2023)
Plantae	Rutaceae	Orixa	japonica	6.46	Luckner and Reisch (1970)
Plantae	Rutaceae	Orixa	japonica	6.122, 6.123, 6.151	Nhoek et al. (2022)
Plantae	Rutaceae	Orixa	japonica	6.46	Yajima et al. (1977)
Plantae	Rutaceae	Orixa	japonica	6.46, 6.147	Funayama et al. (1999)
Plantae	Rutaceae	Paramignya	trimera	6.247	Nguyen et al. (2017)
Plantae	Rutaceae	Platydesma	campanulata	6.144, 6.2, 6.146	Werny and Scheuer (1963)
Plantae	Rutaceae	Raputia	heptaphylla	6.229	Barrera et al. (2011)
Plantae	Rutaceae	Raulinoa	echinata	6.18, 6.36, 6.43	Biavatti et al. (2002)
Plantae	Rutaceae	Ruta	graveolens	6.38	Adams et al. (2004)
Plantae	Rutaceae	Ruta	graveolens	6.36	Ainiwaer et al. (2021)
Plantae	Rutaceae	Ruta	graveolens	6.38	Arthur and Cheung (1960)
Plantae	Rutaceae	Ruta	graveolens	6.38, 6.80, 6.66, 6.43, 6.71, 6.52	Carvalho et al. (2019)
Plantae	Rutaceae	Ruta	chalepensis	6.38, 6.196	El Sayed et al. (2000)

(Continued)

TABLE 6.2 (Continued)
Natural Occurrence of 4-Quinolinone Alkaloids

Kingdom	Family	Genus	Species	Compound(s)	References
Plantae	Rutaceae	Ruta	graveolens	6.38	Ghosh et al. (2013)
Plantae	Rutaceae	Ruta	graveolens	6.43	Grundon and Okely (1979)
Plantae	Rutaceae	Ruta	graveolens	6.38	Hale et al. (2004)
Plantae	Rutaceae	Ruta	angustifolia	6.38	Kamal et al. (2021)
Plantae	Rutaceae	Ruta	graveolens	6.38, 6.21, 6.36, 6.51, 6.66, 6.43, 6.164, 6.116, 6.52	Kostova et al. (1999)
Plantae	Rutaceae	Ruta	graveolens	6.38, 6.67, 6.36, 6.43, 6.91, 6.73	Nakano et al. (2016)
Plantae	Rutaceae	Ruta	graveolens	6.149, 6.80, 6.67, 6.21, 6.36, 6.51, 6.66, 6.43, 6.164, 6.28, 6.52, 6.196, 6.91, 6.234, 6.235, 6.236, 6.237, 6.73, 6.238, 6.239, 6.240, 6.241, 6.242	Oh et al. (2014)
Plantae	Rutaceae	Ruta	graveolens	6.38, 6.67, 6.36, 6.91	Oliva et al. (2003)
Plantae	Rutaceae	Ruta	graveolens	6.67	Reisch (1967)
Plantae	Rutaceae	Ruta	graveolens	6.51	Reisch et al. (1975)
Plantae	Rutaceae	Ruta	graveolens	6.38, 6.80, 6.66, 6.71	Sampaio et al. (2018)
Plantae	Rutaceae	Ruta	graveolens	6.38	Schneider and Immel (1967)
Plantae	Rutaceae	Ruta	montana	6.192, 6.193, 6.194	Touati and Ulubelen (2000)
Plantae	Rutaceae	Ruta	montana	6.146	Ulubelen (1990)
Plantae	Rutaceae	Ruta	angustifolia	6.36	Wahyuni et al. (2014)
Plantae	Rutaceae	Sarcomelicope	megistophylla	6.197, 6.198	Fokialakis et al. (2000a)
Plantae	Rutaceae	Sarcomelicope	megistophylla	6.195	Fokialakis et al. (2000b)
Plantae	Rutaceae	Sarcomelicope	dogniensis	6.186	Mitaku et al. (1995)
Plantae	Rutaceae	Sarcomelicope	megistophylla	6.213	Mitaku et al. (2006)

(Continued)

TABLE 6.2 (Continued)
Natural Occurrence of 4-Quinolinone Alkaloids

Kingdom	Family	Genus	Species	Compound(s)	References
Plantae	Rutaceae	Skimmia	japonica	6.142	Boyd and Grundon (1970)
Plantae	Rutaceae	Skimmia	reevesiana	6.169, 6.170	Wu et al. (1987)
Plantae	Rutaceae	Spathelia	excelsa	6.206, 6.207, 6.208, 6.209, 6.210, 6.211	Lima et al. (2005)
Plantae	Rutaceae	Spiranthera	atlantica	6.142, 6.253	Rocha et al. (2016)
Plantae	Rutaceae	Spiranthera	odoratissima	6.18	Terezan et al. (2010)
Plantae	Rutaceae	Stauranthus	perforatus	6.151	Setzer et al. (2003)
Plantae	Rutaceae	Tetradium	ruticarpum	6.251	Li et al. (2020)
Plantae	Rutaceae	Tetradium	ruticarpum	6.96, 6.75, 6.76, 6.92	Ma et al. (2021)
Plantae	Rutaceae	Tetradium	ruticarpum	6.80, 6.66, 6.78, 6.96	Na et al. (2022)
Plantae	Rutaceae	Tetradium	ruticarpum	6.77, 6.80, 6.66, 6.43, 6.64, 6.97	Pan et al. (2013)
Plantae	Rutaceae	Tetradium	ruticarpum	6.28	To et al. (2021)
Plantae	Rutaceae	Tetradium	ruticarpum	6.77, 6.80, 6.81	Wang et al. (2010)
Plantae	Rutaceae	Vepris	ampody	6.65, 6.153, 6.154	Kan-Fan et al. (1970)
Plantae	Rutaceae	Zanthoxylum	avicennae	6.16	Cho et al. (2012a)
Plantae	Rutaceae	Zanthoxylum	fagara	6.155	Ishi et al. (1972)
Plantae	Rutaceae	Zanthoxylum	avicennae	6.24, 6.256, 6.257, 6.258, 6.259	Ji et al. (2022)
Plantae	Rutaceae	Zanthoxylum	schinifolium	6.28	Liu et al. (2009); Lu et al. (2021); Nguyen et al. (2016); Shu et al. (2019); Wang et al. (2014)
Plantae	Simaroubaceae	Samadera	Bidwillii (SAC-2825)	6.188	Gibbons et al. (1997)

mutilans (centipede) and *Blaps japanensis* (darkling beetle). The 4-quinolinones have valuable ecological and pharmaceutical development prospects. Many species, including those mentioned, develop adaptations reflecting their interactions with nature.

6.2.2 BACTERIAL SOURCES

Fifty-seven 4-quinolinones are found across various bacterial families, each with different chemistry adapted to specific ecological roles within its family. Not all of these 4-quinolinones are bioactive, but their existence in these families suggests a unique biochemistry and evolutionary pressure. The Alteromonadaceae family, with Marinobacterium present, has compounds 6.79 and 6.90. These 4-quinolinones may play an essential role in the marine adaptations of this family, allowing them to thrive and be resilient in their environment. Bacillaceae, which comprises *Bacillus* strains, contains 4-quinolinones like 6.25 and 6.33, which are believed to help living organisms in soil and various environments. Burkholderiaceae, including *Burkholderia* strains (e.g. *B. pseudomallei*, *B. thailandensis*, *B. cepacia*), are listed frequently in Table 6.2, signifying that these microorganisms produce repetitive 4-quinolinones such as 6.21 and 6.27. The same 4-quinolinones are present within different species of this family, probably for important adaptive purposes such as microbial functions or signalling. The family Cytophagaceae, represented by *Cytophaga*, has the unique compound 4-quinolinone (6.31), which may assist them in acquiring nutrients efficiently. The Micrococcaceae (including *Arthrobacter*) and Myxococcaceae (from *Stigmatella*) also form distinctive 4-quinolinones (e.g. 6.53, 6.95, and 6.106) with chemical diversity, which helps them adapt to the soil or decaying organic matter. In Nocardiaceae, the genera *Nocardia* and *Rhodococcus* have been reported to produce 4-quinolinones such as 6.49, 6.56, and 6.107. They are resilient in many diverse habitats. The bacteria Nocardiopsaceae (*Nocardiopsis terrae*) contain the 4-quinolinone 6.4, indicating the genus's adaptation to life in soils. The compound 6.29 shows a 4-quinolinone derived from *Paenibacillus polymyxa* (Paenibacillaceae), which may help improve soil health by interacting with the roots or rhizosphere of plants. The diverse range of 4-quinolinones produced by Pseudomonadaceae (e.g. 6.118, 6.21, 6.36, and 6.50) indicates their chemical versatility. These compounds may help Pseudomonadaceae dominate and adapt to soil-, water-, and plant-associated communities, where they participate in nutrient cycling and compete with microbes. Two 4-quinolinones, 6.50 and 6.51, with algae-killing properties, were isolated from the genus *Alteromonas* (family Alteromonadaceae). Pseudonocardiaceae, the family of *Pseudonocardia* and *Saccharomonospora*, contains the 4-quinolinones 6.87 and 6.152, which may act as signalling compounds; Pseudonocardiaceae is adapted to soil and symbioses. Finally, Streptomycetaceae displays significant variability in 4-quinolinones, with differing *Streptomyces* strains producing different substances (6.2, 6.36, 6.51). The diverse 4-quinolinone profiles exhibited by this group are likely essential to the ecological functions these bacteria perform in soil and as plant-associated bacteria, where they may mediate interactions with other organisms or suppress competing microbes. The types of 4-quinolinones produced by the different bacteria show that these families have evolved chemical forms with varying structures over the years. It also indicates that 4-quinolinones are essential in the metabolism of bacteria.

6.2.3 FUNGAL SOURCES

Fungal species and distributions of families that produce 4-quinolinones are summarized in Table 6.2. The family Aspergillaceae is the most important (Moussa et al. 2024), with species like *Aspergillus sp. SCSIO, FSY-01,* and *FSW-02* yielding compounds 6.5, 6.22, and 6.32. *Penicillium* species, such as *Penicillium citrinum* and *Penicillium steckii*, produce other compounds, including 6.22, 6.32, 6.9, and 6.255. Aspergillaceae gives 6.22 and 6.32 across different species. Other families like Cordycipitaceae (*Cordyceps militaris*), Didymellaceae (*Epicoccum sorghinum*), Onygenaceae (*Auxarthron reticulatum*), Polyporaceae (*Trametes trogii TGC-P-3*), and Trichocomaceae (*Neosartorya hiratsukae*) yield 6.158, 6.5, 6.32, 6.1, and 6.141, respectively. In total, 17 unique 4-quinolinone compounds have been identified. This shows their distribution and universal biochemical pathways amongst various fungal families.

6.2.4 PLANT SOURCES

Table 6.2 shows the isolation of 183 unique 4-quinolinones from various plants. The most common groups of plants include the rue or citrus family, flowering plants, and oak and beech. The genera most represented in the table are *Tetradium, Ruta, Haplophyllum, Euodia,* and Boronia, which belong to the Rutaceae family. The dataset consists of five genera. These five genera are the leading groups in the dataset. It is interesting to note that *Tetradium ruticarpum* and *Ruta graveolens* contain four quinolinones in their chemical composition (Xiao et al. 2023).

In Figure 6.5, the distribution of 4-quinolinones across four kingdoms is depicted: Plantae, Bacteria, Animalia, and Fungi. Plants are the most common, comprising 63%, followed by bacteria at 22%. Animals account for 8%, while fungi represent the smallest portion at 7%.

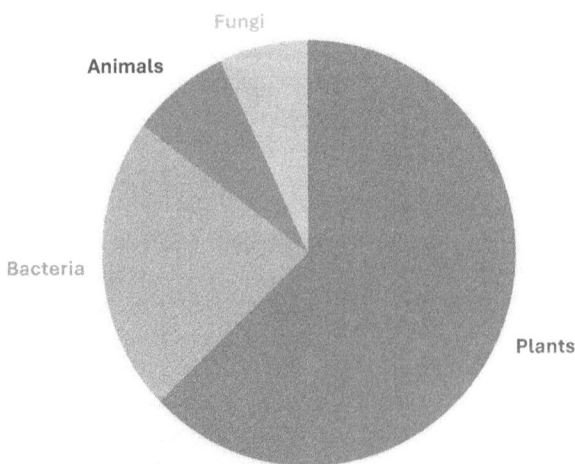

FIGURE 6.5   Distribution of 4-quinolinone alkaloids in various life kingdoms.

6.3 BIOLOGICAL ACTIVITY

Table 6.3 presents the biological activity of 4-quinolinone alkaloids. It describes 115 distinct substances with different biological activities, such as fighting bacteria, fungi, and cancer, reducing inflammation and antioxidant activity. Through various lab tests and animal models, several in vitro and in vivo studies indicate the effectiveness of these compounds in medicine. However, some safety tests suggest that these compounds can be toxic. In a study involving Kunming mice, the LD_{50} varied from 36 mg/kg to 65 mg/kg. Toxicity was also observed in zebrafish embryos at all levels. The most frequently reported biological activity was the antibacterial effects. Thirty-one compounds were found to inhibit the growth of bacteria such as *Staphylococcus aureus*, *Helicobacter pylori*, and Methicillin-resistant Staphylococcus aureus (MRSA), with minimum inhibitory concentration values of 0.38 to 128 µg/mL. The cytotoxicity or cell-killing experiment was observed with 20 compounds on human cancer cell lines MCF-7, HepG2, HeLa, and HL-60. Inhibitory concentration values were in the nanomolar and micromolar range. Thirteen compounds were found to possess antifungal activity against a few pathogens. Research shows that six different compounds can reduce cell growth and inflammation. They do this by blocking a process called nuclear factor kappa B signalling. This blockage helps lower the release of cytokines and reduces nitric oxide production in macrophages. In short, 4-quinolinone alkaloids can serve multipurpose roles and have a bright future in the discovery of drugs.

6.4 DISCUSSION

The 4-quinolineone skeleton possesses great biological significance and diversity in application. These compounds' antimicrobial, antimalarial, antifungal, antialgal, antioxidant, and anticancer activities are extended. The unique molecular characteristics of 4-quinolinone give these agents strong antibacterial activity and allow for chemical modifications that can be used in cancer, inflammatory diseases, and bacterial infection therapy. The fact that there are so many different alkaloids means that there are possibly far more compounds in nature with various biological activities. It was found that certain derivatives, like intervenolin and graveolin, have selective cytotoxicity against gastric and colon cancer cells and melanoma. Their capacity as agents of chemotherapy needs more investigation for optimal effects and fewer side effects. (PQS is used in communication between bacteria, and it is also required for quorum sensing and biofilm formation. PQS and its precursor HHQ influence microbial inter-kingdom interactions. They may even interfere with inter-kingdom communication by hampering the biofilm formation of *Candida albicans*. These findings can help in the development of quorum-sensing inhibitors as antimicrobial agents. 4-Quinolinones are also important ecologically, besides possessing biomedical uses.

Being naturally occurring, signalling most likely evolved to promote conflicts or integrate cooperation amongst microbes. Understanding their ecological importance creates new ways to manage microbes in natural and industrial environments. Although 4-quinolinones possess beneficial properties, their broad application in medicine, agriculture, and aquaculture raises concerns about antibiotic resistance and environmental persistence.

TABLE 6.3
Biological Activity of 4-Quinolinone Alkaloids

Biological Activity	Target	Assay/Method	Results	Compound	References
Acute toxicity	Kunming species mice	Administered by vein in tail	$LD_{50} = 48$ mg/Kg	6.65	Yang et al. (2006)
Acute toxicity	Kunming species mice	Administered by vein in tail	$LD_{50} = 36$ mg/Kg	6.51	Yang et al. (2006)
Acute toxicity	Kunming species mice	Administered by vein in tail	$LD_{50} = 65$ mg/Kg	6.66	Yang et al. (2006)
Acyl glucosaminyl inositol amidase inhibition	Mycothiol-derived S-conjugate amidase (MCA)	In vitro assay for determining amidase activity	Inhibits virulence and reduces antibiotic resistance of mycothiol-producing bacteria; inhibited MCA by 50% at 2 μM concentration	6.112	Fahey et al. (2010)
Algicidal activity	Heterosigma akashiwo, Cochlodinium polykrikoides, Alexandrium tamarense	Inoculation with KNS-16 at Various Growth Phases against Harmful Algal Bloom Species	$LC_{50} = 0.5$, 0.8 and 1.1 μg/mL, respectively	6.50	Cho (2012b)
Algicidal activity	H. akashiwo, C. polykrikoides, A. tamarense	Inoculation with KNS-16 at Various Growth Phases against Harmful Algal Bloom Species	$LC_{50} = 0.7$, 1.2 and 1.4 μg/mL, respectively	6.51	Cho (2012b)
Anti-angiogenic	Human umbilical vascular endothelial cells (HUVECs) Ehrlich ascites carcinoma tumour model in Swiss female albino mice	MTT assay; BrdU incorporation assay; Gelatin zymography; ELISA; Tumour volume evaluation	Suppressed HUVEC proliferation; considerably reduced endothelial cell migration and invasion, and capillary structure formation of HUVEC, downregulated VEGF-A; inhibited MMP-2 & MMP-9 expression in vitro; decreased ascitic fluid (60.94%) & tumour (86.40%) volume in Ehrlich ascites model; reduced peritoneal angiogenesis	6.55	Saraswati et al. (2013)

(Continued)

TABLE 6.3 (Continued)
Biological Activity of 4-Quinolinone Alkaloids

Biological Activity	Target	Assay/Method	Results	Compound	References
Antiarrhythmic	Mice	Aconitine-induced arrhythmia	Mitigated aconitine-induced arrhythmia; no mice died within 3 days after an i.p. dose at 150 mg/kg	6.6	Long et al. (1984)
Antibacterial	Staphylococcus epidermidis DSM 20044, Bacillus subtilis DSM 347, Propionibacterium acnes DSM 1897	Resazurin-based assay	Showed moderate growth inhibition (70%–89% inhibition) at 100 μM against Gram-positive bacteria	6.105	Nachtigall et al. (2010)
Antibacterial	Sa. aureus (209P)	Agar disk-diffusion technique	Inhibits growth with a zone of inhibition of 20 mm at 20 μg	6.45	Bultel-Ponce et al. (1999)
Antibacterial	Escherichia coli, Sa. aureus, Enterobacter aerogenes, P. aeruginosa	—	Minimum inhibitory concentration (MIC) values of 17–45 μmol/L	6.2	Qing-yun et al. (2015)
Antibacterial	Sa. aureus, P. aeruginosa, Enterobacter cloacae	Disk diffusion method	MIC = 0.38, 0.5, 0.45 mg/mL, respectively	6.5	Al-Khalil et al. (1998)
Antibacterial	Helicobacter pylori reference strains (NCTC 1916, NCTC11637, ATCC43504)	Disk method, Agar dilution method	MIC = 2.0 μg/mL [together with compound (76)]	6.77	Hamasaki et al. (2000)
Antibacterial	MRSA (ATCC 33591), Sa. aureus (ATCC 25923)	MIC assay using broth microdilution method	MIC against MRSA = 8 μg/mL; MIC against S. aureus = 8 μg/mL	6.77	Pan et al. (2013)
Antibacterial	H. pylori	Silico molecular docking-based virtual screening	IC_{50} >0.05 μg/mL	6.77	Raj et al. (2021)

(Continued)

TABLE 6.3 (*Continued*)

Biological Activity of 4-Quinolinone Alkaloids

Biological Activity	Target	Assay/Method	Results	Compound	References
Antibacterial	*H. pylori* strains ATCC43504, NCTC11637, NCTC11638, 82516, and 4	MIC assay	MIC = 10–20 µg/mL	6.77	Rho et al. (1999)
Antibacterial	*H. pylori* strain ATCC 43504	*H. pylori* respiration in vitro, eradication of *H. pylori* in the gastric tissues, measurement of myeloperoxidase (MPO) activity	Decreased the number of viable H. pylori by 46.0×10^4, 17.3×10^4, and 8.1×10^4 cfu/stomach, respectively	6.77	Tominaga et al. (2002)
Antibacterial	*Sa. aureus* ATCC25923, *Sa. epidermidis* ATCC12228, *B. subtilis* ATCC6633	MIC assay	MIC = 64,16, and 32 µg/mL, respectively	6.77	Wang et al. (2013)
Antibacterial	*H. pylori* strains 51	Determination of MIC_{50} and MIC_{90} by the broth dilution method	Moderate activity, 58.6% inhibition (MIC_{50} = 99 µM)	6.80	Na et al. (2022)
Antibacterial	*H. pylori* strains ATCC43504, NCTC11637, NCTC11638, 82516, and 4	MIC assay	MIC = 10–20 µg/mL	6.80	Rho et al. (1999)

(Continued)

TABLE 6.3 (Continued)
Biological Activity of 4-Quinolinone Alkaloids

Biological Activity	Target	Assay/Method	Results	Compound	References
Antibacterial	Sa. aureus ATCC25923, Sa. epidernidis ATCC12228, Bacillus subtilis ATCC6633	MIC assay	MIC = 4, 4, and 8 μg/mL, respectively	6.80	Wang et al. (2013)
Antibacterial	Sa. aureus, Vibrio anguil1arum	Standard disc technique on bacteria grown on an agar plate	Strong activity; Inhibition diameter at 100 μg/disk = 8 and 22 mm, respectively	6.21	Debitus et al. (1998)
Antibacterial	Tenacibaculum maritimum	Paper-disc agar diffusion method	At 10 μg: Inhibitory zone = 24 mm	6.21	Li et al. (2018)
Antibacterial	E. coli, Enterobacter aerogeres	—	MIC = 21 μmol/L (for both)	6.36	Liao et al. (2015)
Antibacterial	Sa. aureus, Vibrio anguill1crum	Standard disc technique on bacteria grown on an agar plate	Strong activity; Inhibition diameter at 100 μg/disk = 7 and 9 mm, respectively	6.36	Debitus et al. (1998)
Antibacterial	Sa. aureus, Sarcina lutea	Disc diffusion method	Inhibitory zone of 12 mm against S. aureus and 20 mm against S. lutea	6.36	Nazrul Islam et al. (2002)
Antibacterial	Salmonella typhi sp., B. subtilis, exterotoxigenic E. coli	Disc diffusion method	Inhibitory zones of 10 mm against S. typhi sp., 12 mm against B. subtilis, and 15 mm against exterotoxigenic E. coli	6.3	Nazrul Islam et al. (2002)

(Continued)

TABLE 6.3 (Continued)
Biological Activity of 4-Quinolinone Alkaloids

Biological Activity	Target	Assay/Method	Results	Compound	References
Antibacterial	*H. pylori* strains ATCC43504, NCTC11637, NCTC11638, 82516, 82548, and 4	MIC assay	MIC = 10–20 µg/mL	6.102	Rho et al. (1999)
Antibacterial	*Sa. aureus* ATCC25923, *Sa. epidermidis* ATCC12228	MIC assay	MIC = 16 and 4 µg/mL, respectively	6.102	Wang et al. (2013)
Antibacterial	MRSA (ATCC 33591)	MIC assay using broth microdilution method	MIC against MRSA = 64 µg/mL	6.66	Pan et al. (2014)
Antibacterial	*H. pylori* strains ATCC43504, NCTC11637, NCTC11638, 82516, 82548, and 4	MIC assay	MIC = 10–20 µg/mL	6.66	Rho et al. (1999)
Antibacterial	*Sa. aureus*	–	MIC = 18 µmol/L	6.34	Liao et al. (2015)
Antibacterial	*B. subtilis*	Disc diffusion method	Inhibitory zone of 10 mm against *B. subtilis*	6.43	Nazrul Islam et al. (2002)
Antibacterial	MRSA (ATCC 33591), *Sa. aureus* (ATCC 25923)	MIC assay using broth microdilution method	MIC against MRSA = 64 µg/mL; MIC against *S. aureus* = 64 µg/mL	6.43	Pan et al. (2013)

(Continued)

TABLE 6.3 (Continued)
Biological Activity of 4-Quinolinone Alkaloids

Biological Activity	Target	Assay/Method	Results	Compound	References
Antibacterial	*Sa.aureus* ATCC25923, *Sa.epidermidis* ATCC12228, *B. subtilis* ATCC6633	MIC assay	MIC = 32, 8, and 16 µg/mL, respectively	6.43	Wang et al. (2013)
Antibacterial	*Flavobacterium* 1980, *Sa. aureus* Oxford H strain VI	Agar diffusion method (cup-plate)	Active against *Flavobacterium* 1980 at a concentration as low as 0.1 µg/mL (inhibitory zone diameter: 19.3 mm); Active against *Staphylococcus aureus* Oxford H strain VI at a concentration as low as 24 µg/mL (inhibitory zone diameter: 19.8 mm)	6.31	Evans et al. (1978)
Antibacterial	*Sa. aureus* ATCC25923, *Sa. epidermidis* ATCC12228, *B. subtilis* ATCC6633	MIC assay	MIC = ~ 128 µg/mL	6.37	Wang et al. (2013)
Antibacterial	*P. aeruginosa* ATCC 27853, enteropathogenic *E. coli* (EPEC) O111, *Klebsiella pneumoniae*, *Sa. epidermidis*	Kirby–Bauer (KB) disk diffusion	Size of the inhibition zone = 8.0, 9.0, 8.5, and 8.0 mm, respectively	6.10	Zhang et al. (2005)

(Continued)

TABLE 6.3 (Continued)
Biological Activity of 4-Quinolinone Alkaloids

Biological Activity	Target	Assay/Method	Results	Compound	References
Antibacterial	H. pylori strain 51	Determination of MIC_{50} and MIC_{90} by the broth dilution method	Potent activity against with 94.4% inhibition (MIC_{50} and MIC_{90} values of 22 and 50µM, respectively)	6.78	Na et al. (2022)
Antibacterial	Sa. aureus, E. coli, B. subtilis	-	MIC = 64, 128, and 64 mg/mL, respectively	6.4	Tian et al. (2013)
Antibacterial	B. subtilis, Sa. aureus, Arthrobacter aurescens, Brevibacterium ammoniagenes, Corynebacterium fascians	Serial dilution assay	MIC = 0.15, 0.39, 0.19, 0.05, and 0.78 µg/mL, respectively	6.106	Kunze et al. (1987)
Antibacterial	Sa. epidermidis DSM 20044, B. subtilis DSM 347, Propionibacterium acnes DSM 1897	Resazurin-based assay	Showed moderate growth inhibition (70%–89% inhibition) at 100µM against Gram-positive bacteria	6.106	Nachtigall et al. (2010)
Antibacterial	Sa. aureus CMCC(B)26003, Sr. pyogenes ATCC19615, Bacillus subtilis CICC10283, Micrococcus luteus	-	MIC = 8,8,16, and 8µM, respectively	6.106	Zhang et al. (2017)
Antibacterial	B. subtilis, Sa. aureus, Arthrobacter aurescens, Brevibacterium ammoniagenes, Corynebacterium fascians	Serial dilution assay	MIC = 0.15, 0.39, 0.19, 0.05, and 0.78 µg/mL, respectively	6.95	Kunze et al. (1987)

(Continued)

TABLE 6.3 (Continued)
Biological Activity of 4-Quinolinone Alkaloids

Biological Activity	Target	Assay/Method	Results	Compound	References
Antibacterial	Sa. aureus CMCC(B)26003, Sr. pyogenes ATCC 19615, Bacillus subtilis CICC10283, M.s luteus	-	MIC = 8,4,8, and 4 μM, respectively	6.95	Zhang et al. (2017)
Antibacterial	E. coli, Enterobacter aerogenes, Sa. aureus	MIC assay	MIC = et al. 19 μmol/L (for all 3)	6.29	Liao et al. (2015)
Antibacterial	Sa. aureus, Vibrio anguillarum, Beneckea harveyii B-392	-	Inhibited the growth of all bacteria at 100 μg/disc	6.11	Molinski et al. (1988)
Antibacterial	H. pylori strains ATCC43504, NCTC11637, NCTC11638, 82516, 82548, and 4	MIC assay	MIC = 10–20 μg/mL	6.95	Rho et al. (1999)
Antibacterial	H. pylori strains ATCC43504, NCTC11637, NCTC11638, 82516, 82548, and 4	MIC assay	MIC = 10–20 μg/mL	6.75	Rho et al. (1999)
Antibacterial	Tenacibaculum maritimum	Paper-disc agar diffusion method	At 10 μg: Inhibitory zone = 25 mm	6.41	Li et al. (2018)
Antibacterial	Sa. aureus FDA 209P, Sa. aureus No.5 (MRSA), Sa. epidermidis IID 866, Sa. epidermidis No. 17 (MRSE), B. subtilis ATCC 6633, Mycobacterium smegmatis ATCC607	Serial agar dilution method using the Mueller–Hinton method	MIC = 12.5 μg/mL for M. smegmatis MIC = 6.25 μg/mL for all other bacteria	6.53	Kamigiri et al. (1996)

(Continued)

TABLE 6.3 (*Continued*)
Biological Activity of 4-Quinolinone Alkaloids

Biological Activity	Target	Assay/Method	Results	Compound	References
Antibacterial	*Tenacibaculum maritimum*	Paper-disc agar diffusion method	At 10 µg: Inhibitory zone = 22 mm	6.27	Li et al. (2018)
Antibacterial	*F. oxisporum, F. solani, F. moniliforme, Rhizhoctonia solani*	-	MIC = 32, 64, 64, and 16 µg/mL, respectively	6.25	Dat et al. (2022)
Antibacterial	*T. maritimum*	Paper-disc agar diffusion method	At 10 µg: Inhibitory zone = 55 mm	6.25	Li et al. (2018)
Antibacterial	*Dickeya zeae* (Dz)	Antagonistic activity, agar plug diffusion method, diameter of the antibacterial zone	Antagonistic activity against Dz, Inhibition zone diameter = ~21 mm (after 2 days)	6.25	Toi et al. (2022)
Antibacterial	*Sa. aureus* ATCC25923, *Sa. epidermidis* ATCC12228	Minimum Inhibitory Concentration (MIC) assay	MIC = 64, 16, and 64 µg/mL, respectively	6.71	Wang et al. (2013)
Antibacterial	*H. pylori* reference strains (NCTC 11916, NCTC 11637, ATCC 43504)	Disk method, Agar dilution method	MIC = 2.0 µg/mL [together with Evocarpine (77)]	6.76	Hamasaki et al. (2000)
Antibacterial	*H. pylori*	Silico molecular docking-based virtual screening	IC_{50} = 0.05 µg/mL	6.76	Raj et al. (2021)
Antibacterial	*H. pylori* strain ATCC 43504	*H. pylori* respiration in vitro, eradication of *H. pylori* in the gastric tissues, measurement of myeloperoxidase (MPO) activity	Decreased the number of viable *H. pylori* by 46.0 × 10⁴, 17.3 × 10⁴ and 8.1 × 10⁴ cfu/stomach, respectively	6.76	Tominaga et al. (2002)
Antibacterial	MRSA (ATCC 33591), *S. aureus* (ATCC 25923)	MIC assay using broth microdilution method	MIC against MRSA = 32 µg/mL; MIC against *S. aureus* = 16 µg/mL	6.64	Pan et al. (2013)

(*Continued*)

TABLE 6.3 (Continued)
Biological Activity of 4-Quinolinone Alkaloids

Biological Activity	Target	Assay/Method	Results	Compound	References
Antibacterial	MRSA (ATCC 33591), S. aureus (ATCC 25923)	MIC assay using broth microdilution method	MIC against MRSA = 128 µg/mL; MIC against S. aureus = 128 µg/mL	6.97	Pan et al. (2013)
Antibacterial	H. pylori	Antimicrobial activity using paper disks (inhibition zone)	At 0.5µg/disk, Inhibition zone = 20 mm	6.59	Dekker et al. (1998)
Antibacterial	H. pylori	Antimicrobial activity using paper disks (inhibition zone)	At 0.5µg/disk, Inhibition zone = 17 mm	6.39	Dekker et al. (1998)
Antibacterial	H. pylori	Antimicrobial activity using paper disks (inhibition zone)	At 0.5µg/disk, Inhibition zone = 8 mm	6.60	Dekker et al. (1998)
Antibacterial	H. pylori	Antimicrobial activity using paper disks (inhibition zone)	At 0.5µg/disk, Inhibition zone = 16 mm	6.48	Dekker et al. (1998)
Antibacterial	H. pylori	Antimicrobial activity using paper disks (inhibition zone)	At 0.5µg/disk, Inhibition zone = 24 mm	6.56	Dekker et al. (1998)
Antibacterial	H. pylori	Antimicrobial activity using paper disks (inhibition zone)	At 0.5µg/disk, Inhibition zone = 29 mm	6.68	Dekker et al. (1998)
Antibacterial	H. pylori	Antimicrobial activity using paper disks (inhibition zone)	At 0.5µg/disk, Inhibition zone = 12 mm	6.87	Dekker et al. (1998)
Antibacterial	H. pylori	Antimicrobial activity using paper disks (inhibition zone)	At 0.5µg/disk, Inhibition zone = 23 mm; potent bacteriocidal effect (MBC 10 ng/mL, 0.001% survival); very potent bacteriostatic effect (MIC = 0.1 ng/mL, 0.171% survival)	6.49	Dekker et al. (1998)

(Continued)

TABLE 6.3 (Continued)
Biological Activity of 4-Quinolinone Alkaloids

Biological Activity	Target	Assay/Method	Results	Compound	References
Antibacterial	Sa. aureus ATCC25923, Sa. epidermidis ATCC12228	Minimum Inhibitory Concentration (MIC) assay	MIC = 16, and 4 µg/mL, respectively	6.88	Wang et al. (2013)
Antibacterial	Sinorhizobium meliloti IAM 12611, Deinococcus grandis IAM 13005, B. subtilis IAM 12118, Arthrobacter atrocyaneus IAM 12339, Corynebacterium glutamicum IAM 12435, Nocardia pseudosporangifera IAM 501, Sr. griseus IAM 12311, Rhodococcus erythropolis IAM 12122	Agar diffusion assay	At 10 µg concentration it exhibits a wide and strong antimicrobial spectrum against both high- and low-GC Gram-positive bacteria: MIC = 0.4, 0.2, 3.1, 0.01, 0.01, 0.01, 0.1, and 0.1 µg/mL, respectively	6.107	Kitagawa et al. (2008)
Antibacterial	Enterobacter aerogenes, Sa. aureus	-	MIC = 17 and 87 µmol/L, respectively	6.44	Liao et al. (2015)

(Continued)

TABLE 6.3 (Continued)
Biological Activity of 4-Quinolinone Alkaloids

Biological Activity	Target	Assay/Method	Results	Compound	References
Antibacterial	Sa. aureus ATCC 25923, Sa. epidermidis ATCC 12228, B. subtilis ATCC 6633	MIC assay	MIC = 16, 4, and 16 µg/mL, respectively	6.101	Wang et al. (2013)
Antibacterial	H. pylori strains JCM12093 and JCM12095	MIC determination	Potent and selective, MICs of 16 and 8 µg/mL, respectively	6.110	Kawada et al. (2013)
Antibacterial	Sa. aureus ATCC 25923, Sa. epidermidis ATCC 12228, B. subtilis ATCC 6633	MIC assay	MIC = 64 µg/mL for all 3	6.84	Wang et al. (2013)
Antibacterial	Sa. aureus ATCC 25923, Sa. epidermidis ATCC 12228, B. subtilis ATCC 6633	MIC assay	MIC = 64 µg/mL for all 3	6.85	Wang et al. (2013)

(Continued)

TABLE 6.3 (Continued)
Biological Activity of 4-Quinolinone Alkaloids

Biological Activity	Target	Assay/Method	Results	Compound	References
Antibacterial	Sa. aureus ATCC 25923, Sa. epidermidis ATCC 12228, B. subtilis ATCC 6633	MIC assay	MIC~ 128 µg/mL	6.70	Wang et al. (2013)
Antibacterial	Tenacibaculum maritimum	Paper-disc agar diffusion method	At 10 µg: Inhibitory zone = 25 mm	6.33	Li et al. (2018)
Antibacterial	T. maritimum	Paper-disc agar diffusion method	At 10 µg: Inhibitory zone = 33 mm	6.20	Li et al. (2018)
Antibacterial	Kocuria rhizophila ATCC 9341, Sa. aureus FDA209P JC-1, Rhizobium radiobacter NBRC 14554	MIC determination	MIC = 13, 25, and 50 µg/mL, respectively	6.79	Karim et al. (2022)
Antibacterial	K. rhizophila ATCC 9341, Sa. aureus FDA209P JC-1, Rhizobium radiobacter NBRC 14554	MIC determination	MIC = 6.3, 13, and 50 µg/mL, respectively	6.90	Karim et al. (2022)
Anticancer	Breast cancer cells MCF-7, HER2 (3PPO), ERα (3ERT)	Molecular Docking	The highest affinity for the HER2 receptor but was still below that of the native lapatinib ligand; showed higher affinity for HER2 receptors than ERα.	6.38	Saputra et al. (2024)

(Continued)

TABLE 6.3 (Continued)
Biological Activity of 4-Quinolinone Alkaloids

Biological Activity	Target	Assay/Method	Results	Compound	References
Anticancer	Epidermal Growth Factor Receptor (EGFR) Tyrosine kinase domain	In silico molecular docking	Showed potent inhibition against EGFR and favourable interactions within the active site. Binding energy was lower than the reference	6.55	Saini et al. (2022)
Anticancer	Epidermal growth factor receptor (EGFR) 696-1022 T790M associated with non-small cell lung cancer (NSCLC)	Molecular docking	Efficient drug like property, lowest binding energy = - 6.1 kcal/mol	6.22	Nand et al. (2016)
Antichlamydial	HeLa cells were infected with *Chlamydia trachomatis*	ROS production	Inhibited the formation and growth of *Chlamydia trachomatis* inclusion bodies in a dose-dependent manner, $IC_{50} = 9.54\ \mu M$	6.12	Cheng et al. (2016)
Anticholinesterasic activity	Acetyl cholinesterase	Anticholinesterasic assay, microplate reader based on Ellman's method	High activity, $IC_{50} = 2.5\ \mu M$	6.8	Cardoso-Lopes et al. (2010)
Anticholinesterasic activity	Acetyl cholinesterase	Anticholinesterasic assay, microplate reader based on Ellman's method	Mild activity, $IC_{50} = 0.21\ mM$	6.17	Cardoso-Lopes et al. (2010)
Anticlastogenic	Chromosomes; Micronuclei induced by cyclophosphamide	Micronucleus assay on bone marrow cells of mice	Showed significant reduction (30%–48%) in frequency of micronucleated polychromatic erythrocytes (MNPCE) in bone marrow cells induced by CP, in a dose-dependent manner	6.55	Aher et al. (2013)

(Continued)

TABLE 6.3 (Continued)

Biological Activity of 4-Quinolinone Alkaloids

Biological Activity	Target	Assay/Method	Results	Compound	References
Antifeedant	*Tribolium castaneum*, *Sitophilus zeamais*	Flour disk bioassay	EC_{50} = 47.8 and 85.6 ppm, respectively; Reduced RGR and RCR of both insect species in a concentration-dependent manner	6.28	Liu et al. (2009)
Antifeedant	*Pectinophora gossypiella* (pink bollworm)	Artificial-diet feeding assay	Moderate antifeedant activity	6.13	Nakatsu et al. (1990)
Antifeedant	*P. gossypiella* (pink bollworm)	Artificial-diet feeding assay	Moderate antifeedant activity	6.17	Nakatsu et al. (1990)
Antifeeding activity (Insecticidal)	*Spodoptera litura* (tobacco cutworm)	Feeding inhibitory test	50% Feeding inhibition at 300 ppm concentration	6.46	Yajima et al. (1977)
Antifungal	*Candida albicans*	Bioautography agar overlay assay; Broth microdilution assay	Clear zone of inhibition observed; Minimum Inhibitory Concentration (MIC) = 250 µg/mL	6.38	Kamal et al. (2021)
Antifungal	Isocitrate lyase 1 (ICL1) gene and protein of *C. albicans*	RT-qPCR and western blot analysis to examine the effect on expression of ICL1 gene (CaIcl1) and its encoded protein	Downregulated the expression of both ICL1 gene and protein in *C. albicans*; it inhibits the glyoxylate cycle in *C. albicans* to exert antifungal effects	6.38	Kamal et al. (2021)
Antifungal	*Leucoagaricus gongylophorus*	Fungicidal assay	Growth inhibition at 50 µg/mL = 20% Growth inhibition at 100 µg/mL = 50%	6.36	Biavatti et al. (2002)
Antifungal	*C. albicans*	*Ca. elegans* lifespan assay, *C. albicans* infection model	At 100 mg/L concentration, it increased the lifespan of *Ca. elegans* infected with C. albicans by 75.71% and inhibited *C. albicans* colony formation in the intestinal tract	6.28	Lu et al. (2021)

(Continued)

TABLE 6.3 (Continued)
Biological Activity of 4-Quinolinone Alkaloids

Biological Activity	Target	Assay/Method	Results	Compound	References
Antifungal	C. albicans	Ca. elegans lifespan assay, defecation cycle assay, locomotion behaviour assay, colony-forming units assay, CaSA1::GFP expression assay	Shortened the defecation cycle of Ca. elegans, Increased the locomotion behaviour of Ca. elegans, Increased the lifespan of Ca. elegans after C. albicans infection, Inhibited the colony formation of C. albicans in Ca. elegans intestine	6.28	Shu et al. (2019)
Antifungal	Pyricularia oryzae	-	MIC = 256 mg/mL	6.4	Tian et al. (2013)
Antifungal	F. oxisporum, F. solani, F. moniliforme, Rhizoctonia solani	-	MIC = 16, 64, 64, and 32 µg/mL, respectively	6.41	Dat et al. (2022)
Antifungal	Trichophyon rubrum, Rhizopus oryzae	Paper-disc agar diffusion method	At 10 µg: Inhibitory zone = 30 and 15 mm, respectively	6.41	Li et al. (2018)
Antifungal	Aspergillus niger, Rhizo. solani	Antifungal activity on the Mycelia growth	50% growth inhibition: MIC = 1.25 and 2%, respectively, with 10 µg, a clear zone of 21 mm was observed	6.27	Kilani-Feki et al. (2011a)
Antifungal	A. niger (CTM1099)	Antifungal bioassay, disk diffusion method	25% inhibition at 10 µg concentration	6.27	Kilani-Feki et al. (2011b)
Antifungal	T. rubrum, Rhizop. oryzae	Paper-disc agar diffusion method	At 10 µg: Inhibitory zone = 20 and 30 mm, respectively	6.27	Li et al. (2018)
Antifungal	A. niger, Rhizoc. solani	Antifungal activity on the Mycelia growth	50% Growth inhibition: MIC = 1.25 and 2%, respectively, with 10 µg, a clear zone of 45 mm was observed	6.25	Kilani-Feki et al. (2011)

(Continued)

TABLE 6.3 (Continued)
Biological Activity of 4-Quinolinone Alkaloids

Biological Activity	Target	Assay/Method	Results	Compound	References
Antifungal	A. niger (CTM1099)	Antifungal bioassay, disk diffusion method	45% Inhibition at 10 μg concentration	6.25	Kilani-Feki et al. (2012)
Antifungal	T. rubrum, Rhizop. oryzae	Paper-disc agar diffusion method	At 10 μg: Inhibitory zone = 34 and 10 mm, respectively	6.25	Li et al. (2018)
Antifungal	Cryptococcus neoformans, Candida albicans	–	MIC = 8 and 4 μg/mL, respectively	6.113	Kon et al. (2010)
Antifungal	Cr. neoformans, C. albicans	–	MIC = 4 and 2 μg/mL, respectively	6.111	Kon et al. (2010)
Antifungal	C. albicans CAF2.1	Standard antifungal susceptibility assay (Eucast method)	Showed growth inhibitory activity on both planktonic cells and biofilms: MIC = 8 μg/mL	6.72	Cretton et al. (2016)
Antifungal	T. rubrum, Rhizop. oryzae	Paper-disc agar diffusion method	At 10 μg: Inhibitory zone = 30 and 10 mm, respectively	6.33	Li et al. (2018)
Antifungal	T. rubrum NBRC 5467	MIC determination	MIC = 13 μg/mL	6.79	Karim et al. (2022)
Antifungal	T. rubrum NBRC 5467	MIC determination	MIC = 6.3 μg/mL	6.90	Karim et al. (2022)
Antigenotoxicity	DNA damaged induced by cyclophosphamide (CP)	Comet assay on lymphocytes, liver, spleen, brain, bone marrow cells of mice	Showed significant reduction (26%–52%) in DNA damage induced by CP for all cell types, in a dose-independent manner. Tail length, % DNA in tail and olive moment were significantly lower than CP only treated group	6.55	Aher et al. (2013)

(Continued)

TABLE 6.3 (*Continued*)
Biological Activity of 4-Quinolinone Alkaloids

Biological Activity	Target	Assay/Method	Results	Compound	References
Anti-glioblastoma multiforme (GBM)	Glioblastoma multiforme (brain tumour)	Molecular docking simulation	Strong binding energy with Jun and EGFR tumour-related signalling pathways	6.66	Wu G et al. (2022)
Anti-glioblastoma multiforme (GBM)	Glioblastoma multiforme (brain tumour)	Molecular docking simulation	Strong binding energy with Jun and EGFR tumour-related signalling pathways	6.43	Wu G et al. (2022)
Antihypertensive	Anaesthetized dog	Arterial blood pressure of dogs, lead II electrocardiography on a Narco Biosystem Physiograph recorder	Antihypertensive activity at doses of 50, 100, 200, and 300 mg/kg	6.104	Awaad et al. (2007)
Antihypertensive	Anaesthetized dog	Arterial blood pressure of dogs, lead II electrocardiography on a Narco Biosystem Physiograph recorder	Antihypertensive activity at doses of 50, 100, 200, and 300 mg/kg	6.108	Awaad et al. (2007)
Antihypertensive	Anaesthetized dog	Arterial blood pressure of dogs, lead II electrocardiography on a Narco Biosystem Physiograph recorder	Antihypertensive activity at doses of 50, 100, 200, and 300 mg/kg	6.82	Awaad et al. (2007)
Antihypertensive	Anaesthetized dog	Arterial blood pressure of dogs, lead II electrocardiography on a Narco Biosystem Physiograph recorder	Antihypertensive activity at doses of 50, 100, 200, and 300 mg/kg	6.103	Awaad et al. (2007)

(*Continued*)

TABLE 6.3 (Continued)
Biological Activity of 4-Quinolinone Alkaloids

Biological Activity	Target	Assay/Method	Results	Compound	References
Anti-inflammatory	TNF-α-induced NF-κB activity	Cell culture model with human embryonic kidney (HEK) 293 cells	$IC_{50} = 10.2$ μM	6.74	Liu et al. (2023)
Anti-inflammatory	Lipopolysaccharide (LPS)-induced nitric oxide (NO) production	Cell culture model with RAW 264.7 macrophages	$IC_{50} = 11.4$ μM	6.74	Liu et al. (2023)
Anti-inflammatory	Human neutrophils	Measurement of superoxide anion generation, Measurement of elastase release	Inhibited fMLP/CB-induced elastase release with IC_{50} value 5.10 μg/mL, inhibits superoxide radical anion generation with IC_{50} of 4.46 μg/mL	6.16	Cho et al. (2012a)
Anti-inflammatory	LPS-induced NO production in RAW 264.7 murine macrophages	MTT assay, NO Assay, Reverse Transcription (RT)-PCR, Western Blot Analysis, Enzyme-Linked Immunosorbent Assay (ELISA)	Anti-inflammatory effects on LPS-stimulated macrophage activation via inhibition of ERK, JNK, p38 MAPK phosphorylation, inhibition of NF-κB signalling in LPS-stimulated macrophages, and pro-inflammatory gene expression	6.21	Kim et al. (2017)
Anti-inflammatory	Mouse-Derived RAW264.7 macrophages	MTT assay	Suppressed LPS-induced expression of pro-inflammatory mediators through the NF-κB and MAPK pathways in RAW264.7 cells	6.32	Kim et al. (2014)
Anti-inflammatory	LPS-induced BV-2 cells	Measuring the inhibitory effects of NO production in LPS-induced BV-2 cells	Moderate inhibition towards NO production: $IC_{50} = 55.4$ μM	6.72	Hua et al. (2023)

(Continued)

TABLE 6.3 (Continued)
Biological Activity of 4-Quinolinone Alkaloids

Biological Activity	Target	Assay/Method	Results	Compound	References
Anti-inflammatory	Cytokines IL-1β and IL-6 in macrophages	Gene expression in RAW264.7 cells, Cytokine secretion and LDH release level in BMDMs cells, Western blot analysis	Significantly suppressed the gene expression and secretion of pro-inflammatory cytokines IL-1β and IL-6 in macrophages	6.24	Ji et al. (2022)
Anti-inflammatory, α-Glucosidase inhibition	α-Glucosidase enzyme	In vitro α-glucosidase inhibition assay	Strong α-glucosidase inhibitory effect, $IC_{50} = 82.4\,\mu M$	6.28	Nguyen et al. (2016)
Antimalarial	P. falciparum	In vitro assay using the agar disk-diffusion technique	$ID_{50} = 1\,\mu g/mL$	6.51	Bultel-Ponce et al. (1999)
Anti-melanogenic	Melan-A (murine Melan-A melanocyte) cell line	Cell Viability Assay	Inhibited melanin synthesis in Melan-A cells by 28.2%	6.15	Kim et al. (2016)
Anti-melanogenic	Melan-A (murine Melan-A melanocyte) cell line	Cell Viability Assay	Inhibited melanin synthesis in Melan-A cells by 42.7%, showed the highest inhibition at 8 μg/mL	6.21	Kim et al. (2016)
Anti-melanogenic	Melan-A (murine Melan-A melanocyte) cell line	Cell Viability Assay	Inhibited melanin synthesis in Melan-A cells by 23.0%	6.14	Kim et al. (2016)
Antimycobacterial	M. fortuitum, M. smegmatis, M. phlei	MIC assay	MIC = 2 μg/mL (against all three strains)	6.77	Adams et al. (2005)
Antimycobacterial	M. fortuitum, M. smegmatis, M. phlei	MIC assay	MIC = 32 μg/mL (M. fortuitum), 8 μg/mL (M. smegmatis), 2 μg/mL (M. phlei)	6.66	Adams et al. (2005)
Antimycobacterial	M. fortuitum, M. smegmatis, M. phlei	MIC assay	MIC = 4 μg/mL (against all three strains)	6.95	Adams et al. (2005)

(Continued)

TABLE 6.3 (Continued)

Biological Activity of 4-Quinolinone Alkaloids

Biological Activity	Target	Assay/Method	Results	Compound	References
Antimycobacterial	*M. fortuitum*, *M. smegmatis*, *M. phlei*	MIC assay	MIC = 4 µg/mL (*M. fortuitum*), 4 µg/mL (*M. smegmatis*), 2 µg/mL (*M. phlei*)	6.63	Adams et al. (2005)
Antimycobacterial	*M. fortuitum*, *M. smegmatis*, *M. phlei*	MIC assay	MIC = 4 µg/mL (against all three strains)	6.75	Adams et al. (2005)
Anti-neuroinflammatory	Mouse-Derived BV2 microglia	MTT assay	Suppressed LPS-induced expression of pro-inflammatory mediators through the NF-κB and MAPK pathways in BV2 microglia	6.32	Kim et al. (2014)
Antioxidant	HL-60 cells	Ferric reducing antioxidant power assay, cell-free and cell-based assays	At 10 µM concentration, it reduced the oxidative stress induced by NQO significantly	6.12	Cheng et al. (2016)
Antipathogenic	*Phytophthora capsici*, *Pythium ultimum*	–	MIC = 128 and 64 µg/mL, respectively	6.41	Dat et al. (2022)
Antipathogenic	*Ph. capsica*, *Py. ultimum*	–	MIC = 64 and 32 µg/mL, respectively	6.25	Dat et al. (2022)
Antipathogenic	*Glomerella cingulata* NBRC 5907	MIC determination	MIC = 25 µg/mL	6.79	Karim et al. (2022)
Antipathogenic	*G. cingulata* NBRC 5907	MIC determination	MIC = 13 µg/mL	6.90	Karim et al. (2022)
Antiperistaltic	Rabbit intestines and uteri	–	Stimulates the isolated rabbit's uterus and intestines at 5 ppm concentration	6.55	Steldt and Chen (1943)
Antiplasmodial	*Plasmodium falciparum*, chloroquine-resistant Dd2 and 3D7 strains	Malaria SYBR green I-based fluorescence assay	Good activity: IC_{50} = 330 and 385 nM, respectively	6.26	Presley et al. (2017)

(Continued)

TABLE 6.3 (Continued)
Biological Activity of 4-Quinolinone Alkaloids

Biological Activity	Target	Assay/Method	Results	Compound	References
Antiplasmodial activity	*P. falciparum* 3D7	In vitro antiplasmodial toxicity assay	Moderate activity, $IC_{50} = 2.17\,\mu M$	6.19	Staerk et al. (2009)
Antiplasmodial activity	*P. falciparum* 3D7	In vitro antiplasmodial toxicity assay	Moderate activity, $IC_{50} = 3.79\,\mu M$	6.30	Staerk et al. (2009)
Anti-Platelet Aggregation	Platelet Aggregation Induced by ADP	In vitro assay using an aggregometer	Displayed anti-platelet aggregation activity at $50\,\mu M$ and $100\,\mu M$.	6.93	Qin et al. (2021)
Anti-Platelet Aggregation	Platelet Aggregation Induced by ADP	In vitro assay using an aggregometer	Displayed anti-platelet aggregation activity at $50\,\mu M$ and $100\,\mu M$.	6.94	Qin et al. (2021)
Antiproliferative	HTLV-1-infected T-cell lines MT-1 and MT-2	MTT assay	EC50 values of 150 and 208 μM for MT-1 and MT-2, respectively	6.38	Nakano et al. (2016)
Antiproliferative	HTLV-1-infected T-cell lines MT-1 and MT-2	MTT assay	EC_{50} values of 7.0 and 3.9 μM for MT-1 and MT-2, respectively	6.67	Nakano et al. (2016)
Antiproliferative	HTLV-1-infected T-cell lines MT-1 and MT-2	MTT assay	EC_{50} values of 132 and 134 μM for MT-1 and MT-2, respectively	6.36	Nakano et al. (2016)
Antiproliferative	HTLV-1-infected T-cell lines MT-1 and MT-2	MTT assay	EC_{50} values of 155 and 10.32 μM for MT-1 and MT-2, respectively	6.43	Nakano et al. (2016)
Antiproliferative	HTLV-1-infected T-cell lines MT-1 and MT-2	MTT assay	EC_{50} values of 17.5 and 12.7 μM for MT-1 and MT-2, respectively	6.91	Nakano et al. (2016)
Antiproliferative	HTLV-1-infected T-cell lines MT-1 and MT-2	MTT assay	EC_{50} values of 29.0 and 17.7 μM for MT-1 and MT-2, respectively	6.73	Nakano et al. (2016)
Antitrypanosomal	*Trypanosorta cruzi*, Tulahuen C4 strain	In vitro growth inhibition toward the amastigote form	Potent antichagasic: $IC_{50} = 1.1\,\mu M$ [lower than the reference drug benznidazole ($IC_{50} = 2.9\,\mu M$)]	6.72	Cretton et al. (2014)

(Continued)

TABLE 6.3 (Continued)
Biological Activity of 4-Quinolinone Alkaloids

Biological Activity	Target	Assay/Method	Results	Compound	References
Antitrypanosomatid	*T. cruzi*	Blood from infected Swiss albino mice, counting trypomastigote forms	Moderate activity, $IC_{50} = 144.9 \, \mu M$	6.18	Ambrozin et al. (2008)
Antitrypanosomatid	*T. cruzi*	*In vitro* antitrypanosomal assay	$IC_{50} = 12.9 \, \mu M$, SI (selectivity index) = >1.5	6.109	Cretton et al. (2020)
Antitrypanosomatid	*T. cruzi*	*In vitro* antitrypanosomal assay	$IC_{50} = 10.0 \, \mu M$, SI (selectivity index) = 1.5	6.54	Cretton et al. (2020)
Antitrypanosomatid	*T. cruzi*	*In vitro* antitrypanosomal assay	$IC_{50} = 2.1 \, \mu M$, SI (selectivity index) >12.5	6.98	Cretton et al. (2020)
Antitumour	Human gastric cancer MKN-74 and MKN-7 cells cocultured with Hs738 human gastric stromal cells, Human colorectal cancer HCT-15 cells cocultured with CCD-18Co human colorectal stromal cells	Growth inhibition assay by measuring GFP fluorescence intensity	$IC_{50} = 0.17 \, mg/mL$ (for MKN-74), $IC_{50} = 0.13 \, mg/mL$ (for MKN-7), $IC_{50} = 0.21 \, mg/mL$ (for HCT-15)	6.110	Kawada et al. (2013)
Antitumour	Xenograft model of human colorectal cancer HCT-15 cells in nude mice	*In vivo* antitumour assay by measuring tumour volume	Significantly inhibited tumour growth at 25 mg/kg administered i.v.	6.110	Kawada et al. (2013)

(Continued)

TABLE 6.3 (Continued)
Biological Activity of 4-Quinolinone Alkaloids

Biological Activity	Target	Assay/Method	Results	Compound	References
Antiviral	Hepatitis C virus (HCV)	Time-of-addition experiment, Dose-dependent inhibition of HCV infection, calculated by comparing to the control using SPSS probit analysis	Strong activity: $IC_{50} = 1.4\,\mu g/mL$	6.36	Wahyuni et al. (2014)
Antiviral	HIV-1	In vitro assay	$ID_{50} = 10^{-3}\,\mu g/mL$	6.51	Bultel-Ponce et al. (1999)
Antiviral	Reverse transcriptases (RTs) of Human immunodeficiency viruses type 1 and 2 (HIV-1, HIV-2)	Enzymatic	Strongly inhibits the RNA-directed DNA synthesis of RTs ($IC_{50} = 2.38$ and $6.44\,\mu M$ for HIV-1 and HIV-2)	6.7	Loya et al. (1994)
Apoptosis induction	A375 skin melanoma cells	Annexin V-FITC/PI dual staining and flow cytometry	Progressive increment in Annexin V-positive cell population	6.38	Ghosh et al. (2013)
Apoptosis induction	B16F-10 melanoma cells	Morphological analysis, DNA fragmentation analysis (agarose gel electrophoresis), TUNEL assay; RT-PCR and gene expression analysis; ELISA based transcription factor profiling of NF-κB subunits p65, p50, c-Rel and other transcription factors	Presence of apoptotic bodies observed in cells; Enhanced DNA fragmentation observed in treated cells compared to untreated control; Upregulation of p53 and caspase-3 genes; downregulation of Bcl-2; 71.8%, 78.6% and 80.2% inhibition of p65, p50 and c-Rel nuclear translocation respectively (Decreased nuclear translocation of NF-κB subunits); inhibition of c-Fos, ATF-2 and CREB-1 by 75.8%, 68.9% and 73.7% respectively	6.55	Manu and Kuttan (2009)

(Continued)

TABLE 6.3 (*Continued*)
Biological Activity of 4-Quinolinone Alkaloids

Biological Activity	Target	Assay/Method	Results	Compound	References
Apoptosis induction, Cell cycle arrest	Human lung adenocarcinoma cells (A549)	Flow cytometric analysis with Annexin V-FITC and PI staining	Increased G2/M phase cells from 5.95% to 11.21%, increased early apoptotic cells from 0.33% to 23.37% and apoptotic cells from 0.39% to 25.72% in combination with 4Gy irradiation	6.28	Wang et al. (2014)
Autophagy induction	Skin melanoma A375 cells	Acridine orange staining and fluorescence microscopy; Immunoblot analysis of LC3 and autophagy-related proteins; Confocal microscopy of LC3 and Beclin-1 expression	Detection of acidic vesicular organelles (AVOs); Conversion of LC3-I to LC3-II and upregulation of Atg5, Atg7 expression; downregulation of Bcl-2 and mTOR expression; Increased punctate LC3 and Beclin-1 expression (Increased autophagosome formation, LC3-II expression, and Beclin-1 expression)	6.38	Ghosh et al. (2013)
Cancer chemopreventive	Hepa1c1c7 cells	Luciferase reporter assay in HEK293 cells, quinone reductase (QR) inducing activity	Concentration doubling QR activity: 28.8 μM	6.72	Monteillier et al. (2017)
Cardioprotective, Anti-ischemic	Mice myocardium	*In vivo* ischemia-reperfusion model	Protected against hypoxia and pituitrin-induced acute ischemia, no mice died within 3 days after an i.p. dose at 150 mg/kg	6.6	Long et al. (1984)

(Continued)

TABLE 6.3 (Continued)
Biological Activity of 4-Quinolinone Alkaloids

Biological Activity	Target	Assay/Method	Results	Compound	References
Cell death mechanism	A375 skin melanoma cells	MTT assay with autophagy/apoptosis inhibitors	Cell viability increased when Graveoline-induced autophagy/apoptosis was inhibited, suggesting both pathways contribute to Graveoline-induced cell death.	6.38	Ghosh et al. (2013)
Cerebral and coronary vasodilation	Mice	Blood flow in the brain and heart	Increased the blood flow in the brain and heart of mice; no mice died within 3 days after an i.p. dose at 150 mg/kg	6.6	Long et al. (1984)
Cytochrome P450 inhibitory activity	Cytochrome P450 2D6 (CYP2D6)	In vitro inhibitory activity assay using human liver microsomes and [O-methyl-^{14}C]dextromethorphan as substrate. Radiometric measurement of [^{14}C] formaldehyde formed.	Showed moderate inhibition selective for CYP2D6, with an IC_{50} value of 1.8 μM	6.55	Subehan et al. (2010)
Cytochrome P450 inhibitory activity	CYP2D6	in vitro inhibitory activity assay using human liver microsomes and [O-methyl-^{14}C]dextromethorphan as substrate. Radiometric measurement of [^{14}C] formaldehyde formed.	Showed moderate inhibition selective for CYP2D6, with an IC_{50} value of 4.7 μM	6.47	Subehan et al. (2010)
Cytotoxicity	Lovo, MDA-MB-231, HeLa human tumour cell lines	MTT assay in 96-well microplates	Most potent activity against MDA-MB-231 cell line with IC50 of 7.95 μM	6.92	Ma et al. (2021)

(Continued)

TABLE 6.3 (*Continued*)

Biological Activity of 4-Quinolinone Alkaloids

Biological Activity	Target	Assay/Method	Results	Compound	References
Cytotoxicity	Six Human Tumour Lines: HL-60 (human myeloid leukaemia), Hela (human cervical cancer), HepG2 (human hepatoma), A-549 (human non-small cell lung cancer), AGS (human gastric cancer), MDA-MA-468 (human breast cancer)	MTT Assay	Cytotoxic activity against all six tumour lines.	6.94	Qin et al. (2021)
Cytotoxicity	Tumour cell lines: Lovo, MDA-MB-231, HeLa	MTT method	Moderate cytotoxicity against the three tumour cell lines with IC_{50} values of 6.72 µM, 14.20 µM, and 13.05 µM against Lovo, MDA-MB-231, and HeLa cells, respectively.	6.62	Ma et al. (2018)
Cytotoxicity	Six Human Tumour Lines: HL-60 (human myeloid leukaemia), Hela (human cervical cancer), HepG2 (human hepatoma), A-549 (human non-small cell lung cancer), AGS (human gastric cancer), MDA-MA-468 (human breast cancer)	MTT Assay	Cytotoxic activity against all six tumour lines.	6.93	Qin et al. (2021)

(Continued)

TABLE 6.3 (*Continued*)
Biological Activity of 4-Quinolinone Alkaloids

Biological Activity	Target	Assay/Method	Results	Compound	References
Cytotoxicity	KB cells (cancer cell line)	*In vitro* assay	$IC_{50} < 2\,\mu g/mL$	6.45	Bultel-Ponce et al. (1999)
Cytotoxicity	A375 skin melanoma cells	MTT cell viability assay	$IC_{50} = 22.23\,\mu g/mL$	6.38	Ghosh et al. (2013)
Cytotoxicity	Human cancer cell lines HL-60, N-87, H-460, HepG2	MTT assay	$IC_{50} = 18.12, 17.25, 17.34,$ and $20.38\,\mu M$, respectively	6.77	Huang et al. (2012)
Cytotoxicity	Human colon carcinoma (HT-29), human breast carcinoma (MCF-7), human hepatoblastoma (HepG2)	MTT assay	$IC_{50} = 43, 52,$ and $48\,\mu g/mL$, respectively	6.77	Xu et al. (2006)
Cytotoxicity	Leukaeria HL-60 and prostate cancer PC-3 cell lines	Growth inhibitory assay	$GI_{50} = 10.44$ and $15.11\,\mu M$	6.77	Zhao et al. (2015)
Cytotoxicity	Human cancer cell lines HL-60, N-87, H-460, HepG2	MTT assay	$IC_{50} = 15.41, 18.18, 16.53,$ and $19.90\,\mu M$, respectively	6.80	Huang et al. (2012)
Cytotoxicity	Human hepatoma cells HepG-2, human uterine cervical carcinoma Hela, Human hepatoma cells BEL7402, Human hepatoma cells BEL7403	MTT assay	$IC_{50} = 53.36, 26.98, 37.48,$ and $44.07\,\mu M$, respectively	6.80	Wang et al. (2013)

(Continued)

TABLE 6.3 (Continued)
Biological Activity of 4-Quinolinone Alkaloids

Biological Activity	Target	Assay/Method	Results	Compound	References
Cytotoxicity	Human colon carcinoma (HT-29), human breast carcinoma (MCF-7), human hepatoblastoma (HepG2)	MTT assay	$IC_{50} = 41, 33,$ and $39\,\mu g/mL$, respectively	6.80	Xu et al. (2006)
Cytotoxicity	Leukaemia HL-60 cell lines	Growth inhibitory assay	$GI_{50} = 16.02\,\mu M$	6.80	Zhao et al. (2015)
Cytotoxicity	Human cancer cell lines HL-60, N-87, H-460, Hep G2	MTT assay	$IC_{50} = 17.54, 14.27, 15.79, 15.95\,\mu M$, respectively	6.102	Huang et al. (2012)
Cytotoxicity	Human hepatoma cells HepG-2, human uterine cervical carcinoma Hela, Human hepatoma cells BEL7402, Human hepatoma cells BEL7403	MTT assay	$IC_{50} = 52.30, 23.36, 29.51,$ and $36.86\,\mu M$, respectively	6.102	Wang et al. (2013)
Cytotoxicity	Human cancer cell lines HL-60, N-87, H-460, Hep G2	MTT assay	$IC_{50} = 21.64, 20.52, 21.08,$ and et al. $19.75\,\mu M$, respectively	6.66	Huang et al. (2012)
Cytotoxicity	Leukaemia HL-60 and prostate cancer PC-3 cell lines	Growth inhibitory assay	$GI_{50} = 10.73$ and $17.61\,\mu M$	6.66	Zhao et al. (2015)
Cytotoxicity	Human cancer cell lines HL-60, N-87, H-460, Hep G2	MTT assay	$IC_{50} = 21.39, 23.34, 25.15.$ and $21.92\,\mu M$, respectively	6.43	Huang et al. (2012)

(Continued)

TABLE 6.3 (Continued)
Biological Activity of 4-Quinolinone Alkaloids

Biological Activity	Target	Assay/Method	Results	Compound	References
Cytotoxicity	Human lung adenocarcinoma A549 cells	CCK-8 assay	IC_{50} = 33.7, 21.9, and 16.8 μg/mL after 6, 12, 24 h treatment, respectively	6.28	Wang et al. (2014)
Cytotoxicity	Human cancer cell lines HL-60, N-87, H-460, Hep G2	MTT assay	IC_{50} = 16.06, 12.64, 16.69, and 15.32 μM, respectively	6.95	Huang et al. (2012)
Cytotoxicity	Lovo, MDA-MB-231, HeLa human tumour cell lines	MTT assay in 96-well microplates	Moderate cytotoxic activity against all cell lines	6.95	Ma et al. (2021)
Cytotoxicity	Human cancer cell lines HL-60, N-87, H-460, Hep G2	MTT assay	IC_{50} = 13.81, 14.52, 15.63, and 15.90 μM, respectively [mixture with compound (100)]	6.99	Huang et al. (2012)
Cytotoxicity	Human cancer cell lines HL-60, N-87, H-460, Hep G2	MTT assay	IC_{50} = 13.81, 14.52, 15.63, and 15.90 μM, respectively [mixture with compound (99)]	6.100	Huang et al. (2012)
Cytotoxicity	Human colon carcinoma (HT-29), human breast carcinoma (MCF-7), human hepatoblastoma (Hep G2)	MTT assay	IC_{50} = 59, 56 and 50 μg/mL, respectively	6.100	Xu et al. (2006)
Cytotoxicity	Human cancer cell lines HL-60, N-87, H-460, Hep G2	MTT assay	IC_{50} = 18.80, 18.66, 20.00, and 19.45 μM, respectively	6.63	Huang et al. (2012)
Cytotoxicity	Human cancer cell lines HL-60, N-87, H-460, Hep G2	MTT assay	IC_{50} = 18.50, 17.85, 16.03, and 19.83 μM, respectively	6.75	Huang et al. (2012)

(Continued)

TABLE 6.3 (Continued)
Biological Activity of 4-Quinolinone Alkaloids

Biological Activity	Target	Assay/Method	Results	Compound	References
Cytotoxicity	Tumour cell lines: Lovo, MDA-MB-231, HeLa	MTT method	Moderate cytotoxicity against the three tumour cell lines with IC_{50} values of 18.17 μM, 8.25 μM, and 13.05 μM against Lovo, MDA-MB-231, and HeLa cells, respectively.	6.75	Ma et al. (2018)
Cytotoxicity	Lovo, MDA-MB-231, HeLa human tumour cell lines	MTT assay in 96-well microplates	Moderate cytotoxic activity against all cell lines.	6.75	Ma et al. (2021)
Cytotoxicity	Cancer cell lines A549 (human lung carcinoma), MCF7 (human breast carcinoma), KB (human mouth epidermal carcinoma)	-	IC_{50} = 71.6, 92.0, and 65.1 μg/mL, respectively	6.41	Dat et al. (2022)
Cytotoxicity	Cancer cell lines A549 (human lung carcinoma), MCF7 (human breast carcinoma), KB (human mouth epidermal carcinoma)	-	IC_{50} = 90.9, 95.1, and 79.1 μg/mL, respectively	6.25	Dat et al. (2022)
Cytotoxicity	Human cancer cell lines HL-60, N-87, H-460, Hep G2	MTT assay	$IC50$ = 17.66, 20.82, 18.99, and 16.52 μM, respectively	6.71	Huang et al. (2012)

(Continued)

TABLE 6.3 (Continued)
Biological Activity of 4-Quinolinone Alkaloids

Biological Activity	Target	Assay/Method	Results	Compound	References
Cytotoxicity	Tumour cell lines: Lovo, MDA-MB-231, HeLa	MTT method	Moderate cytotoxicity against the three tumour cell lines with IC_{50} values of 20.78 µM, 15.85 µM, and 15.77 µM against Lovo, MDA-MB-231, and HeLa cells, respectively.	6.76	Ma et al. (2018)
Cytotoxicity	Lovo, MDA-MB-231, HeLa human tumour cell lines	MTT assay in 96-well microplates	Moderate cytotoxic activity against all cell lines.	6.76	Ma et al. (2021)
Cytotoxicity	Human cancer cell lines HL-60, N-87, H-460, Hep G2	MTT assay	IC_{50} = 22.97, 21.69, 21.92, and 18.14 µM, respectively	6.52	Huang et al. (2012)
Cytotoxicity	KB cells (cancer cell line)	In vitro assay	Mild cytotoxicity, IC_{50} = 5 µg/mL	6.50	Bultel-Ponce et al. (1999)
Cytotoxicity	Human cancer cell lines HL-60, N-87, H-460, Hep G2	MTT assay	IC_{50} = 17.72, 16.72, 15.54, and 16.83 µM, respectively	6.88	Huang et al. (2012)
Cytotoxicity	MCF-7 human breast cancer cells, HeLa human cervical cancer cells	MTT assay	Induced cell death: IC_{50} = 240 and 140 nM, respectively	6.35	Mori et al. (2007)
Cytotoxicity	HepG2 cell line	MTT assay	Moderate activity, IC_{50} = 13.2 µM	6.22	Gao et al. (2012)
Cytotoxicity	95-D and HepG2 cell lines	In vitro	Potent cytotoxic, IC_{50} = 0.57 and 6.5 µg/mL, respectively	6.22	Shao et al. (2010)
Cytotoxicity	HepG2, 95-D, MGC832 and HeLa tumour cell lines	MTT assay	Potent activity, IC_{50} = 7, 0.4, 91, and 529 nM, respectively	6.22	Zhu et al. (2013)

(Continued)

TABLE 6.3 (Continued)
Biological Activity of 4-Quinolinone Alkaloids

Biological Activity	Target	Assay/Method	Results	Compound	References
Cytotoxicity	Human hepatoma cells HepG-2, human uterine cervical carcinoma HeLa, Human hepatoma cells BEL7402, Human hepatoma cells BEL7403	MTT assay	IC_{50} = 49.83, 18.53, 15.85, and 35.83 µM, respectively	6.101	Wang et al. (2013)
Cytotoxicity	Human cancer cell lines HL-60, N-87, H-460, Hep G2	MTT assay	IC_{50} = 18.36, 18.04, 20.11, and 21.91 µM, respectively	6.61	Huang et al. (2012)
Cytotoxicity	Hep G2 cell line	MTT assay	Moderate activity, IC_{50} = 11.3 µM	6.32	Gao et al. (2012)
Cytotoxicity	Hep G2, 95-D, MGC832 and HeLa tumour cell lines	MTT assay	Potent activity, IC_{50} = 2.52 µM, 1.54 µM, 13 nM, and 0.110 µM, respectively	6.32	Zhu et al. (2013)
Cytotoxicity	Human cancer cell lines HL-60, N-87, H-460, Hep G2	MTT assay	IC_{50} = 21.67, 17.25, 18.56, and 21.76 µM, respectively	6.40	Huang et al. (2012)
Cytotoxicity	Human cancer cell lines HL-60, N-87, H-460, Hep G2	MTT assay	IC_{50} = 19.56, 16.70, 19.97, and 16.41 µM, respectively	6.57	Huang et al. (2012)
Cytotoxicity	Human cancer cell lines HL-60, N-87, H-460, Hep G2	MTT assay	IC_{50} = 18.26, 16.25, 13.27, and 14.36 µM, respectively	6.89	Huang et al. (2012)
Cytotoxicity	Human cancer cell lines HL-60, N-87, H-460, Hep G2	MTT assay	IC_{50} = 18.12, 17.25, 17.34, and 20.38 µM, respectively	6.83	Huang et al. (2012)

(Continued)

TABLE 6.3 (Continued)
Biological Activity of 4-Quinolinone Alkaloids

Biological Activity	Target	Assay/Method	Results	Compound	References
Cytotoxicity	Human cervical cancer HeLa, human prostate cancer PC-3 and human breast cancer MCF-7 cells	MTT assay	IC_{50} = 50 µM against HeLa and PC-3, not active against MCF-7	6.58	Sinha et al. (2012)
Cytotoxicity	Human hepatoma cells HepG-2, human uterine cervical carcinoma HeLa, Human hepatoma cells BEL7402, Human hepatoma cells BEL7403	MTT assay	IC_{50} = 52.8, 24.25, 2457, and 33.54 µM, respectively	6.69	Wang et al. (2013)
Cytotoxicity	HeLa cells	Standard *in vitro* toxicity protocol (sulforhodamine B procedure)	IC_{50} = 14.4 µg/mL	6.72	Cretton et al. (2016)
Cytotoxicity	Leukaemia HL-60 and prostate cancer PC-3 cell lines	Growth inhibitory assay	GI_{50} = 20.36 and 31.99 µM, respectively	6.114	Zhao et al. (2015)
Cytotoxicity	P388 Murine leukaemia cells	-	IC_{50} = 10 µM	6.79	Karim et al. (2022)
Cytotoxicity	P388 Murine leukaemia cells	-	IC_{50} = 7.3 µM	6.90	Karim et al. (2022)
Cytotoxicity	Leukaemia HL-60 cell lines	Growth inhibitory assay	GI_{50} = 12.07 µM	6.115	Zhao et al. (2015)
Cytotoxicity, P-glycoprotein (p-gp) interaction	CCRF-CEM leukaemia cells; CEM/ADR5000 p-gp overexpressing subline; Porcine brain capillary endothelial cells (PBCECs)	Cell proliferation assay; Calcein accumulation assay using PBCECs	No cytotoxicity at 10 µM, slight activity (13%) at 30 µM; Not transported by p-gp (no increase in fluorescence)	6.77	Adams et al. (2007)

(Continued)

TABLE 6.3 (Continued)
Biological Activity of 4-Quinolinone Alkaloids

Biological Activity	Target	Assay/Method	Results	Compound	References
Cytotoxicity, P-glycoprotein (p-gp) interaction	CCRF-CEM leukaemia cells; CEM/ADR5000 p-gp overexpressing subline; Porcine brain capillary endothelial cells (PBCECs)	Cell proliferation assay; Calcein accumulation assay using PBCECs	IC_{50} against CCRF-CEM: 14.08 μM; IC_{50} against CEM/ADR5000: 33.14 μM; Moderate p-gp modulator (200%–300% increase in fluorescence at 10 μM)	6.80	Adams et al. (2007)
Cytotoxicity, P-glycoprotein (p-gp) interaction	CCRF-CEM leukaemia cells; CEM/ADR5000 p-gp overexpressing subline; Porcine brain capillary endothelial cells (PBCECs)	Cell proliferation assay; Calcein accumulation assay using PBCECs	IC_{50} against CCRF-CEM: 4.56 μM; IC_{50} against CEM/ADR5000 = 17.19 μM; Moderate p-gp modulator (200%–300% increase in fluorescence at 10–25 μM)	6.66	Adams et al. (2007)
Diacylglycerol Acyltransferase (DGAT) inhibition	Microsomes prepared from rat liver	DGAT activity measurement	DGAT activity dose-dependently with $IC_{50} = 23.8$ μM	6.77	Ko et al. (2002)
DGAT inhibition	Microsomes prepared from rat liver	DGAT activity measurement	DGAT activity dose-dependently with $IC_{50} = 13.5$ μM	6.95	Ko et al. (2002)
DGAT inhibition	Microsomes prepared from rat liver	DGAT activity measurement	DGAT activity dose-dependently with $IC_{50} = 20.1$ μM	6.75	Ko et al. (2002)
DGAT) inhibition	Microsomes prepared from rat liver	DGAT activity measurement	DGAT activity dose-dependently with $IC_{50} = 13.5$ μM	6.88	Ko et al. (2002)
Enzyme inhibitory activities against AChE	Acetylcholinesterase (AChE)	AChE inhibitory bioassay, spectrophotometric method	Moderate inhibition, $IC_{50} = 87.3$ μM	6.22	Chen et al. (2021)

(Continued)

TABLE 6.3 (Continued)
Biological Activity of 4-Quinolinone Alkaloids

Biological Activity	Target	Assay/Method	Results	Compound	References
Enzyme inhibitory activities against AChE	AChE	AChE inhibitory bioassay, spectrophotometric method	Moderate inhibition, $IC_{50} = 68.5\,\mu M$	6.9	Chen et al. (2021)
Estrogenic activity	Immature Rats	Uterine and ovaries mass gain	At 50 mg/kg concentration: Uterine mass gain, % of control (a) with fluid 318%; (b) without fluid 243%; Mass gain of ovaries,% of control = 7.6%	6.19	Nazrullaev et al. (2001)
General toxicity	Zebrafish embryos	Zebrafish embryo viability assay	Significant toxicity at all tested concentrations. Embryos died at all concentrations (1 pM to 100 μM)	6.7	Carnovali et al. (2022)
Glycogen synthase kinase 3β inhibition	Glycogen synthase kinase 3β	Luminescent assay	Showed weak inhibitory activity (19% inhibition) at 10μM	6.105	Nachtigall et al. (2010)
Glycogen synthase kinase 3β inhibition	Glycogen synthase kinase 3β	Luminescent assay	Showed weak inhibitory activity (22% inhibition) at 10μM	6.106	Nachtigall et al. (2010)
Human liver microsomal dextromethorphan O-demethylation inhibition	Cytochrome P450 2D6 (CYP2D6)	In vitro enzymatic assay; inhibition assay	Inhibited human liver microsomal dextromethorphan O-demethylation activity, which is a prototype marker of CYP2D6	6.38	Saputra et al. (2024)
Human liver microsomal dextromethorphan O-demethylation inhibition	CYP2D6	In vitro enzymatic assay; inhibition assay	Inhibited human liver microsomal dextromethorphan O-demethylation activity, which is a prototype marker of CYP2D6	6.55	Saputra et al. (2024)
Inhibition of α-Amylase and α-Glucosidase	α-Amylase, α-Glucosidase	–	$IC_{50} = 526.5$ and $794.3\,\mu g/mL$, respectively	6.41	Dat et al. (2022)

(Continued)

TABLE 6.3 (Continued)
Biological Activity of 4-Quinolinone Alkaloids

Biological Activity	Target	Assay/Method	Results	Compound	References
Inhibition of Angiotensin II Receptor Binding	Rat liver plasma membrane	Angiotensin II binding assay	$IC_{50}=43.4\,\mu M$	6.77	Lee et al. (1998)
Inhibition of Angiotensin II Receptor Binding	Rat liver plasma membrane	Angiotensin II binding assay	$IC_{50}=48.2\,\mu M$	6.95	Lee et al. (1998)
Inhibition of Angiotensin II Receptor Binding	Rat liver plasma membrane	Angiotensin II binding assay	$IC_{50}=34.1\,\mu M$	6.75	Lee et al. (1998)
Inhibition of cytochrome oxidase bd	*Azotobacter vinelandii*	Electron transfer studies	Used as a model system to study selectivity	6.106	Nachtigall et al. (2010)
Inhibition of fMLP/CB-induced neutrophil elastase release	Human Neutrophils	Response to N-formyl-l-methionyl-l-leucyl-l-phenylalanine/Cytochalasin B (fMLP/CB)	Elastase release = 2.6 µM	6.77	Wang et al. (2010)
Inhibition of fMLP/CB-induced neutrophil elastase release	Human Neutrophils	Response to N-formyl-l-methionyl-l-leucyl-l-phenylalanine/Cytochalasin B (fMLP/CB)	Elastase release = 3.1 µM	6.81	Wang et al. (2010)
Inhibition of insulin-like growth factors (IGFs)	Insulin-like growth factor I (IGF-I) derived from hematopoietic cell line 32D	Viability and growth of 32D/GR15 cells measured by the WST-8 assay	Induced cell death 32D/GR15 cells with an IC_{50} of 160nM in IGF-I containing medium	6.35	Mori et al. (2007)
Inhibition of mycobacterial enzyme MCA	MCA	Amidase activity assay	Inhibitor of MCA, antimycobacterial activity	6.112	Nicholas et al. (2001)
Inhibition of quinol oxidation sites	Cytochromes bo and bd of *E. coli*	–	Strong inhibition	6.106	Nachtigall et al. (2010)

(Continued)

TABLE 6.3 (*Continued*)
Biological Activity of 4-Quinolinone Alkaloids

Biological Activity	Target	Assay/Method	Results	Compound	References
Inhibition of tyrosine phosphatase 1B (PTP1B) and α-Glucosidase (3W37)	PTP1B and α-Glucosidase (3W37)	*In vitro* enzyme inhibition assay	Significant inhibitory activity: $IC_{50} = 24.3\,\mu M$ against PTP1B and $92.1\,\mu M$ against α-glucosidase	6.28	To et al. (2021)
Inhibition of α-Amylase and α-Glucosidase	α-Amylase, α-Glucosidase	–	$IC_{50} = 649.9$ and $960.4\,\mu g/mL$, respectively	6.25	Dat et al. (2022)
Inhibitory activity on cell adhesion	Cell adhesion molecules (sICAM-1 and LFA-1 in THP-1 cells)	ELISA	$IC_{50} = 109.8\,\mu M$	6.77	Lee et al. (2010)
Inhibitory activity on cell adhesion	Cell adhesion molecules (sICAM-1 and LFA-1 in THP-1 cells)	ELISA	$IC_{50} \sim 150\,\mu M$	6.80	Lee et al. (2010)
Inhibitory activity on cell adhesion	Cell adhesion molecules (sICAM-1 and LFA-1 in THP-1 cells)	ELISA	$IC_{50} \sim 150\,\mu M$	6.66	Lee et al. (2010)
Inhibitory activity on cell adhesion	Cell adhesion molecules (sICAM-1 and LFA-1 in THP-1 cells)	ELISA	$IC_{50} = 40.9\,\mu M$	6.99	Lee et al. (2010)
Inhibitory activity on cell adhesion	Cell adhesion molecules (sICAM-1 and LFA-1 in THP-1 cells)	ELISA	$IC_{50} = 40.9\,\mu M$	6.100	Lee et al. (2010)
Insecticidal	*Bombyx mori* (silkworm) larvae	Effect on moltinism, sleepiness and integument	Slight effect at 400ppm	6.80	Kamikado et al. (1977)

(*Continued*)

TABLE 6.3 (Continued)
Biological Activity of 4-Quinolinone Alkaloids

Biological Activity	Target	Assay/Method	Results	Compound	References
Insecticidal	*Atta sexdens rubropilosa* (leaf-cutting ant)	Ants ingestion bioassay, long-rank test	The ants survival median (S_{50}) was in 7th day, 96% mortality after 25 days	6.18	Terezan et al. (2010)
Insecticidal	*Bo. mori* (silkworm) larvae	Effect on moltinism, sleepiness, and integument	Slight effect at 400ppm	6.102	Kamikado et al. (1977)
Insecticidal	*Bo. mori* (silkworm) larvae	Effect on moltinism, sleepiness and integument	Slight effect at 400ppm	6.66	Kamikado et al. (1977)
Insecticidal	*Aphis gossypii*	Insect mortality and the area of leaf eaten	Strong activity, 100% mortality on a mixed population at 1,000ppm	6.22	Shao et al. (2010)
Leukotriene biosynthesis inhibition	Human polymorphonuclear granulocytes	Ex vivo assay using human granulocytes, LTB4 ELISA immunoassay, probit-log analysis	Dose-dependent inhibitory effect, $IC_{50} = 14.6\,\mu M$	6.77	Adams et al. (2004)
Leukotriene biosynthesis inhibition	Human polymorphonuclear granulocytes	Ex vivo assay using human granulocytes, LTB4 ELISA immunoassay, probit-log analysis	Dose-dependent inhibitory effect, $IC_{50} = 12.1\,\mu M$	6.43	Adams et al. (2004)
Leukotriene biosynthesis inhibition	Human polymorphonuclear granulocytes	Ex vivo assay using human granulocytes, LTB4 ELISA immunoassay, probit-log analysis	Dose-dependent inhibitory effect, $IC_{50} = 12.3\,\mu M$	6.95	Adams et al. (2004)
Leukotriene biosynthesis inhibition	Human polymorphonuclear granulocytes	Ex vivo assay using human granulocytes, LTB4 ELISA immunoassay, probit-log analysis	Dose-dependent inhibitory effect, $IC_{50} = 10.0\,\mu M$	6.63	Adams et al. (2004)
Leukotriene biosynthesis inhibition	Human polymorphonuclear granulocytes	Ex vivo assay using human granulocytes, LTB4 ELISA immunoassay, probit-log analysis	Dose-dependent inhibitory effect, $IC_{50} = 10.1\,\mu M$	6.75	Adams et al. (2004)
Lipase inhibition	Pancreatic lipase	Modified version of a commercially available kit (Lipase Kit S)	$IC_{50} = 3.03\,mM$	6.43	Matsuo et al. (2025)

(Continued)

TABLE 6.3 (Continued)
Biological Activity of 4-Quinolinone Alkaloids

Biological Activity	Target	Assay/Method	Results	Compound	References
Lipase inhibition	Pancreatic lipase	Modified version of a commercially available kit (Lipase Kit S)	$IC_{50} = 3.10$ mM	6.52	Matsuo et al. (2025)
Lipase inhibition	Pancreatic lipase	Modified version of a commercially available kit (Lipase Kit S)	$IC_{50} = 1.40$ mM	6.40	Matsuo et al. (2025)
Lipase inhibition	Pancreatic lipase	Modified version of a commercially available kit (Lipase Kit S)	$IC_{50} = 1.77$ mM	6.63	Matsuo et al. (2025)
Lipase inhibition	Pancreatic lipase	Modified version of a commercially available kit (Lipase Kit S)	$IC_{50} = 3.17$ mM	6.77	Matsuo et al. (2025)
Lipase inhibition	Pancreatic lipase	Modified version of a commercially available kit (Lipase Kit S)	$IC_{50} = 6.85$ mM	6.100	Matsuo et al. (2025)
Lipase inhibition	Pancreatic lipase	Modified version of a commercially available kit (Lipase Kit S)	$IC_{50} = 3.23$ mM	6.264	Matsuo et al. (2025)
Lipase inhibition	Pancreatic lipase	Modified version of a commercially available kit (Lipase Kit S)	$IC_{50} = 7.37$ mM	6.265	Matsuo et al. (2025)
Monoamine Oxidase (MAO) Inhibition	MAO-A and MAO-B	Measured fluorescently using kynuramine as substrate in mouse brain mitochondrial fraction	$IC_{50} = 43.0\,\mu M$ (MAO-B), IC50 > 400 μM (MAO-A)	6.77	Han et al. (2007)
MAO Inhibition	MAO-A and MAO-B	Measured fluorescently using kynuramine as substrate in mouse brain mitochondrial fraction	$IC_{50} > 400\,\mu M$ (MAO-B), IC50 > 400 μM (MAO-A)	6.80	Han et al. (2007)
MAO Inhibition	MAO-A and MAO-B	Measured fluorescently using kynuramine as substrate in mouse brain mitochondrial fraction	$IC_{50} = 19.3\,\mu M$ (MAO-B), $IC_{50} = 398.2\,\mu M$ (MAO-A)	6.66	Han et al. (2007)

(Continued)

TABLE 6.3 (Continued)
Biological Activity of 4-Quinolinone Alkaloids

Biological Activity	Target	Assay/Method	Results	Compound	References
MAO Inhibition	MAO-A and MAO-B	Measured fluorescently using kynuramine as substrate in mouse brain mitochondrial fraction	$IC_{50} = 2.3\,\mu M$ (MAO-B), $IC_{50} = 240.2\,\mu M$ (MAO-A)	6.43	Han et al. (2007)
MAO Inhibition	MAO-A and MAO-B	Measured fluorescently using kynuramine as substrate in mouse brain mitochondrial fraction	$IC_{50} = 3.6\,\mu M$ (MAO-B), $IC50 > 400\,\mu M$ (MAO-A)	6.95	Han et al. (2007)
MAO Inhibition	MAO-A and MAO-B	Measured fluorescently using kynuramine as substrate in mouse brain mitochondrial fraction	$IC_{50} = 13.5\,\mu M$ (MAO-B), $IC_{50} \sim 400\,\mu M$ (MAO-A)	6.63	Han et al. (2007)
MAO Inhibition	MAO-A and MAO-B	Measured fluorescently using kynuramine as substrate in mouse brain mitochondrial fraction	$IC_{50} = 15.3\,\mu M$ (MAO-B), $IC_{50} = 338.2\,\mu M$ (MAO-A)	6.66	Lee et al. (2003)
NADH dehydrogenase inhibition	Beef heart sub-mitochondrial particles	HQNO titration	$IC_{50} = 42\,nM$	6.106	Kunze et al. (1987)
NADH dehydrogenase inhibition	Beef heart sub-mitochondrial particles	HQNO titration	$IC_{50} = 50\,nM$	6.95	Kunze et al. (1987)
Neuroprotective	Cortical neurons in rats (DIV-9)	Sodium nitroprussiateSNP injured primary neuronal model and glutamate induced primary neuron toxicity model; MTT assay	At $10\,\mu M$ concentration: survival status of rat cortical neurons = 90.1%; Effects of compounds on sodium nitroprusside and glutamate-induced apoptosis of cortical neurons in rats = 95.6 and 66.1%, respectively	6.23	Zhuang et al. (2017)

(Continued)

TABLE 6.3 (*Continued*)

Biological Activity of 4-Quinolinone Alkaloids

Biological Activity	Target	Assay/Method	Results	Compound	References
Nuclear factor of activated T cells (NFAT) activity	NFAT-dependent transcription and NF-kB	Comparison with nuclear factor (NF)-kB-dependent transcription using a reporter gene assay	$IC_{50} = 1.09$ and $6.08\,\mu M$ against NFAT and NF-kB, respectively	6.77	Jin et al. (2004)
NFAT activity	NFAT-dependent transcription and NF-kB	Comparison with N-kB-dependent transcription using a reporter gene assay	$IC_{50} = 5.48$ and $9.49\,\mu M$ against NFAT and NF-kB, respectively	6.80	Jin et al. (2004)
NFAT activity	NFAT-dependent transcription and NF-kB	Comparison with NF-kB-dependent transcription using a reporter gene assay	$IC_{50} = 3.44$ and $>100\,\mu M$ against NFAT and NF-kB, respectively	6.36	Jin et al. (2004)
NFAT activity	NFAT-dependent transcription and NF-kB	Comparison with N-kB-dependent transcription using a reporter gene assay	$IC_{50} = 3.29$ and $>100\,\mu M$ against NFAT and NF-kB, respectively	6.51	Jin et al. (2004)
NFAT activity	NFAT-dependent transcription and NF-kB	Comparison with NF-kB-dependent transcription using a reporter gene assay	$IC_{50} = 0.91$ and $11.97\,\mu M$ against NFAT and NF-kB, respectively	6.102	Jin et al. (2004)
NFAT activity	NFAT-dependent transcription and NF-kB	Comparison with NF-kB-dependent transcription using a reporter gene assay	$IC_{50} = 8.22$ and $7.94\,\mu M$ against NFAT and NF-kB, respectively	6.66	Jin et al. (2004)
NFAT activity	NFAT-dependent transcription and NF-kB	Comparison with NF-kB-dependent transcription using a reporter gene assay	$IC_{50} = 15.91$ and $6.08\,\mu M$ against NFAT and NF-kB, respectively	6.43	Jin et al. (2004)
NFAT activity	NFAT-dependent transcription and NF-kB	Comparison with NF-kB-dependent transcription using a reporter gene assay	$IC_{50} = 1.01$ and $6.60\,\mu M$ against NFAT and NF-kB, respectively	6.95	Jin et al. (2004)

(Continued)

TABLE 6.3 (*Continued*)
Biological Activity of 4-Quinolinone Alkaloids

Biological Activity	Target	Assay/Method	Results	Compound	References
NFAT activity	NFAT-dependent transcription and NF-kB	Comparison with NF-kB-dependent transcription using a reporter gene assay	IC_{50} = 1.86 and 10.80 µM against NFAT and NF-kB, respectively	6.75	Jin et al. (2004)
PDE inhibition	PDE10A	Molecular docking, Molecular dynamics (MD) simulation	Stick to the standards of CNS-active drug; was associated with G protein-coupled receptors, emphasizing its significance in Parkinson's disease (PD) therapy; stability in the binding pocket of PDE10A	6.51	Ahmad et al. (2024)
PDE inhibition	PDE10A	Molecular docking, Molecular dynamics (MD) simulation	Stick to the standards of CNS-active drugs which are associated with G protein-coupled receptors, emphasizing their significance in Parkinson's disease (PD) therapy; stability in the binding pocket of PDE10A	6.87	Ahmad et al. (2024)
Photosynthesis inhibition	Photosynthetic electron transport; Photosystem II	Chlorophyll a fluorescence measurements in spinach leaf discs; Polarography assay on noncyclic electron transport in chloroplasts	Inhibited the basal, phosphorylating, and uncoupled electron transport by 40% at 300 µM; increased dV/dt0 and decreased Plabs, both parameters by 60% at 150 µM, indicating damage to the photosynthetic electron transport chain. A J-band was observed in the OJIP curve	6.38	Sampaio et al. (2018)

(*Continued*)

TABLE 6.3 (*Continued*)
Biological Activity of 4-Quinolinone Alkaloids

Biological Activity	Target	Assay/Method	Results	Compound	References
Photosynthesis inhibition	Photosynthetic electron transport; Photosystem II	Chlorophyll a fluorescence measurements in spinach leaf discs; Polarography assay on noncyclic electron transport in chloroplasts	Inhibits energy transfer at 25 µM; Shows slight inhibitory activity on electron-transport reactions at concentrations up to 100 µM; Decreases Plabs by 70% at 150 µM; Reduces PSI0, PHI(E0), ET0/CS0, and ET0/RC by 40, 40, 60, and 40%, respectively, at 150 µM; Reduces ET0/CS0, ET0/RC, and RC/CS0 parameters by 30% at 150 µM; Increases dV/dt0, SmK, Kn, and PHI(D0) at 150 µM	6.80	Sampaio et al. (2018)
Photosynthesis inhibition	Photosynthetic electron transport; Photosystem II	Chlorophyll a fluorescence measurements in spinach leaf discs; Polarography assay on noncyclic electron transport in chloroplasts	Inhibits energy transfer at 25 µM; Shows slight inhibitory activity on electron-transport reactions at concentrations up to 100 µM; Decreases Plabs by 70% at 150 µM; Reduces PSI0, PHI(E0), ET0/CS0, and ET0/RC by 40, 40, 60, and 40%, respectively, at 150 µM; Reduces ET0/CS0, ET0/RC, and RC/CS0 parameters by 30% at 150 µM; Increases dV/dt0, SmK, Kn, and PHI(D0) at 150 µM	6.66	Sampaio et al. (2018)

(Continued)

TABLE 6.3 (Continued)

Biological Activity of 4-Quinolinone Alkaloids

Biological Activity	Target	Assay/Method	Results	Compound	References
Photosynthesis inhibition	Photosynthetic electron transport; Photosystem II	Chlorophyll a fluorescence measurements in spinach leaf discs; Polarography assay on noncyclic electron transport in chloroplasts	Inhibits energy transfer at 25 μM; Shows slight inhibitory activity on electron-transport reactions at concentrations up to 100 μM; Decreases PIabs by 70% at 150 μM; Reduces PSI0, PHI(E0), ET0/CS0, and ET0/RC by 40, 40, 60, and 40%, respectively, at 150 μM; Reduces ET0/CS0, ET0/RC, and RC/CS0 parameters by 30% at 150 μM; Increases dV/dt0, SmK, Kn, and PHI(D0) at 150 μM	6.71	Sampaio et al. (2018)
Phytotoxicity	*Lactuca sativa* (lettuce); *Lemna paucicostata* (duckweed); *Allium cepa* (onion)	Seedling growth inhibition; Chlorophyll content measurement; Frond area measurement; Mitotic indexing	Substantially inhibited growth of *L. sativa* seedlings and reduced chlorophyll content at 100 μM; Inhibited growth of *L. paucicostata* at 100 μM; Substantially reduced cell division in *A. cepa* at or below 100 μM.	6.38	Hale et al. (2004)
Plant Growth promoting	*Capsicum annum* (red pepper)	Seed-germination assay, growth increments in plant height, weight, and number of fruits after foliar applications	Growth-promoting effect at 1 ppm concentration	6.42	Moon et al. (2002)

(Continued)

TABLE 6.3 (Continued)
Biological Activity of 4-Quinolinone Alkaloids

Biological Activity	Target	Assay/Method	Results	Compound	References
Plant Growth promoting	C. annum (red pepper)	Seed-germination assay, growth increments in plant height, weight, and number of fruits after foliar applications	Growth-promoting effect at 1ppm concentration	6.27	Moon et al. (2002)
Plant growth promoting	Capsicum annum (red pepper)	Seed-germination assay, growth increments in plant height, weight, and number of fruits after foliar applications	Increased the number of fruits per plant by 44% at 1ppm concentration	6.25	Moon et al. (2002)
Plant-growth regulator	Lolium perenne	In vivo plant-growth experiments; dry biomass determination in L. perenne plants	Decreases dry biomass by 20% at 150 μM similar to the effect of positive control	6.38	Sampaio et al. (2018)
Plant-growth regulator	L. perenne	In vivo plant-growth experiments; dry biomass determination in L. perenne plants	Decreases dry biomass by 23% at 150 μM similar to the effect of positive control	6.80	Sampaio et al. (2018)
Plant-growth regulator	L. perenne	In vivo plant-growth experiments; dry biomass determination in L. perenne plants	Decreases dry biomass by 23% at 150 μM similar to the effect of positive control	6.66	Sampaio et al. (2018)
Plant-growth regulator	L. perenne	in vivo plant-growth experiments; dry biomass determination in L. perenne plants	Decreases dry biomass by 23% at 150 μM similar to the effect of positive control	6.71	Sampaio et al. (2018)

(Continued)

TABLE 6.3 (Continued)
Biological Activity of 4-Quinolinone Alkaloids

Biological Activity	Target	Assay/Method	Results	Compound	References
Radiosensitizing	Human lung adenocarcinoma cells (A549)	Clonogenic assay, Flow cytometric analysis of cell cycle, Flow cytometric analysis of apoptosis	Increased radiosensitivity of A549 cells in a time- and concentration-dependent manner; Enhanced the effect of radiation on A549 cells, allowing for a reduction in radiation dose without compromising efficacy; Increased the percentage of cells in the G2/M phase of the cell cycle; Increased apoptosis in A549 cells when combined with radiation therapy	6.28	Wang et al. (2014)
ROS generation	A375 skin melanoma cells	H2DCFDA staining, flow cytometry in the presence of ascorbic acid	A marked increase in ROS-positive cell population at 4–6 h post-treatment.	6.38	Ghosh et al. (2013)
Vasorelaxant activity	Wistar rat thoracic aorta	Transient relaxation induced by activation of the Na^+ pump, cyclic nucleotide (cAMP, cGMP) content	Inhibits Ca^{2+} influx through voltage-dependent calcium channels, inhibited the contraction induced by 60 mM K^+ with an IC_{50} of 9.8 μM (21.6% relaxation)	6.77	Yamahara et al. (1988)

The appearance of natural bacterial strains that do not respond to quinolinone treatment is threatening our continued ability to use quinolinone. Therefore, developing new derivatives and combinations that can circumvent the resistance mechanisms is urgent. In addition, the environmental build-up of these substances may harm ecosystems and human health. Therefore, we need to develop safer drugs and improve waste treatment technology. The recurrent global searches for antimicrobial and antiviral agents have become more popular due to drug-resistant pathogens. Exploring the substances as potential antiviral agents may have a greater therapeutic use (Wang et al. 2019). There is a need for further work to optimize synthetic methods, improve pharmacokinetics, and evaluate the toxicity of the compounds for their suitability in clinical use. Advanced drug design techniques, like scaffold hopping, could generate new analogues that give better efficacy and fewer side effects. Studying 4-quinolinones' biosynthetic pathways may help develop insights into their natural production and biotechnological approaches for sustainable synthesis. 4-Quinolinones are valuable models for studying microbe-microbe interaction, antibiotic resistance, and environmental processes. Research on them continues to improve, and we can expect more innovations from them as they lend themselves to novel treatments and global health strategies. Scientists can help keep 4-quinolinones useful in medicine, biotechnology, and ecological research by balancing their therapeutic potential with responsible and efficient use while considering their environmental effects.

REFERENCES

Adams, M., Kunert, O., Haslinger, E., & Bauer, R. (2004). Inhibition of leukotriene biosynthesis by quinolinone alkaloids from the fruits of *Evodia rutaecarpa*. *Planta Medica, 70*(10), 904. 10.1055/s-2004-832614

Adams, M., Wube, A. A., Bucar, F., Bauer, R., Kunert, O., & Haslinger, E. (2005). Quinolinone alkaloids from *Evodia rutaecarpa*: A potent new group of antimycobacterial compounds. *International Journal of Antimicrobial Agents, 26*, 262–264.

Adams, M., Mahringer, A., Kunert, O., Fricker, G., Efferth, T., & Bauer, R. (2007). Cytotoxicity and P-glycoprotein modulating effects of quinolinones and indoloquinazolines from the chinese herb *Evodia rutaecarpa*. *Planta Medica, 73*(15), 1554. 10.1055/s-2007-993743

Agier, C., Bury, M., Aquette, J., Hocquemiller, R., & Waterman, P. G. (2007). New pseudan (2-alkyl-4 (1H)-quinolinone) alkaloids from *Boronia ternata* var. *elongata* and *Boronia alulata* (Rutaceae). *Natural Product Research, 21*(8), 698–703.

Aher, V., Chattopadhyay, P., Goyary, D., & Veer, V. (2013). Evaluation of the genotoxic and antigenotoxic potential of the alkaloid punarnavine from *boerhaavia diffusa*. *Planta Medica, 79*(11), 939. 10.1055/s-0032-1328717

Ahmad, I., Khalid, H., Perveen, A., Shehroz, M., Nishan, U., Rahman, F. U., … & Ojha, S. C. (2024). Identification of novel quinolinone and quinazoline alkaloids as phosphodiesterase 10A inhibitors for Parkinson's disease through a computational approach. *ACS Omega, 9*(14), 16262–16278.

Ahmed, M. S., & Akter, I. (2024). Green strategies for the synthesis of quinolinone derivatives. https://www.qeios.com/read/P5M2Z8

Ahsan, M., Gray, A. I., Leach, G., & Waterman, P. G. (1993). Quinolinone and acridone alkaloids from *Boronia lanceolata*. *Phytochemistry, 33*(6), 1507–1510.

Ahsan, M., Gray, A. I., Waterman, P. G., & Armstrong, J. A. (1994). 4-Quinolinone, 2-quinolinone, 9-acridanone, and furoquinoline alkaloids from the aerial parts of *Boronia bowmanii*. *Journal of Natural Products, 57*(5), 670–672.

Ainiwaer, P., Nueraihemaiti, M., Li, Z., Zang, D., Jiang, L., Li, Y., & Aisa, H. A. (2021). Chemical constituents of *Ruta graveolens* L. and their melanogenic effects and action mechanism. *Fitoterapia.* doi: 15610.1016/j.fitote.2021.105094

Akhmedzhanova, V. I., Angenot, L., & Shakirov, R. S. (2010). Alkaloids from *Haplophyllum leptomerum. Chemistry of Natural Compounds, 46,* 502–503.

Akhmedzhanova, V. I., Bessonova, I. A., & Yunusov, S. Y. (1986). Alkaloids of *Haplophyllum leptomerum.* I. the structure of leptomerine. *Chemistry of Natural Compounds, 22,* 78–79.

Al-Khalil, S., Alkofahi, A., El-Eisawi, D., & Al-Shibib, A. (1998). Transtorine, a new quinoline alkaloid from *Ephedra transitoria. Journal of Natural Products, 61*(2), 262–263.

Aly, A. A., Ramadan, M., Abuo-Rahma, G. E. A., Elshaier, Y. A. M. M., Elbastawesy, M. A. I., Brown, A. B., & Bräse, S. (2021). Quinolinones as prospective drugs: Their syntheses and biological applications. *Advances in Heterocyclic Chemistry, 147.* doi: 10.1016/bs.aihch.2020.08.001

Ambrozin, A. R. P., Vieira, P. C., Fernandes, J. B., Silva, M. F. D. G. F. D., & Albuquerque, S. D. (2008). Piranoflavonas inéditas e atividades tripanocidas das substâncias isoladas de *Conchocarpus heterophyllus. Química Nova, 31,* 740–743.

Arar, G., Gözler, T., Bashir, M., & Shamma, M. (1985). Malatyamine, a 4-quinolinone alkaloid from *Haplophyllum cappadocicum. Journal of Natural Products, 48*(4), 642–643.

Arthur, H. R., & Cheung, H. T. (1960). An examination of the Rutaceae of Hong Kong. VI. graveoline, A new alkaloid from *Ruta graveolens. Australian Journal of Chemistry, 13*(4). doi: 10.1071/ch9600510

Arutiunov, N. A., Zatsepilina, A. M., Aksenova, A. A., Aksenov, N. A., Aksenov, D. A., Leontiev, A. V., & Aksenov, A. V. (2024). One-pot synthesis of N-fused quinolinone-4 tetracyclic scaffolds from 2, 2-disubstituted indolin-3-ones. *ACS Omega, 9*(45), 45501–45517.

Asakawa, A. H., & Manetsch, R. (2021). A comprehensive review of 4 (1H)-quinolinones and 4 (1H)-pyridones for the development of an effective antimalarial. In Tyagi, R. K. (ed.), *Plasmodium Species and Drug Resistance.* DOI: 10.5772/intechopen.97084

Awaad, A. S., Maitland, D. J., & Moneir, S. M. (2007). New alkaloids from *Casimiroa edulis* fruits and their pharmacological activity. *Chemistry of Natural Compounds, 43,* 576–580.

Barrera, C. A. C., Barrera, E. D. C., Falla, D. S. G., Murcia, G. D., & Suarez, L. E. C. (2011). Seco-limonoids and quinoline alkaloids from *Raputia heptaphylla* and their antileishmanial activity. *Chemical and Pharmaceutical Bulletin, 59*(7), 855–859.

Bellete, B. S., Sá, I. C. G. D., Mafezoli, J., Cerqueira, C. D. N., Silva, M. F. D. G. F. D., Fernandes, J. B., … & Pirani, J. R. (2012). Fitoquímica e quimiossistemática de *Conchocarpus marginatus* e *C. inopinatus* (Rutaceae). *Química Nova, 35,* 2132–2138.

Bian, X., Bai, J., Sun, K., Huang, S., Wang, K., Tang, S., Xue, C., Hu, G., Wu, X., Hua, H., & Pei, Y. (2016). Trametramide A, a new pyridone alkaloid from the fungus *Trametes trogii* TGC-P-3. *Magnetic Resonance in Chemistry, 54*(9). doi: 10.1002/mrc.4451

Biavatti, M. W., Vieira, P. C., Silva, M., Fernandes, J. B., Victor, S. R., Pagnocca, F. C., Albuquerque, S., Caracelli, I., & Zukerman-Schpector, J. (2002). Biological activity of quinoline alkaloids from *Raulinoa echinata* and X-ray structure of flindersiamine. *Journal of the Brazilian Chemical Society, 13,* 66–70.

Boteva, A. A., & Krasnykh, O. P. (2009). The methods of synthesis, modification, and biological activity of 4-quinolinones. *Chemistry of Heterocyclic Compounds, 45,* 757–785.

Boyd, D. R., & Grundon, M. F. (1970). Quinoline alkaloids. Part X.(+)-Platydesminium salt and other alkaloids from *Skimmia japonica* Thunb. The synthesis of edulinine. *Journal of the Chemical Society C: Organic, 1970*(4), 556–558.

Buckingham, J. (2023). *Dictionary of natural products, supplement 2.* Boca Raton, FL: Routledge.

Buckingham, J., Baggaley, K. H., Roberts, A. D., & Szabo, L. F. (2010). *Dictionary of alkaloids with CD-ROM*. Boca Raton, FL: CRC Press.

Budzikiewicz, H., Schaller, U., Korth, H., & Pulverer, G. (1979). Bakterieninhaltsstoffe, VI: Alkylchinoline und deren N-oxide aus *Pseudomonas aeruginosa*. *Monatshefte für Chemie-Chemical Monthly*, *110*(4), 947–953.

Bultel-Poncé, V., Berge, J., Debitus, C., Nicolas, J., & Guyot, M. (1999). Metabolites from the sponge-associated bacterium *pseudomonas* species. *Marine Biotechnology, 1*(4), 384–390.

Cabral, R. S., Allard, P., Marcourt, L., Young, M. C. M., Queiroz, E. F., & Wolfender, J. (2016). Targeted isolation of indolopyridoquinazoline alkaloids from *Conchocarpus fontanesianus* based on molecular networks. *Journal of Natural Products, 79*(9), 2270. doi: 10.1021/acs.jnatprod.6b00379

Cardoso-Lopes, E. M., Maier, J. A., Silva, M. R. D., Regasini, L. O., Simote, S. Y., Lopes, N. P., Pirani, J. R., Bolzani, V. D. S., & Young, M. C. M. (2010). Alkaloids from stems of *Esenbeckia leiocarpa* Engl. (Rutaceae) as potential treatment for alzheimer disease. *Molecules, 15*(12), 9205. doi: 10.3390/molecules15129205

Carnovali, M., Ciavatta, M. L., Mollo, E., Roussis, V., Banfi, G., Carbone, M., & Mariotti, M. (2022). Aerophobin-1 from the marine sponge *Aplysina aerophoba* modulates osteogenesis in zebrafish larvae. *Marine Drugs*, 20(2), 135.

Carvalho, L. S. A. D., Queiroz, L. S., Alves Junior, I. J., Almeida, A. D. C., Coimbra, E. S., De Faria Pinto, P., Silva, M. P. N. D., De Moraes, J., & Da Silva Filho, A. A. (2019). In vitro schistosomicidal activity of the alkaloid-rich fraction from *Ruta graveolens* L. (Rutaceae) and its characterization by UPLC-QTOF-MS. *Evidence-Based Complementary and Alternative Medicine, 2019*(1). doi: 10.1155/2019/7909137

Chaudhuri, P. K. (1987). Echinozolinone, an alkaloid from *Echinops echinatus*. *Phytochemistry, 26*(2), 587–589.

Chen, C., Chen, W., Pang, X., Liao, S., Wang, J., Lin, X., Yang, B., Zhou, X., Luo, X., & Liu, Y. (2021). Pyrrolyl 4-quinolinone alkaloids from the mangrove endophytic fungus *Penicillium steckii SCSIO 41025*: Chiral resolution, configurational assignment, and enzyme inhibitory activities. Phytochemistry. doi: 10.1016/j.phytochem.2021.112730

Cheng, C., Othman, E. M., Reimer, A., Grüne, M., Kozjak-Pavlovic, V., Stopper, H., Hentschel, U., & Abdelmohsen, U. R. (2016). Ageloline A, new antioxidant and antichlamydial quinolinone from the marine sponge-derived bacterium *Streptomyces sp*. SBT345. *Tetrahedron Letters, 57*(25), 2786. doi: 10.1016/j.tetlet.2016.05.042

Cheng, W., Pan, Z., Zheng, H., Luo, G., Liu, Z., Xu, S., & Lin, J. (2023). Characterization of phytochemical profile of rhizome of artificial cultured *polygonatum sibiricum* with multiple rhizome buds. Applied Biological Chemistry, *66*(1). doi: 10.1186/s13765-023-00792-4

Cho, J., Bae, S., Kim, H., Lee, M., Choi, Y., Jin, B., Lee, H. J., Jeong, H. Y., Lee, Y. G., & Moon, J. (2015). New quinolinone alkaloids from chestnut (*Castanea crenata* Sieb) honey. *Journal of Agricultural and Food Chemistry, 63*(13), 3587. doi: 10.1021/acs.jafc.5b01027

Cho, J., Hwang, T., Chang, T., Lim, Y., Sung, P., Lee, T., & Chen, J. (2012a). New coumarins and anti-inflammatory constituents from *Zanthoxylum avicennae*. *Food Chemistry, 135*(1), 17. doi: 10.1016/j.foodchem.2012.04.025

Cho, J. Y. (2012b). Algicidal activity of marine Alteromonas sp. KNS-16 and isolation of active compounds. *Bioscience, Biotechnology, and Biochemistry*, 76(8), 1452–1458.

Chuang, W. C., Chu, C. Y., & Sheu, S. J. (1996). Determination of the alkaloids in *Evodiae Fructus* by high-performance liquid chromatography. *Journal of Chromatography A, 727*(2), 317–323.

Chuang, W. C., Cheng, C. M., Chang, H. C., Chen, Y. P., & Sheu, S. J. (1999). Contents of constituents in mature and immature fruits of Evodia species. *Planta Medica*, 65(6), 567–571.

Cimino, G., De Rosa, S., De Stefano, S., Spinella, A., & Sodano, G. (1984). The zoochrome of the sponge *Verongia aerophoba* ("Uranidine"). *Tetrahedron Letters*, *25*(27), 2925–2928.

Clark, B., Capon, R. J., Lacey, E., Tennant, S., & Gill, J. H. (2006). Quinolactacins revisited: From lactams to imide and beyond. Organic & Biomolecular Chemistry, *4*(8). doi: 10.1039/b600959j

Cornforth, J. W., & James, A. T. (1956). Structure of a naturally occurring antagonist of dihydrostreptomycin. *Biochemical Journal*, *63*(1), 124.

Cretton, S., Breant, L., Pourrez, L., Ambuehl, C., Marcourt, L., Ebrahimi, S. N., … & Christen, P. (2014). Antitrypanosomal quinoline alkaloids from the roots of *Waltheria indica*. *Journal of Natural Products*, *77*(10), 2304–2311.

Cretton, S., Dorsaz, S., Azzollini, A., Favre-Godal, Q., Marcourt, L., Ebrahimi, S. N., Voinesco, F., Michellod, E., Sanglard, D., Gindro, K., Wolfender, J., Cuendet, M., & Christen, P. (2016). Antifungal quinoline alkaloids from *Waltheria indica*. *Journal of Natural Products*, *79*(2), 300. doi: 10.1021/acs.jnatprod.5b00896

Cretton, S., Kaiser, M., Karimou, S., Ebrahimi, S. N., Maser, P., Cuendet, M., & Christen, P. (2020). Pyridine-4 (1 H)-one alkaloids from *Waltheria indica* as antitrypanosomatid agents. *Journal of Natural Products*, *83*(11), 3363–3371.

Daneshtalab, M., & Ahmed, A. (2012). Nonclassical biological activities of quinolinone derivatives. *Journal of Pharmacy & Pharmaceutical Sciences*, *15*(1), 52–72.

Dat, T. T. H., Cuong, L. C. V., Ha, D. V., Oanh, P. T. T., Nhi, N. P. K., Anh, H. L. T., Quy, P. T., Bui, T. Q., Triet, N. T., & Nhung, N. T. A. (2022). The study on biological activity and molecular docking of secondary metabolites from *Bacillus* sp. isolated from the mangrove plant *Rhizophora apiculata* Blume. Regional Studies in Marine Science. doi: 10.1016/j.rsma.2022.102583

Datkhayev, U. M. (2017). Phytochemical investigation and technology production of alkaloids in the Kazakh endemic plant *Echinops albicaulis* Kar. Et Kir.(Asteraceae). *International Journal of Green Pharmacy (IJGP)*, *11*(2), S312–S319.

de Medeiros Silva, R., De Castro Lima, M. M., & Cotinguiba, F. (2024). Dereplication of 4-Quinolinone alkaloids from *waltheria indica* (malvaceae) tissues using molecular network tools. *Chemistry & Biodiversity*, *21*(8). doi: 10.1002/cbdv.202400665

Debitus, C., Guella, G., Mancini, I. I., Waikedre, J., Guemas, J., Nicolas, J. L., & Pietra, F. (1998). Quinolinones from a bacterium and tyrosine metabolites from its host sponge, *Suberea creba* from the Coral Sea. *Journal of Marine Biotechnology*, *6*(3), 136–141.

Dekker, K. A., Inagaki, T., Gootz, T. D., Huang, L. H., Kojima, Y., Kohlbrenner, W. E., Matsunaga, Y., McGuirk, P. R., Nomura, E., & Sakakibara, T. (1998). New quinolinone compounds from *Pseudonocardia* sp. with selective and potent anti-*Helicobacter pylori* activity: Taxonomy of producing strain, fermentation, isolation, structural elucidation and biological activities. *The Journal of Antibiotics*, *51*(2), 145–152.

Deng, L., Shi, A., Liu, H., Meruva, N., Liu, L., Hu, H., Yang, Y., Huang, C., Li, P., & Wang, Q. (2016). Identification of chemical ingredients of peanut stems and leaves extracts using UPLC-QTOF-MS coupled with novel informatics UNIFI platform. *Journal of Mass Spectrometry*, *51*(12), 1157. 10.1002/jms.3887

Dhiman, P., Arora, N., Thanikachalam, P. V., & Monga, V. (2019). Recent advances in the synthetic and medicinal perspective of quinolinones: A review. Bioorganic Chemistry. doi: 10.1016/j.bioorg.2019.103291

Di, L., Shi, Y., Yan, Y., Jiang, L., Hou, B., Wang, X., Zuo, Z., Chen, Y., Yang, C., & Cheng, Y. (2015). Nonpeptide small molecules from the insect *Aspongopus chinensis* and their neural stem cell proliferation stimulating properties. *RSC Advances*, *5*(87), 70985. doi: 10.1039/c5ra12920f

Diggle, S. P., Cornelis, P., Williams, P., & Cámara, M. (2006a). 4-quinolinone signalling in pseudomonas aeruginosa: Old molecules, new perspectives. *International Journal of Medical Microbiology*, *296*(2–3), 83. doi: 10.1016/j.ijmm.2006.01.038

Diggle, S. P., Lumjiaktase, P., Dipilato, F., Winzer, K., Kunakorn, M., Barrett, D. A., Chhabra, S. R., Cámara, M., & Williams, P. (2006b). Functional genetic analysis reveals a 2-alkyl-4-quinolinone signaling system in the human pathogen *Burkholderia pseudomallei* and related bacteria. *Chemistry & Biology, 13*(7), 701. doi: 10.1016/j.chembiol.2006.05.006

Dike, C. C., Ezeonu, F. C., Maduka, H. C. C., & Ezeokafor, E. N. (2018). Phytochemical and elemental analysis of *Sida corymbosa* (broom weed or wire weed) leaf and root extracts. *Chemical Science International Journal, 24*(4), 1. doi: 10.9734/csji/2018/44987

Dine, I., Mulugeta, E., Melaku, Y., & Belete, M. (2023). Recent advances in the synthesis of pharmaceutically active 4-quinolinone and its analogues: A review. *RSC Advances, 13*(13), 8657. doi: 10.1039/d3ra00749a

Domagala, J. M. (1994). Structure-activity and structure-side-effect relationships for the quinolinone antibacterials. *Journal of Antimicrobial Chemotherapy*, 33(4), 685–706.

Dube, P. S., Legoabe, L. J., & Beteck, R. M. (2022). Quinolinone: A versatile therapeutic compound class. *Molecular Diversity, 27*(3), 1501. doi: 10.1007/s11030-022-10581-8

Duffield, A. M., & Jefferies, P. R. (1963). The chemistry of the western australian rutaceae. III. The alkaloids of *Boronia ternata* Endl. *Australian Journal of Chemistry, 16*(2), 292–294.

El Sayed, K., Al-Said, M. S., El-Feraly, F. S., & Ross, S. A. (2000). New quinoline alkaloids from *Ruta chalepensis*. *Journal of Natural Products, 63*(7), 995. doi: 10.1021/np000012y

Elsebai, M. F., Rempel, V., Schnakenburg, G., Kehraus, S., Müller, C. E., & König, G. M. (2011). Identification of a potent and selective cannabinoid CB1Receptor antagonist from *Auxarthron reticulatum*. *ACS Medicinal Chemistry Letters, 2*(11), 866. doi: 10.1021/ml200183z

Enenebeaku, U. E., Duru, C. E., Okotcha, E. N., Ogidi, O. I., Mgbemena, I. C., Nwigwe, H. C., & Enenebeaku, C. K. (2022). Phytochemical analysis and antioxidant evaluation of crude extracts from the roots, stem and leaves of *Dictyandra arborescens* (Welw.). *Tropical Journal of Natural Product Research, 6*(1), 62. doi: 10.26538/tjnpr/v6i1.12

Eshonov, M. A., Rasulova, K. A., Turgunov, K. K., & Tashkhodzhaev, B. (2020). New quinoline alkaloid acusine and crystal structures of N-methyl-2-phenylquinolin-4-one and pedicine from *Haplophyllum acutifolium*. *Chemistry of Natural Compounds, 56*(6), 1102. doi: 10.1007/s10600-020-03236-3

Evans, J. R., Napier, E. J., & Fletton, R. A. (1978). G1499-2, a new quinoline compound isolated from the fermentation broth of *Cytophaga johnsonii*. *The Journal of Antibiotics*, 31(10), 952–958.

Fahey, R. C., Newton, G. L., Bewley, C. A., & Nicholas, G. (2010). Inhibitors of acyl glucosaminyl inositol amidase and methods of use. *U.S. Patent* No. 7,642,280.

Felix, D., Ogunka-Nnoka, C. U., & O. Wellington, E. (2022). Phytochemical composition and hepatoprotective potential of ethanolic root extract of *Jatropha curcas* in acetaminophen-induced toxicity in albino wistar rats. *Journal of Applied Life Sciences International, 9*. doi: 10.9734/jalsi/2022/v25i530304

Feng, W. S., Li, Q., Zheng, X. K., & Kuang, H. X. (2007). A new alkaloid from the aerial part of *Rabdosia rubescens*. *Chinese Journal of Natural Medicines, 5*(2), 92–95.

Ferretti, F., Fouad, M. A., Abbo, C., & Ragaini, F. (2023). Effective synthesis of 4-quinolinones by reductive cyclization of 2′-nitrochalcones using formic acid as a CO surrogate. *Molecules, 28*(14), 5424.

Fokialakis, N., Magiatis, P., Mitaku, S., Tillequin, F., & Sevenet, T. (2000a). Two new 3-methoxy-4-quinolinone alkaloids from the bark of *Sarcomelicope megistophylla*. *Chemical and Pharmaceutical Bulletin, 48*(12), 2009–2010.

Fokialakis, N., Magiatis, P., Skaltsounis, A., Tillequin, F., & Sévenet, T. (2000b). The structure of sarcomejine: An application of long-range 1H−15N correlation at natural abundance. *Journal of Natural Products, 63*(7). doi: 10.1021/np000083x

Forghe, B. N., & Nna, P. J. (2020). Phytochemical screening and antimicrobial activities of methanolic extract of *Newbouldia laevis* roots. *World Journal of Pharmaceutical Research, 9*(4), 73–83.

Funayama, S., Noshita, T., Mori, T., Kashiwagura, T., & Murata, K. (1999). Quinoline alkaloids from *Orixa japonica* (Johzan)-isolation of an important biosynthetic intermediate. *Symposium on the Chemistry of Natural Products, 41,* 451–456.

Gach-Janczak, K., Piekielna-Ciesielska, J., Waśkiewicz, J., Krakowiak, K., Wtorek, K., & Janecka, A. (2025). Quinolin-4-ones: Methods of synthesis and application in medicine. *Molecules, 30*(1), 163.

Gao, H., Zhang, L., Zhu, T., Gu, Q., & Li, D. (2012). Unusual pyrrolyl 4-quinolinone alkaloids from the marine-derived fungus *Penicillium* sp. ghq208. *Chemical and Pharmaceutical Bulletin, 60*(11), 1458–1460.

Gao, L. M., Wu, L. J., Huang, J., Sun, B. H., & Gao, H. Y. (2010). Chemical constituents from the flowers of *Castanea mollissima* Blume. *Journal of Shenyang Pharmaceutical University, 27,* 544–547.

Ghosh, P., & Das, S. (2019). Synthesis and functionalization of 4-quinolinones – A progressing story. *European Journal of Organic Chemistry, 2019*(28), 4466. doi: 10.1002/ejoc.201900452

Ghosh, S., Bishayee, K., & Khuda-Bukhsh, A. R. (2013). Graveoline isolated from ethanolic extract of *Ruta graveolens* triggers apoptosis and autophagy in skin melanoma cells: A novel apoptosis-independent autophagic signaling pathway. *Phytotherapy Research, 28*(8), 1153. doi: 10.1002/ptr.5107

Gibbons, S., Craven, L., Dunlop, C., Gray, A. I., Hartley, T. G., & Waterman, P. G. (1997). The secondary metabolites of aff. *Samadera* SAC-2825: An Australian simaroubaceae with unusual chemistry. *Phytochemistry,* 44(6), 1109–1114.

Goodwin, S., Smith, A. F., Velasquez, A. A., & Horning, E. C. (1959). Alkaloids of *Lunasia amara* Blanco. Isolation studies. *Journal of the American Chemical Society, 81*(23), 6209–6213.

Greshoff, M. (1900). Recherches sur l'Echinopsine, nouvel alcaloïde cristallisé. *Recueil Des Travaux Chimiques Des Pays-Bas Et De La Belgique, 19*(11), 360. doi: 10.1002/recl.19000191104

Grundon, M. F., & Okely, H. M. (1979). A new quinoline alkaloid from *Ruta graveolens. Phytochemistry, 18,* 1768–1769.

Guilhon, G. M., Baetas, A. C. S., Maia, J. G. S., & Conserva, L. M. (1994). 2-alkyl-4-quinolinone alkaloids and cinnamic acid derivatives from *Esenbeckia almawillia. Phytochemistry, 37*(4), 1193–1195.

Gulyamova, D. M., Bessonova, I. A., & Yunusov, S. Y. (1971). Alkaloids of *Haplophyllum acutifolium. Chemistry of Natural Compounds, 7*(6), 836–837.

Guo, L., Zhang, Y., He, H., Li, Y., Yu, J., & Hao, X. (2012). A new monoterpenoid indole alkaloid from *Ervatamia chinensis. Chinese Journal of Natural Medicines, 10*(3), 226. doi: 10.3724/SP.J.1009.2012.00226

Guo, M., Wu, D., Yang, H., Zhang, X., Xue, D., & Zhang, W. (2024). Enhanced selectivity in 4-quinolinone formation: A dual-base system for palladium-catalyzed carbonylative cyclization with $Fe(CO)_5$. *Molecules, 29*(4), 850.

Hale, A. L., Meepagala, K. M., Oliva, A., Aliotta, G., & Duke, S. O. (2004). Phytotoxins from the leaves of *Ruta graveolens. Journal of Agricultural and Food Chemistry, 52*(11), 3345. doi: 10.1021/jf0497298

Halim, A. F., Afify, M. S., Ahmed, A. F., Mira, A. S., & Mira, A. S. (2011). The fact about echinopsine and first isolation of echinorine from *Echinops spinosus* L. *Journal of Environmental Sciences, 40*(2), 173–181.

Hamasaki, N., Ishii, E., Tominaga, K., Tezuka, Y., Nagaoka, T., Kadota, S., Kuroki T. & Yano, I. (2000). Highly selective antibacterial activity of novel alkyl quinolinone alkaloids from a Chinese herbal medicine, Gosyuyu (Wu-Chu-Yu), against *Helicobacter pylori in vitro. Microbiology and Immunology,* 44(1), 9–15.

Hamed, A., El-Metwally, M. M., Frese, M., Ibrahim, T. M., El-Haddad, A. F., Sewald, N., & Shaaban, M. (2017). Diverse bioactive secondary metabolites from *Penicillium* sp. 1P. *Journal of Atoms and Molecules, 7*(5), 1121–1132.

Han, X. H., Hong, S. S., Lee, D., Lee, J. J., Lee, M. S., Moon, D. C., ... & Hwang, B. Y. (2007). Quinolinone alkaloids from *Evodiae fructus* and their inhibitory effects on monoamine oxidase. *Archives of Pharmacal Research, 30*, 397–401.

Hart, N. K., Johns, S. R., Lamberton, J. A., & Price, J. R. (1968). Alkaloids of the australian rutaceae: *Lunasia quercifolia*. IV. identification of a minor constituent as 5-hydroxy-1-methyl-2-phenyl-4-quinolinone and the preparation of an angular isomer of (−)-lunine. *Australian Journal of Chemistry, 21*(5), 1389–1391.

Heeb, S., Fletcher, M. P., Chhabra, S. R., Diggle, S. P., Williams, P., & Cámara, M. (2010). Quinolinones: From antibiotics to autoinducers. *FEMS Microbiology Reviews, 35*(2), 247. doi: 10.1111/j.1574-6976.2010.00247.x

Heidarnezhad, Z., & Obidov, Z. (2013). A DFT study on tautomer stability of 4-hydroxyquinoline considering solvent effect and NBO analysis. *Chemical Science Transactions, 2*(4). doi: 10.7598/cst2013.448

Horta, P., Kuş, N., Henriques, M. S. C., Paixão, J. A., Coelho, L., Nogueira, F., O'neill, P. M., Fausto, R., & Cristiano, M. L. S. (2015). Quinolinone–hydroxyquinoline tautomerism in quinolinone 3-esters. preserving the 4-oxoquinoline structure to retain antimalarial activity. *The Journal of Organic Chemistry, 80*(24), 12244. doi: 10.1021/acs.joc.5b02169

Hua, Y., Zeng, K., Liang, H., Liang, H., Jiang, Y., & Tu, P. (2023). Anti-inflammatory quinoline-4(1H)-one derivatives from the aerial parts of *Waltheria indica* Linn. Phytochemistry. doi: 10.1016/j.phytochem.2023.113746

Huang, X., Li, W., & Yang, X. (2012). New cytotoxic quinolinone alkaloids from fruits of *Evodia rutaecarpa*. *Fitoterapia, 83*(4), 709. doi: 10.1016/j.fitote.2012.02.009

Huse, H., & Whiteley, M. (2010). 4-quinolinones: Smart phones of the microbial world. *Chemical Reviews, 111*(1), 152. doi: 10.1021/cr100063u

Hussein, S. A., Barakat, H. H., Nawar, M. A., & Willuhn, G. (1997). Flavonoids from *Ephedra aphylla*. *Phytochemistry, 45*(7), 1529–1532.

Ihenetu, S. C., Ibe, F. C., & Inyamah, P. C. (2021). Comparative study of the properties of yellow and brown *Cyperus esculentus* L. *World News of Natural Sciences, 35*, 25–37.

Iriarte, J., Kincl, F. A., Rosenkranz, G., & Sondheimer, F. (1956). 805. the constituents of *casimiroa edulis* llave *et* lex. Part II. The bark. Journal of the Chemical Society. doi: 10.1039/jr9560004170

Ishi, H., Oida, H., & Haginiwa, J. (1972). Studies on the alkaloids of rutaceous plants. XIX. The chemical constituents of *Xanthoxylum inerme* Koidz. (*Fagara boninensis* Koidz.). 1. Isolation of the chemical constituents from bark and wood. *Yakugaku Zasshi: Journal of the Pharmaceutical Society of Japan, 92*(2), 118–128.

Ito, A., Shamon, L. A., Yu, B., Mata-Greenwood, E., Lee, S. K., Van Breemen, R. B., ... & Kinghorn, A. D. (1998). Antimutagenic constituents of *Casimiroa edulis* with potential cancer chemopreventive activity. *Journal of Agricultural and Food Chemistry, 46*(9), 3509–3516.

Jaiswal, S., Arya, N., Yaduvanshi, N., Devi, M., Jain, S., Jain, S., Dwivedi, J., & Sharma, S. (2023). Current updates on green synthesis and biological properties of 4-quinolinone derivatives. *Journal of Molecular Structure, 1294*. doi: 10.1016/j.molstruc.2023.136565

Ji, K. L., Liu, W., Yin, W. H., Li, J. Y., & Yue, J. M. (2022). Quinoline alkaloids with anti-inflammatory activity from *Zanthoxylum avicennae*. *Organic & Biomolecular Chemistry, 20*(20), 4176–4182.

Jian, J. Y., Fan, Y. M., Liu, Q., Jin, J., Yuan, C. M., Gu, W., ... & Hao, X. J. (2023). Quinoline alkaloids from the roots of *Orixa japonica* with the anti-pathogenic fungi activities. *Chemistry & Biodiversity, 20*(2), e202201097.

Jin, H. Z., Lee, J. H., Lee, D., Lee, H. S., Hong, Y. S., Kim, Y. H., & Lee, J. J. (2004). Quinolinone alkaloids with inhibitory activity against nuclear factor of activated T cells from the fruits of *Evodia rutaecarpa*. *Biological and Pharmaceutical Bulletin, 27*(6), 926–928.

Johnstone, R., Price, J. R., & Todd, A. R. (1958). Alkaloids of the australian rutaceae: *Lunasia quercifolia*. I. 7-methoxy-1-methyl-2–4-quinolinone. *Australian Journal of Chemistry, 11*(4), 562–574.

Kamal, L. Z. M., Adam, M. A. A., Shahpudin, S. N. M., Shuib, A. N., Sandai, R., Hassan, N. M., & Sandai, D. (2021). Identification of alkaloid compounds Arborinine and Graveoline from *Ruta angustifolia* (L.) Pers for their antifungal potential against Isocitrate lyase (ICL 1) gene of *Candida albicans*. Mycopathologia, *186*, 221–236.

Kamigiri, K., Tokunaga, T., Shibazaki, M., Setiawan, B., Morioka, M., Suzuki, K., & Rantiatmodjo, R. M. (1996). YM-30059, a novel quinolinone antibiotic produced by *arthrobacter* sp. *The Journal of Antibiotics, 49*(8), 823–825.

Kamikado, T., Chang, C. F., Murakoshi, S., Sakurai, A., & Tamura, S. (1976). Isolation and structure elucidation of three quinolinone alkaloids from *Evodia rutaecarpa*. *Agricultural and Biological Chemistry, 40*(3), 605–609.

Kamikado, T., Murakoshi, S., & Tamura, S. (1978). Structure elucidation and synthesis of alkaloids isolated from fruits of *Evodia rutaecarpa*. *Agricultural and Biological Chemistry, 42*(8), 1515–1519.

Kan-Fan, C., Das, B. C., Boiteau, P., & Potier, P. (1970). Alcaloïdes de *Vepris ampody* (rutacées). *Phytochemistry, 9*(6), 1283–1291.

Kang, O., Park, S. J., Ahn, H., Jeong, K. C., & Lim, H. J. (2019). Structural assignment of the enol–keto tautomers of one-pot synthesized 4-hydroxyquinolines/4-quinolinones. *Organic Chemistry Frontiers, 6*(2), 183. doi: 10.1039/c8qo00884a

Kapadia, G. J., Shukla, Y. N., & Basak, S. P. (1978). Melovinone, an open chain analogue of melochinone from *Melochia tomentosa*. *Phytochemistry, 17*, 1444–1445.

Karim, M. R., Fukaya, K., In, Y., Sharma, A. R., Harunari, E., Oku, N., Urabe, D., Trianto, A., & Igarashi, Y. (2022). Marinoquinolinones and marinobactoic acid: Antimicrobial and cytotoxic *ortho*-dialkylbenzene-class metabolites produced by a marine obligate gammaproteobacterium of the genus *Marinobacterium*. *Journal of Natural Products, 85*(7), 1763. doi: 10.1021/acs.jnatprod.2c00281

Kawada, M., Inoue, H., Ohba, S., Hatano, M., Amemiya, M., Hayashi, C., Usami, I., Abe, H., Watanabe, T., Kinoshita, N., Igarashi, M., Masuda, T., Ikeda, D., & Nomoto, A. (2013). Intervenolin, a new antitumor compound with anti-*Helicobacter pylori* activity, from *Nocardia* sp. ML96-86F2. *The Journal of Antibiotics, 66*(9), 543. doi: 10.1038/ja.2013.42

Khaleel, A. E. (2002). 2-Phenyl-4-quinolinone alkaloids from *Casimiroa edulis* Llave et Lex (Rutaceae). *Monatshefte für Chemie/Chemical Monthly, 133*, 183–187.

Khan, M. S., Gao, J., Chen, X., Zhang, M., Yang, F., Du, Y., Moe, T. S., Munir, I., Xue, J., & Zhang, X. (2020). Isolation and characterization of plant growth-promoting endophytic bacteria *Paenibacillus polymyxa* SK1 from *Lilium lancifolium*. *BioMed Research International, 2020*, 1. doi: 10.1155/2020/8650957

Kilani-Feki, O., Culioli, G., Ortalo-Magné, A., Zouari, N., Blache, Y., & Jaoua, S. (2011a). Environmental *Burkholderia cepacia* strain Cs5 acting by two analogous alkyl-quinolinones and a didecyl-phthalate against a broad spectrum of phytopathogens fungi. *Current Microbiology, 62*(5), 1490. doi: 10.1007/s00284-011-9892-6

Kilani-Feki, O., Zouari, I., Culioli, G., Ortalo-Magné, A., Zouari, N., Blache, Y., & Jaoua, S. (2011b). Correlation between synthesis variation of 2-alkylquinolinones and the antifungal activity of a *Burkholderia cepacia* strain collection. *World Journal of Microbiology and Biotechnology, 28*(1), 275. doi: 10.1007/s11274-011-0817-0

Kim, D., Lee, H., Ko, W., Lee, D., Sohn, J. H., Yim, J. H., Kim, Y., & Oh, H. (2014). Anti-inflammatory effect of methylpenicinoline from a marine isolate of *Penicillium* sp. (SF-5995): Inhibition of NF-κB and MAPK pathways in lipopolysaccharide-induced RAW264.7 macrophages and BV2 microglia. *Molecules, 19*(11), 18073. doi: 10.3390/molecules191118073

Kim, M. E., Jung, I., Lee, J. S., Na, J., Kim, W. J., Kim, Y., Park, Y., & Lee, J. S. (2017). Pseudane-VII isolated from *Pseudoalteromonas* sp. *M2* ameliorates LPS-induced inflammatory response in vitro and in vivo. Marine Drugs, *15*(11). doi: 10.3390/md15110336

Kim, S., Le, T. C., Han, S., Hillman, P. F., Hong, A., Hwang, S., Du, Y. E., Kim, H., Oh, D., & Cha, S. (2021). Saccharobisindole, neoasterric methyl ester, and 7-chloro-4 (1h)-quinolinone: Three new compounds isolated from the marine bacterium *Saccharomonospora* sp. *Marine Drugs*, 20(1), 35.

Kim, W. J., Kim, Y. O., Kim, J. H., Nam, B., Kim, D., An, C. M., Lee, J. S., Kim, P., Lee, H. M., Oh, J., & Lee, J. S. (2016). Liquid chromatography-mass spectrometry-based rapid secondary-metabolite profiling of marine *Pseudoalteromonas* sp. M2. Marine Drugs, *14*(1). doi: 10.3390/md14010024

Kincl, F. A., Romo, J., Rosenkranz, G., & Sondheimer, F. (1956). The constituents of *Casimiroa edulis* llave *et* lex. part I. The seed. Journal of the Chemical Society, *804*, 4163–4169.

Kitagawa, W., & Tamura, T. (2008). A quinoline antibiotic from *Rhodococcus erythropolis* JCM 6824. *The Journal of Antibiotics, 61*(11), 680–682.

Kiyekbayeva, L., Mohamed, N. M., Yerkebulan, O., Mohamed, E. I., Ubaidilla, D., Nursulu, A., … & Ross, S. A. (2018). Phytochemical constituents and antioxidant activity of *Echinops albicaulis*. *Natural Product Research, 32*(10), 1203–1207.

Ko, J. S., Rho, M. C., Chung, M. Y., Song, H. Y., Kang, J. S., Kim, K., Lee, H.S. & Kim, Y. K. (2002). Quinolinone alkaloids, diacylglycerol acyltransferase inhibitors from the fruits of *Evodia rutaecarpa*. *Planta Medica*, *68*(12), 1131–1133.

Kon, Y., Kubota, T., Shibazaki, A., Gonoi, T., & Kobayashi, J. (2010). Ceratinadins A–C, new bromotyrosine alkaloids from an okinawan marine sponge *Pseudoceratina* sp. *Bioorganic & Medicinal Chemistry Letters, 20*(15), doi: 4569.10.1016/j.bmcl.2010.06.015

Kostova, I., Ivanova, A., Mikhova, B., & Klaiber, I. (1999). Alkaloids and coumarins from *Ruta graveolens*. *Monatshefte für Chemie, 130*, 703–707.

Kozlovskii, A. G., Arinbasarov, M. U., Yakovlev, G. I., Zyakun, A. M., & Adanin, V. M. (1976). 2-(n-Δ1″-heptenyl)-4-quinolinone, a new metabolite of *Pseudomoncs aeruginosa*. *Bulletin of the Academy of Sciences of the USSR, Division of Chemical Science*, *25*, 1115–1119.

Kunze, B., Hofle, G., & Reichenbach, H. (1987). The aurachins, new quinoline antibiotics from myxobacteria: Production, physico-chemical and biological properties. *The Journal of Antibiotics, 40*(3), 258–265.

Kurasawa, Y., Yoshida, K., Yamazaki, N., Iwamoto, K., Hamamoto, Y., Kaji, E., Sasaki, K., & Zamami, Y. (2012). Quinolinone analogues 12: Synthesis and tautomers of 2-substituted 4-quinolinones and related compounds. *Journal of Heterocyclic Chemistry, 49*(6), 1323. doi: 10.1002/jhet.922

Lamberton, J. A. (1966). 1, 2-Dimethyl-4-quinolinone and xanthevodine, minor alkaloids from the leaves of *Acronychia baueri* Schott. *Australian Journal of Chemistry*, *19*(10), 1995–1996.

Larsen, R. D. (2005). 15.4. Product class 4: Quinolinones and related systems. *ChemInform, 36*(24).

Lee, H. S., Oh, W. K., Choi, H. C., Lee, J. W., Kang, D. O., Park, C. S., … & Ahn, J. S. (1998). Inhibition of angiotensin II receptor binding by quinolinone alkaloids from *Evodia rutaecarpa*. *Phytotherapy Research: An International Journal Devoted to Pharmacological and Toxicological Evaluation of Natural Product Derivatives*, *12*(3), 212–214.

Lee, M. C., Chuang, W. C., & Sheu, S. J. (1996). Determination of the alkaloids in *Evodiae fructus* by capillary electrophoresis. *Journal of Chromatography A, 755*(1), 113–119.

Lee, M. K., Hwang, B. Y., Lee, S. A., Oh, G. J., Choi, W. H., Hong, S. S. & Ro, J. S. (2003). 1-methyl-2-undecyl-4 (1*H*)-quinolinone as an irreversible and selective inhibitor of type B monoamine oxidase. *Chemical and Pharmaceutical Bulletin, 51*(4), 409–411.

Lee, S., Chang, J., Lim, J., Kim, M., Park, S., Jeong, H., Kim, M. S., Lee, W., & Rho, M. (2010). Quinolinone alkaloids from *Evodiae fructus* inhibit LFA-1/ICAM-1-mediated cell adhesion. *Bulletin of the Korean Chemical Society, 31*(1), 64. doi: 10.5012/bkcs.2010.31.01.064

Lee, W., Lee, J., Kulkarni, R., Kim, M., Hwang, J. S., Na, M., & Bae, J. (2016). Antithrombotic and antiplatelet activities of small-molecule alkaloids from *Scolopendra subspinipes mutilans*. *Scientific Reports, 6*(1). doi: 10.1038/srep21956

Lee, Y. G., Cho, J., Kim, C., Jeong, H., Lee, D. I., Kim, S. R., Lee, S., Kim, W., Park, K., & Moon, J. (2013). Isolation and identification of 3 low-molecular compounds from pear (*Pyrus pyrifolia* Nakai cv. Chuhwangbae) fruit peel. *Korean Journal of Food Science and Technology, 45*(2), 174. doi: 10.9721/kjfst.2013.45.2.174

Li, D., Oku, N., Hasada, A., Shimizu, M., & Igarashi, Y. (2018). Two new 2-alkylquinolinones, inhibitory to the fish skin ulcer pathogen *Tenacibaculum maritimum*, produced by a rhizobacterium of the genus *Burkholderia* sp. *Beilstein Journal of Organic Chemistry, 14*, 1446. doi: 10.3762/bjoc.14.122

Li, J., Sun, W., Saalim, M., Wei, G., Zaleta-Pinet, D. A., & Clark, B. R. (2020). Isolation of 2-alkyl-4-quinolinones with unusual side chains from a chinese *Pseudomonas aeruginosa* isolate. *Journal of Natural Products, 83*(7), 2294. doi: 10.1021/acs.jnatprod.0c00026

Li, W., Sun, X., Liu, B., Zhang, L., Fan, Z., & Ji, Y. (2016). Screening and identification of hepatotoxic component in *evodia rutaecarpa* based on spectrum–effect relationship and UPLC-Q-TOFMS. *Biomedical Chromatography, 30*(12), 1975. doi: 10.1002/bmc.3774

Li, Y., He, J., Li, Y., Wu, X., Peng, L., Du, R., Cheng, X., Zhao, Q., & Li, R. (2014). Evollionines A–C, three new alkaloids isolated from the fruits of *Evodia rutaecarpa*. *Helvetica Chimica Acta, 97*(11), 1481. doi: 10.1002/hlca.201300449

Li, Y. Y., Feng, J. L., Li, Z., Zang, X. Y., & Yang, X. W. (2022). Separation and enrichment of alkaloids from *Coptidis rhizoma* and *Euodiae fructus* by macroporous resin and evaluation of the effect on bile reflux gastritis rats. *Molecules, 27*(3), 724.

Li, Y. H., Liu, X., Yin, M., Liu, F., Wang, B., Feng, X., & Wang, Q. Z. (2020). Two new quinolinone alkaloids from the nearly ripe fruits of *Tetradium ruticarpum*. *Natural Product Research, 34*(13), 1868–1873.

Liao, Q. Y., Wang, Z. Y., Li, W. J., & Zhu, W. M. (2015). Secondary metabolites of alkalitolerant Streptomyces sindenensis OUCMDZ-1368. *Chinese Journal of Antibiotics, 40*, 19–27.

Lien, T. P., Sung, T. V., & Adam, G. (2002). Alkaloids from *Evodia sutchuenensis* dode. *Tap Chi Hoa Hoc. 40*(4), 79–83.

Lima, M. D. P., Rosas, L. V., Da Silva, M Fátima Das G F, Ferreira, A. G., Fernandes, J. B., & Vieira, P. C. (2005). Alkaloids from *Spathelia excelsa*: Their chemosystematic significance. *Phytochemistry, 66*(13), 1560. doi: 10.1016/j.phytochem.2005.05.019

Ling, Y., Hu, P., Zhang, L., Jin, H., Chen, J., Tao, Z., Huang, L., & Ren, R. (2016). Identification and structural characterization of acylgluconic acids, flavonol glycosides, limonoids and alkaloids from the fruits of *Evodia rutaecarpa* by high performance liquid chromatography coupled to electrospray ionization and quadrupole time-of-flight mass spectrometry. *Journal of Chromatographic Science, 54*(9), 1593. doi: 10.1093/chromsci/bmw109

Liu, F., O'Donnell, T. J., Park, E. J., Kovacs, S., Nakamura, K., Dave, A., … & Chang, L. C. (2023). Anti-inflammatory Quinoline Alkaloids from the Roots of *Waltheria indica*. *Journal of Natural Products, 86*(2), 276–289.

Liu, R., Chu, X., Sun, A., & Kong, L. (2005). Preparative isolation and purification of alkaloids from the Chinese medicinal herb *Evodia rutaecarpa* (Juss.) Benth by high-speed counter-current chromatography. *Journal of Chromatography A, 1074*(1–2), 139. doi: 10.1016/j.chroma.2005.03.099

Liu, Y., Li, H., Chen, L., Zhao, H., Liu, J., Gong, S., Ma, D., Chen, C., Zeng, S., Long, H., & Ren, W. (2023). Mechanism and pharmacodynamic substance basis of raw and wine-processed *Evodia rutaecarpa* on smooth muscle cells of dysmenorrhea mice. *Pain Research and Management, 2023*, 1. doi: 10.1155/2023/7711988

Liu, Z. L., Chu, S. S., & Jiang, G. H. (2009). Feeding deterrents from *Zanthoxylum schinifolium* against two stored-product insects. *Journal of Agricultural and Food Chemistry, 57*(21), 10130. doi: 10.1021/jf9012983

Liu, Z., Li, J., & Wu, H. (1990). A study on chemical constituents of South China Sea soft coral *Sinularia microclavate. Chinese Journal of Organic Chemistry, 10*(3), 277.

Long, K., Ju, Z., Lin, Y., Xu, S., Xie, Q., & Xie, R. (1984). Studies on the chemical constituents of the Chinese soft corals collected from the South China Sea (X). Chemical structure and physiological activity of a new quinolinone derivative. *Journal of Central South University (Science and Technology), (4)*, 85–92.

Loya, S., Rudi, A., Tal, R., Kashman, Y., Loya, Y., & Hizi, A. (1994). 3, 5, 8-Trihydroxy-4-quinolinone, a novel natural inhibitor of the reverse transcriptases of human immunodeficiency viruses type 1 and type 2. *Archives of Biochemistry and Biophysics, 309*(2), 315–322.

Lu, L., Li, Z., Shan, C., Ma, S., Nie, W., Wang, H., Chen, G., Li, S., & Shu, C. (2021). Whole transcriptome analysis of schinifoline treatment in *Caenorhabditis elegans* infected with *Candida albicans. Molecular Immunology, 135*, 312. doi: 10.1016/j.molimm.2021.04.019

Luckner, M., & Reisch, J. (1970). Japonin, 1-methyl-2-phenyl-3, 6-dimethoxy-chinolon-(4), ein neues alkaloid aus *Orixa japonica. Phytochemistry, 9*(10), 2199–2208.

Luo, Y., Zhang, R., & Chen, F. (2022). Chemical constituents from the fruits of *Vitex rotundifolia* and their chemotaxonomic significance. *Biochemical Systematics and Ecology.* doi: 10310.1016/j.bse.2022.104440

Ma, C., Liu, X., Shan, Y., Xu, S., Feng, X., & Wang, Q. Z. (2021). A new quinolinone alkaloid from the fruits of *Tetradium ruticarpum. Natural Product Research, 35*(2), 222–227.

Ma, C., Liu, X., Shan, Y., Xu, S., Su, X. L., Feng, X., & Wang, Q. Z. (2018). A new quinolinone alkaloid with cytotoxic activity from the fruits of *Euodia Rutaecarpa. Natural Product Communications, 13*(3), 1934578X1801300317.

Ma, X., Liang, X., Huang, Z. H., & Qi, S. H. (2020). New alkaloids and isocoumarins from the marine gorgonian-derived fungus *Aspergillus sp.* SCSIO 41501. *Natural Product Research, 34*(14), 1992–2000.

Makowicz, E., Kafarski, P., & Jasicka-Misiak, I. (2018). Chromatographic fingerprint of the volatile fraction of rare hedera helix honey and biomarkers identification. *European Food Research and Technology, 244*(12), 2169. doi: 10.1007/s00217-018-3127-z

Mandal, A., & Khan, A. T. (2024). Recent advancement in the synthesis of quinoline derivatives via multicomponent reactions. *Organic & Biomolecular Chemistry*, 22, 2339–2358.

Manu, K. A., & Kuttan, G. (2009). Punarnavine induces apoptosis in B16F-10 melanoma cells by inhibiting NF-kB signaling. *Asian Pacific Journal of Cancer Prevention, 10*(6), 1031–1038.

Matsuo, Y., Nozaki, T., Kamewada, Y., Nakagawa, M., Nakamura, Y., Inaba, N., & Mimaki, Y. (2025). New quinolinone alkaloids from euodia fruit, and their pancreatic lipase inhibitory and PPAR-γ ligand-binding activities. *Fitoterapia, 180*, 106322.

Mitaku, S., Fokialakis, N., Magiatis, P., & Tillequin, F. (2006). Alkaloids from *Sarcomelicope megistophylla. Fitoterapia, 78*(2), 169. doi: 10.1016/j.fitote.2006.10.003

Mitaku, S., Skaltsounis, A., Tillequin, F., Koch, M., Pusset, J., & Sevenet, T. (1995). New alkaloids from *Sarcomelicope dogniensis. Natural Product Letters, 7*(3), 219. doi: 10.1080/10575639508043214

Molinski, T. F., & Faulkner, D. J. (1988). An antibacterial pigment from the sponge *Dendrilla membranosa. Tetrahedron Letters, 29*(18), 2137–2138.

Monteillier, A., Cretton, S., Ciclet, O., Marcourt, L., Ebrahimi, S. N., Christen, P., & Cuendet, M. (2017). Cancer chemopreventive activity of compounds isolated from *Waltheria indica. Journal of Ethnopharmacology, 203*, 214. doi: 10.1016/j.jep.2017.03.048

Moon, S., Kang, P. M., Park, K. S., & Kim, C. H. (1996). Plant growth promoting and fungicidal 4-quinolinones from *Pseudomonas cepacia. Phytochemistry*, 42(2), 365–368.

Moon, S., Myung, E., Cho, S., Park, J., & Chung, B. (2002). Plant growth promoting effect of 4-quinolinone metabolites from *Pseudomonas cepacia* and 4-quinolinone-3-carboxylate derivatives on red pepper plant (*Capsicum annum*). *The Korean Journal of Pesticide Science, 6*(2), 64–71.

Mori, T., Yamashita, T., Furihata, K., Nagai, K., Suzuki, K., Hayakawa, Y., & Shin-Ya, K. (2007). Burkholone, a new cytotoxic antibiotic against IGF-I dependent cells from *Burkholderia* sp. *The Journal of Antibiotics, 60*(11), 713–716.

Moussa, A. Y., Albelbisy, M. A. K., & Singab, A. N. B. (2024). The underrepresented quinolinone alkaloids in genera *Penicillium* and *Aspergillus*: Structure, biology, and biosynthetic machinery. Chemistry & Biodiversity. doi: 10.1002/cbdv.202402218

Mphahlele, M. J., & El-Nahas, A. M. (2004). Tautomeric 2-arylquinolin-4(1H)-one derivatives-spectroscopic, X-ray and quantum chemical structural studies. *Journal of Molecular Structure, 688*(1–3), 129. doi: 10.1016/j.molstruc.2003.10.003

Mukai, H., Kubota, T., Aoyama, K., Mikami, Y., Fromont, J., & Kobayashi, J. (2009). Tyrokeradines A and B, new bromotyrosine alkaloids with an imidazolyl-quinolinone moiety from a verongid sponge. *Bioorganic & Medicinal Chemistry Letters, 19*(5), 1337. doi: 10.1016/j.bmcl.2009.01.056

Na, M. W., Jeong, S. Y., Ko, Y. J., Kang, D. M., Pang, C., Ahn, M. J., & Kim, K. H. (2022). Chemical investigation of *Tetradium Ruticarpum* fruits and their antibacterial activity against *Helicobacter Pylori*. ACS *Omega, 7*(27), 23736–23743.

Nachtigall, J., Schneider, K., Nicholson, G., Goodfellow, M., Zinecker, H., Imhoff, J. F., Süssmuth, R. D., & Fiedler, H. (2010). Two new aurachins from *Rhodococcus* sp. Acta 2259. *The Journal of Antibiotics, 63*(9), 567. doi: 10.1038/ja.2010.79

Naeem, A., Badshah, S. L., Muska, M., Ahmad, N., & Khan, K. (2016). The current case of quinolinones: Synthetic approaches and antibacterial activity. *Molecules, 21*(4), 268.

Nakano, D., Ishitsuka, K., Matsuda, N., Kouguchi, A., Tsuchihashi, R., Okawa, M., Okabe, H., Tamura, K., & Kinjo, J. (2016). Screening of promising chemotherapeutic candidates from plants against human adult T-cell leukemia/lymphoma (V): Coumarins and alkaloids from *Boenninghausenia japonica* and *Ruta graveolens. Journal of Natural Medicines, 71*(1), 170. doi: 10.1007/s11418-016-1046-5

Nakano, H., Cantrell, C. L., Mamonov, L. K., Kustova, T. S., Fronczek, F. R., & Ross, S. A. (2012). Chemical constituents from *Echinops nanus* and *Echinops transiliensis. Biochemical Systematics and Ecology, 45*, 127. doi: 10.1016/j.bse.2012.07.008

Nakatsu, T., Johns, T., Kubo, I., Milton, K., Sakai, M., Chatani, K., Saito, K., Yamagiwa, Y. & Kamikawa, T. (1990). Isolation, structure, and synthesis of novel 4-quinolinone alkaloids from *Esenbeckia leiocarpa. Journal of Natural Products*, 53(6), 1508–1513.

Nammalwar, B., & Bunce, R. A. (2013). Recent syntheses of 1,2,3,4-tetrahydroquinolines, 2,3-dihydro-4(1H)-quinolinones and 4(1H)-quinolinones using domino reactions. *Molecules, 19*(1), 204. doi: 10.3390/molecules19010204

Nand, M., Maiti, P., Pant, R., Kumari, M., Chandra, S., & Pande, V. (2016). Virtual screening of natural compounds as inhibitors of EGFR 696-1022 T790M associated with non-small cell lung cancer. *Bioinformation, 12*(6), 311.

Nasiri, H. R., Bolte, M., & Schwalbe, H. (2006). Tautomerism of 4-hydroxy-4 (1H) quinolon. *Heterocyclic Communications, 12*(5), 319–322.

Nawwar, M., Barakat, H. H., Buddrust, J., & Linscheidt, M. (1985). Alkaloidal, lignan and phenolic constituents of *Ephedra alata. Phytochemistry*, 24(4), 878–879.

Nazrul Islam, S. K., Gray, A. I., Waterman, P. G., & Ahasan, M. (2002). Screening of eight alkaloids and ten flavonoids isolated from four species of the genus *Boronia* (Rutaceae) for antimicrobial activities against seventeen clinical microbial strains. *Phytotherapy Research, 16*(7), 672. doi: 10.1002/ptr.999

Nazrullaev, S. S., Bessonova, I. A., & Akhmedkhodzhaeva, K. S. (2001). Estrogenic activity as a function of chemical structure in *Haplophyllum quinoline* alkaloids. *Chemistry of Natural Compounds, 37*, 551–555.

Nery, A. C. S., De Souza, K. R., & Domingos, J. L. D. O. (2023). Highlights on quinolonic compounds with changes on the basic structure as promising molecules to new drugs. *Brazilian Journal of Health Review, 6*(6), 28702. doi: 10.34119/bjhrv6n6-167

Nguyen, N. T., Dang, P. H., Vu, N. X., Le, T. H., & Nguyen, M. T. (2017). Quinoliniumolate and 2 H-1, 2, 3-triazole derivatives from the stems of *Paramignya trimera* and their α-glucosidase inhibitory activities: In vitro and in silico studies. *Journal of Natural Products, 80*(7), 2151–2155.

Nguyen, P., Zhao, B. T., Kim, O., Lee, J. H., Choi, J. S., Min, B. S., & Woo, M. H. (2016). Anti-inflammatory terpenylated coumarins from the leaves of *Zanthoxylum schinifolium* with α-glucosidase inhibitory activity. *Journal of Natural Medicines, 70*(2), 276. doi: 10.1007/s11418-015-0957-x

Nhoek, P., Ahn, S., Pel, P., Kim, Y. M., Huh, J., Kim, H. W., … & Chin, Y. W. (2022). Alkaloids and coumarins with adiponectin-secretion-promoting activities from the leaves of *orixa japonica. Journal of Natural Products*, 86(1), 138–148.

Nicholas, G. M., Newton, G. L., Fahey, R. C., & Bewley, C. A. (2001). Novel bromotyrosine alkaloids: Inhibitors of mycothiol S-conjugate amidase. *Organic Letters*, 3(10), 1543–1545.

Nwankwo, N. E., Ezeako, E. C., Nworah, F. N., Ogara, A. L., Oka, S. A., Aham, E. C., Joshua, P. E., Nwiloh, B. I., Ezike, T. C., Ashiakpa, N. P., Ngozi, H. C., Ezeugwu, C. P., Obiora, O. M., Nwadike, G. C., Ezeh, T. C., Alotaibi, S. S., Albogami, S. M., & Batiha, G. E. (2022). Bioactive compounds, anti-inflammatory, anti-nociceptive and antioxidant potentials of ethanolic leaf fraction of *Sida linifolia* L. (malvaceae). Arabian Journal of Chemistry, *16*(1). doi: 10.1016/j.arabjc.2022.104398

Oh, E., Lee, J., Kim, C., & Keum, Y. (2014). Alkylquinolinone alkaloid profiles in *Ruta graveolens. Biochemical Systematics and Ecology, 57*, 384. doi: 10.1016/j.bse.2014.09.011

Okonkwo-Uzor, N. J., Obi, C., Enwelum, A., Okezie, U. M., Eze, P. M., Okoye, F. B. C., & Esimone, C. O. (2022). Antimicrobial and antioxidant activities of secondary metabolites of an endophytic fungus of *Azadirachta indica. Journal of Drug Delivery and Therapeutics, 12*(5-S), 112. doi: 10.22270/jddt.v12i5-s.5642

Oliva, A., Meepagala, K. M., Wedge, D. E., Harries, D., Hale, A. L., Aliotta, G., & Duke, S. O. (2003). Natural fungicides from *Ruta graveolens* L. leaves, including a new quinolinone alkaloid. *Journal of Agricultural and Food Chemistry, 51*(4), 890. doi: 10.1021/jf0259361

Oliveira, F. M., Conserva, L. M., Maia, J. S., & Guilhon, G. M. (1996). Alkaloids and coumarins from *esenbeckia* species. *Phytochemistry, 41*(2), 647–649.

Organic Chemistry Portal. (2025). Synthesis of 4-quinolinones. https://www.organic-chemistry.org/synthesis/heterocycles/benzo-fused/4-quinolinones.shtm

Osuagwu, O. L., Igwe, C. U., Nzebude, C. P., Njoku, O. C., & Ejiofor, J. C. (2023). Phenolic profile and antioxidant potential of aqueous extracts of selected traditional anti-cough plants. *Journal of Drug Delivery and Therapeutics, 13*(7), 22. doi: 10.22270/jddt.v13i7.6124

Paluka, J., Kanokmedhakul, K., Soytong, M., Soytong, K., Yahuafai, J., Siripong, P., & Kanokmedhakul, S. (2020). Meroterpenoid pyrones, alkaloid and bicyclic brasil-iamide from the fungus *Neosartorya hiratsukae.* Fitoterapia, *142*10. doi: 1016/j. fitote.2020.104485

Pan, X., Bligh, S. W. A., & Smith, E. (2013). Quinolinone alkaloids from fructus euodiae show activity against methicillin-resistant *Staphylococcus aureus. Phytotherapy Research, 28*(2), 305. doi: 10.1002/ptr.4987

Pang, X., Cai, G., Lin, X., Salendra, L., Zhou, X., Yang, B., … & Liu, Y. (2019). New alkaloids and polyketides from the marine sponge-derived fungus *Penicillium sp.* SCSIO41015. *Marine Drugs, 17*(7), 398.

Panthong, K., Srisud, Y., Rukachaisirikul, V., Hutadilok-Towatana, N., Voravuthikunchai, S. P., & Tewtrakul, S. (2013). Benzene, coumarin and quinolinone derivatives from roots of *citrus hystrix. Phytochemistry, 88*, 79. doi: 10.1016/j.phytochem.2012.12.013

Parimah Parhoodeh, P. P., Mawardi Rahmani, M. R., Najihah Mohd. Hashim, N. M. H., Mohd. Aspollah Sukari, M. A. S., & Ee ChengLian [Ee, C. (2012). Alkaloid constituents of *Haplophyllum laeviusculum* (Rutaceae). *Sains Malaysiana 41*(1), 47–52.

Paulsen, S. S., Isbrandt, T., Kirkegaard, M., Buijs, Y., Strube, M. L., Sonnenschein, E.C., Larsen, T. O., & Gram, L. (2020). Production of the antimicrobial compound tetrabro-mopyrrole and the pseudomonas quinolinone system precursor, 2-heptyl-4-quinolinone, by a novel marine species *Pseudoalteromonas galatheae* sp. nov. Scientific Reports, *10*(1). doi: 10.1038/s41598-020-78439-3

Pesci, E. C., Milbank, J. B., Pearson, J. P., McKnight, S., Kende, A. S., Greenberg, E. P., & Iglewski, B. H. (1999). Quinolinone signaling in the cell-to-cell communication system of *Pseudomonas aeruginosa. Proceedings of the National Academy of Sciences, 96*(20), 11229–11234.

Peterson, L. R. (2001). Quinolinone molecular structure-activity relationships: What we have learned about improving antimicrobial activity. *Clinical Infectious Diseases, 33*(Supplement_3), S180–S186.

Pham, T. D. M., Ziora, Z. M., & Blaskovich, M. A. T. (2019). Quinolinone antibiotics. *MedChemComm, 10*(10), 1719. doi: 10.1039/c9md00120d

Pitchai, P., Ulagi, R., Mohan, S., & Gengan, R. M. (2012). A novel alkaloid, four alkaloid precursors and a coumarin from *Feronia limonia. Indian Journal of Chemistry, 51*(B), 1771–1775.

Presley, C. C., Du, Y., Dalal, S., Merino, E. F., Butler, J. H., Rakotonandrasana, S., Rasamison, V. E., Cassera, M. B., & Kingston, D. G. I. (2017). Isolation, structure elucidation, and synthesis of antiplasmodial quinolinones from *Crinum firmifolium. Bioorganic & Medicinal Chemistry, 25*(15), 4203. doi: 10.1016/j.bmc.2017.06.017

Qin, J., Liao, C., Chen, W., Li, H., Su, J., Wu, X., He, J., & Zhang, G. (2021). New limonoids and quinolinone alkaloids with cytotoxic and anti-platelet aggregation activities from *Evodia rutaecarpa* (Juss.) Benth. Fitoterapia, *152*. doi: 1016/j.fitote.2021.104875

Qing-yun, L., Zhi-ying, W., Wen-jun, L., & Wei-ming, Z. (2015). Secondary metabolites of alkali tolerant *Streptomyces sindenensis* OUCMDZ-1368. *Chinese Journal of Antibiotics, 40*(1), 19–27 and 50.

Rahim, A., Saito, Y., Fukuyoshi, S., Miyake, K., Goto, M., Chen, C. H., … & Nakagawa-Goto, K. (2020). Paliasanines A–E, 3, 4-Methylenedioxyquinoline alkaloids fused with a phe-nyl-14-oxabicyclo [3.2. 1] octane unit from *Melochia umbellata* var. *deglabrata. Journal of natural products, 83*(10), 2931–2939.

Raj, R., Agarwal, N., Raghavan, S., Chakraborti, T., Poluri, K. M., Pande, G., & Kumar, D. (2021). Epigallocatechin gallate with potent anti-*Helicobacter pylori* activity binds effi-ciently to its histone-like DNA binding protein. ACS *Omega, 6*(5), 3548–3570.

Razakova, D. M., & Bessonova, I. A. (1981). Hapovine, a new alkaloid from *Haplophyllum popovii* [aerial parts]. *Khimiya Prirodnykh Soedinenii, 4*, 528–529.

Razakova, D. M., Bessonova, I. A., & Yunusov, S. Y. (1979). Alkaloids of *Haplophyllum dubium*. *Chemistry of Natural Compounds, 15*(6), 716–718.

Razakova, D. M., Bessonova, I. A., & Yunusov, S. Y. (1986). Components of *Haplophyllum acutifolium*. *Khimiya Prirodnykh Soedinenii, 1986*(3), 384–385.

Razzakova, D. M., Bessonova, I. A., & Yunusov, S. Y. (1972). The structure of folimidine. *Chemistry of Natural Compounds, 8*, 737–740.

Razzakova, D. M., Bessonova, I. A., & Yunusov, S. Y. (1973). Alkaloids of *Haplophyllum acutifolium*. *Chemistry of Natural Compounds, 9*, 199–202.

Reisch, J. (1967). 2-4′-(3″, 4″-methylenedioxyphenyl)-n-butyl-quinolinone (4). An alkaloid from *Ruta graveolens*. *Naturwissenschaften, 54*(19), 517.

Reisch, J., Rózsa, Z., Szendrei, K., & Körösi, J. (1975). 2-(N-Undecyl)-chinolon-(4) aus den blüten von *Ptelea trifoliata* und den wurzeln von *Ruta graveolens*. *Phytochemistry, 14*(3), 840–841.

Rho, T. C., Bae, E., Kim, D., Oh, W. K., Kim, B. Y., Ahn, J. S., & Lee, H. S. (1999). Anti-helicobacter pylori acticvity of quinolinone alkaloids from *Evodiae fructus*. *Biological and Pharmaceutical Bulletin, 22*(10), 1141–1143.

Rizvi, S. H., Kapil, R. S., & Shoeb, A. (1985). Alkaloids and coumarins of *Casimiroa edulis*. *Journal of Natural Products, 48*(1), 146–146.

Robertson, L. P., Hall, C. R., Forster, P. I., & Carroll, A. R. (2018). Alkaloid diversity in the leaves of Australian *Flindersia* (Rutaceae) species driven by adaptation to aridity. *Phytochemistry, 152*, 71. doi: 10.1016/j.phytochem.2018.04.011

Rocha, M. R. E., Da Cunha, C. P., Filho, R. B., & Vieira, I. J. C. (2016). A novel alkaloid isolated from *Spiranthera atlantica* (Rutaceae). *Natural Product Communications, 11*(3), 1934578X1601100318.

Rodriguez, J. P., Bernardi, D. I., Gubiani, J. R., Magalhães de Oliveira, J., Morais-Urano, R. P., Bertonha, A. F., … & Berlinck, R. G. (2020). Water-soluble glutamic acid derivatives produced in culture by *Penicillium solitum IS1-A* from King George Island, Maritime Antarctica. *Journal of Natural Products, 83*(1), 55–65.

Roitman, J. N., Mahoney, N. E., Janisiewicz, W. J., & Benson, M. (1990). A new chlorinated phenylpyrrole antibiotic produced by the antifungal bacterium *Pseudomonas cepacia*. *Journal of Agricultural and Food Chemistry, 38*(2), 538–541.

Saalim, M., Villegas-Moreno, J., & Clark, B. R. (2020). Bacterial alkyl-4-quinolinones: Discovery, structural diversity and biological properties. *Molecules, 25*(23). doi: 10.3390/molecules25235689

Sahashi, Y. (1925). The occurrence of dihydroxyquinoline carboxylic acid (the β-acid of suzuki) in rice bran. *Biochemische Zeitschrift, 159*, 221–234.

Saini, N., Grewal, A. S., Lather, V., & Gahlawat, S. K. (2022). Natural alkaloids targeting EGFR in non-small cell lung cancer: Molecular docking and ADMET predictions. *Chemico-Biological Interactions*. doi: 10.1016/j.cbi.2022.109901

Salama, H. M. H. (2012). Alkaloids and flavonoids from the air dried aerial parts of *Citrullus colocynthis*. *Journal of Medicinal Plants Research, 6*(38), 5150. doi: 10.5897/jmpr12.406

Sampaio, O. M., Vieira, L. C. C., Bellete, B. S., King-Diaz, B., Lotina-Hennsen, B., Da Silva, Maria Fátima Das Graças Fernandes, & Veiga, T. A. M. (2018). Evaluation of alkaloids isolated from *Ruta graveolens* as photosynthesis inhibitors. *Molecules, 23*(10) doi: 10.3390/molecules23102693

Sandmann, A., Dickschat, J., Jenke-kodama, H., Kunze, B., Dittmann, E., & Müller, R. (2007). A type II polyketide synthase from the gram-negative bacterium *Stigmatella aurantiaca* is involved in aurachin alkaloid biosynthesis. *Angewandte Chemie International Edition, 46*(15), 2712. doi: 10.1002/anie.200603513

Santosh, J.S. & Sunil, C.N. (2019). Phytochemical investigation of methanolic extracts of fruit of *Luffa acutangula* var. *amara*. *Journal of Pharmacognosy and Phytochemistry 8*(2), 489–491.

Saputra, A., Wientarsih, I., Rafi, M., Sutardi, L. N., & Mariya, S. (2024). LC-HRMS metabolite profiling of *Lunasia amara* stem bark and in silico study in breast cancer receptors. *Indonesian Journal of Pharmacy, 35*(1), 116–125.

Saraswati, S., Alhaider, A. A., & Agrawal, S. S. (2013). Punarnavine, an alkaloid from *Boerhaavia diffusa* exhibits anti-angiogenic activity via downregulation of VEGF in vitro and in vivo. *Chemico-Biological Interactions, 206*(2), 204. doi: 10.1016/j.cbi.2013.09.007

Sarker, S. D., Armstrong, J. A., & Waterman, P. G. (1995). An alkaloid, coumarins and a triterpene from *Boronia algida. Phytochemistry, 39*(4), 801–804.

Sartor, C. F. P., da Silva, M. F. D. G., Fernandes, J. B., Vieira, P. C., Fo, E. R., & Cortez, D. A. G. (2003). Alkaloids from *Dictyoloma vandellianum*: Their chemosystematic significance. *Phytochemistry, 63*(2), 185–192.

Schneider, G., & Immel, D. (1967). Die struktur des alkaloids "rutamin" aus *Ruta graveolens* L. *Archiv Der Pharmazie, 300*(11), 953. doi: 10.1002/ardp.19673001111

Seixas, R. S. G. R., Silva, A. M. S., Alkorta, I., & Elguero, J. (2011). An experimental NMR and computational study of 4-quinolinones and related compounds. *Monatshefte Für Chemie - Chemical Monthly, 142*(7), 731. doi: 10.1007/s00706-011-0473-y

Selim, N. (2012). Chemical and bioactivity studies of the red sea soft corals: *Sinularia polydactyla* (Ehrenberg) and *Lobophytum crassum* (Marenzellar)(Fam. Alcyoniidae) (PhD Thesis), University of Cairo, Egypt.

Setzer, W. N., Vogler, B., Bates, R. B., Schmidt, J. M., Dicus, C. W., Nakkiew, P., & Haber, W. A. (2003). HPLC-NMR/HPLC-MS analysis of the bark extract of *Stauranthus perforatus. Phytochemical Analysis, 14*(1), 54. doi: 10.1002/pca.687

Shakirov, R., Telezhenetskaya, M. V., Bessonova, I. A., Aripova, S. F., Israilov, I. A., Sultankhodzhaev, M. N., ... & Tel'nov, V. A. (1996). Alkaloids. Plants, structure, properties. Dedicated to the memory of academician S. Yu. Yunosov. 102–173.

Shao, C., Wang, C., Gu, Y., Wei, M., Pan, J., Deng, D., She, Z., & Lin, Y. (2010). Penicinoline, a new pyrrolyl 4-quinolinone alkaloid with an unprecedented ring system from an endophytic fungus *penicillium* sp. *Bioorganic & Medicinal Chemistry Letters, 20*(11), 3284. doi: 10.1016/j.bmcl.2010.04.043

Sharma, V., Das, R., Kumar Mehta, D., Gupta, S., Venugopala, K. N., Mailavaram, R., Nair, A. B., Shakya, A. K., & Kishore Deb, P. (2022). Recent insight into the biological activities and SAR of quinolinone derivatives as multifunctional scaffold. Bioorganic & Medicinal Chemistry. doi: 10.1016/j.bmc.2022.116674

Shen, C., Wang, A., Xu, J., An, Z., Loh, K. Y., Zhang, P., & Liu, X. (2019). Recent advances in the catalytic synthesis of 4-quinolinones. *Chem, 5*(5), 1059. doi: 10.1016/j.chempr.2019.01.006

Shrestha, S., Park, J., Cho, J., Lee, D., Jeong, R., Han, J., Cho, S. K., Lee, D., & Baek, N. (2016). Phytochemical constituents of the *Urena lobata* fruit. *Chemistry of Natural Compounds, 52*(1), 178. doi: 10.1007/s10600-016-1586-7

Shu, C., Zhang, M., Zhang, W., Ma, S., Chen, G., & Shi, B. (2019). Antifungal activity of schinifoline against *Candida albicans* in *Caenorhabditis elegans. Phyton, 88*(3), 317. doi: 10.32604/phyton.2019.07766

Silva, V. L. M., & Silva, A. M. S. (2019). Palladium-catalysed synthesis and transformation of quinolinones. Molecules, 24(2). doi: 10.3390/molecules24020228

Simpson, D. S., & Jacobs, H. (2005). Alkaloids and coumarins from *Esenbeckia pentaphylla* (rutaceae). *Biochemical Systematics and Ecology, 33*(8), 841. doi: 10.1016/j.bse.2004.12.022

Sinha, S., Mishra, P., Amin, H., Rah, B., Nayak, D., Goswami, A., Kumar, N., Vishwakarma, R., & Ghosal, S. (2012). A new cytotoxic quinolinone alkaloid and a pentacyclic steroidal glycoside from the stem bark of *Crataeva nurvala*: Study of anti-proliferative and apoptosis inducing property. *European Journal of Medicinal Chemistry, 60*, 490. doi: 10.1016/j.ejmech.2012.12.017

Siuda, J. F. (1974). Chemical defense mechanisms of marine organisms. identification of 8-hydroxy-4-quinolinone from the ink of the giant octopus, *Octopus dofleini* martini. *Lloydia, 37*(3), 501–503.

Staerk, D., Kesting, J. R., Sairafianpour, M., Witt, M., Asili, J., Emami, S. A., & Jaroszewski, J. W. (2009). Accelerated dereplication of crude extracts using HPLC–PDA–MS–SPE–NMR: Quinolinone alkaloids of *Haplophyllum acutifolium*. *Phytochemistry, 70*(8), 1055. doi: 10.1016/j.phytochem.2009.05.004

Steldt, F. A., & Chen, K. K. (1943). The alkaloids of *lunasia amara*. *Journal of the American Pharmaceutical Association, 32*(4), 107–111.

Su, H., Yang, H., Meng, C., Peng, C., Guo, L., & Dai, O. (2016). Study on chemical constituents of seeds of *croton tiglium* and their cytotoxicities. *Zhongguo Zhong Yao Za Zhi = Zhongguo Zhongyao Zazhi = China Journal of Chinese Materia Medica, 41*(19), 3620–3623.

Subehan, Takahashi, N., Kadota, S., & Tezuka, Y. (2010). Cytochrome P450 2D6 inhibitory constituents of *lunasia amara*. *Phytochemistry Letters, 4*(1), 30. doi: 10.1016/j. phytol.2010.10.006

Sugimoto, T., Miyase, T., Kuroyanagi, M., & Ueno, A. (1988). Limonoids and quinolinone alkaloids from *Evodia rutaecarpa* Bentham. *Chemical and pharmaceutical bulletin, 36*(11), 4453–4461.

Sun, J., Jin, M., Zhou, W., Diao, S., Zhou, Y., Li, S., … & Li, G. (2017). A new ribonucleotide from *Cordyceps militaris*. *Natural Product Research, 31*(21), 2537–2543.

Sunarpi, H., Sunarwidhi, E., Ariyana, M., Nikmatulla, A., Zulkifli, L., Yoshie, S., Miyake, M., Kobayashi, D., & Hazama, A. (2018). Cytotoxicity and antiproliferative activity of indonesian red algae *Acanthophora muscoides* crude ethanol extracts. *Journal of Biological Sciences, 18*(8), 425. doi: 10.3923/jbs.2018.425.433

Tang, W., Ding, X., & Xin, Y. (2004). A new lignan glycoside from the flower of *Castanea mollissima* blume. *Yao Xue Xue Bao = Acta Pharmaceutica Sinica, 39*(7), 531–533.

Tang, Y. Q., Feng, X. Z., & Huang, L. (1996). Quinolinone alkaloids from *Evodia rutaecarpa*. *Phytochemistry, 43*(3), 719–722.

Teeyapant, R., Kreis, P., Wray, V., Witte, L., & Proksch, P. (1993). Brominated secondary compounds from the marine sponge *Verongia aerophoba* and the sponge feeding gastropod *tylodina perversa*. *Zeitschrift Für Naturforschung* C, *48*(7–8), 640–644.

Terezan, A. P., Rossi, R. A., Almeida, R. N., Freitas, T. G., Fernandes, J. B., Silva, M. F. G. F., … & Pirani, J. R. (2010). Activities of extracts and compounds from *Spiranthera odoratissima* St. Hil.(Rutaceae) in leaf-cutting ants and their symbiotic fungus. *Journal of the Brazilian Chemical Society, 21*, 882–886.

Thoms, C., Wolff, M., Padmakumar, K., Ebel, R., & Proksch, P. (2004). Chemical defense of mediterranean sponges *Aplysina cavernicola* and *Aplysina aerophoba*. *Zeitschrift Für Naturforschung C, 59*(1–2), 113–122.

Tian, S., Yang, Y., Liu, K., Xiong, Z., Xu, L., & Zhao, L. (2014). Antimicrobial metabolites from a novel halophilic actinomycete *Nocardiopsis terrae YIM 90022*. *Natural Product Research, 28*(5), 344–346.

Tian, X., Hai-Feng, T., Yu-Shan, L., Hou-Wen, L., Zhang, X., Jun-Tao, F., & Zhang, X. (2015). Studies on the chemical constituents from marine bryozoan *Cryptosula pallasiana*. *Records of Natural Products, 9*(4), 628.

Tillequin, F., Koch, M., & Sevenet, T. (1980). Alcaloïdes des Ecorces de Tronc de Flindersia fournieri. *Planta Medica, 39*(08), 383–387.

To, D. C., Bui, T. Q., Nhung, N. T. A., Tran, Q. T., Do, T. T., Tran, M. H., … & Nguyen, P. H. (2021). On the inhibitability of natural products isolated from *Tetradium ruticarpum* towards tyrosine phosphatase 1B (PTP1B) and α-glucosidase (3W37): An in vitro and in silico study. *Molecules, 26*(12), 3691.

Toi, N. D., Mai, D. T. N., Thu, N. T. H., Dat, N. T., Thao, N. K. N., & Hang, D. T. (2022). Isolation of rice endophytic bacterial strain VY81 and study on its bioactive compound antagonizing the phytopathogen *Dickeya zeae*. *Vietnam Journal of Biotechnology, 20*(4), 683–691.

Tominaga, K., Higuchi, K., Hamasaki, N., Hamaguchi, M., Takashima, T., Tanigawa, T., Watanabe, T., Fujiwara, Y., Tezuka, Y., Nagaoka, T. and Kadota, S., Ishii, E., Kobayashi, K. & Arakawa, T. (2002). In vivo action of novel alkyl methyl quinolinone alkaloids against *Helicobacter pylori*. *Journal of Antimicrobial Chemotherapy, 50*(4), 547–552.

Touati, D., & Ulubelen, A. (2000). Alkaloids from *Ruta montana*. *Phytochemistry, 53*(2), 277–279.

Truchado, P., Martos, I., Bortolotti, L., Sabatini, A. G., Ferreres, F., & Tomas-Barberan, F. A. (2009). Use of quinoline alkaloids as markers of the floral origin of chestnut honey. *Journal of Agricultural and Food Chemistry, 57*(13), 5680. doi: 10.1021/jf900766v

Tschesche, R., & Werner, W. (1967). Evocarpin, ein neues Alkaloid aus *Evodia rutaecarpa*. *Tetrahedron, 23*(4), 1873–1881.

Tummanapalli, S., Gulipalli, K. C., Bodige, S., Pommidi, A. K., Boya, R., Choppadandi, S., Bakangari, M. R., Punna, S. K., Medaboina, S., Mamindla, D. Y., Kanuka, A., Endoori, S., Ganapathi, V. K., Kottam, S. D., Kalbhor, D., & Valluri, M. (2024). Cu-catalyzed tandem C–N and C–C bond formation leading to 4(1H)-quinolinones: A scaffold with diverse biological properties from totally new raw materials in a single step. *The Journal of Organic Chemistry, 89*(3), 1609. doi: 10.1021/acs.joc.3c02215

Ulubelen, A. (1990). A new alkaloid, montanine, from *Ruta montana*. *Journal of Natural Products, 53*(1), 207–208.

Vivero-Gomez, R. J., Mesa, G. B., Higuita-Castro, J., Robledo, S. M., Moreno-Herrera, C. X., & Cadavid-Restrepo, G. (2021). Detection of quorum sensing signal molecules, particularly N-acyl homoserine lactones, 2-alky-4-quinolinones, and diketopiperazines, in gram-negative bacteria isolated from insect vector of leishmaniasis. Frontiers in Tropical Diseases. doi: 10.3389/fitd.2021.760228

Wahyuni, T. S., Widyawaruyanti, A., Lusida, M. I., Fuad, A., Soetjipto, Fuchino, H., Kawahara, N., Hayashi, Y., Aoki, C., & Hotta, H. (2014). Inhibition of hepatitis C virus replication by chalepin and pseudane IX isolated from *Ruta angustifolia* leaves. *Fitoterapia, 99*, 276. doi: 10.1016/j.fitote.2014.10.011

Wakana, D., Hosoe, T., Itabashi, T., Okada, K., De Campos Takaki, G. M., Yaguchi, T., Fukushima, K., & Kawai, K. (2006). New citrinin derivatives isolated from *Penicillium citrinum*. *Journal of Natural Medicines, 60*(4), 279. 10.1007/s11418-006-0001-2

Wang, C., Fan, L., Tian, M., Qi, X., Liu, J., Feng, J., Du, S., Su, X., & Wang, Y. (2014). Radiosensitizing effect of schinifoline from *Zanthoxylum schinifolium* Sieb et Zucc on human non-small cell lung cancer A549 cells: A preliminary in vitro investigation. *Molecules, 19*(12), 20128. doi: 10.3390/molecules191220128

Wang, R., Xu, K., & Shi, W. (2019). Quinolinone derivatives: Potential anti-HIV agent – development and application. *Archiv Der Pharmazie, 352*(9). doi: 10.1002/ardp.201900045

Wang, T., Wu, J., Hwang, T., Kuo, Y., & Chen, J. (2010). A new quinolinone and other constituents from the fruits of *tetradium ruticarpum*: Effects on neutrophil pro-inflammatory responses. *Chemistry & Biodiversity, 7*(7), 1828. doi: 10.1002/cbdv.200900289

Wang, X., Zan, K., Shi, S., Zeng, K., Jiang, Y., Guan, Y., Xiao, C., Gao, H., Wu, L., & Tu, P. (2013). Quinolinone alkaloids with antibacterial and cytotoxic activities from the fruits of *Evodia rutaecarpa*. *Fitoterapia, 89*, 1. doi: 10.1016/j.fitote.2013.04.007

Werny, F., & Scheuer, P. J. (1963). Hawaiian plant studies – IX: The alkaloids of *Platydesma campanulata* mann. *Tetrahedron, 19*(8), 1293–1305.

Wu, B., & Ge, Y. (2023). Extraction of polyhydroxycyclohexane derivatives from *Epicoccum sorghinum* fungi. *China Patent* No. CN116003316.

Wu, G., Hao, Q., Liu, B., Zhou, J., Fan, C., & Liu, R. (2022). Network pharmacology-based screening of the active ingredients and mechanisms of *Evodiae fructus* anti-glioblastoma multiforme. *Medicine, 101*(39). doi: 10.1097/md.0000000000030853

Wu, T. S. (1987). Alkaloids and coumarins of *Skimmia reevesiana. Phytochemistry, 26*(3), 873–875.

Xiao, S., Xu, X., Chen, W., Xin, J., Yuan, W., Zu, X., & Shen, Y. (2023). Traditional Chinese medicine *Euodiae fructus*: Botany, traditional use, phytochemistry, pharmacology, toxicity and quality control. *Natural Products and Bioprospecting, 13*(1). doi: 10.1007/s13659-023-00369-0

Xu, K., Jiang, S., Sun, H., Zhou, Y., Xu, X., Peng, S., & Ding, L. (2012). New alkaloids from the seeds of *Notopterygium incisum. Natural Product Research, 26*(20), 1898. doi: 10.1080/14786419.2011.628177

Xu, M., Li, G., Moon, D., Lee, C., Woo, M. H., Lee, E. S., Jahng, Y., Chang, H., Lee, S., & Son, J. (2006). Cytotoxicity and DNA topoisomerase inhibitory activity of constituents isolated from the fruits of *Evodia officinalis. Archives of Pharmacal Research, 29*, 541–547.

Yajima, T., Kato, N., & Munakata, K. (1977). Isolation of insect anti-feeding principles in *Orixa japonica* Thunb. *Agricultural and Biological Chemistry, 41*(7), 1263–1268.

Yamahara, J., Kobayashi, G., Matsuda, H., & Fujimura, H. (1988). The vasorelaxant effect of evocarpine in isolated aortic strips: Mode of action. *European Journal of Pharmacology, 155*(1–2), 139–143.

Yan, Q., Shan, Y., Yin, M., Xu, S., Ma, C., Tong, H., Feng, X., & Wang, Q. (2019). Phytochemical and chemotaxonomic study on *Evodia rutaecarpa var. officinalis. Biochemical Systematics and Ecology.* doi: 10.1016/j.bse.2019.103961

Yan, Y., Zhu, H., Zhou, F., Tu, Z., & Cheng, Y. (2019). Phenolic compounds from the insect *blaps japanensis* with inhibitory activities towards cancer cells, COX-2, ROCK1 and JAK3. *Tetrahedron, 75*(8), 1029. doi: 10.1016/j.tet.2018.12.015

Yang, X., Zhang, H., Li, M., Du, L., Yang, Z., & Xiao, S. (2006). Studies on the alkaloid constituents of *Evodia rutaecarpa* (Juss) Benth var. *bodinaieri* (Dode) Huang and their acute toxicity in mice. *Journal of Asian Natural Products Research, 8*(8), 697. doi: 10.1080/10286020412331286425

Ye, S., Liu, Y., Wang, S., Guo, M., Wu, J., Hao, Z., Pan, J., Guan, W., Kuang, H., & Yang, B. (2024). Two new alkaloids isolated from *Dictamnus dasycarpus* turcz. Natural Product Research, 1. doi: 10.1080/14786419.2024.2305207

Yin, S., Boyle, G. M., Carroll, A. R., Kotiw, M., Dearnaley, J., Quinn, R. J., & Davis, R. A. (2010). Caelestines A–D, brominated quinolinecarboxylic acids from the Australian ascidian *Aplidium caelestis. Journal of Natural Products, 73*(9), 1586. doi: 10.1021/np100329w

Yong, X., Wang, B., Wang, M., Lyu, H., Yin, M., Jin, T., … & Wang, Q. (2024). Comprehensive analysis of 11 species of *Euodia* (Rutaceae) by untargeted LC-IT-TOF/MS metabolomics and in vitro functional methods. *Molecules, 29*(5), 1059.

Youn, U. J., Han, S. J., Kim, I. C., & Yim, J. H. (2018). Quinolinone alkaloids from the Arctic Bacterium, *Pseudomonas aeruginosa. Korean Journal of Pharmacognosy, 49*(2), 108–112.

Yu, J., Song, X., Yang, P., Wang, X., & Wang, X. (2018). Alkaloids from *Scindapsus officinalis* (Roxb.) Schott. and their biological activities. *Fitoterapia, 129*, 54. doi: 10.1016/j.fitote.2018.06.006

Zhang, B. (2019). Quinolinone derivatives and their antifungal activities: An overview. *Archiv Der Pharmazie, 352*(5). doi: 10.1002/ardp.201800382

Zhang, F., Li, B., Wen, Y., Liu, Y., Liu, R., Liu, J., Liu, S., & Jiang, Y. (2022a). An integrated strategy for the comprehensive profiling of the chemical constituents of *Aspongopus chinensis* using UPLC-QTOF-MS combined with molecular networking. *Pharmaceutical Biology, 60*(1), 1349–1364.

Zhang, M., Yang, C. L., Xiao, Y. S., Zhang, B., Deng, X. Z., Yang, L., Shi, J., Wang, Y. S., Li, W., Jiao, R. H., Tan, R. X., & Ge, H. M. (2017). Aurachin SS, a new antibiotic from *Streptomyces* sp. NA04227. *The Journal of Antibiotics, 70*(7), 853. doi: 10.1038/ja.2017.50

Zhang, S., Hou, Y., Liu, S., Guo, S., Ho, C., & Bai, N. (2022b). Exploring active ingredients, beneficial effects, and potential mechanism of *Allium tenuissimum* L. flower for treating T2DM mice based on network pharmacology and gut microbiota. Nutrients, *14*(19). doi: 10.3390/nu14193980

Zhang, S., Meng, L., Gao, W., Jia, W., & Duan, H. (2005). A new quinolinone alkaloid with antibacterial activity from *Lappula echinata. Chinese Traditional and Herbal Drugs, 36*(4), 490–492.

Zhang, W., Liu, W. K., & Che, C. (2003). Polyhydroxylated steroids and other constituents of the soft coral *Nephthea chabroli. Chemical and Pharmaceutical Bulletin, 51*(8), 1009–1011.

Zhao, N., Li, Z., Li, D., Sun, Y., Shan, D., Bai, J., Pei, Y., Jing, Y., & Hua, H. (2015). Quinolinone and indole alkaloids from the fruits of *Euodia rutaecarpa* and their cytotoxicity against two human cancer cell lines. *Phytochemistry, 109*, 133. doi: 10.1016/j.phytochem.2014.10.020

Zhu, F., Chen, G., Wu, J., & Pan, J. (2013). Structure revision and cytotoxic activity of marinamide and its methyl ester, novel alkaloids produced by co-cultures of two marine-derived mangrove endophytic fungi. Natural *Product Research, 27*(21), 1960–1964.

Zhu, W., He, H., Fan, L., Shen, Y., Zhou, J., & Hao, X. (2005). Components of stem barks of *Winchia calophylla* A. DC. and their bronchodilator activities. *Journal of Integrative Plant Biology, 47*(7), 892. doi: 10.1111/j.1744-7909.2005.00042.x

Zhuang, P., Wang, X., Chen, J., Zhang, D., Lin, X., & Yang, Y. (2017). Extraction of novel quinolinones alkaloid in *Evodia rutaecarpa* useful as neuroprotector. *China Patent* No. CN106810495.

7 2,4-Quinolinediones

7.1 INTRODUCTION

As explained in Chapter 2, a partially hydrogenated quinoline, 1,2,3,4-tetrahydroquinoline(THQ), is an especially pertinent heterocyclic system that serves as a critical structural feature in many natural and synthetic compounds with potent biological properties (Khadem and Marles 2025a). 2,4(1*H*,3*H*)-Quinolinedione or 2,4-quinolinedione (2,4-QD) (7.1) is a dicarbonyl derivative of THQ (Figure 7.1).

Various tautomeric forms of 2,4-QD exist theoretically between the carbonyl groups, C-3, and NH (Aly et al. 2020; Abdou 2017). These forms have been analysed through different chemical reactivity, spectral, thermochemical, and computational methods. The stability and predominance of these forms depend on several factors, including the state (solid, liquid, and gas), polarity, and pH of the solution. For instance, 4-hydroxy-2(1*H*)-quinolinone [(1a) in Figure 7.2] was found to be the major tautomer in the gas phase (Hebanowska et al. 1986; Elshaier et al. 2022). However, many 2,4-QDs can be sufficiently stable for isolation, characterization, and biological evaluation. 2-quinolinone derivatives, such as (1a), have been reviewed comprehensively (Hamama et al. 2014; Abdou 2018; Proisl et al. 2017; Khadem and Marles 2025b, 2025c, 2025d).

Depicted in Figure 7.3, 2,4-QD contains the structurally interesting elements of a benzylic ketone, a 1,3-dicarbonyl moiety, and a lactam, showing a bioactivity potential for interaction with enzymes and disruption of cellular processes (Shakhatreh et al. 2016). The distinctive reactivity of the benzylic ketone yields antibiotic and antifungal properties (Vásquez-Martínez et al. 2019). The chiral properties of compounds

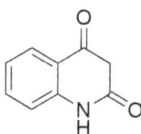

FIGURE 7.1 Chemical structure of 2,4-quinolinedione (2,4-QD).

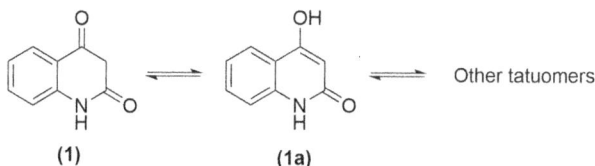

FIGURE 7.2 Tautomeric forms of 2,4-QD.

DOI: 10.1201/9781003534600-7

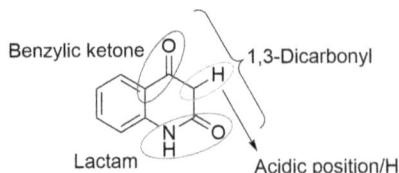

FIGURE 7.3   Structural features of 2,4-QD.

with these characteristics enable selective oxidation and asymmetric transformations during chiral drug synthesis (Lubov et al. 2020). The 1,3-dicarbonyl moiety with a pKa of 9–11 forms stabilized enolate intermediates, which modulate biological interactions via keto-enol tautomerism and metal chelation, which in turn affects solubility, membrane permeability, and covalent modification of biomolecules (Shokova et al. 2015; Caine et al. 2020; Bērziņa and Mieriņa 2023). The lactam contributes to metabolic stability through hydrogen bonding and π–π stacking with pharmaceutical targets. Several proteases resist proteolytic cleavage, thereby enhancing ADMET properties (Tashima 2015; Meiring et al. 2018). The lactam non-planar geometry due to two sp³-hybridized atoms predisposes it to stable and biologically favourable interactions, indicating that 2,4-QD is a possible candidate for pharmaceutical studies (Liu et al. 2013; Han et al. 2015). Our focus is to investigate the presence and bioactivity of 2,4-QDs, which have not been studied comprehensively.

7.1.1   CHEMICAL IDENTITY

The chemical identity (structure, name, and CAS number) of 2,4-QD alkaloids in this chapter is shown in Table 7.1. Some of the reliable literature sources were reviewed to verify the accuracy of these data (e.g. Buckingham et al. 2010; Buckingham 2023).

7.1.2   SYNTHESIS

Most synthetic efforts for 2,4-QD have focused on the 3,3-disubstituted version (Figure 7.4) because of its skeletal abundance. A summary of the comprehensive review by Proisl et al. (2017) will be provided here. Additionally, we will discuss several studies published after 2017. The general strategy for obtaining these compounds (3,3-disubstituted 2,4-QD) involves halogenation, alkylation, the incorporation of aryl groups, or cyclization reactions of suitable precursor molecules such as 4-hydroxy-2-quinolinone. They enable the preparation of both symmetric and asymmetric disubstituted products.

A common method for generating 3-halo derivatives involves chlorinating or brominating 4-hydroxy-2-quinolinones. For example, using sulfuryl chloride in dioxane at mildly elevated temperatures allows selective mono- or dihalogenation at the 3-position. Chlorination can also be performed using chlorine gas, chlorine generated in situ from hydrogen peroxide and hydrochloric acid, or chloride-chlorate mixtures in diluted sulfuric acid. The bromination analogues are synthesized similarly using

TABLE 7.1
The Chemical Identity of 2,4-Quinolinedione Alkaloids

Structure (Number)	Systematic Name	Common Name	CAS #
(7.1)	2,4(1*H*,3*H*)-Quinolinedione	2,4-Quinolinedione (Quinoline-2,4-dione)	52851-41-9
(7.2)	1-Methyl-3,3-bis(3-methyl-2-buten-1-yl)-2,4(1*H*,3*H*)-quinolinedione	3,3-diisopentenyl-N-methyl-2,4-quinoldione	57931-85-8
(7.3)	(+)-3-Heptyl-3-hydroxy-2,4(1*H*,3*H*)-quinolinedione	MY 12-62A	69808-30-6
(7.4)	3-Benzoyl-3-(phenylmethyl)-2,4(1*H*,3*H*)-quinolinedione	–	70611-42-6
(7.5)	3-(1,1-Dimethyl-2-propen-1-yl)-3-(3-methyl-2-buten-1-yl)-2,4(1*H*,3*H*)-quinolinedione	Buchapine	84017-97-0
(7.6)	3,3-Bis[(2E)-3,7-dimethyl-2,6-octadien-1-yl]-6-hydroxy-2,4(1*H*,3*H*)-quinolinedione	Severibuxine	219998-24-0
(7.7)	rel-(2*S*,3*S*)-3-[(1*R*)-1-(1-Hydroxy-1-methylethyl)-4-methyl-3-penten-1-yl] spiro[oxirane-2,3′(2′*H*)-quinoline]-2′,4′(1′*H*)-dione	Haplotubinone	352320-42-4
(7.8)	(+)-3-Hydroxy-2,5′,6-trimethoxy-7′-methylspiro [2,5-cyclohexadiene-1,3′(2′*H*)-quinoline]-2′,4,4′(1′*H*)-trione	Aspertumoid	848073-02-9
(7.9)	Methyl 1,2,3,4-tetrahydro-3-hydroxy-1-methyl-2,4-dioxo-3-quinolineacetate	–	1159999-39-9

(Continued)

TABLE 7.1 (*Continued*)
The Chemical Identity of 2,4-Quinolinedione Alkaloids

Structure (Number)	Systematic Name	Common Name	CAS #
(7.10)	3-Hydroxy-3-(2-oxopropyl)-2,4(1*H*,3*H*)-quinolinedione	–	1989563-15-6
(7.11)	3-Hydroxy-1-methyl-3-(2-oxopropyl)-2,4(1*H*,3*H*)-quinolinedione	–	1159999-40-2
(7.12)	(3*S*)-3-Hydroxy-1-methyl-3-(3-methyl-2-buten-1-yl)-2,4(1*H*,3*H*)-quinolinedione	Clausenaside G	2011753-43-6
(7.13)	(3*R*)-3-Methoxy-1-methyl-3-(3-methyl-2-buten-1-yl)-2,4(1*H*,3*H*)-quinolinedione	Dunnine E	2229853-41-0
(7.14)	(3*S*)-3-Hydroxy-3-methyl-2,4(1*H*,3*H*)-quinolinedione	(−)-(3S)-Isatindinoline D	2449214-31-5
(7.15)	(3*R*)-3-Hydroxy-3-methyl-2,4(1*H*,3*H*)-quinolinedione	(+)-(3R)-Isatindinoline D	2449214-30-4
(7.16)	Methyl (3*S*)-1,2,3,4-tetrahydro-3-hydroxy-2,4-dioxo-3-quinolineacetate	(+)-(3S)-Isatindinoline C	2449214-29-1
(7.17)	Methyl (3*R*)-1,2,3,4-tetrahydro-3-hydroxy-2,4-dioxo-3-quinolineacetate	(−)-(3R)-Isatindinoline C	2449214-28-0
(7.18)	6,8-dimethoxy-3-methyl-3-(3′-methylbut-2′-enyl)-1*H*-quinoline-2,4-dione	–	–
(7.19)	3-dihydro-spiro[2(1*H*), 3′(1′*H*)-diquinoline]-3′,4,4′-trione	Spirodiquinolinone	–

(Continued)

TABLE 7.1 (*Continued*)
The Chemical Identity of 2,4-Quinolinedione Alkaloids

Structure (Number)	Systematic Name	Common Name	CAS #
(7.20)	3,3-diisopropyl-1-methylquinoline-2,4(1*H*,3*H*)-dione	–	–
(7.21)	3-n-Nonyl-3-hydroxy-1,2,3,4-tetrahydroquinoline-2,4-dione	–	–
(7.22)	1,3-dimethyl-3-(3-methylbut-2-en-1-yl)quinoline-2,4(1*H*,3*H*)-dione	–	–

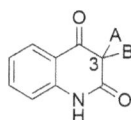

FIGURE 7.4   3,3-Disubstituted 2,4-QD.

elemental bromine or various brominating agents such as *N*-bromosuccinimide or pyridinium bromide perbromide. Solvents have been employed for this transformation, including acetic acid, formic acid, aqueous dioxane, benzene, or carbon tetrachloride. These 3-halo derivatives serve as versatile intermediates for nucleophilic substitution reactions. Halogen atoms at the 3-position can be displaced with nucleophiles such as amines, azides, or fluoride, leading to various functionalized 2,4-QD structures. Typically, chlorinated precursors are treated with potassium fluoride in the presence of a phase-transfer catalyst such as 18-crown-6 ether to achieve fluorination. 3-Alkyl-substituted 4-hydroxy-2-quinolinones were successfully iodinated with iodine/potassium iodide solutions (Lugol's reagent) under basic aqueous conditions. Halogenated quinolinone compounds have the potential to be used as synthetic intermediates, and subsequently, more complex and functional derivatives can be made.

The oxidation reactions of 4-hydroxy-2-quinolinones can be utilized to synthesize 3-hydroxy-2,4-QD. The oxidizing agents are hydrogen peroxide, peroxyacetic acid, and 3-chloroperoxybenzoic acid. Alternatively, a nitration reaction followed by hydrolysis leads to the desired hydroxy diones.

Treating 3-unsubstituted 4-hydroxy-2-quinolinone with an alkyl halide and a base like lithium hydroxide or sodium hydroxide afforded 3,3-dialkyl 2,4-QD. However, in some cases, the proposed reaction may be hindered due to side reactions. For instance, in the case of propargyl bromide, there could be complex mixtures obtained.

Also, 3,3-dialkyl-, 3-alkyl-, and 3-aryl-2,4-QD have been analogously synthesized through alkylation of monosubstituted 4-hydroxy-2-quinolinones with alkyl halides using copper catalysis in aqueous media. Cyclization strategies like Dieckmann condensation with anthranilate derivatives and strong bases [e.g. lithium bis(trimethylsilyl)amide] allow disubstituted quinolinediones to be formed.

Another synthetic method involves the reaction of isatoic anhydride with 2-alkyl-2-(diethoxyphosphoryl)acetates under basic conditions for carboxylated derivatives. *N*-Methylisatoic anhydride reacts with silyl ketene acetals in a Mukaiyama-type reaction with titanium (IV) chloride catalyst to give trimethyl-substituted diones. These methods offer further opportunities for Friedel-Crafts-type side-chain functionalization.

A novel multicomponent and radical cascade strategy has been introduced for constructing disubstituted 2,4-QD. One-pot reactions involving diphenylamine, diethyl malonate, and substituted benzaldehydes proceed under mild conditions with zirconium oxide catalysis. Radical additions starting from acrylamidobenzonitrile derivatives have also been demonstrated using sulfinates, aldehydes, hydrazides, or alcohols as radical sources. These reactions proceed in aqueous conditions, tolerate various functional groups, and afford products in moderate to excellent yields. However, they are limited by challenges in preparing certain derivatives, such as 1-unsubstituted or 1-acetyl derivatives.

At atmospheric pressure, a multicomponent palladium catalyst reaction was conducted with o-alkynylanilines, aryl iodides, and carbon dioxide. This reaction, as reported by Wang et al. in (2017), produced 3,3-diaryl 2,4-quinazolinones (2,4-QD). The reaction involved several steps: first, sequential carboxylation took place, followed by trans-oxopalladation of the alkyne bond (C≡C) by ArPdX species, and finally, a reductive elimination process. This sequence led to the formation of benzoxazine-2-ones that contained tetrasubstituted vinyl fragments. Subsequently, a rearrangement of these benzoxazine-2-ones occurred, resulting in the formation of 3,3-diaryl 2,4-QD. This process demonstrated the simultaneous formation of four bonds in a single reaction vessel in moderate to excellent yields under mild conditions.

An approach to the synthesis of 3,3-dialkyl 2,4-QD was developed based on a methyl triflate (TfOMe)-promoted intramolecular Houben–Hoesch reaction of α,α-dialkyl-substituted cyanoacetanilides (Zhong et al. 2018).

Recently, structurally diverse 2,4-QD compounds were synthesized under metal-free and photocatalyst-free conditions through a visible light-promoted cascade sulfonylation and cyclization reaction, using selenosulfonates (such as TsSePh) as sources of sulfonyl-centred radicals (Zhang et al. 2023). The reactions were performed at room temperature, and various substituents (halogen, alkyl, and aryl) and substituted products were obtained.

The deprotonation of a few benzo[*e*][1,4]diazepine-2,5-diones resulted in a ring contraction to the corresponding 2,4-QD substances with high enantioselectivity (Antolak et al. 2014). This discovery is exciting because benzo[*e*][1,4]diazepine-2,5-diones are significant in medicinal chemistry, and several synthetic strategies have been developed.

A great example of the synthesis of other 2,4-QD derivatives was provided by Liu et al. (2013). They prepared a series of imine derivatives (diketone enamine):

3-(1-aminoethylidene) 2,4-QD. Treatment of methyl 2-aminobenzoate with a diketene afforded an ester amide. Condensation of this ester amide in the presence of sodium gave the key intermediate 3-acetyl 4-hydroxy-2-quinolinone, which reacted with alkyl amines to afford 3-(1-aminoethylidene) 2,4-QD derivatives. The bioassay results indicated that most compounds displayed good to excellent herbicidal activities.

7.1.3 STRUCTURE–ACTIVITY RELATIONSHIP (SAR)

2,4-QD features a benzylic ketone, a 1,3-dicarbonyl system, and a lactam ring, making it capable of forming various intermolecular interactions with biological targets (Khadem and Marles 2024). Its aromatic ring offers a stable scaffold for substitution, allowing precise tuning of electronic and steric properties that affect biological binding. These characteristics make the scaffold suitable for broad pharmacological applications. Structure–Activity Relationship (SAR) studies use rational design, synthesis, and evaluation strategies supported by computational tools like comparative molecular field analysis, comparative molecular similarity indices analysis, and molecular docking. Experimental models such as zebrafish embryos and encephalomyelitis mice help validate therapeutic potential (Han et al. 2015). These integrated approaches reveal how specific structural modifications influence bioactivity.

Initially explored for antimalarial activity (Smeyne 2020), the 2,4-QD scaffold has since been examined for diverse therapeutic uses. It includes modulation of cannabinoid receptors, anticancer effects, and antimicrobial and antiviral activity (Han et al. 2015). Substitution patterns on the quinoline core strongly influence receptor interactions. For example, C5 and C8 substitution favours Cannabinoid Receptor 2 (CB2) receptor agonism, with 8-substituted analogues showing greater selectivity and potency, particularly when bulky groups like adamantyl or long alkyl chains are added. These compounds show efficacy in reducing inflammation and demyelination in multiple sclerosis models. In contrast, C6 and C7 substitutions yield CB2 receptor antagonists, with 7-methyl derivatives showing the highest potency (Han et al. 2015).

The scaffold also demonstrates anticancer potential by inhibiting kinases and affecting microtubule dynamics (Aref et al. 2023). Derivatives such as pyrimidoquinolinediones (Dou et al. 2013) and pyranoquinolinones (Elshaier et al. 2022) inhibit targets such as epidermal growth factor receptor, vascular endothelial growth factor receptor, and Aurora kinases. Hybrid molecules, including those combining quinolinedione with naphthoquinone (Mancini et al. 2022) or spiro-quinazoline motifs (Divar et al. 2023), have improved activity profiles and are being explored for oncological applications. Antimicrobial activity is achieved by inhibiting bacterial enzymes like DNA gyrase and topoisomerase IV, particularly in quinazoline-dione and 4-aminoquinoline hybrids (Zayed 2023; Ubaid et al. 2024). Substituents such as triazoles and oxadiazoles at nitrogen sites enhance antibacterial effects (Boshta et al. 2022). Some derivatives are also active against fungi, with efficacy influenced by substituent electronics (Kadela-Tomanek et al. 2019).

Quinolinedione-based antiviral compounds show efficacy against HIV, influenza, and coronaviruses (Han et al. 2015). Buchapine and its analogues display high anti-HIV activity, which is lost upon aromatic ring substitution (Proisl et al. 2017), indicating strict structural requirements. Other active structures include 3-aryl-quinolinones for influenza (Ahmed et al. 2010) and aminoquinazolinones for SARS-CoV-2 (Lee et al. 2021). Hybrids bearing pyrazole, triazole, or oxadiazole groups enhance antiviral spectra, including activity against tobacco mosaic virus (Liao et al. 2024). Additionally, biological effects include anticonvulsant activity, sleep modulation, and selective serotonin receptor antagonism (Proisl et al. 2017), suggesting utility in neuropsychiatric conditions. Some compounds also serve non-therapeutic roles as UV absorbers, antioxidants, or antifungal agents in materials science (Han et al. 2015).

SAR analyses emphasize the importance of substitution patterns. C8-substitution and N1-pentyl chains enhance cannabinoid receptor type 2 agonism, while C6/C7 substitutions favour antagonist behaviour (Han et al. 2015). Heterocyclic additions, such as triazoles or oxadiazoles, improve antibacterial properties (Boshta et al. 2022). Sulfur-containing (Dou et al. 2013) or arylamino groups (Kadela-Tomanek et al. 2019) boost anticancer effects. Methoxy and halogen substitutions can enhance or impair activity depending on position and target (Chen et al. 2023; Proisl et al. 2017). For example, chlorination improves sleep-potentiating effects but can abolish anti-HIV activity. Functional groups that enable hydrogen bonding, such as hydroxyl and amino groups, are also important, particularly in antiviral and anticancer settings. Maintaining the 2,4-dione motif is critical, as carbonyl changes generally reduce activity (Han et al. 2015).

As mentioned above, hybridization with other bioactive scaffolds remains an effective strategy for enhancing biological profiles. Examples include antibacterial and antifungal 4-aminoquinoline-benzohydrazide hybrids (Ubaid et al. 2024), anti-tumour naphthoquinone conjugates (Mancini et al. 2022), and antiviral triazole-quinolinones (Liao et al. 2024). These hybrids allow multiple mechanisms of action and can improve pharmacokinetics or expand target coverage. Selectivity is enhanced through precise functional group placement, as seen in selective CB2 agonists and 5-HT$_6$ antagonists. However, inappropriate substitutions, such as altering critical side chains or replacing optimal groups, can diminish activity or introduce toxicity (Qian et al. 2017), highlighting the necessity of detailed SAR understanding.

In neurology, 2,4-QD shows promise in treating multiple sclerosis, epilepsy, and possibly Alzheimer's and Parkinson's diseases. Its anticancer applications span kinase inhibition and microtubule interference. Its antimicrobial and antiviral derivatives address resistant bacterial strains and emerging viruses like SARS-CoV-2 (Lee et al. 2021). Moreover, 2,4-QD physicochemical properties support broader industrial and material science use (Han et al. 2015).

7.2   NATURAL OCCURRENCE

Woad, *Isatis tinctoria* L. (Brassicaceae), has a history of use as a medicinal plant and a source of blue dye. It has been cultivated for ornamental purposes and used as animal feed. In recent years, it has also gained popularity in the cosmetic industry. Numerous studies conducted in the past have provided substantial evidence supporting the traditional uses of *I. tinctoria*. These studies have shown that extracts or compounds derived from different parts of the plant have a wide range of biological activities (Speranza et al. 2020). Xi et al. (2019) isolated several pairs of alkaloid enantiomers from the leaves of *I. tinctoria*. Among those enantiomers, (−)-(3S)-isatindinoline D (7.14), (+)-(3R)-isatindinoline D (7.15), (+)-(3S)-isatindinoline C (7.16), and (−)-(3R)-isatindinoline C (7.17) are 2,4-QD alkaloids. These alkaloids serve as valuable chemotaxonomic markers for identifying this species (Qiu et al. 2024). 3-Hydroxy-3-(2-oxopropyl)-2,4(1*H*,3*H*)-quinolinedione (7.10) has been isolated from the bulbils of the Chinese yam (*Dioscorea opposita* Thunb., a synonym of *D. oppositifolia* L., Dioscoreaceae) (Chen et al. 2022). Limeberry (*Micromelum falcatum* (Lour.) Tanaka, a synonym of *M. minutum* (G.Forst.) Wight & Arn., Rutaceae) is a citrus species of the Southern Hemisphere tropics whose leaves are used in folk medicine, are well-studied chemically and pharmacologically (Thant et al. 2020). 3-Hydroxy-1-methyl-3-(2-oxopropyl)-2,4(1*H*,3*H*)-quinolinedione (7.11) and methyl 1,2,3,4-tetrahydro-3-hydroxy-1-methyl-2,4-dioxo-3-quinolineacetate (7.9) have been isolated from the stem bark of *M. falcatum* (Luo et al. 2009; Zhang et al. 2011). In a study to investigate the essential oil composition of chaste tree fruits (*Vitex agnus-castus* L., Lamiaceae), 1,3-dimethyl-3-(3-methylbut-2-en-1-yl) quinoline-2,4(1*H*,3*H*)-dione (7.22) was found in a concentration of about 0.9% (Tin et al. 2017). Baja California Jopoy (*Esenbeckia flava* Brandegee, Rutaceae) contains several alkaloids, limonoids, phenolic derivatives, steroids, and terpenoids, which highlights its potential as a source of bioactive metabolites (Carvalho et al. 2022). The 2,4-QD alkaloid 3,3-diisopropyl-1-methylquinoline-2,4(1*H*,3*H*)-dione (7.20) was found as a major component in the wood extracts of *E. flava* (Dreyer 1980). Clausenaside G (7.12) was obtained from the stems of wampee (*Clausena lansium* (Lour.) Skeels, Rutaceae) (Liu et al. 2015). Within the Rutaceae plant family, dunnine E (7.13) is present in *Clausena dunniana* H.Lév. (synonym of *C. anisata* (Willd.) Hook.f. ex Benth.), *Glycosmis craibii* var. glabra (Craib) *Yu.Tanaka* (synonym of *G. ovoidea* Pierre), and the curry leaf tree (*Murraya koenigii* (L.) Spreng., synonym of *Bergera koenigii* L.) (Cao et al. 2018; Chen et al. 2022; Wei et al. 2020). 3-Heptyl-3-hydroxy-2,4(1*H*,3*H*)-quinolinedione (7.3) is sourced from various strains of bacteria, primarily from the Pseudomonadaceae family. The sources include different strains of *Pseudomonas* sp., *Pseudomonas methanica*, *Pseudomonas azotoformans*, and *Pseudomonas aeruginosa*, as well as *Amycolatopsis saalfeldensis* and *Streptomyces sp. strain MBTG13*. Compound 7.21 was first identified in 1945 by Hays and co-workers. Japanese Chestnut (*Castanea crenata* Siebold & Zucc., Fagaceae) is native to Korea and Japan, and an interesting heterocyclic spiro compound, spirodiquinolinone (7.19), was isolated from chestnut honey by Cho and co-workers in 2015. It consists of a 2,4-QD/4-oxo-THQ bicycle with only one common carbon. Buchapine (7.5) is a 2,4-QD alkaloid found in plants from the Rutaceae family, such as *Haplophyllum bucharicum* Litv., *Haplophyllum tuberculatum* (Forssk.) A.Juss., and *Euodia roxburghiana* (Cham.) Benth. A recent study by Chóez-Guaranda et al. (2023) analysed the chemical properties of extracts from four fungal species of *Trichoderma* (Hypocreaceae) in Ecuador. The alkaloid 6,8-dimethoxy-3-methyl-3-(3′-methylbut-2′-enyl)-1*H*-quinoline-2,4-dione (7.18) was isolated from *Trichoderma harzianum* Rifai, known for its mycoparasitic properties. Another alkaloid, 3-*n*-nonyl-3-hydroxy-1,2,3,4-tetrahydroquinoline-2,4-dione (7.21), was isolated from a bacterial culture of *Pseudomonas aeruginosa* (Pseudomonadaceae) and the ethanolic extract of the aerial parts of Orange Jessamine (*Murraya paniculata* (L.) Jack, Rutaceae). From the trunk bark, wood, essential oil, and roots of *Esenbeckia almawillia* Kaastra (Rutaceae), the 2,4-QD alkaloid 1-methyl-3,3-bis(3-methyl-2-buten-1-yl)-2,4(1*H*,3 *H*)-quinolinedione (7.2) was isolated. This compound is an important chemotaxonomic marker for the genus *Esenbeckia*. Haplotubinone (7.7) was isolated from *Haplophyllum tuberculatum* (Forssk.) A.Juss. and *Fissistigma oldhamii* (Hemsl.) Merr. The relative stereochemistry of (7.7) was confirmed by X-ray crystallography. 3-Benzoyl-3-(phenylmethyl)-2,4(1*H*,3*H*)-quinolinedione (7.4) was isolated from the essential oil of Rabdosia (*Rabdosia rubescens* (Hemsl.) H.Hara), volatile constituents of *Corydalis speciosa* Maxim., Diospyros lotus Lour., and *Diospyros kaki* Thunb., as well as Noni fruit extract (*Morinda citrifolia* L.). Another compound, asperfumoid (7.8), has been found in several fungi, including *Penicillium* spp., *Aspergillus fumigatus, and Myrothecium roridum IFB-E091*. Severibuxine (7.6) was discovered in the root bark of Chinese Box-orange (*Severinia buxifolia* (Poir.) Ten.).

Table 7.2 presents the list of natural sources of 2,4-QD s derived from bacteria, fungi, and plants.

TABLE 7.2
Natural Occurrence of 2,4-Quinolinedione Alkaloids

Kingdom	Family	Genus	Species	Compound(s)	References
Bacteria	Pseudomonadaceae	*Pseudomonad*	–	7.3	Debitus et al. (1998)
Bacteria	Pseudomonadaceae	*Pseudomonas*	*azotoformans* strain UICC B-91	7.3	Pratiwi et al. (2022)
Bacteria	Pseudomonadaceae	*Pseudomonas*	*methanica* KY4634	7.3	Kitamura et al. (1986)
Bacteria	Pseudomonadaceae	*Pseudomonas*	*sp. 20#-5*	7.3	Lan (1992)
Bacteria	Pseudomonadaceae	*Pseudomonas*	*aeruginosa*	7.3	Martínez-Luis et al. (2019); Neuenhaus et al. (1979); Hays et al. (1945)
Bacteria	Pseudomonadaceae	*Pseudomonas*	*aeruginosa* BCC76810	7.3	Supong et al. (2016)
Bacteria	Pseudomonadaceae	*Pseudomonas*	*aeruginosa*	7.21	Neuenhaus et al. (1979); Hays et al. (1945)
Bacteria	Pseudonocardiaceae	*Amycolatopsis*	*saalfeldensis (LL2-4B)*	7.3	Zhu et al. (2021)
Bacteria	Streptomycetaceae	*Streptomyces*	*sp. strain MBTG13*	7.3	Kim et al. (2019)
Fungi	Aspergillaceae	*Aspergillus*	*fumigatus CY018*	7.8	Liu et al. (2004)
Fungi	Aspergillaceae	*Penicillium*	*spp.*	7.8	Wang et al. (2008)

(Continued)

TABLE 7.2 (*Continued*)
Natural Occurrence of 2,4-Quinolinedione Alkaloids

Kingdom	Family	Genus	Species	Compound(s)	References
Fungi	Hypocreaceae	*Trichoderma*	*spp.*	7.18	Chóez-Guaranda et al. (2023)
Fungi	Stachybotryaceae	*Myrothecium*	*roridum IFB-E091*	7.8	Shen et al. (2015)
Plantae	Annonaceae	*Fissistigma*	*oldhamii*	7.7	Hu et al. (2021)
Plantae	Brassicaceae	*Isatis*	*tinctoria*	7.14, 7.15, 7.16, 7.17	Xi et al. (2019)
Plantae	Dioscoreaceae	*Dioscorea*	*opposite Thunb.*	7.10	Chen et al. (2022)
Plantae	Ebenaceae	*Diospyros*	*lotus*	7.4	Dekebo et al. (2022)
Plantae	Ebenaceae	*Diospyros*	*kaki*	7.4	Dekebo et al. (2022)
Plantae	Fagaceae	*Castanea*	*crenata*	7.19	Cho et al. (2015)
Plantae	Lamiaceae	*Rabdosia*	*rubescens*	7.4	Wu et al. (2010)
Plantae	Papaveraceae	*Corydalis*	*speciosa*	7.4	Dekebo et al. (2022)
Plantae	Rubiaceae	*Morinda*	*citrifolia*	7.4	Moh et al. (2024)
Plantae	Rutaceae	*Esenbeckia*	*almawillia*	7.2	Barros-Filho et al. (2004); Nunes et al. (2006)
Plantae	Rutaceae	*Haplophyllum*	*bucharicum*	7.5	Ahmed et al. (2010); Nesmelova et al. (1982a, 1982b)
Plantae	Rutaceae	*Haplophyllum*	*tuberculatum*	7.5	Ahmed et al. (2010); Sheriha et al. (1987)
Plantae	Rutaceae	*Euodia*	*roxburghiana*	7.5	Ahmed et al. (2010); McCormick et al. (1996)
Plantae	Rutaceae	*Severinia*	*buxifolia*	7.6	Ito et al. (2013); Wu et al. (1998)
Plantae	Rutaceae	*Haplophyllum*	*tuberculatum*	7.7	Al-Rehaily et al. (2001)
Plantae	Rutaceae	*Micromelum*	*falcatum*	7.9	Luo et al. (2009); Zhang et al. (2011)
Plantae	Rutaceae	*Micromelum*	*falcatum*	7.11	Luo et al. (2009)
Plantae	Rutaceae	*Clausena*	*lansium*	7.12	Liu et al. (2015)
Plantae	Rutaceae	*Clausena*	*dunniana*	7.13	Cao et al. (2018)
Plantae	Rutaceae	*Glycosmis*	*craibii var. glabra*	7.13	Chen et al. (2022)
Plantae	Rutaceae	*Murraya*	*koenigii*	7.13	Wei et al. (2020)
Plantae	Rutaceae	*Esenbeckia*	*flava*	7.20	Dreyer et al. (1980)
Plantae	Rutaceae	*Murraya*	*paniculata*	7.21	Shah et al. (2014)
Plantae	Verbenaceae	*Vitex*	*agnus-castus*	7.22	Tin et al. (2017)

7.3   BIOLOGICAL ACTIVITY

The protective effects of compound 7.10 on renal cells and nerve cells were studied using an adriamycin-induced normal rat kidney 52E and corticosterone-induced rat pheochromocytoma (PC-12) cell injury model, respectively. Researchers reported the cell viability for normal rat kidney-52E and PC-12 as 69% and 85%, indicating moderate renal and nerve cell protection (Chen et al. 2022). The toxicity of compounds 7.9 and 7.11 was tested towards brine shrimp larvae; their LD_{50} values were 355 and 143 µg/mL (Luo et al. 2009). Scientists tested the activity of these compounds against HNNY-LQ-8601 breast cancer cells by the thiazolyl blue (MTT) method. They found that compound 7.9 has an average half-inhibitory concentration IC_{50} value of 357 µg/mL (Zhang et al. 2011). In addition, the MTT method was used to evaluate the activity of these compounds against HNNY-LQ-8602 lung cancer cells, and compound 7.11 has an IC_{50} value of 66 µg/mL. Investigators measured the anti-adhesion activity (antifouling) of compounds 7.9 and 7.11 against barnacle larvae, giving IC_{50} values of 121.5 and 13.1 µg/mL, respectively (Zhang et al. 2011). *Vitex agnus-castus* [containing compound 7.22] has been approved by the German health authorities for use in treating menstrual cycle irregularities, premenstrual syndrome, and mastalgia (breast pain) due to its significant effects on hormones (Niroumand et al. 2018). Health Canada (2024) approved using natural health products with *Vitex agnus-castus* fruit for purposes including relief of premenstrual symptoms, to stabilize menstrual cycle irregularities, and to help relieve symptoms associated with menopause, such as hot flashes. Although several alkaloids from *C. lansium* showed moderate inhibitory effects on LPS-induced NO production in murine microglial BV2 cells, compound 7.12 had an IC_{50} value of over 10 µM. Wampee contains several bioactive compounds that may offer various health benefits, including antioxidative properties, neuroprotection, hepatoprotective, antidiabetic, anticancer, anti-inflammatory, and antimicrobial activities (Huang et al. 2023). Dunnine E (7.13) demonstrated a strong neuroprotective effect against 6-hydroxydopamine (6-OHDA)-induced apoptosis in neuron-like PC12 cells, with an IC_{50} value of 16.3 µM (Cao et al. 2018). It was also tested for its hepatoprotective activity using an MTT colourimetric assay to assess its effect on cytotoxicity induced by D-galactosamine in the human hepatic cell line (HL-7702). At a concentration of 10 µM, it exhibited moderate hepatoprotective activity, resulting in a cell survival rate of 79.5% compared to 53.2% inhibition in the control. Bicyclol, a drug with hepatoprotective activity, was used as the positive control (Wei et al. 2020). Compound 7.3 was discovered in 1945 by Hays and co-workers, making it the first 2,4-QD compound to be identified. Despite the inability to determine the exact structure of compound 7.3 at that time, it was found to have an antibacterial effect against several Gram-positive organisms. 5-Lipoxygenase (5-LO) inhibitors prevent the 5-LO enzyme from producing inflammatory leukotrienes. Compound 7.3 was identified as a moderate and selective inhibitor of 5-LO in rat basophilic leukaemia cells in a dose-dependent manner, with an IC_{50} value of 19 µM (Kitamura et al. 1986).

Additionally, it exhibited inhibitory activity against hyphal growth induction in the dimorphic fungus *Candida albicans* by inhibiting the expression of mRNAs related to the cAMP-Efg1 pathway, which is a key regulator of cell wall dynamics and hyphal growth in *C. albicans* (Kim et al. 2019). A study by Martínez-Luis et al. (2019) found that compound 7.3 displayed significant antiparasitic activity against *Plasmodium falciparum*, the parasite responsible for malaria, with an IC_{50} value of 3.47 µg/mL. Buchapine (7.5) exhibited moderate anti-HIV activity in an XTT-tetrazolium assay against HIV-1 in cultured human lymphoblastoid (CEM-SS) cells, with an EC_{50} value of 0.94 µM and an IC_{50} of 29.0 µM. Its potential as an anti-HIV agent was further emphasized by its inhibitory activity ($IC_{50} = 12$ µM) in an HIV-1 reverse transcriptase (RT) assay (McCormick et al. 1996). HIV-1 is known to cause the loss of CD4+T lymphocytes, either by directly killing the cells or indirectly by impairing their function, which ultimately results in cell destruction by apoptosis (Oxenius et al. 2001). The anti-HIV potential of buchapine was tested on a human CD4+T cell line (CEM-GFP) infected with HIV-1NL4.3 virus, using p24 antigen capture ELISA assay. The results showed that buchapine had a potent inhibitory activity against the virus, with an IC_{50} value of 2.99 µM (Ahmed et al. 2010). The extract containing compound 7.18 inhibited the growth of *Moniliophthora perniciosa*, a fungus causing Witches' Broom disease, maximally at 10 µg/mL (39.04% ± 2.43%). The same extract at a concentration of 1 µg/mL slightly inhibited the growth of another parasite, *Moniliophthora roreri*, the fungus causing Frosty Pod Rot disease, one of the most serious problems for cacao. However, at 1000 µg/mL that extract conversely slightly stimulated the growth of *M. roreri*. The authors explained that this was due to the metabolites in the extract at higher concentrations favouring the growth of the pathogen (Chóez-Guaranda et al. 2023). Compound 7.21 exhibited antibacterial efficacy against several Gram-positive bacteria but demonstrated less activity against Gram-negative bacteria (Hays et al. 1945). Compound 7.2 is cytotoxic in a few tumour cell lines, as determined by the MTT assay. Alkaloid 7.2 has shown activity against the HL-60 (human leukaemia), CEM (human leukaemia), HCT-8 (human colon), and MCF-7 (human breast) cell lines, with IC_{50} values of 9.2, 13.9, 21.3, and 20.6 µg/mL, respectively (Nunes et al. 2006). Asperfumoid (7.8) was tested in an *in vitro* antifungal activity assay against three human pathogenic fungi, *Candida albicans*, *Trichophyton rubrum*, and *Aspergillus niger*. Compound 7.8 showed activity against *C. albicans* with a Minimum Inhibitory Concentration (MIC) of 75.0 µg/mL (Liu et al. 2004). In another study, asperfumoid inhibited the growth of *C. albicans* with an MIC of 20.0 µg/mL (Wang et al. 2008). This compound exhibited cytotoxic activity against the human nasopharyngeal epidermoid tumour KB cell line with an IC_{50} value of 20.0 µg/mL and against the human liver cancer HepG2 cell line with an IC_{50} of 15.0 µg/mL.

The summarized biological activities of 2,4-QD alkaloids are presented in Table 7.3.

TABLE 7.3
Biological Activity of 2,4-Quinolindione Alkaloids

Biological Activity	Target	Assay/Method	Results	Compound	References
Antibacterial	Gram-positive bacteria	Growth inhibition	Effective	7.3	Hays et al. (1945)
Antibacterial	Gram-positive bacteria	Growth inhibition	Effective	7.21	Hays et al. (1945)
Antibacterial	Gram-negative bacteria	Growth inhibition	Less effective	7.21	Hays et al. (1945)
Anticancer	HNNY-LQ-8601 breast cancer cells	MTT assay	$IC_{50} = 357\ \mu g/mL$	7.9	Zhang et al. (2011)
Anticancer	HNNY-LQ-8602 lung cancer cells	MTT assay	$IC_{50} = 66\ \mu g/mL$	7.11	Zhang et al. (2011)
Antifouling	*Barnacle larvae*	Adhesion inhibition assay	$IC_{50} = 121.5\ \mu g/mL$	7.9	Zhang et al. (2011)
Antifouling	*Barnacle larvae*	Adhesion inhibition assay	$IC_{50} = 13.1\ \mu g/mL$	7.11	Zhang et al. (2011)
Antifungal	*Candida albicans*	Hyphal growth inhibition	Inhibits cAMP-Efg1 pathway	7.3	Kim et al. (2019)
Antifungal	*Moniliophthora perniciosa*	Growth inhibition	$39.04\% \pm 2.43\%$ at 10 μg/mL	7.18	Chóez-Guaranda et al. (2023)
Antifungal	*Moniliophthora roreri*	Growth modulation	Inhibition at 1 μg/mL, stimulation at 1,000 μg/mL	7.18	Chóez-Guaranda et al. (2023)
Antifungal	*Candida albicans*	MIC determination	$MIC = 20.0–75.0\ \mu g/mL$	7.8	Liu et al. (2004); Wang et al. (2008)
Anti-HIV	CEM-SS cells	XTT assay	$EC_{50} = 0.94\ \mu M; IC_{50} = 29\ \mu M$	7.5	McCormick et al. (1996)
Anti-HIV	CEM-GFP cells	p24 ELISA	$IC_{50} = 2.99\ \mu M$	7.5	Ahmed et al. (2010)
Anti-inflammatory	BV2 microglia	LPS-induced NO production	$IC_{50} > 10\ \mu M$	7.12	Huang et al. (2023)
Anti-inflammatory	5-LO enzyme	Leukotriene inhibition assay (RBL cells)	$IC_{50} = 19\ \mu M$	7.3	Kitamura et al. (1986)

(Continued)

TABLE 7.3 (Continued)

Biological Activity of 2,4-Quinolindione Alkaloids

Biological Activity	Target	Assay/Method	Results	Compound	References
Antimalarial	*Plasmodium falciparum*	Growth inhibition assay	IC_{50} = 3.47 µg/mL	7.3	Martinez-Luis et al. (2019)
Cytotoxic	HL-60 cells	MTT assay	IC_{50} = 9.2 µg/mL	7.2	Nunes et al. (2006)
Cytotoxic	CEM cells	MTT assay	IC_{50} = 13.9 µg/mL	7.2	Nunes et al. (2006)
Cytotoxic	HCT-8 cells	MTT assay	IC_{50} = 21.3 µg/mL	7.2	Nunes et al. (2006)
Cytotoxic	MCF-7 cells	MTT assay	IC_{50} = 20.6 µg/mL	7.2	Nunes et al. (2006)
Cytotoxic	KB (nasopharyngeal) cells	MTT assay	IC_{50} = 20.0 µg/mL	7.8	Liu et al. (2004)
Cytotoxic	HepG2 (liver) cells	MTT assay	IC_{50} = 15.0 µg/mL	7.8	Liu et al. (2004)
Hepatoprotective	HL-7702 cells	D-galactosamine-induced injury	79.5% cell survival at 10 µM	7.13	Wei et al. (2020)
HIV-1 RT inhibition	Reverse transcriptase	RT inhibition assay	IC_{50} = 12 µM	7.5	McCormick et al. (1996)
Hormonal regulation	PMS, menopause, mastalgia	Clinical use (Germany, Canada)	Approved	7.22	Niroumand et al. (2018)
Nerve cell protection	PC-12 cells	Corticosterone-induced injury	85% cell viability	7.10	Chen et al. (2022)
Neuroprotective	PC-12 cells	6-OHDA-induced apoptosis	IC_{50} = 16.3 µM	7.13	Cao et al. (2018)
Renal cell protection	NRK-52E cells	Adriamycin-induced injury	69% cell viability	7.10	Chen et al. (2022)
Toxicity	Brine shrimp larvae	In vivo bioassay	LD_{50} = 355 µg/mL	7.9	Luo et al. (2009)
Toxicity	Brine shrimp larvae	In vivo bioassay	LD_{50} = 143 µg/mL	7.11	Luo et al. (2009)

7.4 DISCUSSION

This chapter showed that many organisms produce 2,4-QD, which is a crucial part of many natural products. These natural products show various pharmacological activities and promise, including cytotoxic (anticancer), antiviral, and neuroprotective effects. The 2,4-QD skeleton significantly contributes to the bioactivity of these small molecules, acting via different mechanisms. The future exploration of dunnine E, buchapine, and severibuxine is especially warranted due to their high potency and selective bioactivities. In the coming times, natural products with a 2,4-QD moiety can unlock drug discovery opportunities. The diverse structures of these compounds might also allow them to become preclinical candidates via SAR and optimization. The 2,4-QD moiety can be a chemical probe for applications beyond small-molecule therapeutics. We might discover many more natural products with medicinally interesting 2,4-QD scaffolds as sequencing technologies and other approaches further improve our capacity to explore microbial and marine biodiversity. The occurrence of several 2,4-QD compounds in a broad array of unrelated families and even kingdoms is of interest to evolutionary and chemical ecology. The widespread presence of these components suggests they may have appeared early in plant evolution. Alternatively, they could have developed independently as bioactive secondary metabolites that help protect plants against pathogens and herbivores. Interdisciplinary research at the interface of chemistry, biology, and natural product biosynthesis should help not only uncover more sources of 2,4-QD analogues but also provide insights into how to produce and derivatize these compounds efficiently.

Studies showed that 2,4-QD derivatives are versatile, as the biological activity is affected by the position and nature of substituents. Whether the quinoline ring acts as a CB2 receptor agonist or antagonist depends on the variation. This platform's use with other pharmacophores enhances bioactivity and multi-target interactions. Even with progress made, we need to understand the molecular interactions further. Understanding binding mechanisms through tools like molecular dynamics, new substitutes, and hybrid scaffolds can optimize pharmacological properties, and the mechanistic and in vivo studies are crucial to confirm effectiveness. The results support the further development of 2,4-QD-based drugs targeting neurological, oncological, and infectious diseases, which could lead to good cannabinoid and serotonin receptor ligands, kinase inhibitors, and antimicrobial agents.

REFERENCES

Abdou, M. M. (2017). Chemistry of 4-hydroxy-2 (1H)-quinolinone. Part 1: Synthesis and reactions. *Arabian Journal of Chemistry, 10*, S3324–S3337.

Abdou, M. M. (2018). Chemistry of 4-hydroxy-2 (1H)-quinolinone. Part 2. As synthons in heterocyclic synthesis. *Arabian Journal of Chemistry, 11*(7), 1061–1071.

Ahmed, N., Brahmbhatt, K. G., Sabde, S., Mitra, D., Singh, I. P., & Bhutani, K. K. (2010). Synthesis and anti-HIV activity of alkylated quinoline 2, 4-diols. *Bioorganic & Medicinal Chemistry, 18*(8), 2872–2879.

Al-Rehaily, A. J., Al-Howiriny, T. A., Ahmad, M. S., Al-Yahya, M. A., El-Feraly, F. S., Hufford, C. D., & McPhail, A. T. (2001). Alkaloids from *Haplophyllum tuberculatum*. *Phytochemistry, 57*(4), 597–602.

Aly, A. A., El-Sheref, E. M., Mourad, A. E., Bakheet, M. E., & Bräse, S. (2020). 4-Hydroxy-2-quinolinones: Syntheses, reactions and fused heterocycles. *Molecular Diversity, 24*, 477–524.

Antolak, S. A., Yao, Z., Richoux, G. M., Slebodnick, C., & Carlier, P. R. (2014). Enantioselective deprotonative ring contraction of N 1-methyl-N 4-boc-benzo [e][1, 4] diazepine-2, 5–diones. *Organic Letters, 16*(19), 5204–5207.

Aref, M. M., Mohamed, A. A., Dahab, M. A., & El-Zahabi, M. A. A. (2023). An overview of quinoline derivatives as anti-cancer agents. *Al-Azhar Journal of Pharmaceutical Sciences, 68*(2), 130–158.

Barros-Filho, B. A., Nunes, F. M., de Oliveira, M. C., Mafezoli, J., Andrade-Neto, M., Silveira, E. R., & Pirani, J. R. (2004). Volatile constituents from *Esenbeckia almawillia* (Rutaceae). *Biochemical Systematics and Ecology, 32*(9), 817–821.

Bērziṇa, L., & Mieriṇa, I. (2023). Antiradical and antioxidant activity of compounds containing 1, 3-dicarbonyl moiety: An overview. *Molecules, 28*(17), 6203.

Boshta, N. M., El-Essawy, F. A., Alshammari, M. B., Noreldein, S. G., & Darwesh, O. M. (2022). Discovery of quinazoline-2, 4 (1 H, 3 H)-dione derivatives as potential antibacterial agent: Design, synthesis, and their antibacterial activity. *Molecules, 27*(12), 3853.

Buckingham, J. (2023). *Dictionary of natural products, supplement 2*. Boca Raton, FL: Routledge.

Buckingham, J., Baggaley, K. H., Roberts, A. D., & Szabo, L. F. (2010). *Dictionary of alkaloids with CD-ROM*. Boca Raton, FL: CRC Press.

Caine, B. A., Bronzato, M., Fraser, T., Kidley, N., Dardonville, C., & Popelier, P. L. (2020). Aqueous pKa prediction for tautomerizable compounds using equilibrium bond lengths. *Communications Chemistry, 3*(1), 21.

Cao, N., Chen, Y., Ma, X., Zeng, K., Zhao, M., Tu, P., Li, J., & Jiang, Y. (2018). Bioactive carbazole and quinoline alkaloids from *Clausena dunniana*. *Phytochemistry, 151*, 1–8.

Carvalho, J. C., Pirani, J. R., & Ferreira, M. J. (2022). *Esenbeckia* (Pilocarpinae, Rutaceae): Chemical constituents and biological activities. *Brazilian Journal of Botany, 45*(1), 41–65.

Chen, H., Zhu, S., Tu, P., & Jiang, Y. (2022). Chemical constituents from the stems and leaves of *Glycosmis craibii* var. glabra (Craib) Tanaka and their chemotaxonomic significance. *Biochemical Systematics and Ecology, 105*, 104492.

Chen, Y., Lawal, B., Huang, L., Kuo, S., Sumitra, M. R., Mokgautsi, N., Lin, H., & Huang, H. (2023). In vitro and in silico biological studies of 4-phenyl-2-quinolinone (4-PQ) derivatives as anticancer agents. *Molecules, 28*(2), 555.

Chen, X., Guo, M., Cao, Y., Wang, M., Mi, W., Zheng, X., & Feng, W. (2022). Cyclopeptides chemical constituents of bulbils of *Dioscorea opposita* Thunb. and their bioactivities. *Chinese Journal of Pharmaceuticals, 57*(12), 1002–1007.

Cho, J., Bae, S., Kim, H., Lee, M., Choi, Y., Jin, B., Lee, H. J., Jeong, H. Y., Lee, Y. G., & Moon, J. (2015). New quinolinone alkaloids from chestnut (*Castanea crenata* Sieb) honey. *Journal of Agricultural and Food Chemistry, 63*(13), 3587–3592.

Chóez-Guaranda, I., Espinoza-Lozano, F., Reyes-Araujo, D., Romero, C., Manzano, P., Galarza, L., & Sosa, D. (2023). Chemical characterization of *Trichoderma* spp. extracts with antifungal activity against cocoa pathogens. *Molecules, 28*(7), 3208.

Debitus, C., Guella, G., Mancini, I. I., Waikedre, J., Guemas, J., Nicolas, J. L., & Pietra, F. (1998). Quinolinones from a bacterium and tyrosine metabolites from its host sponge, *Suberea creba* from the Coral Sea. *Journal of Marine Biotechnology, 6*(3), 136–141.

Dekebo, A., Kim, M., Son, M., & Jung, C. (2022). Comparative analysis of volatile organic compounds from flowers attractive to honey bees and bumblebees. *Journal of Ecology and Environment, 46*(1), 62–75.

Divar, M., Edraki, N., Damghani, T., Moosavi, F., Mohabbati, M., Alipour, A., Pirhadi, S., Saso, L., Khabnadideh, S., & Firuzi, O. (2023). Novel spiroindoline quinazolinedione derivatives as anticancer agents and potential FLT3 kinase inhibitors. *Bioorganic & Medicinal Chemistry, 90*, 117367.

Dou, X., Li, X., Tao, L., Hu, C., Zhang, L., He, Q., … & Hu, Y. (2013). Synthesis and biological evaluation of novel pyrimido [4, 5-b] quinoline-2, 4-dione derivatives as MDM2 ubiquitin ligase inhibitors. *Medicinal Chemistry, 9*(4), 581–587.

Dreyer, D. L. (1980). Alkaloids, limonoids and furocoumarins from three Mexican *Esenbeckia* species. *Phytochemistry, 19*(5), 941–944.

Elshaier, Y. A., Aly, A. A., El-Aziz, M. A., Fathy, H. M., Brown, A. B., & Ramadan, M. (2022). A review on the synthesis of heteroannulated quinolinones and their biological activities. *Molecular Diversity, 26*, 2341–2370.

Hamama, W. S., Hassanien, A. E. E., & Zoorob, H. H. (2014). Studies on quinolinedione: Synthesis, reactions, and applications. *Synthetic Communications, 44*(13), 1833–1858.

Han, S., Zhang, F., Qian, H., Chen, L., Pu, J., Xie, X., & Chen, J. (2015). Development of quinoline-2, 4 (1 H, 3 H)-diones as potent and selective ligands of the cannabinoid type 2 receptor. *Journal of Medicinal Chemistry, 58*(15), 5751–5769.

Hays, E. E., Wells, I. C., Katzman, P. A., Cain, C. K., Jacobs, F. A., Thayer, S. A., Doisy, E. A., Gaby, W. L., Roberts, E. C., & Muir, R. D. (1945). Antibiotic substances produced by *Pseudomonas aeruginosa. Biological Chemistry, 159*(3), 725–750.

Hebanowska, E., Tempczyk, A., Łobocki, L., Szafranek, J., Szafranek, A., & Urbanek, Z. H. (1986). Gas phase prototropic equilibrium studies of 2, 4-dihydroxyquinoline by CNDO/2 calculations, mass spectrometry and derivatization. *Journal of Molecular Structure, 147*(3–4), 351–361.

Hu, H., Lee-Fong, Y., Peng, J., Hu, B., Li, J., Li, Y., & Huang, H. (2021). Comparative research of chemical profiling in different parts of fissistigma oldhamii by ultra-high-performance liquid chromatography coupled with hybrid quadrupole-orbitrap mass spectrometry. *Molecules, 26*(4), 960.

Huang, X., Wang, M., Zhong, S., & Xu, B. (2023). Comprehensive review of phytochemical profiles and health-promoting effects of different portions of wampee (*Clausena lansium*). *ACS Omega, 8*(30), 26699–26714.

Ito, C., Murata, T., Kato, M., Suzuki, N., Wu, T., Kaneda, N., Furukawa, H., & Itoigawa, M. (2013). Severibuxine, isolated from *Severinia buxifolia*, induces apoptosis in HL-60 leukemia cells. *Natural Product Communications, 8*(6), 1934578X1300800623.

Kadela-Tomanek, M., Bębenek, E., Chrobak, E., & Boryczka, S. (2019). 5, 8-quinolinedione scaffold as a promising moiety of bioactive agents. *Molecules, 24*(22), 4115.

Khadem, S., & Marles, R. J. (2025a). Tetrahydroquinoline-containing natural products discovered within the last decade: Occurrence and bioactivity. *Natural Product Research, 39*(1), 182–194.

Khadem, S., & Marles, R. J. (2025b). Natural 3,4-dihydro-2(1H)-quinolinones- Part I: Plant sources. *Natural Product Research, 39*(3), 593–608.

Khadem, S., & Marles, R. J. (2025c). Natural 3,4-dihydro-2(1H)-quinolinones- Part II: Animal, bacterial, and fungal sources. *Natural Product Research, 39*(2), 374–387.

Khadem, S., & Marles, R. J. (2025d). Natural 3,4-dihydro-2(1H)-quinolinones- Part III: Biological activities. *Natural Product Research*, 39(8), 2252–2259.

Khadem, S., & Marles, R. J. (2024). 2, 4-Quinolinedione alkaloids: Occurrence and biological activities. Natural Product Research, 1–12. https://doi.org/10.1080/14786419.2024. 2390611

Kim, H., Hwang, J., Chung, B., Cho, E., Bae, S., Shin, J., & Oh, K. (2019). 2-Alkyl-4-hydroxyquinolines from a marine-derived *Streptomyces* sp. inhibit hyphal growth induction in *Candida albicans. Marine Drugs, 17*(2), 133.

Kitamura, S., Hashizume, K., Iida, T., Miyashita, E., Shirahata, K., & Kase, H. (1986). Studies on lipoxygenase inhibitors II. KF8940 (2-n-heptyl-4-hydroxyquinoline-N-oxide), a potent and selective inhibitor of 5-lipoxygenase, produced by *Pseudomonas methanica. The Journal of Antibiotics, 39*(8), 1160–1166.

Lan, X. (1992). Isolation, structural identification of active components from metabolite of *Pseudomonas* sp. 20~#-5. Microbiology, *32*(6), 16–19.

Lee, J. Y., Shin, Y. S., Jeon, S., Lee, S. I., Noh, S., Cho, J., Jang, M. S., Kim, S., Song, J. H., & Kim, H. R. (2021). Design, synthesis and biological evaluation of 2-aminoquinazolin-4 (3H)-one derivatives as potential SARS-CoV-2 and MERS-CoV treatments. *Bioorganic & Medicinal Chemistry Letters, 39*, 127885.

Liao, Y., Cheng, L., Luo, R., Guo, Q., Shao, W., Feng, Y., Zhou, X., Liu, L., & Yang, S. (2024). Discovery of new 1, 2, 4-triazole/1, 3, 4-oxadiazole-decorated quinolinones as agrochemical alternatives for controlling viral infection by inhibiting the viral replication and self-assembly process. *Journal of Agricultural and Food Chemistry, 72*(50), 27750–27761.

Liu, J. Y., Song, Y. C., Zhang, Z., Wang, L., Guo, Z. J., Zou, W. X., & Tan, R. X. (2004). *Aspergillus fumigatus* CY018, an endophytic fungus in *Cynodon dactylon* as a versatile producer of new and bioactive metabolites. *Journal of Biotechnology, 114*(3), 279–287.

Liu, J., Li, C., Ni, L., Yang, J., Li, L., Zang, C., Bao, X., Zhang, D., & Zhang, D. (2015). Anti-inflammatory alkaloid glycoside and quinoline alkaloid derivates from the stems of *Clausena lansium*. *RSC Advances, 5*(98), 80553–80560.

Liu, Y., Zhao, H., Wang, Z., Li, Y., Song, H., Riches, H., Beattie, D., Gu, Y., & Wang, Q. (2013). The discovery of 3-(1-aminoethylidene) quinoline-2, 4 (1 H, 3 H)-dione derivatives as novel PSII electron transport inhibitors. *Molecular Diversity, 17*, 701–710.

Lubov, D. P., Talsi, E. P., & Bryliakov, K. P. (2020). Methods for selective benzylic C–H oxofunctionalization of organic compounds. *Russian Chemical Reviews, 89*(6), 587.

Luo, X. M., Qi, S. H., Yin, H., Gao, C. H., & Zhang, S. (2009). Alkaloids from the stem bark of *Micromelum falcatum*. *Chemical and Pharmaceutical Bulletin, 57*(6), 600–602.

Mancini, I., Vigna, J., Sighel, D., & Defant, A. (2022). Hybrid molecules containing naphthoquinone and quinolinedione scaffolds as antineoplastic agents. *Molecules, 27*(15), 4948.

Martínez-Luis, S., Cherigo, L., Spadafora, C., & Gutiérrez, M. (2019). Antiparasitic compounds from the Panamanian marine bacterium *Pseudomonas aeruginosa*. *Natural Product Communications, 14*(1), 1934578X1901400109.

McCormick, J. L., McKee, T. C., Cardellina, J. H., & Boyd, M. R. (1996). HIV inhibitory natural products. 26. Quinoline alkaloids from *Euodia roxburghiana*. *Journal of Natural Products, 59*(5), 469–471.

Meiring, L., Petzer, J. P., & Petzer, A. (2018). A review of the pharmacological properties of 3,4-dihydro-2(1H)- quinolinones. *Mini Reviews in Medicinal Chemistry, 18*(10), 828–836. 10.2174/1389557517666170927141323

Moh, J. H. Z., Okomoda, V. T., Mohamad, N., Waiho, K., Noorbaiduri, S., Sung, Y. Y., Manan, H., Fazhan, H., Ma, H., & Abualreesh, M. H. (2024). *Morinda citrifolia* fruit extract enhances the resistance of *Penaeus vannamei* to vibrio parahaemolyticus infection. *Scientific Reports, 14*(1), 5668.

Nesmelova, E. F., Bessonova, I. A., & Yunusov, S. Y. (1982a). Buchapine—A new alkaloid from *Haplophyllum bucharicum*. *Chemistry of Natural Compounds, 18*(4), 508–509.

Nesmelova, E. F., Bessonova, I. A., & Yunusov, S. Y. (1982b). Components of *Haplophyllum bucharicum*. *Chemistry of Natural Compounds, 18*(4), 507–508.

Neuenhaus, W., Budzikiewicz, H., Korth, H., & Pulverer, G. (1979). Bakterieninhaltsstoffe, III 3-alkyl-tetrahydrochinolinderivate aus pseudomonas/bacterial constituents, III 3-alkyl-tetrahydroquinoline derivatives from *Pseudomonas*. *Zeitschrift für Naturforschung B, 34*(2), 313–315.

Niroumand, M. C., Heydarpour, F., & Farzaei, M. H. (2018). Pharmacological and therapeutic effects of *Vitex agnus-castus* L.: A review. *Pharmacognosy Reviews, 12*(23), 103–114.

Nunes, F. M., Barros-Filho, B. A., de Oliveira, M. C., de Mattos, M. C., Andrade-Neto, M., Barbosa, F. G., Mafezoli, J., Montenegro, R. C., Pessoa, C., & de Moraes, M. O. (2006). 3, 3-diisopentenyl-N-methyl-2, 4-quinoldione from *esenbeckia almawillia*: The antitumor activity of this alkaloid and its derivatives. *Natural Product Communications, 1*(4), 1934578X0600100409.

Oxenius, A., Fidler, S., Brady, M., Dawson, S. J., Ruth, K., Easterbrook, P. J., Weber, J. N., Phillips, R. E., & Price, D. A. (2001). Variable fate of virus-specific CD4 T cells during primary HIV-1 infection. *European Journal of Immunology, 31*(12), 3782–3788.

Pratiwi, R. H., Oktarina, E., Mangunwardoyo, W., Hidayat, I., & Saepudin, E. (2022). Antimicrobial compound from endophytic *Pseudomonas azotoformans* UICC B-91 of *Neesia altissima* (Malvaceae). *Pharmacognosy Journal, 14*(1), 172–181.

Proisl, K., Kafka, S., & Kosmrlj, J. (2017). Chemistry and applications of 4-hydroxyquinolin-2-one and quinoline-2, 4-dionebased compounds. *Current Organic Chemistry, 21*(19), 1949–1975.

Qian, H., Wang, Z., Pan, Y., Chen, L., Xie, X., & Chen, J. (2017). Development of quinazoline/pyrimidine-2, 4 (1 H, 3 H)-diones as agonists of cannabinoid receptor type 2. *ACS Medicinal Chemistry Letters, 8*(6), 678–681.

Qiu, Y., Jia, Q., Song, S., & Wang, X. (2024). Alkaloids in *isatis indigotica* and their chemotaxonomic significance. *Biochemical Systematics and Ecology, 113*, 104800.

Shah, S., Saied, S., Mahmood, A., & Malik, A. (2014). Phytochemical screening of volatile constituents from aerial parts of *murraya paniculata*. *Pakistan Journal of Botany, 46*(6), 2051–2056.

Shakhatreh, M. A. K., Al-Smadi, M. L., Khabour, O. F., Shuaibu, F. A., Hussein, E. I., & Alzoubi, K. H. (2016). Study of the antibacterial and antifungal activities of synthetic benzyl bromides, ketones, and corresponding chalcone derivatives. Drug Design, Development and Therapy, *10*, 3653–3660.

Shen, L., Li, L. Y., Zhang, X. J., Li, M., & Song, Y. C. (2015). A new indole derivative from endophyte *Myrothecium roridum* IFB-E091 in *Artemisia annua. Yao Xue Xue Bao= Acta Pharmaceutica Sinica, 50*(10), 1305–1308.

Sheriha, G. M., Abouamer, K., Elshtaiwi, B. Z., Ashour, A. S., Abed, F. A., & Alhallaq, H. H. (1987). Quinoline alkaloids and cytotoxic lignans from *Haplophyllum tuberculatum*. *Phytochemistry, 26*(12), 3339–3341.

Shokova, E. A., Kim, J. K., & Kovalev, V. (2015). 1, 3–diketones. Synthesis and properties. *Russian Journal of Organic Chemistry, 51*, 755–830.

Smeyne, D. (2020). Structure activity relationship (SAR) studies of neurotoxin quinolinederivatives (Honors College Thesis), Georgia Southern University.

Speranza, J., Miceli, N., Taviano, M. F., Ragusa, S., Kwiecień, I., Szopa, A., & Ekiert, H. (2020). Isatis tinctoria L. (Woad): A review of its botany, ethnobotanical uses, phytochemistry, biological activities, and biotechnological studies. *Plants, 9*(3), 298.

Supong, K., Thawai, C., Supothina, S., Auncharoen, P., & Pittayakhajonwut, P. (2016). Antimicrobial and anti-oxidant activities of quinoline alkaloids from *Pseudomonas aeruginosa* BCC76810. *Phytochemistry Letters, 17*, 100–106.

Tashima, T. (2015). The structural use of carbostyril in physiologically active substances. *Bioorganic & Medicinal Chemistry Letters, 25*(17), 3415–3419.

Thant, T. M., Aminah, N. S., Kristanti, A. N., Ramadhan, R., Aung, H. T., & Takaya, Y. (2020). Phytoconstituents of genus *micromelum* and their bioactivity—A review. *Natural Product Communications, 15*(5), 1934578X20927124.

Tin, B., Kurtoğlu, C., & Sevindik, E. (2017). Evaluation of chemical composition of *Vitex agnus-castus* (Verbenaceae) fruits essential oils grown in Aydın/Turkey. *Turkish Journal of Life Sciences, 2*(2), 171–174.

Ubaid, A., Shakir, M., Ali, A., Khan, S., Alrehaili, J., Anwer, R., & Abid, M. (2024). Synthesis and Structure–Activity relationship (SAR) studies on new 4-aminoquinoline-hydrazones and isatin hybrids as promising antibacterial agents. *Molecules, 29*(23), 5777.

Vásquez-Martínez, Y., Torrent, C., Toledo, G., Cabezas, F., Espinosa, V., Montoya -K, M., Mejias, S., Cortez-San Martín, M., Sepúlveda-Boza, S., & Mascayano, C. (2019). New antibacterial and 5-lipoxygenase activities of synthetic benzyl phenyl ketones: Biological and docking studies. *Bioorganic Chemistry, 82*, 385–392.

Wang, B., Sun, S., Yu, J., Jiang, Y., & Cheng, J. (2017). Palladium-catalyzed multicomponent reactions of O-alkynylanilines, aryl iodides, and CO_2 toward 3, 3-diaryl 2, 4-quinolinediones. *Organic Letters, 19*(16), 4319–4322.

Wang, F. W., Hou, Z. M., Wang, C. R., Li, P., & Shi, D. H. (2008). Bioactive metabolites from *Penicillium* sp., an endophytic fungus residing in *hopea hainanensis*. *World Journal of Microbiology and Biotechnology, 24*, 2143–2147.

Wei, R., Ma, Q., Zhong, G., Su, Y., Yang, J., Wang, A., Ji, T., Guo, H., Wang, M., & Jiang, P. (2020). Structural characterization, hepatoprotective and antihyperlipidemic activities of alkaloid derivatives from *Murraya koenigii*. *Phytochemistry Letters, 35*, 135–140.

Wu, T., Leu, Y., Chan, Y., Lin, F., Li, C., Shi, L., Kuo, S., Chen, C., & Wu, Y. (1998). Severibuxine, a new quinolin-2, 4-dione and other constituents from *Severinia buxifolia*. *Phytochemistry, 49*(5), 1467–1470.

Wu, Y., Dong, L., & Yuan, K. (2010). Analysis of the volatile constituents of *Rabdosia rubescens* by gas chromatography-mass spectrometry using headspace solid-phase micro-extraction. *Asian Journal of Chemistry, 22*(5), 3903–3909.

Xi, Y., Lou, L., Xu, Z., Hou, Z., Wang, X., Huang, X., & Song, S. (2019). Alkaloid enantiomers from *Isatis tinctoria* with neuroprotective effects against H_2O_2-induced SH-SY5Y cell injury. *Planta Medica, 85*(17), 1374–1382.

Zayed, M. F. (2023). Medicinal chemistry of quinazolines as anticancer agents targeting tyrosine kinases. *Scientia Pharmaceutica, 91*(2), 18.

Zhang C, Luo X, Yin H, Qi S. (2011). 2,4-diketoquinoline alkaloids, and preparation methods and applications thereof. *Patent* No. CN101429158 (Chinese), Application No. CN2008-10219869. South China Sea Institute of Oceanology, Chinese Academy of Sciences.

Zhang, Y., Qiu, G., Liu, F., Zhao, D., Tian, M., & Sun, K. (2023). Visible light-induced cascade sulfonylation/cyclization to produce quinoline-2, 4-diones under metal-free conditions. *Molecules, 28*(7), 3137.

Zhong, S., Huang, P., Wang, X., Lin, M., & Ge, C. (2018). Synthesis of quinoline-2, 4-diones from cyanoacetanilide derivatives. *Chinese Journal of Organic Chemistry, 38*(5), 1199.

Zhu, C., Lew, C. I., Neuhaus, G. F., Adpressa, D. A., Zakharov, L. N., Kaweesa, E. N., Plitzko, B., & Loesgen, S. (2021). Biodiversity, bioactivity, and metabolites of high desert derived oregonian soil bacteria. *Chemistry & Biodiversity, 18*(4), e2100046.

8 Other Quinoline Derivatives

8.1 INTRODUCTION

In Chapter 1, we looked into how semi-aromatic quinoline derivatives maintain the aromatic characteristics of the benzene ring while adding saturation or partial saturation to the pyridine ring (Figure 8.1). This structural adjustment significantly alters the three-dimensional shape and properties of these compounds compared with fully aromatic systems. Increasing the sp^3 character of the molecule is increasingly seen as beneficial in drug discovery. Compounds with higher sp^3 content often show improved selectivity, reduced toxicity, and better pharmacokinetic profiles than highly planar sp^2-rich aromatic systems (Twigg et al. 2016).

Chapters 2–7 focus on exploring the primary natural semi-aromatic quinoline derivatives. While various other structural frameworks were identified as falling under the category of semi-aromatic quinoline derivatives, not all occur naturally. Further research to uncover the remaining natural semi-aromatic quinoline derivatives led to identifying additional skeletons (Figure 8.2): 1,2-dihydroquinoline, 1,4-dihydroquinoline, 2,3-dihydroquinolin-4(1H)-imine, and 2,3-dihydroquinolin-4(1H)-one. This concise chapter aims to bridge the final gap by analysing the presence of these remaining quinoline derivatives and their biological activities.

The chemical identities of other quinoline derivatives presented in this chapter are shown in Table 8.1.

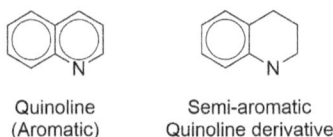

Quinoline
(Aromatic)

Semi-aromatic
Quinoline derivative

FIGURE 8.1 Chemical structure of quinoline and semi-aromatic quinoline derivatives.

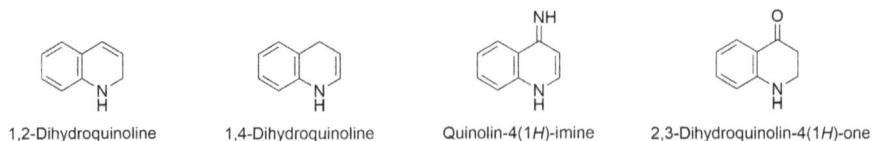

1,2-Dihydroquinoline   1,4-Dihydroquinoline   Quinolin-4(1H)-imine   2,3-Dihydroquinolin-4(1H)-one

FIGURE 8.2 Structural skeleton for the remaining natural semi-aromatic quinoline derivatives.

DOI: 10.1201/9781003534600-8

TABLE 8.1
The Chemical Identity of Other Quinoline Derivatives

Structure (Number)	Systematic Name	Common Name	CAS #
(8.1)	2,3-Dihydro-4(1*H*)-quinolinone	–	4295-36-7
(8.2)	2,3-Dihydro-1-methyl-4(1*H*)-quinolinone	NSC 123397	1198-15-8
(8.3)	2,3-Dihydro-8-hydroxy-4(1*H*)-quinolinone	–	28884-04-0
(8.4)	1,2-Dihydro-8-methoxyquinoline	–	874498-34-7
(8.5)	1-Methyl-4(1*H*)-quinolinimine	Echinopsidine	2400-75-1
(8.6)	4(1*H*)-Quinolinimine, 1-methyl-, hydriodide (1:1)	Echinopsidine. HI (Adepren)	24667-93-4

(*Continued*)

TABLE 8.1 (*Continued*)
The Chemical Identity of Other Quinoline Derivatives

Structure (Number)	Systematic Name	Common Name	CAS #
(8.7)	Quinoline, 1,2-dihydro-2,2,4-trimethyl-	1,2-Dihydro-2,2,4-trimethyl-quinoline	147-47-7
(8.8)	N-[[(2S,3R)-2-(1,1-Dimethyl-2-propen-1-yl)-1,2,3,4-tetrahydro-4-oxo-3-quinolinyl]methyl]acetamide	Brocaeloid A	1620765-89-0
(8.9)	(4R)-1,4-Dihydro-4-methoxy-1,4-dimethyl-3-(3-methyl-2-buten-1-yl)-2,7-quinolinediol	–	1588847-76-0
(8.10)	4-Quinolinepropanoic acid, α-amino-β-(aminocarbonyl)-1,4-dihydro-4-methoxy-1-methyl-	Echinoramine I	59669-20-4
(8.11)	N-[[(2S,3R)-2-(1,1-Dimethyl-2-propen-1-yl)-1,2,3,4-tetrahydro-4-oxo-3-quinolinyl]methyl]-L-glutamine	Solitumine A	2587177-68-0

8.2   NATURAL OCCURRENCE

Plants, fungi, and animals have been identified as sources of the quinoline derivatives mentioned above (Table 8.2).

Several species of *Echinops* (Asteraceae family) contain a wide range of compounds. For instance, *Echinops ritro* has compounds 8.5 and 8.10. Unidentified *Echinops* species also share these two compounds in other studies. This genus produces many similar metabolites. Also, *Ganoderma resinaceum* and *Saussurea medusa* contain compound 8.1, suggesting that unrelated species may make the compound similar. In contrast, the chemical varieties of *Neopetrosia* sp., *Leonurus japonicus* (an animal species), and *Ruta graveolens* are fewer as they are linked to one compound only. The appearance of compound 8.1 in various unrelated species shows that these species may produce the same compound through a similar mechanism or that the species have evolved to make the compound independently.

TABLE 8.2
Natural Occurrence of Other Quinoline Derivatives

Kingdom	Family	Genus	Species	Compound(s)	References
Animalia	Petrosiidae	*Neopetrosia*	*sp.*	8.1	Sorek et al. (2007)
Fungi	Aspergillaceae	*Penicillium*	*Brocae MA-192*	8.8	Zhang et al. (2014)
Fungi	Aspergillaceae	*Penicillium*	*Solitum MCCC 3A00215*	8.11	He et al. (2021)
Fungi	Aspergillaceae	*Penicillium*	*Solitum IS1-A*	8.11	Rodriguez et al. (2020)
Fungi	Ganodermataceae	*Ganoderma*	*cochlear*	8.1, 8.3	Lei et al. (2015)
Fungi	Ganodermataceae	*Ganoderma*	*resinaceum*	8.1	Yang (2019)
Plantae	Acanthaceae	*Strobilanthes*	*cusia*	8.1	Huang et al. (2024)
Plantae	Asteraceae	*Echinops*	*ritro*	8.5	Ulubelen and Kurucu (1991)
Plantae	Asteraceae	*Echinops*	*spp.*	8.5	Kurucu (1991)
Plantae	Asteraceae	*Echinops*	*echinatus*	8.5	Chaudhuri (1987)
Plantae	Asteraceae	*Echinops*	*niveus*	8.5	Bhakuni et al. (1990)
Plantae	Asteraceae	*Echinops*	*sphaerocephalus*	8.5	Horn et al. (2008)
Plantae	Asteraceae	*Echinops*	*heterophyllous*	8.2	Khadim et al. (2014)
Plantae	Asteraceae	*Echinops*	*ritro*	8.10	Ulubelen and Kurucu (1991)
Plantae	Asteraceae	*Echinops*	*spp.*	8.10	Pham Thanh et al. (1976); Kurucu (1991)
Plantae	Asteraceae	*Saussurea*	*medusa*	8.1	Cao et al. (2023)
Plantae	Apiaceae	*Bupleurum*	*chinense*	8.7	Lei et al. (2012)
Plantae	Brassicaceae	*Isatis*	*Indigotica (tinctoria)*	8.4	Wang et al. (2021)
Plantae	Lamiaceae	*Leonurus*	*japonicus*	8.1	Bu et al. (2022)
Plantae	Onagraceae	*Ludwigia*	*adscendens*	8.1	Chen et al. (2023)
Plantae	Plumbaginaceae	*Limonium*	*gmelini*	8.7	Ikhsanov et al. (2019)
Plantae	Pteridaceae	*Adiantum*	*lunulatum*	8.7	Jerom et al. (2023)
Plantae	Rutaceae	*Ruta*	*graveolens*	8.9	Salib et al. (2014)

The species included sponges, for example, *Neopetrosia*, fungi (like *Ganoderma*), and plants (such as Strobilanthes cusia, *Leonurus japonicus, Ludwigia adscendens*, and *Saussurea medusa*). The comparison reveals that both *Echinops* and *Ganoderma* are rich sources of these unique natural substances; hence, they deserve more attention in terms of phytochemistry and pharmacology.

8.3 BIOLOGICAL ACTIVITY

Echinopsidine (8.5) and its hydroiodide salt 8.6 have been identified as bioactive among the other quinoline derivatives. Echinopsidine is best known for its antidepressant properties (Avramova et al. 1970). It was initially developed in Bulgaria (named Adepren) to treat depression. A monoamine oxidase inhibitor (MAOI) is an antidepressant at the monoamine neurotransmitter level. Echinopsidine inhibits the MAO enzyme, leading to increased brain neurotransmitters, a common mechanism of antidepressant action (Tripathi et al. 2018). MAOIs are important in the treatment of neuropsychiatric diseases with altered expression of MAO enzymes, which may be involved in neurodegenerative diseases like Alzheimer's and Parkinson's. Echinopsidine's labelling as a second-generation MAOI means that it is more selective, with fewer side effects when compared to the first-generation MAOIs (Hong and Li 2019). It is hard to know the subtype-specific effects of MAO-A and MAO-B. Because it affects neurotransmitters, echinopsidine may have other uses for different neuropsychiatric disorders apart from depression, if it is selective and crosses the blood-brain barrier.

Compound 8.7 was assessed for interactions with cyclooxygenase-2, which is involved in inflammation, and with the transient receptor potential cation channel V member 1 and transient receptor potential channel 3, which are involved in pain. Its positive interactions with this enzyme's amino acids and receptors reduced inflammatory pathogenesis (Jerom et al. 2023). Interestingly, compound 8.7 also showed a positive antioxidant activity for protecting the polyunsaturated fatty acids in fish meal (de Koning 2002).

The effects of compound 8.10 (Echinoramine I) on the motor behaviour of mice were measured using an electronic device that produces movement profiles (Bekemeier and Wenzel 1989). The range of movements was classified by direction and intensity, showing an approximately normal distribution in control mice. Generally, depressing drugs reduce the incidence of strong movements while increasing weaker movements. In contrast, stimulating drugs initially increase strong movements, but this effect eventually leads to a state of exhaustion and depression. Echinoramine I exhibited depressant effects at low doses and stimulant effects at high doses.

8.4 DISCUSSION

Alkaloids are present in bacteria. However, the types of alkaloids being dealt with in this chapter are not encountered in bacteria. This is probably due to limits in evolution and metabolism. Bacteria do not possess the pathways required to manufacture steroidal alkaloids. These compounds likely rely on ingredients similar to steroids, which prokaryotes do not possess. Bacteria do not have complex enzymes that perform all the steps of the changes that plant or fungal alkaloids undergo. Instead, bacteria usually make secondary metabolites that cost less energy, like β-lactams or non-ribosomal peptides. By getting these compounds, bacteria can compete better and save on the high energy costs of making complex alkaloids. Besides that, bacteria serve different functions in the ecosystem than other organisms. They are less likely to face predation that would require them to produce neurotoxic or hormone-mimicking alkaloids. This indicates bacteria frequently generate quick-acting toxins or substances that interfere with biofilms. The absence of quinolinone derivatives within bacteria reflects their different evolutionary paths and biochemical limitations. *Echinops* is an important source of plant-defence chemistry. *Echinops* species contain a lot of alkaloids, which are essential to their role and how they are made (Bitew and Hymete 2019). These plants produce quinolone-type alkaloids, which are concentrated in the seed and aerial parts of the plant, like leaves, stems, and flowers. The chemical defence against plant eaters and disease may happen due to this alkaloid concentration. Their biosynthesis pathways efficiently produce nitrogen-containing heterocycles, resulting in high yields of alkaloids, up to 2.2% in seeds. Alkaloid diversity may be influenced by ecological pressures in arid or high-altitude environments, while providing potentially beneficial medicinal properties, such as cardiotonic effects. Further research will be directed towards developing green approaches for synthesizing these alkaloids, evaluating their potential as therapeutic agents, and synthesizing new molecules that combine quinoline derivatives with other active compounds. Scientific progress is moving faster because of technologies like AI in medicine design and better evaluation methods. The better we understand how the structure affects the activity of these versatile compounds, the more likely we are to develop new drugs that are safer and more effective.

REFERENCES

Avramova, B., Zhelyazkov, L., Daleva, L., & Stefanova, D. (1970). 1-Substituted quinolin-4-one imines with biological activity. I. *Chemistry of Natural Compounds, 6*(1), 92–94.

Bekemeier, H., & Wenzel, U. (1989). The effect of drugs on electronically determined movement profile (movement profile analysis). [Beeinflussung des elektronisch erfassten Bewegungsprofils von Mausen durch Pharmaka (Bewegungsprofilanalyse)]. *Die Pharmazie, 44*(11), 781–783.

Bhakuni, R. S., Shukla, Y. N., & Thakur, R. S. (1990). Alkaloids and lipid constituents of *Echinops niveus*. *Phytochemistry, 29*(8), 2697–2698.

Bitew, H., & Hymete, A. (2019). The genus *Echinops*: Phytochemistry and biological activities: A review. *Frontiers in Pharmacology*, 10, 1234.

Bu, L., Peng, C., Liu, F., Meng, C., Guo, L., Zhou, Q., & Xiong, L. (2022). A novel diterpene glycoside from *Leonurus japonicus*. *Chinese Traditional and Herbal Drugs, 53*(1), 8–13.

Cao, J., Yu, R., & Tao, Y. (2023). Chemical constituents of the whole plants of *Saussurea medusa*. *Chemistry of Natural Compounds, 59*(3), 616–618.

Chaudhuri, P. K. (1987). Echinozolinone, an alkaloid from *Echinops echinatus*. *Phytochemistry, 26*(2), 587–589.

Chen, L., Liao, G., Zeng, Y., Zhan, X., & Lu, R. (2023). Study on the chemical constituents of *Polyporus zhuangyao*. *Journal of Guangxi Normal University – Natural Science Edition, 41*(2), 131.

de Koning, A. J. (2002). The antioxidant ethoxyquin and its analogues: A review. *International Journal of Food Properties, 5*(2), 451–461.

He, Z., Wu, J., Xu, L., Hu, M., Xie, M., Hao, Y., Li, S., Shao, Z., & Yang, X. (2021). Chemical constituents of the deep-sea-derived *Penicillium solitum*. *Marine Drugs, 19*(10), 580.

Hong, R., & Li, X. (2019). Discovery of monoamine oxidase inhibitors by medicinal chemistry approaches. *MedChemComm, 10*(1), 10–25.

Horn, G., Kupfer, A., Kalbitz, J., Gerdelbracht, H., Kluge, H., Eder, K., & Dräger, B. (2008). Great globe thistle fruit (*Echinops sphaerocephalus* L.), a potential new oil crop. *European Journal of Lipid Science and Technology, 110*(7), 662–667.

Huang, S., He, C., He, Q., Lang, G., Chen, H., Teng, L., & Zhou, Z. (2024). Chemical constituents from *Strobilanthes cusia* and their chemotaxonomic significance. *Biochemical Systematics and Ecology, 114*, 104822.

Ikhsanov, Y. S., Zhussupova, A. I., Kasymova, D. T., Ross, S. A., & Zhusupova, G. E. (2019). Study of the hexane fraction isolated from the substance obtained from the roots *limonium gmelinii* by GC-MS. *International Journal of Biology and Chemistry, 12*(1), 159–163.

Jerom, J. P., Nair, R. H., Sajan, A. L., Manirajan, B. A., & Mohammed, S. (2023). GC-MS screening of *Adiantum lunulatum* Burm. F phytochemicals and interaction with COX-2, TRPV1, and TRPC3 proteins-bioinformatics approach. *Current Bioactive Compounds, 19*(3), 35–49.

Khadim, E. J., Abdulrasool, A. A., & Awad, Z. J. (2014). Phytochemical investigation of alkaloids in the Iraqi *Echinops heterophyllus* (Compositae). *Iraqi Journal of Pharmaceutical Sciences, 23*(1), 26–34.

Kurucu, S. (1991). Researches on the alkaloids of *Echinops* species growing in Turkey-I. Section: Oligolepis and ritrodes. *FABAD Journal of Pharmaceutical Sciences, 16*, 1–7.

Lei, T., Xin-long, W., Yan-zhi, W., & Yong-xian, C. (2015). Two new alkaloids from *Ganoderma cochlear*. *Natural Product Research and Development, 27*(8), 1325.

Lie, X., Wang, Q., Yang, Y., & Li, X. (2012). Analysis of *Bupleurum chinense* obtained from different harvest period in supercritical fluid extraction by GC-MS spectroscopy. *Chinese Journal of Experimental Traditional Medical Formulae, 18*(2), 69–71.

Pham Thanh Ky, P. T. K., & Schroder, P. (1976). Studies on the physiology of the quinoline alkaloids in *echinops* species. *Biochemie Und Physiologie Der Pflanzen, 169*(5), 461–470.

Rodriguez, J. P., Bernardi, D. I., Gubiani, J. R., Magalhães de Oliveira, J., Morais-Urano, R. P., Bertonha, A. F., Bandeira, K. F., Bulla, J. I., Sette, L. D., & Ferreira, A. G. (2020). Water-soluble glutamic acid derivatives produced in culture by *Penicillium solitum* IS1-A from King George Island, Maritime Antarctica. *Journal of Natural Products, 83*(1), 55–65.

Salib, J. Y., El-Toumy, S. A., Hassan, E. M., Shafik, N. H., Abdel-Latif, S. M., & Brouard, I. (2014). New quinoline alkaloid from *Ruta graveolens* aerial parts and evaluation of the antifertility activity. *Natural Product Research, 28*(17), 1335–1342.

Sorek, H., Rudi, A., Benayahu, Y., & Kashman, Y. (2007). Njaoamines G and H, two new cytotoxic polycyclic alkaloids and a tetrahydroquinolone from the marine sponge *Neopetrosia* sp. *Tetrahedron Letters, 48*(43), 7691–7694.

Tripathi, A. C., Upadhyay, S., Paliwal, S., & Saraf, S. K. (2018). Privileged scaffolds as MAO inhibitors: Retrospect and prospects. *European Journal of Medicinal Chemistry, 145*, 445–497.

Twigg, D. G., Kondo, N., Mitchell, S. L., Galloway, W. R., Sore, H. F., Madin, A., & Spring, D. R. (2016). Partially saturated bicyclic heteroaromatics as an sp^3-enriched fragment collection. *Angewandte Chemie, 128*(40), 12667–12671.

Ulubelen, A., & Kurucu, S. (1991). Sesquiterpene acids from *Echinops ritro*. *Fitoterapia*, *62*(3), 280.

Wang, F., Bi, J., He, L., Chen, J., Zhang, Q., Hou, X., & Xu, H. (2021). The indole alkaloids from the roots of *Isatidis radix*. *Fitoterapia, 153*, 104950.

Yang, Q. (2019). A new meroterpenoid from *Ganoderma resinaceum*. Zhongcaoyao, *2019*, 1902–1905.

Zhang, P., Meng, L., Mándi, A., Kurtán, T., Li, X., Liu, Y., Li, X., Li, C., & Wang, B. (2014). Brocaeloids A–C, 4-Oxoquinoline and indole alkaloids with C-2 reversed prenylation from the Mangrove-Derived endophytic fungus *Penicillium brocae*. *European Journal of Organic Chemistry, 2014*(19), 4029–4036.

9 Concluding Remarks and Future Perspectives on Semi-Aromatic Quinoline Derivatives

9.1 FINDINGS

In the previous chapters, we discussed the distribution and pharmacological potential of 730 alkaloids, including tetrahydroquinolines (THQs), tetrahydronaphthoquinoline-diones, 2-oxo-tetrahydroquinolines (2O-THQs), 2-quinolinones, 4-quinolinones, 2,4-quinolinediones, and others across multiple biological sources. Figure 9.1 illustrates that plant-derived alkaloids constitute the majority at 61.8%, followed by bacterial (19.0%) and fungal (15.2%) origins, while those derived from animals represent only 3.9%. This distribution confirms the dominance of plants as the principal source of these alkaloids.

Among the structural subclasses, 2-quinolinones and 4-quinolinones are the most commonly identified, particularly within plant-derived extracts (Figure 9.2).

Pharmacological analysis of these identified alkaloids reveals that approximately 41% report bioactivity, a high proportion compared to the natural products' overall annotated bioactivity rate. The *Dictionary of Natural Products* is currently catalogued with more than ~300,000 compounds, including only 3,882 (approximately 1.3%) active with known or predicted pharmacological activity (Chassagne et al. 2019). This low rate highlights the unexplored diversity in natural product chemistry and the need for expanded screening efforts. Its biological origin and structural class influence a natural product's likelihood of displaying bioactivity. Bacterial metabolites, particularly those from *Streptomyces* spp., tend to exhibit more known bioactivities. At the same time, plant-derived compounds, despite their numerical prevalence, are often less well-characterized due to historical limitations in bioassay coverage.

Antibacterial activity is the most frequently reported among bioactive quinoline alkaloids. It suggests robust and broad-spectrum potential against pathogenic bacterial strains. Antibacterial activity may arise from mechanisms such as the disruption of cell wall synthesis, inhibition of nucleic acid replication, or interference with protein biosynthesis. Cytotoxic effects represent the second most prevalent activity. It indicates potential for anticancer applications through apoptosis induction, cell cycle arrest, or mitochondrial disruption. Antifungal activity is also frequently noted, with several compounds demonstrating efficacy by disrupting fungal membrane integrity or inhibiting ergosterol biosynthesis.

DOI: 10.1201/9781003534600-9

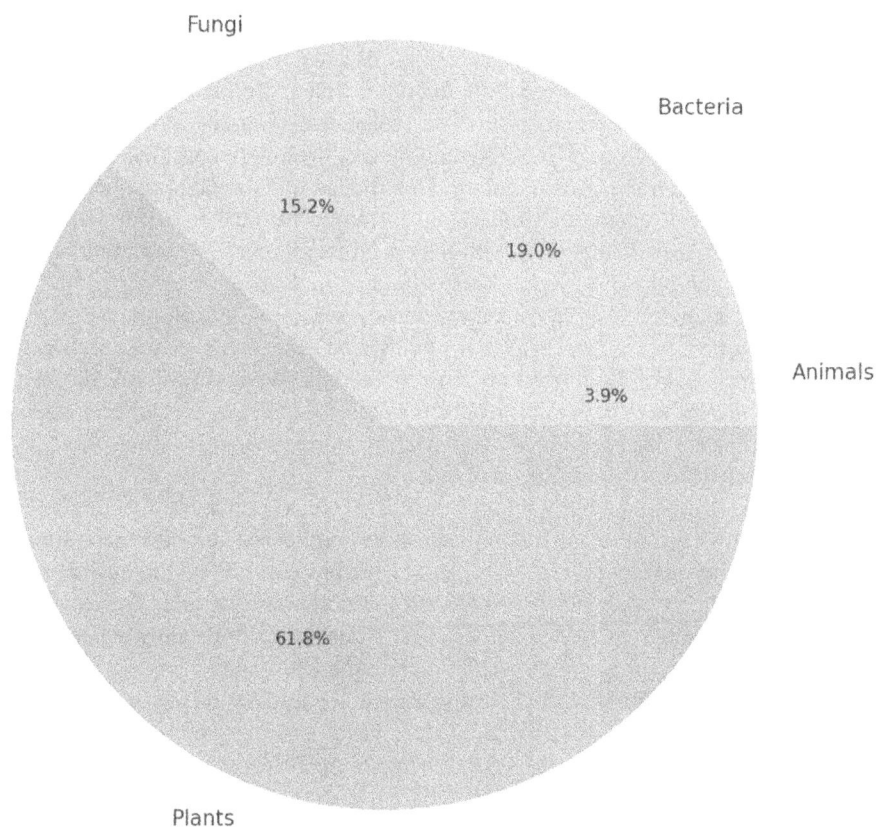

FIGURE 9.1 Contribution of each natural source to the total alkaloids.

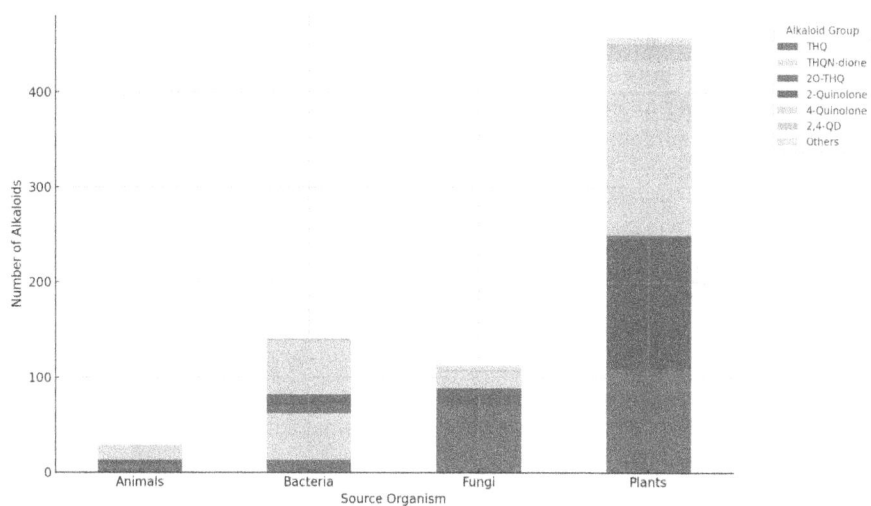

FIGURE 9.2 Distribution of alkaloid groups by natural sources.

Semi-aromatic quinoline derivatives represent a chemically and biologically diverse group characterized by a benzene ring fused to a partially saturated pyridine ring. This partial saturation profoundly alters the electronic distribution, aromatic stability, and overall conformation of these molecules compared to fully aromatic quinolines. Classes such as THQs, 2-quinolinones, 4-quinolinones, 2,4-quinoline-diones, 2O-THQs, tetrahydronaphthoquinoline-diones, and dihydroquinolines (both 1,2- and 1,4-derivatives) demonstrate distinct physicochemical properties that significantly impact their interactions with biological targets. Partial saturation brings flexibility and non-planarity, significantly affecting molecular recognition, binding affinity, and selectivity for proteins, enzymes, or receptors. For example, THQs' sp³ hybridized carbon atoms disrupt planar conjugation, enhancing their fit into non-planar biological pockets and improving specificity and potency. Similarly, the presence of keto groups in quinolinones results in keto-enol tautomeric equilibria, which critically influence hydrogen-bonding capacity, polarity, and pharmacokinetics, thereby affecting their therapeutic potential.

Studies on structure–activity relationship (SAR) have provided crucial insights into how slight structural modifications significantly influence quinoline derivatives' biological activity, selectivity, and pharmacokinetic properties. Key pharmacophores have been identified at specific positions on the quinoline scaffold. For example, electron-withdrawing substituents at the 2- and 4-positions significantly enhance the antibacterial potency of quinolinones by stabilizing DNA gyrase and topoisomerase IV interactions. Similarly, SAR investigations focused on quinoline derivatives have elucidated the role of electron-withdrawing and donating groups in modulating neuropharmacological effects, particularly regarding Alzheimer's disease targets. Substitution patterns significantly influence enzyme inhibition, antioxidant properties, and lipophilicity-driven brain penetration. Structure-based pharmacophore alignment and 3D-QSAR modelling contribute to identifying lead compounds with optimized activity profiles and therapeutic relevance in neurodegenerative disorders (Li et al. 2023). For instance, changes to THQs can affect how flexible their rings are or how bulky their side groups are. These changes can significantly influence how well they bind to neurotransmitter receptors and enzymes. As a result, this can impact their abilities to protect nerve cells and fight cancer.

Antimicrobial quinolinones primarily inhibit bacterial DNA gyrase and topoisomerase IV, disrupting DNA replication and transcription. Structural modifications to the quinolinone core influence target selectivity, potency, and spectrum of activity. Resistance mechanisms arise from target mutations, efflux pumps, and reduced permeability, necessitating the design of next-generation derivatives that evade resistance pathways (Aldred et al. 2014; Hooper and Jacoby 2015). Several quinoline analogues exhibit non-antibiotic pharmacological profiles, including antimalarial, antiplasmodial, anti-tumour, and neuroprotective effects. These effects are mediated through diverse molecular targets such as kinases, heme detoxification pathways, and neurotransmitter enzymes. Scaffold flexibility and substitution at key positions on the quinoline ring are critical in modulating bioavailability and receptor affinity (Daneshtalab and Ahmed 2012; Sharma et al. 2023).

Comprehensive reviews of quinoline bioactivity highlight their role as privileged scaffolds in anti-infective and anticancer domains. Mechanistic insights support quinoline-mediated generation of reactive oxygen species, metal chelation, and mitochondrial disruption in targeted cells. Their multifunctionality underscores the continued interest in tailoring these structures for therapeutic selectivity and minimal toxicity (El-Mrabet et al. 2025; Millanao et al. 2021).

Hybrid molecules are synthetic constructs that combine two or more pharmacophoric elements into a single chemical entity. It often results in synergistic or dual-mode mechanisms of action. This strategy simultaneously facilitates targeting multiple pathways or biological sites, improving efficacy and reducing the probability of resistance development. Utilizing hybrid molecules has successfully designed ligands for complex diseases such as cancer, malaria, and multidrug-resistant infections, where single-target therapeutics may be insufficient (Meunier 2008; Morphy and Rankovic 2005). Rationally designed hybrids often retain key structural motifs responsible for binding affinity while introducing new functional elements to optimize pharmacodynamics (Tanaka et al. 2025). Applications span antimalarial agents, anti-tumour therapies, neuroactive ligands, parasitic diseases, and complex diseases, such as Alzheimer's, where dual or complementary activity profiles are essential for therapeutic success (Viegas-Junior et al. 2007; Ivasiv et al. 2019). Hybridization/combination of quinoline moieties with other bioactive scaffolds such as indoles, triazoles, benzothiazoles, hydrazones, chalcones, benzodiazepines, or other heterocycles has increasingly been pursued to achieve multi-target therapeutic profiles. These hybrids show more substantial effects against bacteria, parasites, and cancer cells. They work by causing oxidative stress, breaking down cell membranes, and blocking enzymes. Structural tailoring improves lipophilicity and target specificity, with some candidates achieving activity against resistant clinical strains (Abdullah et al. 2024; Vinindwa et al. 2021).

Their chelating ability allows quinolines to form stable complexes with metal ions. They are useful as catalysts or building blocks in metal-organic frameworks, which highlights their multifunctional nature across scientific disciplines (Kumar et al. 2024). Quinoline derivatives act as fluorescent probes or chemical sensors, responding sensitively to environmental changes or analyte binding through photophysical modifications (Luo et al. 2022). Semi-aromatic quinolines may have applications beyond medicine, including roles in materials science and supramolecular chemistry. Their electronic characteristics make them valuable in organic electronics, specifically organic light-emitting diodes, organic photovoltaic cells, and organic semiconductors (Lewinska et al. 2021).

Natural products, including quinoline derivatives, that function as specific, small-molecule protein-binding agents and modulate protein–protein interactions are complex and exhibit diverse three-dimensional structures with chiral functional groups, crucial for specificity. Developing high-throughput methods for creating complex polycyclic compounds with limited asymmetric diversity is essential for understanding protein interactions in chemical biology (Nandy et al. 2009). To generate a library using a polycyclic derivative with an enamide group, an enantioselective synthesis of a THQ derivative was achieved, utilizing its phenolic hydroxyl as an anchoring site for solid-phase synthesis (Khadem et al. 2004). A ring-closing metathesis approach produced the desired polycyclic product in five steps. This compound is a novel scaffold for generating small molecules as chemical probes. These findings provided insights into the conformations of bicyclic scaffolds (quinolines) based on ring size. Modular synthesis using solid support successfully generated various macrocyclic derivatives. The growing interest in small molecules for deciphering protein interactions underscores the need for structurally complex and diverse small-molecule libraries, as they are vital for investigating intricate protein networks and cell-signalling pathways. The synthesis of THQ derivatives with unique functional groups was also achieved through solid-phase techniques (Khadem et al. 2010). These methods can synthesize alkaloid-like compounds from various semi-aromatic quinoline derivatives for biological activity testing.

9.2   CASE STUDY

2O-THQs and 1-oxo-tetrahydroisoquinolines (1O-THIQs), Figure 9.3, are structurally related nitrogen-containing heterocycles with significant bioactive potential. Despite their isomeric relationship and shared bicyclic frameworks, they diverge in their structural geometry, chemical reactivity, biological origin, bioactivity, and biosynthesis, which leads to distinct functional properties and applications in medicinal chemistry.

Structurally, 2O-THQs are characterized by a 3,4-dihydroquinolin-2(1H)-one scaffold. The oxo group resides at the C2 position, and nitrogen at N1 (Chapter 4). In contrast, the 3,4-dihydroisoquinolin-1(2H)-one core is the basis for 1O-THIQs. The oxo group is found at C1 and nitrogen at N2 (Khadem and Marles 2025). Electronic and steric differences are imparted by the differing orientation of the fused benzene ring. The oxo group at C2 in 2O-THQs perturbs aromatic resonance in the adjacent ring, resulting in reduced electron density. In 1O-THIQs, the aromaticity of the benzene ring is preserved by the C1 oxo group, leading to increased electronic stability. Additionally, envelope or half-chair conformations are often adopted by 2O-THQs due to ring flexibility. Twisted boat-like geometries are frequently adopted by 1O-THIQs, as crystallographic data confirms.

Hydrogen bonding and substitution patterns are influenced by these structural characteristics. Intermolecular N–H···O hydrogen bonding is exhibited by both classes. Strong intramolecular hydrogen bonds can also be formed by 1O-THIQs, especially in derivatives bearing amide groups. Biosynthetic origin is reflected by substitution patterns. 2O-THQs typically feature prenyl, hydroxyl, or methoxy groups at C4, C6, and C8. 1O-THIQs more commonly exhibit methoxy or hydroxyl substitutions at C6 and C7, mimicking tyrosine residues. Distinct chemical shifts, reflecting the oxo group's deshielding effects and differences in conjugation, are revealed by spectroscopic data, especially NMR.

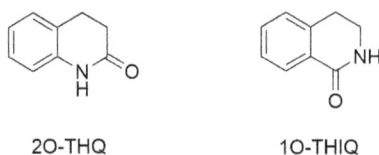

2O-THQ   1O-THIQ

FIGURE 9.3   Chemical structures of 2O-THQ and 1O-THIQ.

Regarding chemical reactivity, the electrophilic nature of the carbonyl group influences both compounds. The 2-oxo group in 2O-THQs is adjacent to sp³-hybridized nitrogen. This increases reactivity towards nucleophiles and enables transformations such as enolization, conjugate addition, and reduction. In contrast, resonance with the nitrogen lone pair stabilizes the 1-oxo moiety in 1O-THIQs. This reduces its electrophilicity and makes it less prone to nucleophilic attack. However, oxidative transformations are undergone by 1O-THIQs, especially hydroxylation at the B-ring and N-dealkylation. Tautomerism is exhibited by both compounds. Enamine–keto tautomers are more readily formed by 2O-THQs. Lactim–lactam tautomers can be formed by 1O-THIQs, although less frequently.

Their susceptibility to hydrolytic cleavage also differs. 2O-THQs can be hydrolysed to anthranilic acid derivatives under strongly acidic or basic conditions. However, 1O-THIQs hydrolyse to 2-aminobenzoyl acetic acids. Microbial degradation and biotransformation find these pathways relevant. Electron-rich positions of the benzene ring are typically where electrophilic aromatic substitution is observed. The electronic influence of substituents and the lactam moiety guide this process. For example, hydroxylation or methoxylation frequently occurs at C6 or C8 in both scaffolds. Their electronic environments influence this pattern.

Marked differences in biological source and ecological distribution are revealed by the natural occurrence of these alkaloids. 2O-THQs are primarily found in organisms such as fungi (e.g. *Aspergillus*, *Penicillium*) and actinobacteria (e.g. *Streptomyces*). They are also found in some plants (notably Rutaceae) and even in insects. Defensive secondary metabolites typically serve this function. In contrast, higher plants contain 1O-THIQs widely, particularly within Papaveraceae, Berberidaceae, Lauraceae, and Ranunculaceae. Examples include northalifoline, corydaldine, and thalifoline. Ecological roles are played by these alkaloids, such as deterring herbivores and modulating microbial communities. Their significance in traditional medicine is highlighted by the broad distribution of 1O-THIQs in medicinal plants.

Biologically, antimicrobial, anti-inflammatory, and cytotoxic properties are known for 2O-THQs. Activity against Gram-positive bacterial and fungal pathogens is exhibited by specific derivatives. Others demonstrate antimalarial and anticancer activities via mechanisms such as DNA intercalation and inhibition of topoisomerases. Aripiprazole, a synthetic 2O-THQ, functions as a dopaminergic modulator. Acetylcholinesterase is also inhibited by some 2O-THQs, or they suppress inflammatory mediators like nitric oxide. Broader bioactivities are exhibited by 1O-THIQs. These include cholinesterase and poly(ADP-ribose) polymerase 1 (PARP1) inhibition, antioxidant, hepatoprotective, antifibrotic, and anti-neoplastic effects. Certain derivatives interact with neurodegenerative disease targets such as beta-site APP-cleaving enzyme 1 and acetylcholinesterase (AChE). This makes them candidates for Alzheimer's therapy. Vasorelaxation is induced by derivatives like thalifoline. Others exhibit antifungal or weak antiviral effects. DNA intercalation in some cases is also permitted by the relatively planar geometry of 1O-THIQs.

The biosynthetic origins of these alkaloids differ significantly. Anthranilic acid or tryptophan, through intermediates of the kynurenine pathway, are typically the source of 2O-THQs. Cyclization is undergone by prenylated anthranilates in plants such as those in the Rutaceae family. In fungi, hybrid non-ribosomal peptide synthetase-polyketide synthase pathways are involved in biosynthesis. Anthranilate-containing benzodiazepines are precursors to the quinolinone core in these pathways. Key enzymes include AsqI, a haemocyanin-like oxidase that catalyses ring contraction. Other tailoring enzymes like prenyltransferases and cytochrome P450s are also involved. The biosynthetic gene clusters (e.g. asq cluster) for fungal 2O-THQs have been partially elucidated. On the other hand, 1O-THIQs, as members of the benzylisoquinoline alkaloid family, originate from L-tyrosine. Tyrosine is converted into dopamine. Pictet–Spengler condensation with aldehydes is then undergone to form the THIQs skeleton. The C1-oxo functionality is introduced via oxidative deamination or P450-mediated transformations of intermediates such as N-methylated THIQs. Iminium intermediates that are hydrolysed to the lactam form are often yielded by these reactions. Sometimes, end-products or metabolic shunts of broader benzylisoquinoline alkaloid pathways are what 1O-THIQs represent. Oxidative coupling catalysed by laccases or peroxidases forms dimeric structures like berbanine. Methylation, glycosylation, and acylation are included in the tailoring steps.

Looking forward, underexploited resources in natural product research are represented by both scaffolds. For 2O-THQs, future directions include the discovery of novel analogues from extremophiles. Complete biosynthetic pathway elucidation and heterologous expression of gene clusters are also included. Unexplored ecological niches and enzymatic machinery with biotechnological potential are suggested by their relative scarcity. For 1O-THIQs, multifunctional analogues targeting neurodegenerative and inflammatory diseases should be the focus of drug development efforts. Modular biosynthesis of complex derivatives may be enabled by synthetic biology approaches. Additionally, both classes may play roles in chemical ecology. Pest deterrence and microbiome modulation are included in these roles.

9.3   SEARCH IN PROGRESS

A primary source of pharmacologically active compounds has historically been served by natural products. Structural diversity is derived from evolutionary biochemical processes. The identification and classification of natural scaffolds across plant, microbial, and marine sources have been enhanced by advances in isolation techniques, dereplication strategies, and metabolomic profiling. Broad-spectrum bioactivities are exhibited by natural products, including antimicrobial, anti-inflammatory, and antineoplastic effects. They often function through novel, inaccessible mechanisms via synthetic small molecules (Atanasov et al. 2015). A curated collection of structurally annotated natural molecules is offered by the *Dictionary of Natural Products*. This resource enables bioactivity mining and cheminformatics analyses. As quantitative database evaluations reveal, biosynthetic origin strongly influences structural diversity. Dominant classes comprise polyketides, alkaloids, and terpenoids (Figure 9.4). Drug discovery is enhanced by integrating DNP data into virtual screening pipelines. Target-specific pharmacophore matching and scaffold enrichment are allowed by this integration (Chassagne et al. 2019).

The structural rigidity of the quinoline ring is combined with chemically diverse moieties by quinoline-based hybrids to enhance biological activity and molecular

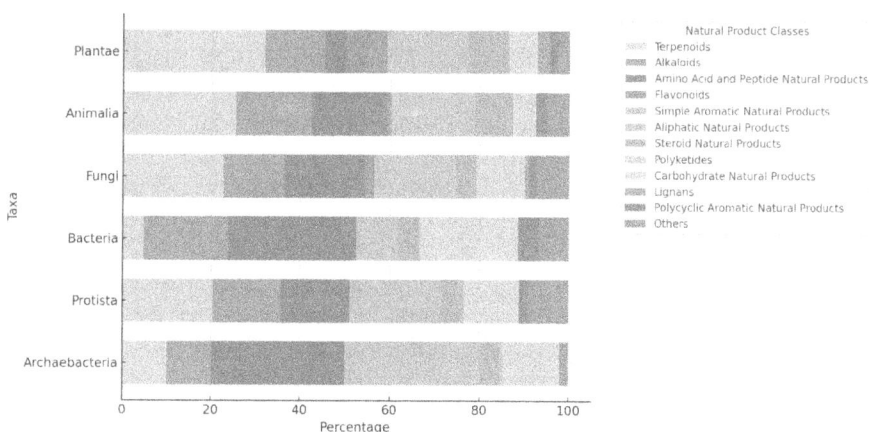

FIGURE 9.4 Comparative chemical biodiversity of natural product classes among major taxa. (Adopted from Chassagne et al. 2019.)

selectivity. Broad-spectrum cytotoxicity is exhibited by these compounds through DNA intercalation, enzyme inhibition, or redox modulation. Green chemistry approaches to their synthesis are emphasized by recent developments. Atom economy is improved, and reaction complexity is reduced by these approaches. Their viability in medicinal chemistry pipelines is supported by this improvement (Abdullah et al. 2024; Ahmed and Akter 2024).

Stereochemical complexity and functional group diversity are introduced by dearomatization of aromatic natural products, such as quinolines, to semi-aromatic quinolines. These features are critical for synthesizing architecturally elaborate scaffolds. Non-planar intermediates that facilitate selective functionalization at previously inaccessible positions are generated by such transformations. A key role in the total synthesis of bioactive compounds is played by these reactions. They also create libraries for structure–activity relationship exploration (Roche and Porco Jr 2011).

Historically, classical reactions such as Skraup and Friedländer syntheses were heavily relied upon for synthetic access to quinoline derivatives. Harsh conditions were typically required by all of these methods. Moderate yields and limited structural diversity were produced by these approaches. Greener and more efficient approaches have been shifted towards by modern synthetic chemistry to overcome these limitations. The one-step construction of diverse quinoline derivatives is allowed by multicomponent reactions, notably the Povarov reaction. Readily available substrates are combined by this approach. Atom economy, reaction selectivity, and throughput are improved by this method. The direct functionalization of quinoline cores has been enabled by recent advancements in transition metal-catalysed C–H activation methods. Several synthetic steps that were previously required for pre-functionalization are eliminated by this approach. Additionally, emerging methodologies such as photocatalytic and electrochemical syntheses offer milder and more environmentally sustainable conditions. Selective transformations that conventional methods cannot easily achieve are enabled by these conditions.

Increasingly significant roles are now played by asymmetric catalysis and biocatalysis, particularly in synthesizing enantiomerically pure THQs. These compounds are crucial for biological activity. Reliance on chiral auxiliaries or costly resolution procedures is reduced by these methods. More economically viable routes to chiral quinoline derivatives are offered by this approach. Reproducibility and scalability are further enhanced by continuous-flow chemistry and automated synthesis. Rapid optimization and large-scale production for potential pharmaceutical applications are facilitated by these methods.

A predictive framework for estimating untested compounds' toxicity and other biological properties based on their structural attributes is offered by quantitative structure–activity relationship (QSAR) models. These models are increasingly adopted in chemical safety evaluations to reduce reliance on in vivo experimentation. Standardized validation procedures, applicability domain analysis, and transparent performance metrics facilitate regulatory integration. Due to these features, QSAR is a practical alternative in preliminary risk assessments (Benfenati et al. 2013). The adoption of ensemble modelling strategies, expanded descriptor spaces, and machine learning(ML) algorithms that accommodate nonlinear associations between chemical features and bioactivity is included in recent advancements in QSAR. Central roles in building robust models have become occupied by data curation standards, descriptor selection protocols, and reproducibility practices. Efforts are underway to unify these methodologies through interoperable platforms and standardized modelling guidelines. Broader acceptance in pharmaceutical and toxicological sciences is promoted by this unification (Cherkasov et al. 2014). Computational strategies that involve selecting relevant structural descriptors, calibrating regression or classification algorithms, and conducting internal and external validation are heavily relied upon by QSAR development (Jäntschi and Bolboaca 2007). Topological indices, quantum-chemical parameters, and 3D spatial characteristics may be included in descriptor classes. Error minimization, generalization capacity, and resistance to overfitting are emphasized by model development cycles. These features are crucial for reliable predictive applications (Dudek et al. 2006). Globally harmonized frameworks that integrate artificial intelligence(AI) with cheminformatics are now encompassed by the evolution of QSAR modelling. Consensus modelling and mechanistic interpretation of results strengthen predictive reliability. The applicability of QSAR beyond academic research into regulatory and clinical decision-making is expanded by these features, which are essential (Muratov et al. 2020; Spiegel and Senderowitz 2020). An alternative to traditional hazard assessment is provided by using QSAR-based toxicological models in regulatory science. Endpoints such as mutagenicity, hepatotoxicity, and carcinogenicity are predicted from molecular structure by these models. Validation standards such as those established by Organisation for Economic Co-operation and Development (OECD) guidelines are applied to these models. Regulatory frameworks such as Registration, Evaluation, Authorisation and Restriction of Chemicals (REACH) and the Environmental Protection Agency (EPA)'s predictive tools increasingly incorporate them. The reliability of predictions for specific chemical classes is ensured by incorporating applicability domain analysis (Benfenati et al. 2013).

Multidimensional chemical-bioactivity landscapes for quinoline derivatives are functioned as by SAR maps. The graphical identification of structural analogues associated with favourable or deleterious biological profiles is enabled by these maps. Molecular similarity and potency data are incorporated by these visual tools. Precise structural trends within a chemical series are distinguished by medicinal chemists through their use. Compound prioritization is streamlined by their application. Scaffold replacement decisions are aided, and rational lead optimization workflows are enhanced by these tools (Agrafiotis et al. 2007).

Integral roles in contemporary drug discovery have been assumed by AI and ML techniques. Tools for prediction, classification, and optimization of bioactive molecules are offered by these systems. High-dimensional chemical and biological data are processed by these systems to uncover patterns that inform target identification, lead optimization, and de novo molecular design. The development of dynamic models that adapt to new information is allowed by integration with cheminformatics (chemoinformatics) pipelines. The experimental burden is reduced, and decision-making during early-stage screening is enhanced by this integration (Dara et al. 2022). Algorithmic advances such as random forests, support vector machines, neural networks, and gradient boosting are leveraged by cheminformatics applications based on ML. Curated datasets are used to train these models to forecast biological activity, toxicity, solubility, and other drug-like attributes. Recent efforts have focused on improving generalizability, addressing data imbalance, and enhancing interpretability of predictions through model-agnostic explanations and attention-based frameworks (Niazi and Mariam 2023).

The development of predictive frameworks that combine pharmacological data with disease-specific omics signatures has been led by cross-disciplinary ML applications. Assistance in identifying targets, discovering biomarkers, and clarifying their mechanisms is provided by these methods. Minimal prior knowledge is required by these approaches. Multi-modal data is increasingly used to train ML models for phenotypic screening, virtual patient simulations, and drug repurposing strategies across therapeutic areas (Vamathevan et al. 2019).

Cheminformatics modelling of electronic structure, reaction kinetics, and compound permeability has been extended by advances in AI. First-principles physics and data-derived heuristics are incorporated by these advances. Quantitative structure–property relationships are integrated with physiologically-based pharmacokinetic models by hybrid mechanistic-AI frameworks to simulate drug absorption and distribution. Mechanistic understanding of drug behaviour is improved by these tools. In silico evaluation of formulation strategies and dosage regimens is facilitated by these tools (Arav 2024).

The rapid evaluation of large chemical libraries is allowed by virtual screening. Compounds are ranked based on predicted binding affinity or pharmacophoric alignment by this method. Active compounds in candidate pools are identified more effectively by techniques such as docking simulations, pharmacophore filtering, and similarity scoring. Accurate target modelling, scoring function calibration, and post-processing filters to eliminate artefacts and frequent hitters are dependent upon successful applications (Lavecchia and Di Giovanni 2013). Hit discovery without full structural knowledge of the target protein is enabled by pharmacophore-based virtual screening. Important features, like hydrogen bond acceptors and donors, hydrophobic areas, and aromatic rings from known active compounds are looked at. New structures with similar biological effects are found by examining these features. Scaffold diversity is improved by this approach. Early-stage discovery is supported by this method, particularly for allosteric or orphan targets (Giordano et al. 2022).

Cases where small structural changes in a molecule lead to disproportionately large changes in biological activity are referred to as activity cliffs. Challenges in predictive modelling are presented by these discontinuities. Insight into critical pharmacophoric elements is offered by them. SAR models are refined by the identification and analysis of activity cliffs. Structural regions for optimization during lead development are prioritized by this process (Cruz-Monteagudo et al. 2014).

Specific psychotropic and neurological drug discovery applications have been tailored by computational pipelines. Pharmacophore models, docking tools, and ML-based scoring functions are integrated by these systems (Sun et al. 2025). Complex CNS-relevant properties, such as blood–brain barrier penetration, receptor subtype selectivity, and neurotoxicity prediction, are accommodated by these systems. Prioritization of ligands with multi-target activity in neurodegenerative and psychiatric disease contexts is enabled by them (Dorahy et al. 2023).

Absorption, distribution, metabolism, excretion, and toxicity(ADMET) profiling is essential in early drug development to evaluate candidate compounds' pharmacokinetic behaviour and safety. Quantitative descriptors and ML algorithms are utilized by in silico ADMET prediction models to estimate bioavailability, metabolic pathways, and clearance rates. Reliance on animal testing is reduced by these models. The prioritization of drug-like molecules with favourable human pharmacokinetics is enabled by them (Jung et al. 2024). Key drug properties, including their ADMETs, have been enhanced in their prediction by recently improved machine-learning methods. Drug candidates with good characteristics are identified early on by this advancement. The chances of failure during clinical trials are lowered by this improvement.

A standard practice in early discovery phases has become the filtering of compounds based on known structural alerts associated with promiscuity or assay interference. False-positive biological signals can be caused by substructures known as pan-assay interference compounds. Downstream development is confounded by this issue. The likelihood of identifying clean, target-specific leads is increased by applying such filters. The cost of late-stage failures due to toxicity or off-target effects is reduced by this approach (Senger et al. 2016).

The prediction of idiosyncratic and mechanistically complex toxicities is being enhanced by hybrid models combining toxicogenomics, cheminformatics, and Bayesian frameworks. Chemical structure is integrated with biological response markers, such as gene expression profiles, by these systems. The characterization of mode-of-action and pathway-level effects is enabled by this integration. Improved mechanistic interpretability compared to traditional black-box predictors is offered by such models. Biomarker discovery and risk assessment strategies may be informed by them (Benfenati et al. 2013; Jung et al. 2024).

A computational strategy employed to identify structurally distinct cores that maintain essential biological activity is scaffold hopping. The discovery of novel chemotypes with improved pharmacokinetic or pharmacodynamic properties is enabled by this approach. The repositioning of ligands within bioactive space is facilitated by combining focused libraries and pharmacophore constraints. Known resistance or selectivity issues are bypassed by this method. The development of efflux pump inhibitors and anti-virulence agents against microbial targets is included in applications. Key binding motifs are preserved while chemical backbones are diversified by these applications (Cedraro et al. 2021). Data-driven prioritization of chemical cores using similarity metrics that surpass conventional substructure matching is incorporated by modern developments in scaffold hopping. Topological fingerprints, three-dimensional alignment, and target-specific scoring functions to guide scaffold substitution are included in this approach. Chemical novelty in drug discovery campaigns is enhanced by these techniques. Pharmacophoric fidelity is maintained, and biological relevance across compound series is preserved by them (Hu et al. 2017). Molecular complexity, binding site structure and bioisosteric tolerance influence the practical limits of scaffold hopping. Receptor affinity may be preserved or disrupted by small scaffold changes depending on local interaction networks, as empirical analyses have demonstrated. A central question in medicinal chemistry optimization remains understanding the extent to which scaffold modifications can alter receptor recognition without loss of function (Schneider et al. 2006).

The basic physical and chemical features needed for a substance to have biological activity are found by pharmacophore modelling. Known chemicals that connect with biological targets are looked at to accomplish this. Large chemical databases are used to identify candidate molecules from these models in virtual screening. Defined spatial arrangements of functional groups are matched by these candidate molecules. Reliance on full structural similarity is reduced by pharmacophore-based approaches. Ligand discovery when little structural data about the target is available is enabled by these approaches (Giordano et al. 2022). The intentional incorporation of pharmacophores targeting distinct binding domains or biological pathways within a single molecule is involved in the design of multiple ligands. The modulation of polypharmacological profiles, synergistic action, and reduced likelihood of resistance is supported by this strategy. Broadened activity spectra are often exhibited by the resulting compounds when coupled with scaffold hopping. Acceptable physicochemical and ADMET properties are maintained by these compounds (Morphy and Rankovic 2005).

A rich source of drug leads continues to be served by natural resources. High-throughput technologies and cheminformatics are combined by modern screening strategies to prioritize candidates for development. Ethnopharmacological knowledge, in vitro bioassays, and in silico modelling are integrated to enhance lead identification. Privileged interactions with biological macromolecules, including kinases, proteases, and membrane receptors, are often demonstrated by the resulting compounds (Xu et al. 2022). Fine-tuning of pharmacokinetic and pharmacodynamic properties is enabled by the synthetic modification of natural compounds. The inherent biological activity of the parent molecule is retained by this process. Semi-synthesis, bioisosteric replacement, and prodrug design to enhance solubility, metabolic stability, and target specificity are included in strategies. Derivatives with improved therapeutic indices and expanded utility in treating complex diseases like cancer, infection, and inflammation have been yielded by this approach (Guo 2017).

The biological roles of bioactive compounds derived from nutraceuticals, functional foods, and cosmeceuticals are also elucidated using computational approaches. Health-promoting agents from botanicals, microbial fermentation, and dietary metabolites are evaluated for immunomodulatory, antioxidant, and anti-inflammatory properties. The identification of bioactivity signatures is facilitated by databases and cheminformatics platforms. Repositioning and formulation of compounds with functional benefits for metabolic disorders, dermatological health, and neuroprotection are enabled by this identification (Carpio et al. 2021).

9.4   FUTURE DIRECTIONS

Future research strategies must focus on sustainable and efficient synthetic protocols. These protocols include solvent-free or minimal-solvent conditions, microwave-assisted synthesis, and biodegradable catalyst utilization. In addition, researchers must enhance targeted drug delivery through prodrug approaches or nanoparticle carriers. Scientists must establish comprehensive and accessible SAR databases. The field must encourage interdisciplinary collaboration among synthetic chemists, computational scientists, pharmacologists, and environmental researchers.

Despite extensive research, scientists have incompletely discovered the semi-aromatic quinolines' potential due to significant challenges. The complex synthesis of highly substituted or chiral quinoline derivatives creates a considerable barrier. This complexity often limits accessibility and scalability. Additionally, researchers partly understand metabolic pathways, bioactivation potentials, long-term toxicity profiles, and environmental persistence. This situation requires further comprehensive *in vivo* studies and predictive modelling. Moreover, resistance development, particularly to antimicrobials, continues to pose severe threats. This issue underscores the necessity for new combination therapies and alternative patient treatment approaches.

The combination of high-throughput screening, cheminformatics, and synthetic biology offers the potential for discovering new bioactive scaffolds from these underexplored sources. These data collectively highlight the pharmacological versatility of quinoline-derived alkaloids. This versatility also reinforces their importance as lead compounds in anti-infective and anticancer drug discovery.

Hybrid pharmacophores allow fine-tuning of ligand orientation and binding conformation within diverse receptor environments. Functional diversity within the hybrid structure contributes conformational flexibility. This flexibility enables interactions with multiple binding domains. Researchers advantageously design multi-target-directed ligands using such molecular configurations. These prove particularly useful in treating neurodegenerative and polygenic disorders (Elebiju et al. 2023). In addition to enhancing efficacy, hybrid molecule strategies contribute to improved pharmacokinetics. These strategies modulate metabolic stability, absorption, and membrane permeability. Hybrid molecules reduce off-target toxicity while maintaining therapeutic concentrations within the desired biological compartment. Optimization balances molecular weight, hydrogen bond donors/acceptors, and polar surface area. This balance satisfies Lipinski's and Veber's criteria for drug-likeness (Shehab et al. 2023).

Scaffold-hopping strategies enhance chemical diversity by substituting the quinoline core with structurally similar frameworks. These strategies provide advantages for patentability. They may improve essential physicochemical properties crucial for clinical translation. These approaches substantially enhance the therapeutic potential of quinoline-based medicinal agents. Computational modelling and high-throughput screening aid these approaches.

Bioprinting enhances the bioavailability and precision of alkaloid-based treatments' evaluation. Three-dimensional bioprinting represents an advanced technique in pharmaceutical sciences. This technique enables the creation of tissue-like structures, personalized medications, and organ-on-chip models. This process deposits layers of biomaterials and cell-containing inks. It supports controlled drug delivery, wound healing, and *in vitro* toxicity screening. Three-dimensional bioprinting offers more accurate predictive models and reduces reliance on animal testing. The progression to four-dimensional bioprinting incorporates time as a functional parameter. This approach uses stimuli-responsive materials (e.g., shape-memory, thermoresponsive, or pH-sensitive polymers). These materials enable post-printing transformations and dynamic drug release. These systems allow precise temporal and spatial control of therapeutics. Targeted cancer treatments and gastrointestinal drug delivery particularly benefit from this control. Advanced bioprinting also provides for the integration of living cells into bioactive scaffolds. This integration supports regenerative medicine and developing disease models replicating human tissue environments. Advanced bioprinting enhances drug efficacy and toxicity assessment accuracy. This enhancement particularly benefits from patient-specific or rare disease contexts where conventional formulations prove inadequate. Bioprinting has the potential to enable on-demand drug manufacturing in remote or resource-limited settings, utilizing portable or modular printers. Patient data and computer-aided design guide these systems (Mahapatra and Karuppasamy 2022). They enable localized, waste-reducing production tailored to individual needs. This approach aligns with the broader goal of decentralized pharmaceutical care (Mihaylova et al. 2024).

Finally, standardized regulatory and safety protocols must account for ecological impacts alongside clinical safety, which proves imperative. Implementing these strategies will significantly advance semi-aromatic quinoline derivatives' therapeutic and technological applications. This advancement will position them as promising entities in drug development and beyond.

REFERENCES

Abdullah, A. H., Alarareh, A. K., Al-Sha'er, M. A., Habashneh, A. Y., Awwadi, F. F., & Bardaweel, S. K. (2024). Docking, synthesis, and anticancer assessment of novel quinoline-amidrazone hybrids. *Pharmacia, 71*, 1–12.

Agrafiotis, D. K., Shemanarev, M., Connolly, P. J., Farnum, M., & Lobanov, V. S. (2007). SAR maps: A new SAR visualization technique for medicinal chemists. *Journal of Medicinal Chemistry, 50*(24), 5926–5937.

Ahmed, M. S., & Akter, I. (2024). Green strategies for the synthesis of quinolinone derivatives. https://www.qeios.com/read/P5M2Z8

Aldred, K. J., Kerns, R. J., & Osheroff, N. (2014). Mechanism of quinolinone action and resistance. *Biochemistry, 53*(10), 1565–1574.

Arav, Y. (2024). Advances in modeling approaches for oral drug delivery: Artificial intelligence, physiologically-based pharmacokinetics, and first-principles models. *Pharmaceutics, 16*(8), 978.

Atanasov, A. G., Waltenberger, B., Pferschy-Wenzig, E., Linder, T., Wawrosch, C., Uhrin, P., Temml, V., Wang, L., Schwaiger, S., & Heiss, E. H. (2015). Discovery and resupply of pharmacologically active plant-derived natural products: A review. *Biotechnology Advances, 33*(8), 1582–1614.

Benfenati, E., Pardoe, S., Martin, T., Diaza, R. G., Lombardo, A., Manganaro, A., & Gissi, A. (2013). Using toxicological evidence from QSAR models in practice. *ALTEX-Alternatives to Animal Experimentation, 30*(1), 19–40.

Carpio, L. E., Sanz, Y., Gozalbes, R., & Barigye, S. J. (2021). Computational strategies for the discovery of biological functions of health foods, nutraceuticals and cosmeceuticals: A review. *Molecular Diversity, 25*, 1425–1438.

Cedraro, N., Cannalire, R., Astolfi, A., Mangiaterra, G., Felicetti, T., Vaiasicca, S., Cernicchi, G., Massari, S., Manfroni, G., & Tabarrini, O. (2021). From quinoline to quinazoline-based *S. aureus* NorA efflux pump inhibitors by coupling a focused scaffold hopping approach and a pharmacophore search. *ChemMedChem, 16*(19), 3044–3059.

Chassagne, F., Cabanac, G., Hubert, G., David, B., & Marti, G. (2019). The landscape of natural product diversity and their pharmacological relevance from a focus on the dictionary of natural products®. *Phytochemistry Reviews, 18*, 601–622.

Cherkasov, A., Muratov, E. N., Fourches, D., Varnek, A., Baskin, I. I., Cronin, M., Dearden, J., Gramatica, P., Martin, Y. C., & Todeschini, R. (2014). QSAR modeling: Where have you been? Where are you going to? *Journal of Medicinal Chemistry, 57*(12), 4977–5010.

Cruz-Monteagudo, M., Medina-Franco, J. L., Pérez-Castillo, Y., Nicolotti, O., Cordeiro, M. N. D., & Borges, F. (2014). Activity cliffs in drug discovery: Dr Jekyll or Mr Hyde? *Drug Discovery Today, 19*(8), 1069–1080.

Daneshtalab, M., & Ahmed, A. (2012). Nonclassical biological activities of quinolinone derivatives. *Journal of Pharmacy & Pharmaceutical Sciences, 15*(1), 52–72.

Dara, S., Dhamercherla, S., Jadav, S. S., Babu, C. M., & Ahsan, M. J. (2022). Machine learning in drug discovery: A review. *Artificial Intelligence Review, 55*(3), 1947–1999.

Dorahy, G., Chen, J. Z., & Balle, T. (2023). Computer-aided drug design towards new psychotropic and neurological drugs. *Molecules, 28*(3), 1324.

Dudek, A. Z., Arodz, T., & Gálvez, J. (2006). Computational methods in developing quantitative structure-activity relationships (QSAR): A review. *Combinatorial Chemistry & High Throughput Screening, 9*(3), 213–228.

Elebiju, O. F., Ajani, O. O., Oduselu, G. O., Ogunnupebi, T. A., & Adebiyi, E. (2023). Recent advances in functionalized quinoline scaffolds and hybrids—Exceptional pharmacophore in therapeutic medicine. *Frontiers in Chemistry, 10*, 1074331.

El-mrabet, A., Haoudi, A., Kandri-Rodi, Y., & Mazzah, A. (2025). An overview of quinolinones as potential drugs: Synthesis, reactivity and biological activities. *Organics, 6*(2), 16.

Giordano, D., Biancaniello, C., Argenio, M. A., & Facchiano, A. (2022). Drug design by pharmacophore and virtual screening approach. *Pharmaceuticals, 15*(5), 646.

Guo, Z. (2017). The modification of natural products for medical use. *Acta Pharmaceutica Sinica B, 7*(2), 119–136.

Hooper, D. C., & Jacoby, G. A. (2015). Mechanisms of drug resistance: Quinolinone resistance. *Annals of the New York Academy of Sciences, 1354*(1), 12–31.

Hu, Y., Stumpfe, D., & Bajorath, J. (2017). Recent advances in scaffold hopping: Miniperspective. *Journal of Medicinal Chemistry, 60*(4), 1238–1246.

Ivasiv, V., Albertini, C., Gonçalves, A. E., Rossi, M., & Bolognesi, M. L. (2019). Molecular hybridization as a tool for designing multitarget drug candidates for complex diseases. *Current Topics in Medicinal Chemistry*, *19*(19), 1694–1711.

Jäntschi, L., & Bolboaca, S. (2007). Results from the use of molecular descriptors family on structure property/activity relationships. *International Journal of Molecular Sciences*, *8*(3), 189–203.

Jung, W., Goo, S., Hwang, T., Lee, H., Kim, Y., Chae, J., Yun, H., & Jung, S. (2024). Absorption distribution metabolism excretion and toxicity property prediction utilizing a pre-trained natural language processing model and its applications in early-stage drug development. *Pharmaceuticals*, *17*(3), 382.

Khadem, S., Joseph, R., Rastegar, M., Leek, D. M., Oudatchin, K. A., & Arya, P. (2004). A solution-and solid-phase approach to tetrahydroquinoline-derived polycyclics having a 10-membered ring. *Journal of Combinatorial Chemistry*, *6*(5), 724–734.

Khadem, S., Udachin, K. A., & Arya, P. (2010). Solution-and solid-phase synthesis of tetra-hydroquinoline-based polycyclics having α, β-unsaturated γ-lactam and δ-lactone functionalities. *Synlett*, *2010*(02), 199–202.

Khadem, S., & Marles, R. J. (2025). Biological activities of selected 1-Oxo-tetrahydroisoquinolinone alkaloids. *Natural Product Research*, *39*(6), 1658–1671.

Kumar, R., Thakur, A., Chandra, D., Dhiman, A. K., Verma, P. K., & Sharma, U. (2024). Quinoline-based metal complexes: Synthesis and applications. *Coordination Chemistry Reviews*, *499*, 215453.

Lavecchia, A., & Di Giovanni, C. (2013). Virtual screening strategies in drug discovery: A critical review. *Current Medicinal Chemistry*, *20*(23), 2839–2860.

Lewinska, G., Sanetra, J., & Marszalek, K. W. (2021). Application of quinoline derivatives in third-generation photovoltaics. *Journal of Materials Science: Materials in Electronics*, *32*(14), 18451–18465.

Li, Z., Yin, L., Zhao, D., Jin, L., Sun, Y., & Tan, C. (2023). SAR studies of quinoline and derivatives as potential treatments for alzheimer's disease. *Arabian Journal of Chemistry*, *16*(2), 104502.

Luo, M., Sun, B., Zhou, C., Pan, Q., Hou, Y., Zhang, H., … & Zou, C. (2022). A novel quinoline derivative as a highly selective and sensitive fluorescent sensor for Fe^{3+} detection. *RSC Advances*, *12*(36), 23215–23220.

Mahapatra, M. K., & Karuppasamy, M. (2022). Fundamental considerations in drug design. *Computer aided drug design (CADD): From ligand-based methods to structure-based approaches* (pp. 17–55). Elsevier.

Meunier, B. (2008). Hybrid molecules with a dual mode of action: Dream or reality? *Accounts of Chemical Research*, *41*(1), 69–77.

Mihaylova, A., Shopova, D., Parahuleva, N., Yaneva, A., & Bakova, D. (2024). (3D) Bioprinting—Next dimension of the pharmaceutical sector. *Pharmaceuticals*, *17*(6), 797.

Millanao, A. R., Mora, A. Y., Villagra, N. A., Bucarey, S. A., & Hidalgo, A. A. (2021). Biological effects of quinolinones: A family of broad-spectrum antimicrobial agents. *Molecules*, *26*(23), 7153.

Morphy, R., & Rankovic, Z. (2005). Designed multiple ligands. An emerging drug discovery paradigm. *Journal of Medicinal Chemistry*, *48*(21), 6523–6543.

Muratov, E. N., Bajorath, J., Sheridan, R. P., Tetko, I. V., Filimonov, D., Poroikov, V., Oprea, T. I., Baskin, I. I., Varnek, A., & Roitberg, A. (2020). QSAR without borders. *Chemical Society Reviews*, *49*(11), 3525–3564.

Nandy, J. P., Prakesch, M., Khadem, S., Reddy, P. T., Sharma, U., & Arya, P. (2009). Advances in solution-and solid-phase synthesis toward the generation of natural product-like libraries. *Chemical Reviews*, *109*(5), 1999–2060.

Niazi, S. K., & Mariam, Z. (2023). Recent advances in machine-learning-based chemoinformatics: A comprehensive review. *International Journal of Molecular Sciences, 24*(14), 11488.

Roche, S. P., & Porco Jr, J. A. (2011). Dearomatization strategies in the synthesis of complex natural products. *Angewandte Chemie International Edition, 50*(18), 4068–4093.

Schneider, G., Schneider, P., & Renner, S. (2006). Scaffold-hopping: How far can you jump? *QSAR & Combinatorial Science, 25*(12), 1162–1171.

Senger, M. R., Fraga, C. A., Dantas, R. F., & Silva Jr, F. P. (2016). Filtering promiscuous compounds in early drug discovery: Is it a good idea? *Drug Discovery Today, 21*(6), 868–872.

Sharma, B., Chowdhary, S., Legac, J., Rosenthal, P. J., & Kumar, V. (2023). Quinoline-based heterocyclic hydrazones: Design, synthesis, anti-plasmodial assessment, and mechanistic insights. *Chemical Biology & Drug Design, 101*(4), 829–836.

Shehab, W. S., Amer, M. M., Elsayed, D. A., Yadav, K. K., & Abdellattif, M. H. (2023). Current progress toward synthetic routes and medicinal significance of quinoline. *Medicinal Chemistry Research, 32*(12), 2443–2457.

Spiegel, J., & Senderowitz, H. (2020). Evaluation of QSAR equations for virtual screening. *International Journal of Molecular Sciences, 21*(21), 7828.

Sun, J., Li, Z., Yang, Y., & Zhang, S. (2025). Machine learning for active sites prediction of quinoline derivatives. *Artificial Intelligence Chemistry, 3*(1), 100082.

Tanaka, M., Szatmári, I., & Vécsei, L. (2025). Quinoline quest: Kynurenic acid strategies for next-generation therapeutics via rational drug design. *Pharmaceuticals, 18*(5), 607.

Vamathevan, J., Clark, D., Czodrowski, P., Dunham, I., Ferran, E., Lee, G., Li, B., Madabhushi, A., Shah, P., & Spitzer, M. (2019). Applications of machine learning in drug discovery and development. *Nature Reviews Drug Discovery, 18*(6), 463–477.

Viegas-Junior, C., Danuello, A., da Silva Bolzani, V., Barreiro, E. J., & Fraga, C. A. M. (2007). Molecular hybridization: A useful tool in the design of new drug prototypes. *Current Medicinal Chemistry, 14*(17), 1829–1852.

Vinindwa, B., Dziwornu, G. A., & Masamba, W. (2021). Synthesis and evaluation of chalcone-quinoline based molecular hybrids as potential anti-malarial agents. *Molecules, 26*(13), 4093.

Xu, Z., Eichler, B., Klausner, E. A., Duffy-Matzner, J., & Zheng, W. (2022). Lead/drug discovery from natural resources. *Molecules, 27*(23), 8280.

Index

Note: **Bold** page numbers refer to tables; *Italic* page numbers refer to figures.

For Product Safety Concerns and Information please contact our EU
representative GPSR@taylorandfrancis.com
Taylor & Francis Verlag GmbH, Kaufingerstraße 24, 80331 München, Germany

www.ingramcontent.com/pod-product-compliance
Lightning Source LLC
Chambersburg PA
CBHW060743220326
41598CB00022B/2310